LAND SURVEYOR REFERENCE MANUAL

Andrew L. Harbin, P.E.

PROFESSIONAL PUBLICATIONS, INC.
San Carlos, CA 94070

DISCLAIMER: This book is written specifically for engineers preparing for the L-S-I-T and L.S. examinations. Because it is a reference manual, summarizing and simplifying related theory, this book should not be used for design.

Data presented in this book are representative and they are not intended to be exhaustive, precise, or useful for every application. By using this book, the purchaser assumes all responsibility for its use. The author, publisher, distributors, and other interested entities do not assume or accept any responsibility or liability, including liability for negligence, for errors or oversight, or for the use of this book in preparing plans or designs.

In the *ENGINEERING REVIEW MANUAL SERIES*

Engineer-In-Training Review Manual
Quick Reference Cards for the E-I-T Exam
Mini-Exams for the E-I-T Exam
Civil Engineering Review Manual
Seismic Design for the Civil P.E. Exam
Timber Design for the Civil P.E. Exam
Structural Engineering Practice Problem Manual
Mechanical Engineering Review Manual
Electrical Engineering Review Manual
Chemical Engineering Review Manual
Chemical Engineering Practice Exam Set
Land Surveyor Reference Manual
Expanded Interest Tables
Engineering Law, Design Liability, and Professional Ethics

Distributed by: Professional Publications, Inc.
Post Office Box 199
Department 77
San Carlos, CA 94070
(415) 593-9119

LAND SURVEYOR REFERENCE MANUAL

Copyright © 1985 by Professional Publications, Inc.
All rights are reserved. No part of this publication may be reproduced, stored in a retrieval system, or transmitted, in any form or by any means, electronic, mechanical, photocopying, recording, or otherwise, without the prior written permission of the publisher.

Printed in the United States of America

ISBN: 0-932276-46-6

Professional Publications, Inc.
Post Office Box 199, San Carlos, CA 94070

Current printing of this edition (last number) 6 5 4 3 2 1

TABLE OF CONTENTS

PROFESSIONAL PUBLICATIONS, INC. • P.O. Box 199, San Carlos, CA   94070

PROFESSIONAL PUBLICATIONS, INC. • P.O. Box 199, San Carlos, CA 94070

TABLE OF CONTENTS

PROFESSIONAL PUBLICATIONS, INC. • P.O. Box 199, San Carlos, CA 94070

PROFESSIONAL PUBLICATIONS, INC. • P.O. Box 199, San Carlos, CA 94070

TABLE OF CONTENTS

PROFESSIONAL PUBLICATIONS, INC. • P.O. Box 199, San Carlos, CA 94070

PROFESSIONAL PUBLICATIONS, INC. • P.O. Box 199, San Carlos, CA 94070

TABLE OF CONTENTS

PREFACE

This book was written as a complete review of important techniques unique to the profession of land surveying. It is intended to appeal to anyone wishing to gain exposure to the breadth of land surveying. The contents, examples, and problems in this book have been extensively tested in actual classroom courses, as well as in continuing education correspondence courses.

With this book, you will also be able to prepare effectively for the national and state Land Surveyor licensing examinations. An important requirement for passing those examinations is adequate exposure to the subjects covered. This book will provide you with basic exposure to subjects new to you.

This edition is consistent with the new examination structure. The introductory chapter fully describes this new structure, as well as the new scoring method introduced in November 1981. In addition, the new problem categories resulting from NCEE's 1983 task analysis are covered.

Andrew L. Harbin, P.E.
Waco, TX
May 1985

PROFESSIONAL PUBLICATIONS, INC. • P.O. Box 199, San Carlos, CA   94070

ACKNOWLEDGMENTS

Appreciation is extended to Seth S. Searcy III, attorney, my son-in-law, for guiding this work to publication; to James W. Strawn, attorney, my friend and counselor, for advice and for furnishing many legal documents; and to Kenneth G. Gold, surveyor extraordinaire, for a thorough review of part of this work.

Special thanks are extended to Barbara Blancett, my former student, for the illustrations, and to Linda Loepf, Librarian, and her staff at Texas State Technical Institute, for cooperation over many years.

Special thanks belong to Michael R. Lindeburg, P.E., President, and the staff at Professional Publications, Inc. for providing the chapters on Astronomical Observations, Restoration of Lost Corners, Contract Law, Statistics, and Economic Analysis.

From a technical standpoint, I would like to acknowledge the contributions of Rhonda A. Jones, Art Director, who reworked many of the illustrations contained in my chapters, Susan A Madden, Marijo Jamie Brown, and Wendy Nelson for their able editing, and Yasuko Kitajima of Aldine Press for her skill as a $\TeX$ typsetter.

And last but not least, thanks to the many fine students of the Civil Engineering Technology Department of Texas State Techincal Institute for teaching me how to teach.

Andrew L. Harbin

INTRODUCTION

Purpose of Registration

As a surveyor, you may have to obtain your surveying license through procedures which have been established by the state in which you reside. These procedures are designed to protect the public by preventing unqualified individuals from legally practicing as surveyors.

There are many reasons for wanting to become a licensed land surveyor. Among them are the following:

- You may wish to become an independent consultant. By law, consultants must be registered.

- Your company may require a license as a requirement for employment or advancement.

- Your state may require registration if you use the title *land surveyor*.

The Registration Procedure

The registration procedure is similar in most states. You probably will take two 8-hour written examinations. The first examination is the *Land Surveyor in Training* examination, also known as the *Fundamentals of Land Surveying* exam. The initials LSIT, SIT, and FLS are also used. The second examination is the *Professional Land Surveyor* (LS) exam, which differs from the LSIT exam in format and content. (The LS exam is also known as the *Principles and Practice of Land Surveying* exam.)

Actual details of registration, experience requirements, minimum education levels, fees, and examination schedules vary from state to state. You should contact your state's Board of Registration for Professional Engineers and Land Surveyors.

Reciprocity Among States

All but a few states use the NCEE examinations.[1] If you take and pass the examination in one state, all or part of your certificate probably will be honored by other states which have used the same NCEE examination. It will not be necessary to retake the LSIT examination.

The simultaneous administration of identical examinations in multiple states has led to the term *Uniform Examination*. However, each state is free to choose its own minimum passing score or to add special questions to the NCEE examination. Therefore, this Uniform Examination does not automatically ensure reciprocity among states.

Of course, you may apply for and receive a license from another state. However, a license from one state will not permit you to practice land surveying in another state. You must have a land surveying license from each state in which you work.

Applying for the Examination

Each state charges different fees, requires different qualifications, and uses different forms. Therefore, it will be necessary for you to request an application and an information packet from the state in which you reside or in which you plan to take the exam. It generally is sufficient to phone for this information. Telephone numbers for all of the U.S. state boards of registration are given here. (The District of Columbia does not license land surveyors.)

Phone numbers of
State Boards of Registration

Alabama	(205)261-5568
Alaska	(907)465-2540
Arizona	(602)255-4053
Arkansas	(501)371-2517
California	(916)445-5544
Colorado	(303)866-2396
Connecticut	(203)566-3386
Delaware	(302)656-7311
District of Columbia	(202)727-7454
Florida	(904)488-9912
Georgia	(404)656-3926
Guam	(671)646-8643
Hawaii	(808)548-4100
Idaho	(208)334-3860

---

[1] The National Council of Engineering Examiners (NCEE) in Seneca, South Carolina, produces, distributes, and grades the national examinations used by most states. It does not distribute applications to take the examination.

Illinois	(217)785-0872
Indiana	(317)232-1840
Iowa	(515)281-6566
Kansas	(913)296-3053
Kentucky	(502)564-2680
Louisiana	(504)568-8450
Maine	(207)289-3236
Maryland	(301)659-6322
Massachusetts	(617)727-3088
Michigan	(517)373-3880
Minnesota	(612)296-2388
Mississippi	(601)354-7241
Missouri	(314)751-2334
Montana	(406)444-3737
Nebraska	(402)471-2021
Nevada	(702)329-1955
New Hampshire	(603)271-2219
New Jersey	(201)648-2660
New Mexico	(505)827-9940
New York	(518)474-3833
North Carolina	(919)781-9499
North Dakota	(701)258-0786
Ohio	(614)466-8948
Oklahoma	(405)521-2874
Oregon	(503)378-4180
Pennsylvania	(717)783-7049
Puerto Rico	(809)722-2121
Rhode Island	(401)277-2565
South Carolina	(803)758-2855
South Dakota	(605)394-2510
Tennessee	(615)741-3221
Texas	(512)452-9427
Utah	(801)530-6632
Vermont	(802)828-2363
Virgin Islands	(809)774-1301
Virginia	(804)257-8512
Washington	(206)753-6966
West Virginia	(304)348-3554
Wisconsin	(608)266-1397
Wyoming	(307)777-6156

Examination Format–LSIT Exam

The NCEE Land-Surveyor-In-Training (LSIT) examination consists of two four-hour sessions separated by a one-hour lunch period. Most states use both four-hour parts, although some may substitute their own examination if testing of concepts unique to that state is required.

Part I of the NCEE LSIT exam contains 100 multiple choice problems, with five answers labeled A to E for each problem. There is no penalty for guessing, and a calculator may be used. However, no references are permitted in this part of the exam.

Part II of the LSIT exam contains 25 multiple choice problems, with five answers labeled A to E for each problem. The problems in this part of the examination are more difficult than in part I. There is no penalty for guessing. Bound references and calculators are allowed.

In 1983, NCEE completed a task analysis of land surveying activities. The study was intended to determine all the important activities in which land surveyor's engage, and then to suggest changes in the NCEE examinations. At the current time, the examination can be expected to contain the following types of problems.

Typical LSIT
Examination Format

subject	% of exam
mathematics	16%
measurement techniques	26
legal precedent and principles	15
field techniques	18
data integration and analysis	21
record sources	4

Parts of the examination are open book. Most states do not limit the number and types of books you can bring into the exam. Loose-leaf papers and writing tablets are usually forbidden, although you may be able to bring in loose reference pages in a three-ring binder.

Any battery-powered, silent calculator may be used. Depending on the state and the examination section being worked, there may be restrictions on programmable and preprogrammed calculators. Printers must not be used unless they are totally silent.

You will not be permitted to share books, calculators, or any other items with other examinees. You will receive the results of your examination by mail. Allow 12–14 weeks for notification. Your score may or may not be revealed to you depending on your state's procedure.

Grading the LSIT Exam

Each problem in Part I is worth one point, and each problem in Part II is worth four points. The maximum possible score is, therefore, 200 points. There are no mandatory problems.

The scores from both sessions are added together to determine your total score. Both sessions are given equal weight. It is not necessary to achieve any minimum score on either the morning or afternoon sessions.

The minimum passing score is established by each individual State Board of Registration for Professional Engineers and Surveyors, although most boards use the cut-off score recommended by NCEE. A modification of the *Angoff procedure* is used to develop the cut-off score.[2] With this method, experts estimate the fraction of minimally-qualified surveyors who will be able to answer each question correctly. The summation over all test questions of these estimated fractions becomes the cut-off score. Grading on a curve and the concept of a 70% passing rate are not employed.

The use of the Angoff method of establishing minimum passing score, while being more defensible than other methods, is not expected to change the passing rate appreciably. The passing score for a recent examination was determined to be between 120 points (out of 200) and 128 points when the Angoff method was used. Future examinations are expected to have passing rates in this same general range.

All grading is done by computer optical sensing. Choices to the multiple-choice questions must be recorded on score sheets with number 2 pencils. No credit is given for answers recorded in ink.

Examination Format—LS Exam

The Land Surveyor professional exam consists of two four-hour sessions separated by a one-hour lunch break. Most states supply the second part of this exam, since testing of state laws is required. However, NCEE supplies most states with the first four-hour examination.

Part III (i.e., the first part of the LS exam) of the examination process is four hours in length. It consists of ten professional situations requiring calculation and evaluation. Each of the ten situations contains five independent multiple-choice problems with possible answers labeled A to E. Thus, there are a total of 50 problems during the initial four hours. There is no penalty for guessing, and books and calculators are allowed.

Part IV of the examination is unique to the state in which you take the exam. Although you should check with your state, it is likely that this session will consist of problems requiring calculated and essay solutions. The problems are not usually multiple choice. Calculators and bound references should be permitted, but

again, you should verify the rules of the exam with your state.

Typical Part III Examination Format

subject	% of exam
measurement techniques	18%
standards and ethics	24
legal precedent and principles	20
data analysis and integration	38

NCEE has suggested the following format for the part IV session of the LS exam.

Typical Part IV Examination Format

subject	% of exam
local history	38%
standards and ethics	8
legal precedent and principles	38
field techniques	8
record sources	8

Grading the LS Exam

Each problem in Part III is worth one point, with a maximum of 50 points. There is no penalty for guessing, nor are there any mandatory problems.

As with the LSIT exam, NCEE has adopted the Angoff method of establishing the minimum passing score for the Part III exam. For a recent examination, that score was 32 points.

Details of grading Part IV of the exam vary from state to state.

Examination Dates

The national examinations are administered on the same weekend in all states. Each state decides independently whether to offer the examinations on Thursday, Friday, or Saturday of the examination period. The upcoming examination dates are given in the accompanying table.

---

[2] NCEE calls this method *criterion referencing*, to distinguish it from the older *norm referencing* method.

PROFESSIONAL PUBLICATIONS, INC. • P.O. Box 199, San Carlos, CA 94070

Future Dates
NCEE Uniform Examinations

Year	Spring Dates	Fall Dates
1985	April 18–20	October 24–26
1986	April 10–12	October 23–25
1987	April 9–11	October 29–31
1988	April 14–16	October 27–29
1989	April 13–15	October 26–28
1990	April 19–21	October 25–27
1991	April 11–13	October 24–26
1992	April 9–11	October 29–31
1993	April 15–17	October 28–30
1994	April 14–16	October 27–29
1995	April 6–8	October 26–28
1996	April 18–20	October 24–26

Preparing for the Examination

Since all problems must be worked to get full credit, it will be necessary for you to prepare in all of the subjects. Do not make the mistake of studying only a few subjects in hopes of finding *enough* questions to pass with it. You must work in all subjects to pass the exam.

More important than strategy are fast recall and stamina. You must be able to quickly recall solution procedures, formulas, and important data—and, this sharpness must be maintained for eight hours.

It is imperative that you develop and adhere to a review outline and schedule if you are not taking a classroom review course where the order of your preparation is determined by the lectures.

It is unnecessary to bring a large quantity of books to the examination. This book, a dictionary, and a few other references of your choice should be sufficient.[3] The examination is very fast-paced. You will not have time to look up solution procedures, data, or equations with which you are not familiar. Although parts of the examination are open-book, there is insufficient time to use books with which you are not thoroughly familiar.

What to Do Before the Exam

Here are some suggestions for making your examination experience comfortable and successful.

---

[3] Check with your state to see if review books can be brought into the examination. Most states do not have any restrictions. Some states ban only collections of solved problems, such as Schaum's Outline Series. A few prohibit all review books.

- Keep a copy of your examination application. Send the original application by certified mail and request a receipt of delivery. Tape your delivery receipt on the first page of this book.

- Visit the exam site the day before your examination. This is especially important if you are not familiar with the area. Find the examination room, the parking area, and the rest rooms.

- Plan on arriving at least 30 minutes before the examination starts. This will assure you a convenient parking place and adequate time for site, room, and seating changes.

- If you live a considerable distance from the examination site, consider getting a hotel room in which to spend the night before.

- Take off the day before the examination to relax. Don't cram the last night. Rather, get a good night's sleep.

- Be prepared to find that the examination room is not ready at the designated time. Bring an interesting novel or magazine to read in the interim and at lunch.

- If you make arrangements for babysitters or transportation, allow for a delayed completion.

- Prepare your examination kit the day before. Here is a checklist of items to bring with you to the examination.

  ▫ this and your other reference books

  ▫ course notes in a binder

  ▫ calculator and a spare

  ▫ spare calculator batteries or battery pack

  ▫ battery charger and 20' extension cord

  ▫ chair cushions. A large, thick bath mat works well.

  ▫ earplugs

  ▫ desk expander. If you are taking the exam in theater chairs with tiny, fold-up writing surfaces, you should bring a long, wide board to place across the arm rests.

  ▫ a cardboard box cut to fit your references

  ▫ twist-to-advance pencils

  ▫ extra leads

  ▫ machine-scoring pencils

  ▫ snacks such as raisin, nuts, or trail mix

  ▫ thermos filled with hot chocolate

□ a light lunch

□ a collection of graph paper

□ scissors, stapler, and staple puller

□ construction paper for stopping drafts and sunlight

□ scotch and masking tape

□ sunglasses

□ extra prescription glasses if you wear them

□ aspirin

□ travel pack of Kleenex

□ Webster's dictionary

□ dictionary of scientific and engineering terms

□ $2 in change

□ a light comfortable sweater

□ comfortable shoes or slippers for the exam room

□ raincoat, boots, gloves, hat, and umbrella

□ local street maps

□ photographic identification

□ letter admitting you to the examination

□ your copy of the original application and delivery receipt

□ note to the parking patrol for your windshield

□ pad of scrach paper with holes for 3-ring binder

□ straight-edge, ruler, compass, protractor, and French curves

□ battery-powered desk lamp

□ watch

□ wire coat hanger

What to Do in the Exam

• Do not spend excessive time on any difficult problem. If you have not finished a question in a reasonable amount of time, make a note of it and continue on.

• Stop five minutes before the end of each session and guess at all remaining unsolved problems. Do not work up to the end. You will be lucky with about 20% of your guesses. These points will more than make up for the few points you would have earned by working during the last five minutes.

• Record the details of any problem for which you cannot find a correct response. Being able to point out an error may later give you the margin needed to pass.

• Make sure all of your responses on the answer sheet are dark and that they completely fill the 'bubbles.'

• Follow these guidelines when solving essay and other Part IV problems:

Do not rewrite the problem statement.

Do not unnecessarily redraw any figures.

Use pencil only.

Be neat. (Print all text. Use a straightedge or template where possible.)

Draw a box around each answer.

Label each answer with a symbol.

Give the units.

List your sources whenever you use obscure solution methods or data.

Write on one side of the page only.

Use one page per problem, no matter how short the solution is.

Go through all calculations a second time and check for mathematical errors.

Recommended References

The books listed below have been reported by previous examinees to be the most valuable resources they brought with them into the examination. Depending on your background, you may need additional books to learn subjects new to you.

Where possible, book titles, authors, and publishers have been listed. However, dates of publication and editions have been omitted since these are subject to frequent change.

• Brown, BOUNDARY CONTROL AND LEGAL PRINCIPLES (John Wiley & Sons)

• Brown and Eldridge, EVIDENCE AND PROCEDURES FOR BOUNDARY LOCATION (John Wiley & Sons)

• Bureau of Land Management, MANUAL OF INSTRUCTIONS FOR THE SURVEY OF THE PUBLIC LANDS OF THE UNITED STATES

• Hickerson, ROUTE LOCATION AND DESIGN (McGraw-Hill, Inc.)

- ACSM and ASCE, DEFINITIONS OF SURVEYING AND ASSOCIATED TERMS

- Wattles, WRITING LEGAL DESCRIPTIONS

- In addition to these references, it is essential that you bring with you a solar ephemeris and plane coordinate projection tables from your state.

PROFESSIONAL PUBLICATIONS, INC. • P.O. Box 199, San Carlos, CA 94070

1 ARITHMETIC AND MEASUREMENTS

1 COMMON FRACTIONS

A common fraction is the division of one number by another. In the fraction $\frac{2}{3}$, the number above the line is known as the numerator, and the number below the line is known as the denominator. The line between the numbers indicates division.

A fraction whose numerator is smaller than its denominator is known as a proper fraction. A fraction whose numerator is equal to or larger than its denominator is known as an improper fraction.

2 CHANGING THE FORM OF FRACTIONS

If the numerator and denominator of a fraction are multiplied by the same number, the value of the fraction is not changed. Thus, if we multiply the numerator and denominator of the fraction $\frac{1}{2}$ by 2, we get the fraction $\frac{2}{4}$; if we multiply the numerator and denominator of $\frac{2}{3}$ by 3, we get $\frac{6}{9}$. The fractions $\frac{1}{2}$ and $\frac{2}{4}$ are known as equivalent fractions, as are $\frac{2}{3}$ and $\frac{6}{9}$.

Example 1.1

(a) $\dfrac{1}{4} \times \dfrac{2}{2} = \dfrac{2}{8}$

(b) $\dfrac{1}{3} \times \dfrac{3}{3} = \dfrac{3}{9}$

(c) $\dfrac{7}{8} \times \dfrac{4}{4} = \dfrac{28}{32}$

The numerator and denominator of a fraction may be divided by the same number without changing the value of the fraction. Thus, $\frac{3}{6}$ is equivalent to $\frac{1}{2}$ and $\frac{2}{8}$ is equivalent to $\frac{1}{4}$. In each case the numerator and denominator have been divided by 2. If the numerator and denominator have no common divisor except 1, the fraction is said to be in its lowest terms. Examples of fractions in their lowest terms are: $\frac{1}{2}$, $\frac{2}{9}$, $\frac{3}{8}$, $\frac{4}{11}$ and $\frac{5}{16}$.

3 MIXED NUMBERS

A number which consists of a whole number and a fraction, such as $5\frac{3}{8}$, is known as a mixed number. An improper fraction can be changed to a whole number or to a whole number and a fraction. A mixed number can be changed to an improper fraction.

To change an improper fraction to a mixed number, divide the numerator by the denominator.

To change a mixed number to an improper fraction, multiply the whole number by the denominator of the fraction and add the improper fraction and the proper fraction.

Example 1.2

(a) Change $\frac{15}{4}$ to a mixed number.

(b) Change $5\frac{3}{8}$ to an improper fraction.

Solutions

(a) $\dfrac{15}{4} = 15 \div 4 = 3\dfrac{3}{4}$

(b) $5\dfrac{3}{8} = \dfrac{5}{1} + \dfrac{3}{8} = \dfrac{5 \times 8}{1 \times 8} + \dfrac{3}{8} = \dfrac{40}{8} + \dfrac{3}{8} = \dfrac{43}{8}$

Example 1.3

Change each fraction to an equivalent fraction having the higher denominator.

(a) $\dfrac{1}{2} = \dfrac{x}{8}$

(b) $\dfrac{1}{4} = \dfrac{x}{32}$

(c) $\dfrac{5}{16} = \dfrac{x}{64}$

(d) $\dfrac{7}{12} = \dfrac{x}{72}$

PROFESSIONAL PUBLICATIONS, INC. • P.O. Box 199, San Carlos, CA 94070

Solutions

(a) $\dfrac{1}{2} = \dfrac{4}{8}$

(b) $\dfrac{1}{4} = \dfrac{8}{32}$

(c) $\dfrac{5}{16} = \dfrac{20}{64}$

(d) $\dfrac{7}{12} = \dfrac{42}{72}$

Example 1.4

Reduce the fractions to the lowest terms.

(a) $\dfrac{3}{9}$

(b) $\dfrac{12}{16}$

(c) $\dfrac{15}{35}$

(d) $\dfrac{9}{24}$

Solutions

(a) $\dfrac{3}{9} = \dfrac{1}{3}$

(b) $\dfrac{12}{16} = \dfrac{3}{4}$

(c) $\dfrac{15}{35} = \dfrac{3}{7}$

(d) $\dfrac{9}{24} = \dfrac{3}{8}$

Example 1.5

Change the improper fractions to equivalent mixed numbers reduced to the lowest terms.

(a) $\dfrac{21}{8}$

(b) $\dfrac{37}{7}$

(c) $\dfrac{51}{16}$

(d) $\dfrac{455}{64}$

Solutions

(a) $\dfrac{21}{8} = 2\dfrac{5}{8}$

(b) $\dfrac{37}{7} = 5\dfrac{2}{7}$

(c) $\dfrac{51}{16} = 3\dfrac{3}{16}$

(d) $\dfrac{455}{64} = 7\dfrac{7}{64}$

Example 1.6

Change the mixed numbers to improper fractions.

(a) $7\dfrac{8}{9}$

(b) $3\dfrac{3}{5}$

(c) $3\dfrac{5}{16}$

(d) $4\dfrac{5}{7}$

Solutions

(a) $7\dfrac{8}{9} = \dfrac{71}{9}$

(b) $3\dfrac{3}{5} = \dfrac{18}{5}$

(c) $3\dfrac{5}{16} = \dfrac{53}{16}$

(d) $4\dfrac{5}{7} = \dfrac{33}{7}$

4 ADDITION AND SUBTRACTION OF FRACTIONS

Before fractions can be added or subtracted, they must have the same denominator. The fractions $\frac{1}{2}$, $\frac{2}{3}$, and $\frac{3}{4}$ cannot be added directly, but if they are changed to the equivalent fractions $\frac{6}{12}$, $\frac{8}{12}$, and $\frac{9}{12}$, they can be added.

5 PRIME FACTORS

When we write $12 = 3 \times 4$, we have factored 12 into a product of two numbers. The numbers 3 and 4 are factors of 12. Also, 6 and 2 are factors of 12, and 12 and 1 are factors of 12.

A prime factor, or prime number, is a number which is divisible only by 1 and itself. The prime numbers are 2, 3, 5, 7, 11, 13, ...

The prime factors of numbers can be found by dividing the number by the lowest prime, 2; dividing the result by 2 until it is not divisible by 2; then dividing by the next higher prime until the quotient is 1.

Example 1.7

Find the prime factors of each number.

(a) 252

(b) 1575

Solutions

(a)

$$
\begin{array}{r|r}
2 & 252 \\
\hline
2 & 126 \\
\hline
3 & 63 \\
\hline
3 & 21 \\
\hline
7 & 7 \\
\hline
& 1
\end{array}
$$

$$252 = 2 \times 2 \times 3 \times 3 \times 7$$

(b)

$$
\begin{array}{r|r}
3 & 1575 \\
\hline
3 & 525 \\
\hline
5 & 175 \\
\hline
5 & 35 \\
\hline
7 & 7 \\
\hline
& 1
\end{array}
$$

$$1575 = 3 \times 3 \times 5 \times 5 \times 7$$

6 LEAST COMMON MULTIPLE

The least common multiple of two or more numbers is the least number which has each of the numbers as a factor. The least common multiple (LCM) of two or more numbers is found by finding the prime factors of each number and the product of all the prime factors using each prime the greatest number of times it appears in either number.

Example 1.8

Find the LCM of 12, 15, and 18.

Solution

$$12 = 2 \times 2 \times 3$$
$$15 = 3 \times 5$$
$$18 = 2 \times 3 \times 3$$

The LCM is $2 \times 2 \times 3 \times 3 \times 5 = 180$

7 LEAST COMMON DENOMINATOR

The least common denominator of two or more fractions, the LCD, is the least common multiple, the LCM, of the denominators. It is the smallest number that can be exactly divided by the denominators.

To add or subtract fractions with different denominators, first find the LCD for the fractions and change the fractions to equivalent ones using the LCD. Then add the numerators and place the sum over the common denominator.

Example 1.9

Add

(a) $\dfrac{3}{4} + \dfrac{2}{3} + \dfrac{5}{9}$

(b) $\dfrac{11}{16} + \dfrac{12}{14} + \dfrac{13}{21}$

Solutions

(a) $4 = 2 \times 2$
$\qquad 3 = 3$
$\qquad 9 = 3 \times 3$

$LCD = 2 \times 2 \times 3 \times 3 = 36$

$$\dfrac{3}{4} + \dfrac{2}{3} + \dfrac{5}{9} = \dfrac{27}{36} + \dfrac{24}{36} + \dfrac{20}{36} = \dfrac{71}{36} = 1\dfrac{35}{36}$$

(b) $16 = 2 \times 2 \times 2 \times 2$
$\qquad 14 = 2 \times 7$
$\qquad 21 = 3 \times 7$

$LCD = 2 \times 2 \times 2 \times 2 \times 3 \times 7 = 336$

$$\dfrac{11}{16} + \dfrac{12}{14} + \dfrac{13}{21} = \dfrac{231}{336} + \dfrac{288}{336} + \dfrac{208}{336} = \dfrac{727}{336} = 2\dfrac{55}{336}$$

8 MULTIPLICATION OF FRACTIONS

The product of fractions is the product of the numerators over the product of the denominators.

Example 1.10

(a) $\dfrac{3}{8} \times \dfrac{2}{3} \times \dfrac{1}{4} = \dfrac{6}{96} = \dfrac{1}{16}$

(b) $2\dfrac{7}{8} \times 3\dfrac{2}{3} \times 4\dfrac{1}{8} = \dfrac{23}{8} \times \dfrac{11}{3} \times \dfrac{33}{8} = \dfrac{8349}{192} = 43\dfrac{83}{192}$

9 DIVISION OF FRACTIONS

To divide a number by a fraction, to divide a fraction by a number, or to divide a fraction by a fraction, invert the second number and multiply by the first.

Example 1.11

(a) $4 \div \dfrac{3}{8} = \dfrac{4}{1} \times \dfrac{8}{3} = \dfrac{32}{3} = 10\dfrac{2}{3}$

(b) $\dfrac{3}{4} \div 5 = \dfrac{3}{4} \div \dfrac{5}{1} = \dfrac{3}{4} \times \dfrac{1}{5} = \dfrac{3}{20}$

(c) $\dfrac{3}{4} \div \dfrac{2}{3} = \dfrac{3}{4} \times \dfrac{3}{2} = \dfrac{9}{8} = 1\dfrac{1}{8}$

It will be noted that when a number is divided by a proper fraction, the quotient will be larger than the number.

PROFESSIONAL PUBLICATIONS, INC. • P.O. Box 199, San Carlos, CA   94070

10 CANCELLATION

A fraction containing the product of several numbers in the numerator and the product of several numbers in the denominator may be simplified by performing as many divisions as possible by cancellation.

Example 1.12

Simplify by cancellation:

(a) $\dfrac{(15)(6)}{(12)(5)}$

(b) $\dfrac{(6)(10)(8)}{(4)(3)(5)}$

(c) $\dfrac{(16)(15)(14)}{(30)(28)(8)}$

(d) $\dfrac{(15)(7)(8)(9)}{(21)(30)(5)}$

Solutions

(a) $\dfrac{(\overset{3}{\cancel{15}})(\overset{1}{\cancel{6}})}{(\underset{2}{\cancel{12}})(\underset{1}{\cancel{5}})} = \dfrac{3}{2} = 1\dfrac{1}{2}$

(b) $\dfrac{(\overset{2}{\cancel{6}})(\overset{2}{\cancel{10}})(\overset{2}{\cancel{8}})}{(\underset{1}{\cancel{4}})(\underset{1}{\cancel{3}})(\underset{1}{\cancel{5}})} = 8$

(c) $\dfrac{(\overset{1}{\underset{2}{\cancel{16}}})(\overset{1}{\cancel{15}})(\overset{1}{\cancel{14}})}{\underset{\underset{1}{2}}{\cancel{30}})(\underset{2}{\cancel{28}})(\underset{1}{\cancel{8}})} = \dfrac{1}{2}$

(d) $\dfrac{(\overset{1}{\cancel{15}})(\overset{1}{\cancel{7}})(\overset{4}{\cancel{8}})(\overset{3}{\cancel{9}})}{(\underset{\underset{1}{3}}{\cancel{21}})(\underset{\underset{1}{2}}{\cancel{30}})(5)} = \dfrac{12}{5} = 2\dfrac{2}{5}$

11 READING DECIMAL FRACTIONS

A decimal fraction is a fraction whose denominator is 10 or some power of 10. The fraction $\frac{5}{10}$ is written 0.5, and the fraction $\frac{75}{100}$ is written 0.75. The period preceding 75 is called the decimal point. The number 25,376.4921 is read twenty-five thousand, three hundred seventy-six and four thousand nine hundred twenty-one ten-thousandths. The "and" is used for the decimal point. Decimal terminology for this number is illustrated in Fig. 1.1.

A decimal fraction which is greater than one is called a mixed decimal. An integer, that is a whole number such as 325, is understood to have a decimal point to the right of the units digit (325.).

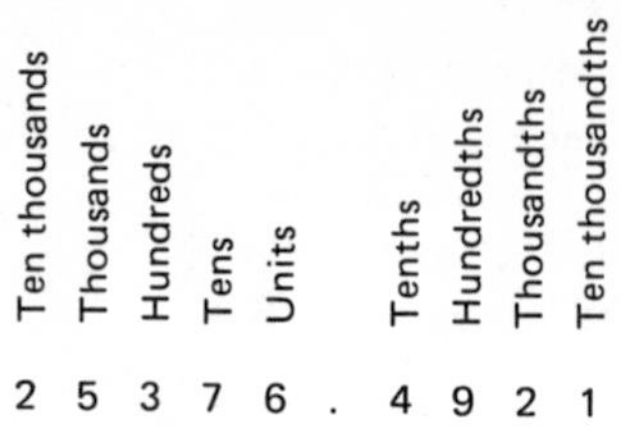

Ten thousands	Thousands	Hundreds	Tens	Units		Tenths	Hundredths	Thousandths	Ten thousandths
2	5	3	7	6	.	4	9	2	1

Figure 1.1

Example 1.13

Write in words:

(a) 388.152

(b) 72.006

(c) 0.00005

(d) 326.2200

Solutions

(a) Three hundred eighty-eight and one hundred fifty-two thousandths

(b) Seventy-two and six thousandths

(c) Zero and five hundred-thousandths

(d) Three hundred twenty-six and two thousand two hundred ten-thousandths

12 MULTIPLYING AND DIVIDING DECIMAL FRACTIONS

Multiplying a number by 10 moves the decimal point in the number one place to the *right*. Multiplying by 100 moves the decimal point two places to the right. In general, multiplying a number by 10 or a multiple of 10 moves the decimal to the right a number of places equivalent to the number of zeros in the multiplier.

Example 1.14

$$537.26 \times 10 = 5372.6$$
$$421.38 \times 100 = 42138$$
$$6.7382 \times 1000 = 6738.2$$

Dividing a number by 10 or a multiple of 10 moves the decimal point to the *left* a number of places equivalent to the number of zeros in the divisor.

Example 1.15

$$537.26 \div 10 = 53.726$$
$$421.38 \div 100 = 4.2138$$
$$6.7382 \div 1000 = 0.0067382$$

Multiplying a number by 0.1 is the same as dividing by 10 and moves the decimal point to the *left*.

Example 1.16

$$4762 \times 0.1 = 4762 \times \frac{1}{10} = 476.2$$
$$378 \times 0.01 = 378 \times \frac{1}{100} = 3.78$$
$$9843 \times 0.0001 = 9843 \times \frac{1}{10,000} = 0.9843$$

Dividing a number by 0.1 is the same as multiplying by 10 and moves the decimal point to the *right*.

Example 1.17

$$\frac{537.26}{0.1} = \frac{537.26}{\frac{1}{10}} = 537.26 \times \frac{10}{1} = 5372.6$$
$$\frac{421.38}{0.01} = \frac{421.38}{\frac{1}{100}} = 421.38 \times \frac{100}{1} = 42138$$

13 PERCENT

Our word percent comes from a Latin word meaning *by the hundred*. The symbol "%" is used for the word *percent*. A fraction having 100 as a denominator expresses a percent. Thus, $\frac{6}{100}$ is the same as 6% as is 0.06.

14 CHANGING A DECIMAL FRACTION TO A PERCENT

To change a decimal fraction to a percent, move the decimal point two places to the right and place the percent sign, %, after the number.

Example 1.18

$$0.07 = 7\%$$
$$0.25 = 25\%$$
$$8.39 = 839\%$$
$$0.005 = 0.5\%$$

15 CHANGING A PERCENT TO A DECIMAL FRACTION

To change a percent to a decimal fraction, remove the % sign and move the decimal point two places to the left.

Example 1.19

$$4\% = 0.04$$
$$75\% = 0.75$$
$$150\% = 1.50$$
$$0.7\% = 0.007$$

16 FINDING A PERCENT OF A NUMBER

When we say 25% of a number we mean 25% times the number. To find a percent of a number, change the percent to a decimal fraction and multiply the number by this decimal fraction.

Example 1.20

Find the percent of the number as indicated.

(a) 10% of 156
(b) 1% of 7823
(c) 25% of 200
(d) 68% of 300

Solutions

(a) 10% of 156 = (0.10)(156) = 15.6
(b) 1% of 7823 = (0.01)(7823) = 78.23
(c) 25% of 200 = (0.25)(200) = 50
(d) 68% of 300 = (0.68)(300) = 204

17 FINDING WHAT PERCENT ONE NUMBER IS OF ANOTHER

To find what percent one number is of another, divide the first by the second and express as a decimal fraction. Then move the decimal point two places to the right and add the percent sign.

Example 1.21

(a) 15 is what percent of 750?
(b) What percent of 64 is 16?

Solutions

(a) $\frac{15}{750} = 0.02 = 2\%$

(b) $\frac{16}{64} = 0.25 = 25\%$

PROFESSIONAL PUBLICATIONS, INC. • P.O. Box 199, San Carlos, CA 94070

18 FINDING A NUMBER WHEN A PERCENT OF THE NUMBER IS KNOWN

In the example under "Finding a Percent of a Number," 35% of 200 = 0.35 × 200 = 70. Now if we know that 70 is 35% of some number, then 70 ÷ 35% = the number = 200.

Example 2.22

(a) 25% of what number is 10?

(b) 40 is 80% of what number?

Solutions

(a) $\dfrac{10}{25\%} = \dfrac{10}{0.25} = 40$

(b) $\dfrac{40}{80\%} = \dfrac{40}{0.80} = 50$

By converting percent to a common fraction, many problems can be solved more rapidly. If 30 is 75% of some number and we want to find that number, we divide 30 by 75% to find the number 40 as has been explained.

$$\frac{30}{75\%} = \frac{30}{0.75} = 40$$

But knowing that $75\% = 0.75 = \frac{3}{4}$, we divide 30 by $\frac{3}{4}$.

$$\frac{30}{\frac{3}{4}} = \frac{30}{1} \times \frac{4}{3} = 40$$

By relating the common-fraction equivalent to certain percentages in solving problems involving percent, we make the solution much simpler. The first step in using this method is to memorize the following equivalents.

$$\frac{1}{100} = 1\% \qquad \frac{1}{3} = 33\frac{1}{3}\%$$
$$\frac{1}{10} = 10\% \qquad \frac{1}{2} = 50\%$$
$$\frac{1}{8} = 12.5\% \qquad \frac{2}{3} = 66\frac{2}{3}\%$$
$$\frac{1}{4} = 25\% \qquad \frac{3}{4} = 75\%$$
$$\frac{3}{8} = 37.5\% \qquad 1 = 100\%$$

Example 1.23

(a) 6 is 50% of what number?

(b) 12 is 25% of what number?

(c) 45 is $33\frac{1}{3}\%$ of what number?

(d) 120 is 75% of what number?

(e) 12 is $12\frac{1}{2}\%$ of what number?

(f) 6.5 is 25% of what number?

(g) $\dfrac{3}{8}$ is $37\frac{1}{2}\%$ of what number?

(h) $\dfrac{2}{3}$ is 75% of what number?

Solutions

(a) $\dfrac{6}{50\%} = \dfrac{6}{\frac{1}{2}} = \dfrac{6}{1} \times \dfrac{2}{1} = 12$

(b) $\dfrac{12}{25\%} = \dfrac{12}{\frac{1}{4}} = \dfrac{12}{1} \times \dfrac{4}{1} = 48$

(c) $\dfrac{45}{33\frac{1}{3}\%} = \dfrac{45}{\frac{1}{3}} = \dfrac{45}{1} \times \dfrac{3}{1} = 135$

(d) $\dfrac{120}{75\%} = \dfrac{120}{\frac{3}{4}} = \dfrac{120}{1} \times \dfrac{4}{3} = 160$

(e) $\dfrac{12}{12\frac{1}{2}\%} = \dfrac{12}{\frac{1}{8}} = \dfrac{12}{1} \times \dfrac{8}{1} = 96$

(f) $\dfrac{6.5}{25\%} = \dfrac{6.5}{\frac{1}{4}} = \dfrac{6.5}{1} \times \dfrac{4}{1} = 26$

(g) $\dfrac{\frac{3}{8}}{37.5\%} = \dfrac{3}{8} \div \dfrac{3}{8} = \dfrac{3}{8} \times \dfrac{8}{3} = 1$

(h) $\dfrac{\frac{2}{3}}{75\%} = \dfrac{2}{3} \div \dfrac{3}{4} = \dfrac{2}{3} \times \dfrac{4}{3} = \dfrac{8}{9}$

19 ROUNDING-OFF NUMBERS

We often hear someone say: "in round numbers." We might estimate the crowd at a football game at 50,000, meaning that it is between 40,000 and 60,000. Or, we might say that a certain university has 12,000 students enrolled, meaning that there are between 11,000 and 13,000 students. The exact figure is often unimportant in conveying an idea of size or quantity, as is shown in the two examples: the size of a crowd and the enrollment of a university.

In like manner, the use of approximate numbers—rounded-off numbers—can make locating the decimal point in computations relatively simple.

For instance, in finding the product of 310 and 21 we know it is approximately 6000 because 300 × 20 is equal to 6000. We have rounded-off 310 to 300 and 21 to 20. Mentally we have said 3 × 2 is 6, and seeing three zeros in the approximation 300 × 20 we add three zeros to the right of 6 and get 6000. Thus, we are able to estimate, immediately, the approximate product. This system is one of those used in finding the decimal point when approximate answers are needed quickly.

In rounding-off a number so that it contains a certain number of significant digits, the following procedure should be observed:

1. When the digit to be dropped is less than 5, no change is made in the number retained. Thus, 58.463 becomes 58.46.

2. When the digit to be dropped is greater than 5, the digit preceding it is increased by 1. Thus, 58.467 becomes 58.47.

3. When the digit to be dropped is exactly 5, the digit preceding it is increased by 1 if it is odd but left unchanged if it is even. Thus, 3.55 becomes 3.6, and 15.45 becomes 15.4.

The procedure in 1 and 2 is universally accepted; however, the procedure in 3 is not always followed. Some computers prefer to take the next higher number when the digit to be dropped is 5.

Example 1.24

Round-off each number to the nearest tenth.

(a) 3.66
(b) 6.74
(c) 5.45
(d) 5.35
(e) 2.751
(f) 2.749
(g) 7.049
(h) 7.051

Solutions

(a) 3.7
(b) 6.7
(c) 5.4
(d) 5.4
(e) 2.8
(f) 2.7
(g) 7.0
(h) 7.1

Example 1.24

Round-off each number to the nearest hundredth.

(a) 8.344
(b) 7.465
(c) 5.295
(d) 2.355
(e) 6.114
(f) 9.455
(g) 1.006
(h) 3.015

Solutions

(a) 8.34
(b) 7.46
(c) 5.30
(d) 2.36
(e) 6.11
(f) 9.46
(g) 1.01
(h) 3.02

Example 1.25

Round-off each number to the nearest hundred.

(a) 28232
(b) 32008
(c) 41500
(d) 8501
(e) 9552
(f) 25962
(g) 60499
(h) 1990

Solutions

(a) 28200
(b) 32000
(c) 41500
(d) 8500
(e) 9600
(f) 26000
(g) 60500
(h) 2000

20 EXACT AND APPROXIMATE NUMBERS

Many beginning students in engineering technology experience some confusion in dealing with approximate numbers. Mathematics, to these students, has dealt largely with exact numbers. Counting numbers are exact numbers. The number of students in a classroom is an exact number. The number of inches in a foot is an exact number. Thus, when the student has been asked to multiply 2.3×12.78, he finds the result to be 29.394. But if the numbers 2.3 and 12.78 represent measured quantities, the result 29 might better represent the actual value. It is difficult for the student to know "how many places to carry the answer."

A measured quantity is an approximate number and can never be an exact number. Let us consider measuring the distance between two points with two different tapes. The first tape is graduated into feet and tenths of a foot. The second tape is graduated into feet, tenths, and hundredths of a foot.

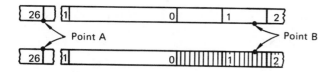

Figure 1.2

Suppose we record the distance as 26.18 feet using the first tape. This means that we can be sure that the first *three* digits are exact or positive digits, but that the last digit is estimated. We have determined the last digit, 8, by estimating that one of the points lies along our tape between the 1 and 2 tenth marks, and we estimate that it lies 8 tenths of the way between the 1 and the 2.

Now we use the other tape, graduated into hundredths, and record the distance as 26.185. We can say that the first *four* digits are exact or positive numbers and that the last digit is estimated. Thus, it can be seen that the number of digits recorded for a measurement is an indication of the accuracy of the measurement. There will always be an indeterminable amount of error in a measured quantity, and we will never reach an exact number.

21 SIGNIFICANT DIGITS

The accuracy of computed solutions can be no greater than the accuracy of the data used in the computations.

A method commonly used to indicate the degree of accuracy of computations is the use of significant digits. The number of decimal places in a number does *not* indicate the number of significant digits in the number.

All the digits in a number are significant except zero, which *may* or *may not* be significant.

A zero is *not* significant if it is used as a placeholder following a decimal, as in .004, .062, and .0002

A zero is *not* significant if it occurs at the end of a measured number unless information is available which indicates that it is. For example, if a distance is recorded as 260 feet, the zero is significant if the distance is measured to the nearest foot, but not significant if measured to the nearest 10 feet.

Zeros following a decimal point are usually considered significant. For example, a measurement recorded as 216.00 feet would indicate that the zeros are significant.

A zero is always significant when it occurs between two other digits as in 306.

In general, the following rules can be applied.

1. Initial zeros *are not* significant, as in 0.0065.

2. Final zeros *are not* significant unless they follow a decimal point, as 22.00.

3. Zeros between two other digits *are* significant, as 30,605.

Example 1.26

Write the number of significant digits in each of the following numbers.

   (a) 369.5

   (b) 0.036

   (c) 603

   (d) 300

   (e) 16.50

   (f) 0.0006

   (g) 0.0020

   (h) 0.0301

Solutions

   (a) 4

   (b) 2

   (c) 3

   (d) 1

   (e) 4

(f) 1

(g) 2

(h) 3

Example 1.27

Round-off each number to three significant digits.

(a) 325,496

(b) 36,374

(c) 0.05078

(d) 2.465

(e) 286,501

(f) 2.002

(g) 0.00606

(h) 4.255

Solutions

(a) 325,000

(b) 36,400

(c) 0.0508

(d) 2.46

(e) 287,000

(f) 2.00

(g) 0.00606

(h) 4.26

Example 1.28

Round-off each number to one significant digit.

(a) 0.449

(b) 56,325

(c) 0.0056

(d) 3.448

(e) 0.0719

(f) 100,788

(g) 251

(h) 16.32

Solutions

(a) 0.4

(b) 60,000

(c) 0.006

(d) 3

(e) 0.07

(f) 100,000

(g) 300

(h) 20

22 COMPUTATIONS WITH APPROXIMATE DATA

Suppose we measure a rectangle and find it to be 14.2 feet long and 12.6 feet wide. The student not familiar with computations using measured quantities might be tempted to say that the area of the rectangle is $(14.2)(12.6) = 178.92$ square feet, but he would indicate false accuracy. The result should be rounded-off to the same number of significant digits as in the measured quantities. A more reasonable result would be 179 square feet.

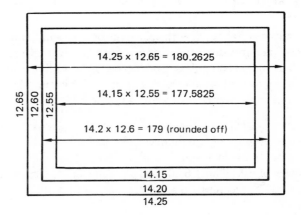

Figure 1.3

Let us examine this more closely. The length was measured to the nearest tenth of a foot and indicates that the exact distance lies between 14.15 feet and 14.25 feet, remembering that the last digit recorded is an estimate. Likewise, the exact distance for the width would lie between 12.55 feet and 12.65 feet. Now, using the lower limits of the measurements, we find that the area is $(14.15)(12.55) = 177.5825$ square feet, and using the upper limits, we find that the area is $(14.25)(12.65) = 180.2625$ square ft. The more accurate result lies somewhere between the two measurements.

In multiplying (or dividing) two approximate numbers, the product (or quotient) should not contain more significant digits than the number containing the fewer significant digits.

Example 1.29

Multiply the approximate numbers.

$$246.34 \times 3.12$$

Solution

$$246.34 \times 3.12 = 768.5808; \text{ therefore,}$$
$$246.34 \times 3.12 = 769$$

In adding numbers, the accuracy of values is governed by the number of places of digits rather than the number of significant digits. However, when two quantities are added together which were measured to a different degree of accuracy, the accuracy of the result cannot be greater than that of the quantity of least accuracy. For instance, in summing up earthwork, a quantity of 26.5 cubic yards added to a quantity of 156 cubic yards would give a result of 182 cubic yards and not 182.5 cubic yards.

23 SCIENTIFIC NOTATION

Scientists often work with very large or very small numbers. For instance, the distance from the earth to the sun is 93,000,000 miles, and the coefficient of linear expansion of steel is 0.0000065. In scientific notation these numbers are written 9.3×10^7 and 6.5×10^{-6}.

A number expressed in scientific notation is expressed as the product of two numbers, the first number having a value between 1 and 10 and the second number as a power of 10. Thus, 6,500,000 becomes 6.5×10^6, and 0.0000078 becomes 7.8×10^{-6}.

Example 1.30

Express 128,000 in scientific notation.

Solution

Place a caret to the right of the first significant digit as follows: $1_\wedge 28000$. Starting at the caret, count the number of places to the decimal point. Since we count 5 places to the right to the decimal point, the number 1.28000 must be multiplied by 10^5 to equal 128,000 Therefore, $128,000 = 1.28 \times 10^5$.

Example 1.31

Express 0.000433 in scientific notation.

Solution

Place a caret to the right of the first significant digit as follows: $0.0004_\wedge 33$. Starting at the caret, we count four places to the left to the decimal point. Then 4.33 must be multiplied by 10^{-4} to equal 0.000433. Therefore, $0.000433 = 4.33 \times 10^{-4}$ in scientific notation.

Where a number is expressed in scientific notation, the number of digits in the first factor should equal the number of significant digits of the number. For example, the number 4.23×10^6 has three significant digits.

Example 1.32

Express each of the following numbers in scientific notation.

   (a) 856,000

   (b) 0.00042

   (c) 0.563

   (d) 1,500,000

   (e) 55200

   (f) 7.73

Solutions

   (a) 8.56×10^5

   (b) 4.2×10^{-4}

   (c) 5.63×10^{-1}

   (d) 1.5×10^6

   (e) 5.52×10^4

   (f) 7.73×10^0

24 SQUARES AND SQUARE ROOTS

When we say 3 squared equals 9, we mean that 3 multiplied by itself equals 9.

When we say the square root of 9 is 3, we mean that 3 is the number which produces 9 when multiplied by itself. We use the symbol, $\sqrt{}$, to represent "the square root of." The symbol is known as the radical sign, or simply, the radical. Thus,

$$
\begin{aligned}
1^2 &= 1, &\text{and} &&\sqrt{1} &= 1 \\
2^2 &= 4, &\text{and} &&\sqrt{4} &= 2 \\
3^2 &= 9, &\text{and} &&\sqrt{9} &= 3 \\
4^2 &= 16, &\text{and} &&\sqrt{16} &= 4 \\
5^2 &= 25, &\text{and} &&\sqrt{25} &= 5 \\
6^2 &= 36, &\text{and} &&\sqrt{36} &= 6 \\
7^2 &= 49, &\text{and} &&\sqrt{49} &= 7 \\
8^2 &= 64, &\text{and} &&\sqrt{64} &= 8 \\
9^2 &= 81, &\text{and} &&\sqrt{81} &= 9 \\
10^2 &= 100, &\text{and} &&\sqrt{100} &= 10
\end{aligned}
$$

PROFESSIONAL PUBLICATIONS, INC. • P.O. Box 199, San Carlos, CA   94070

25 FINDING THE SQUARE ROOT OF A NUMBER LONGHAND

The square root, or the approximate square root, of a number can be quickly found on an electronic calculator, but the ability to find the square root with only pencil and paper still has value.

The first step in finding the square root of a number longhand is to pair the digits in the number by placing a line, or bar, over each pair of digits. The digits are paired by starting at the decimal point and pairing the first two digits to the left of the decimal point, then the next two, if there are two more, et cetera, until all the digits to the left of the decimal point are in pairs. If there is an odd number of digits to the left of the decimal point, the last digit on the left will be considered to be a pair. After digits to the left have been paired, digits to the right are paired in the same way, by moving to the right. If there is an odd number of digits to the right, a zero is added on the right to make the pair.

Example 1.33

Place bars over pairs of digits of each number in preparation for finding the square root of the number.

(a) 356

(b) 5874

(c) 72365

(d) 847.2

(e) 4469.35

(f) 289.396

Solutions

(a) $\overline{3}\,\overline{56}$

(b) $\overline{58}\,\overline{74}$

(c) $\overline{7}\,\overline{23}\,\overline{65}$

(d) $\overline{8}\,\overline{47}.\overline{20}$

(e) $\overline{44}\,\overline{69}.\overline{35}$

(f) $\overline{2}\,\overline{89}.\overline{39}\,\overline{60}$

After the digits are paired, the square root is found as explained in the following examples.

Example 1.34

Find the square root of 3136.

Solution

1. Group the digits in pairs as explained above.

$\sqrt{\overline{31}\,\overline{36}}$

2. Find the largest number whose square is equal to or less than the number under the left bar and place it over the left bar. The number is 5 ($5^2 = 25$ is less than 31).

$\begin{array}{c} 5 \\ \sqrt{\overline{31}\,\overline{36}} \end{array}$

3. Square the number above the left bar (5) and write it under the left pair of digits.

$\begin{array}{c} 5 \\ \sqrt{\overline{31}\,\overline{36}} \\ 25 \end{array}$

4. Subtract the number (25) from the left pair of digits. The remainder is 6.

$\begin{array}{c} 5 \\ \sqrt{\overline{31}\,\overline{36}} \\ \underline{25} \\ 6 \end{array}$

5. Bring down the next pair of digits (36).

$\begin{array}{c} 5 \\ \sqrt{\overline{31}\,\overline{36}} \\ \underline{25} \\ 636 \end{array}$

6. Find the trial divisor for the remainder (636) by doubling the number in the answer (5) and multiplying that number by ten ($2 \times 5 = 10$; $10 \times 10 = 100$).

$\begin{array}{c} 5 \\ \sqrt{\overline{31}\,\overline{36}} \\ 25 \\ 100)\overline{636} \end{array}$

7. Divide the dividend (636) by this trial divisor and find a one-significant-digit quotient ($636 \div 100 = 6$); then place this quotient (6) over the next bar and replace the last digit in the trial divisor (0) with this quotient (6).

$\begin{array}{c} 5\;\;6 \\ \sqrt{\overline{31}\,\overline{36}} \\ 25 \\ 106)\overline{636} \end{array}$

8. Multiply the correct divisor (106) by the second number in the answer (6) and place the product under the remainder (636). It can be seen that there will be no remainder when this product is subtracted, so we can say that the square root of 3136 is 56, which should be proved by multiplying 56×56.

$\begin{array}{c} 5\;\;6 \\ \sqrt{\overline{31}\,\overline{36}} \\ 25 \\ 106)\overline{636} \\ \underline{636} \end{array}$

Example 1.35

Find the approximate square root of 736.412 to two decimal places.

Solution

$$\overline{7}\,\overline{36}.\overline{41}\,\overline{20}\,\overline{00}$$

1. Pair the digits. There will be a digit in the answer to match each pair in the number. To find the answer to 2 decimals rounded, we add enough decimals to find 3 decimals in the answer and then round the last digit.

$$\frac{2}{\sqrt{\overline{7}\,\overline{36}.\overline{41}\,\overline{20}\,\overline{00}}}$$

2. The number 2 is the largest number which, when squared, can be written under the left pair (7). Write it above the left pair (7).

$$\begin{array}{c} 2 \\ \sqrt{\overline{7}\,\overline{36}.\overline{41}\,\overline{20}\,\overline{00}} \\ 4 \end{array}$$

3. Square the number 2 and write it (4) under the left pair of digits (7).

$$\begin{array}{c} 2 \\ \sqrt{\overline{7}\,\overline{36}.\overline{41}\,\overline{20}\,\overline{00}} \\ 4 \\ \hline 336 \end{array}$$

4. Subtract and bring down the next pair.

$$\begin{array}{c} 2 \\ \sqrt{\overline{7}\,\overline{36}.\overline{41}\,\overline{20}\,\overline{00}} \\ 4 \\ 40)\overline{336} \end{array}$$

5. To find a trial divisor for 336, double the digit in the answer (2) and multiply by 10. ($2 \times 2 \times 10 = 40$).

$$\begin{array}{c} 2\ \ 7 \\ \sqrt{\overline{7}\,\overline{36}.\overline{41}\,\overline{20}\,\overline{00}} \\ 4 \\ 47)\overline{336} \end{array}$$

6. Divide 336 by 40 to one-significant-digit quotient ($336 \div 40 = 7$). Write 7 above next pair of digits and change last digit in 40 to 7 (47).

$$\begin{array}{c} 2\ \ 7 \\ \sqrt{\overline{7}\,\overline{36}.\overline{41}\,\overline{20}\,\overline{00}} \\ 4 \\ 47)\overline{336} \\ \underline{329} \\ 7 \end{array}$$

7. Multiply 47 by 7, place product under remainder (336) and subtract. Note: If we had tried 8 instead of 7, we would have gotten the product $8 \times 48 = 364$, which is greater than 336.

$$\begin{array}{c} 2\ \ 7 \\ \sqrt{\overline{7}\,\overline{36}.\overline{41}\,\overline{20}\,\overline{00}} \\ 4 \\ 47)\overline{336} \\ \underline{329} \\ 7\ 41 \end{array}$$

8. Bring down the next pair of digits (41). Ignore decimal for now.

$$\begin{array}{c} 2\ \ 7 \\ \sqrt{\overline{7}\,\overline{36}.\overline{41}\,\overline{20}\,\overline{00}} \\ 4 \\ 47)\overline{336} \\ \underline{329} \\ 540)\ 7\ 41 \end{array}$$

9. Find a trial divisor by doubling the answer to this point (27) and multiplying by 10. ($27 \times 2 \times 10 = 540$).

$$\begin{array}{c} 2\ \ 7\ \ 1 \\ \sqrt{\overline{7}\,\overline{36}.\overline{41}\,\overline{20}\,\overline{00}} \\ 4 \\ 47)\overline{336} \\ \underline{329} \\ 541)\ 7\ 41 \end{array}$$

10. Divide dividend (741) by trial divisor to one-significant-digit quotient ($741 \div 540 = 1$). Place this quotient in answer above next pair and change last digit in trial divisor (0) to 1.

$$\begin{array}{c} 2\ \ 7.\ 1\ \ 3\ \ 6 \\ \sqrt{\overline{7}\,\overline{36}.\overline{41}\,\overline{20}\,\overline{00}} \\ 4 \\ 47)\overline{336} \\ \underline{329} \\ 541)\ 7\ 41 \\ \ \ \ \ \ 5\ 41 \\ 5423)\overline{2\ 0\ 0\ 2\ 0} \\ \ \ \ \ 1\ 6\ 2\ 6\ 9 \\ 54266)\overline{3\ 7\ 5\ 1\ 0\ 0} \\ \ \ \ \ \ 3\ 2\ 5\ 5\ 9\ 6 \\ \ \ \ \ \ \ 4\ 9\ 5\ 0\ 4 \end{array}$$

11. Continue the procedure.

12. The approximate square root of 736.412 is 27.14, rounded to 2 decimal places. Note that the decimal point in the answer is directly above the decimal point in the number. There are two pairs (bars) to the left of the decimal point in the number and two digits to the left of the decimal point in the answer. Each bar in the number is matched by a digit in the answer.

With very little practice, an estimated square root of most any number can be made mentally. Let us consider the square root of the following numbers.

(a) $\overline{1}\,\overline{28}$

(b) $\overline{9}\,\overline{75}$

(c) $\overline{11}\,\overline{33}$

(d) $\overline{98}\,\overline{68}$

Solution

(a) The square root has two digits; the first digit is 1.

(b) The square root has two digits; the first digit is 3.

Considering $\sqrt{100} = 10$ and $\sqrt{900} = 30$, a good guess for (a) and (b) would be: $\sqrt{128} = 11$ and $\sqrt{975} = 32$.

(c) The square root has two digits; the first digit is 3; so the square root must be less than 40. Considering $\sqrt{1600} = 40$, a good guess for (c) would be: $\sqrt{1133} = 35$.

(d) The square root has two digits; the first digit is 9. Considering that $\sqrt{10,000} = 100$, a good guess for (d) would be: $\sqrt{9868} = 99$. Now consider larger numbers.

(a) $\overline{1}\,\overline{23}\,\overline{46}$

(b) $\overline{9}\,\overline{67}\,\overline{41}$

(c) $\overline{13}\,\overline{72}\,\overline{42}$

Solution

(a) The square root has three digits; the first digit is 1. Considering $\sqrt{10,000} = 100$, a good guess is 110.

(b) The square root has three digits; the first digit is 3. Considering $\sqrt{160,000} = 400$, a good guess is 310.

(c) The square root has three digits; the first digit is 3. Considering (b) and that the square root must be less than 400, a good guess is: $\sqrt{137,242} = 375$. Now let us consider smaller numbers.

(a) 0.25

(b) 0.348

(c) 0.0056

(d) 0.000084

Solution

(a) $0.\overline{25}$ (b) $0.\overline{34}\overline{80}$ (c) $0.\overline{00}\overline{56}$ (d) $0.\overline{00}\overline{00}\overline{84}$

(a) There is a digit for each bar: $\sqrt{0.25} = 0.5$

(b) There is a digit for each bar; the first digit is 5. A good guess is: $\sqrt{0.3480} = 0.58$.

(c) There is a digit for each bar; the first digit must be 0. The second digit is 7. A good guess is $\sqrt{0.0056} = 0.07$.

(d) There is a digit for each bar; the first two digits must be 0. The third digit is 9. A good guess: $\sqrt{0.000084} = 0.009$.

The idea can be expanded to make numbers more interesting. Notice that the square root of a decimal fraction is larger than the decimal fraction.

26 THE PYTHAGOREAN THEOREM

The Greek mathematician Pythagoras, who lived some 500 years before Christ, is given credit for a very important theorem which is widely used today. The Pythagorean Theorem states: In a right triangle, the square of the hypotenuse is equal to the sum of the squares of the other two sides.

If we call the base b, the altitude a and the hypotenuse c, we have

$$c^2 = a^2 + b^2$$
$$a^2 = c^2 - b^2$$
$$b^2 = c^2 - a^2$$
$$c = \sqrt{a^2 + b^2}$$
$$a = \sqrt{c^2 - b^2}$$
$$b = \sqrt{c^2 - a^2}$$

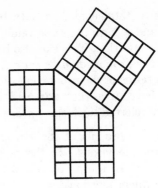

Figure 1.4

Example 1.36

Find the hypotenuse of the triangle shown.

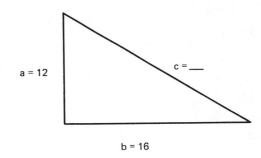

$a = 12$ $c = \underline{\quad}$ $b = 16$

Figure 1.5

Solution

$$c^2 = a^2 + b^2$$
$$c = \sqrt{a^2 + b^2} = \sqrt{12^2 + 16^2}$$
$$= \sqrt{144 + 256} = \sqrt{400} = 20$$

Example 1.37

A city lot forms a right triangle with the hypotenuse 100 feet and one side 60 feet. Find the other side.

Solution

$$b = \sqrt{c^2 - a^2}$$
$$= \sqrt{100^2 - 60^2} = 80 \text{ feet}$$

27 RATIO AND PROPORTION

A ratio is a comparison of two quantities. It is the quotient of the first quantity divided by the second. The ratio of 1 to 2 is $\frac{1}{2}$. The ratio of 2 to 1 is $\frac{2}{1}$. A ratio can be stated in four ways: The ratio of 2 to 4, $2 : 4$, $2 \div 4$, or $\frac{2}{4}$. Thus, a ratio is a fraction and can be treated as a fraction.

PROFESSIONAL PUBLICATIONS, INC. ● P.O. Box 199, San Carlos, CA   94070

A proportion is a statement of equality between two ratios. Since $\frac{2}{4}$ and $\frac{1}{2}$ have the same value, we can say that $\frac{2}{4} = \frac{1}{2}$ is a proportion. This can also be written as $2 : 4 = 1 : 2$. In the proportion $a : b = c : d$, we refer to a, b, c, and d as terms of the proportion. The first and last terms are referred to as the "*extremes*" and the two middle terms are referred to as the "*means*."

In a proportion, the product of the extremes is equal to the product of the means. In the proportion $a : b = c : d$, $ad = bc$. In the proportion $3 : 4 = 6 : 8$, $3 \times 8 = 4 \times 6$. In the proportion $\frac{1}{2} = \frac{5}{10}$, $2 \times 5 = 1 \times 10$. (This is sometimes called cross-multiplying.)

In the proportion $x : 3 = 4 : 6$, $6x = 12$, and $x = 2$.

Example 1.38

Find the value of x.

(a) $x : 4 = 2 : 8$

(b) $2 : x = 3 : 9$

(c) $3 : 6 = x : 2$

(d) $\dfrac{4}{12} = \dfrac{x}{6}$

(e) $\dfrac{9}{6} = \dfrac{12}{x}$

Solutions

(a) $x : 4 = 2 : 8$; $8x = 8$; $x = 1$

(b) $2 : x = 3 : 9$; $3x = 18$; $x = 6$

(c) $3 : 6 = x : 2$; $6x = 6$; $x = 1$

(d) $\dfrac{4}{12} = \dfrac{x}{6}$; $12x = 24$; $x = 2$

(e) $\dfrac{9}{6} = \dfrac{12}{x}$; $9x = 72$; $x = 8$

PRACTICE PROBLEMS FOR CHAPTER 1

1. Change the given fraction to an equivalent fraction having the higher denominator indicated.

Examples: (1) $\dfrac{2}{5} = \dfrac{10}{25}$ (2) $\dfrac{2}{3} = \dfrac{6}{9}$

(a) $\dfrac{1}{4} = \dfrac{x}{16}$ (b) $\dfrac{1}{6} = \dfrac{x}{24}$ (c) $\dfrac{3}{5} = \dfrac{x}{15}$ (d) $\dfrac{3}{4} = \dfrac{x}{16}$

(e) $\dfrac{3}{8} = \dfrac{x}{32}$ (f) $\dfrac{1}{3} = \dfrac{x}{15}$ (g) $\dfrac{2}{3} = \dfrac{x}{18}$ (h) $\dfrac{2}{7} = \dfrac{x}{21}$

(i) $\dfrac{5}{8} = \dfrac{x}{24}$ (j) $\dfrac{8}{3} = \dfrac{x}{9}$ (k) $\dfrac{7}{8} = \dfrac{x}{32}$ (l) $\dfrac{1}{5} = \dfrac{x}{25}$

(m) $\dfrac{4}{5} = \dfrac{x}{30}$ (n) $\dfrac{5}{8} = \dfrac{x}{32}$ (o) $\dfrac{1}{6} = \dfrac{x}{30}$ (p) $\dfrac{3}{9} = \dfrac{x}{81}$

(q) $\dfrac{2}{9} = \dfrac{x}{45}$ (r) $\dfrac{3}{16} = \dfrac{x}{48}$ (s) $\dfrac{5}{12} = \dfrac{x}{72}$ (t) $\dfrac{3}{7} = \dfrac{x}{56}$

2. Reduce the fractions to lowest terms.

Examples: (1) $\dfrac{6}{8} = \dfrac{3}{4}$ (2) $\dfrac{8}{12} = \dfrac{2}{3}$

(a) $\dfrac{2}{8}$ (b) $\dfrac{12}{16}$ (c) $\dfrac{21}{42}$ (d) $\dfrac{4}{10}$

(e) $\dfrac{12}{32}$ (f) $\dfrac{27}{45}$ (g) $\dfrac{6}{54}$ (h) $\dfrac{9}{16}$

(i) $\dfrac{30}{45}$ (j) $\dfrac{18}{54}$ (k) $\dfrac{36}{72}$ (l) $\dfrac{48}{64}$

(m) $\dfrac{14}{28}$ (n) $\dfrac{9}{27}$ (o) $\dfrac{12}{16}$ (p) $\dfrac{24}{32}$

(q) $\dfrac{54}{81}$ (r) $\dfrac{17}{51}$ (s) $\dfrac{27}{81}$ (t) $\dfrac{25}{35}$

3. Change the improper fractions to equivalent mixed numbers reduced to lowest terms.

Examples: (1) $\dfrac{16}{3} = 5\dfrac{1}{3}$ (2) $\dfrac{8}{6} = 1\dfrac{2}{6} = 1\dfrac{1}{3}$

(a) $\dfrac{15}{12}$ (b) $\dfrac{25}{4}$ (c) $\dfrac{19}{3}$ (d) $\dfrac{48}{15}$

(e) $\dfrac{30}{9}$ (f) $\dfrac{15}{4}$ (g) $\dfrac{50}{7}$ (h) $\dfrac{71}{12}$

(i) $\dfrac{77}{44}$ (j) $\dfrac{51}{12}$ (k) $\dfrac{64}{15}$ (l) $\dfrac{71}{56}$

(m) $\dfrac{71}{12}$ (n) $\dfrac{43}{5}$ (o) $\dfrac{71}{9}$ (p) $\dfrac{70}{8}$

(q) $\dfrac{77}{16}$ (r) $\dfrac{84}{13}$ (s) $\dfrac{78}{5}$ (t) $\dfrac{87}{15}$

4. Change the mixed numbers to improper fractions.

Examples: (1) $4\dfrac{2}{3} = \dfrac{14}{3}$ (2) $3\dfrac{3}{8} = \dfrac{27}{8}$

(a) $2\dfrac{1}{2}$ (b) $4\dfrac{7}{8}$ (c) $2\dfrac{4}{5}$ (d) $3\dfrac{1}{8}$

(e) $2\dfrac{5}{8}$ (f) $5\dfrac{3}{7}$ (g) $3\dfrac{7}{9}$ (h) $5\dfrac{2}{7}$

(i) $3\dfrac{5}{9}$ (j) $5\dfrac{3}{8}$ (k) $7\dfrac{8}{9}$ (l) $4\dfrac{5}{6}$

(m) $8\dfrac{3}{5}$ (n) $3\dfrac{4}{7}$ (o) $8\dfrac{1}{3}$ (p) $5\dfrac{2}{7}$

(q) $9\dfrac{1}{8}$ (r) $3\dfrac{3}{16}$ (s) $4\dfrac{7}{9}$ (t) $7\dfrac{7}{64}$

5. Find the prime factors.

Examples:

(1)

```
2)180
2) 90
3) 45
3) 15
5)  5
    1
```
$180 = 2 \times 2 \times 3 \times 3 \times 5$

(2)

```
2)2310
3)1155
5) 385
7)  77
11) 11
     1
```
$2310 = 2 \times 3 \times 5 \times 7 \times 11$

(a) 24   (b) 210   (c) 126
(d) 1323   (e) 1125   (f) 3234
(g) 5775   (h) 4851   (i) 15,015

PROFESSIONAL PUBLICATIONS, INC. ● P.O. Box 199, San Carlos, CA 94070

6. Find the Least Common Multiple (LCM) of the following sets of numbers.

Example: 18, 24, 45

solution: $18 = 2 \times 3 \times 3$

$\quad\quad\quad\quad 24 = 2 \times 2 \times 2 \times 3$

$\quad\quad\quad\quad 45 = 3 \times 3 \times 5$

$\quad\quad\quad\quad \text{LCM} = 2 \times 2 \times 2 \times 3 \times 3 \times 5 = 360$

(a) 4, 6, 12   (b) 10, 15, 20
(c) 24, 35, 60   (d) 16, 45, 75
(e) 45, 54, 75   (f) 30, 42, 60

7. Find the Least Common Denominator (LCD) of the fractions and add or subtract as indicated. Express answer in lowest terms.

Examples:

(1) $\dfrac{3}{8} + \dfrac{5}{8} + \dfrac{7}{8} = \dfrac{15}{8} = 1\dfrac{7}{8}$

(2) $\dfrac{1}{4} + \dfrac{1}{6} + \dfrac{1}{9} = \dfrac{9}{36} + \dfrac{6}{36} + \dfrac{4}{36} = \dfrac{19}{36}$

(a) $\dfrac{1}{2} + \dfrac{2}{3} + \dfrac{3}{5}$

(b) $\dfrac{3}{8} + \dfrac{1}{4} + \dfrac{2}{3}$

(c) $\dfrac{9}{16} + \dfrac{5}{12} - \dfrac{1}{3}$

(d) $\dfrac{1}{4} + \dfrac{3}{8} + \dfrac{2}{3} + \dfrac{5}{6}$

(e) $\dfrac{5}{8} + \dfrac{3}{16} + \dfrac{3}{4} - \dfrac{5}{6}$

(f) $6\dfrac{2}{3} + 4\dfrac{5}{8}$

(g) $5\dfrac{7}{8} - 3\dfrac{1}{9}$

(h) $5\dfrac{4}{9} + 2\dfrac{7}{12} - 3\dfrac{1}{8}$

(i) $2\dfrac{2}{3} + 3\dfrac{5}{16} - 1\dfrac{1}{16}$

(j) $2\dfrac{5}{8} + 3\dfrac{1}{4} - \dfrac{15}{16}$

8. Multiply and express answer in lowest terms.

Examples:

(1) $\dfrac{2}{3} \times \dfrac{1}{4} = \dfrac{2}{12} = \dfrac{1}{6}$

(2) $\dfrac{3}{5} \times \dfrac{1}{3} \times \dfrac{15}{16} = \dfrac{3}{16}$

(a) $\dfrac{3}{8} \times \dfrac{2}{3}$

(b) $\dfrac{3}{5} \times \dfrac{2}{3} \times \dfrac{3}{8}$

(c) $\dfrac{3}{4} \times \dfrac{2}{5} \times \dfrac{1}{2}$

(d) $\dfrac{3}{8} \times \dfrac{1}{3} \times \dfrac{3}{4}$

(e) $\dfrac{2}{3} \times \dfrac{4}{5} \times \dfrac{1}{4} \times \dfrac{5}{8} \times \dfrac{9}{16}$

(f) $\dfrac{3}{5} \times \dfrac{7}{9} \times \dfrac{20}{21} \times \dfrac{3}{8} \times \dfrac{1}{12}$

(g) $5\dfrac{3}{5} \times 2\dfrac{3}{4} \times 3\dfrac{3}{7}$

(h) $3\dfrac{1}{4} \times 2\dfrac{2}{5} \times 4\dfrac{2}{3}$

(i) $1\dfrac{15}{16} \times 2\dfrac{5}{8} \times 3\dfrac{1}{2}$

(j) $2\dfrac{4}{9} \times 1\dfrac{8}{9} \times 3\dfrac{1}{3}$

9. Divide and show answers in lowest terms.

Examples:

(1) $3 \div \dfrac{2}{3} = 3 \times \dfrac{3}{2} = 4\dfrac{1}{2}$

(2) $\dfrac{2}{3} \div 4 = \dfrac{2}{3} \times \dfrac{1}{4} = \dfrac{1}{6}$

(a) $3 \div \dfrac{3}{5}$

(b) $5 \div \dfrac{5}{6}$

(c) $3 \div \dfrac{5}{8}$

(d) $\dfrac{3}{4} \div \dfrac{5}{8}$

(e) $\dfrac{5}{6} \div \dfrac{2}{3}$

(f) $\dfrac{3}{8} \div 5$

(g) $\dfrac{3}{8} \div 3$

(h) $\dfrac{1}{8} \div 5$

(i) $\dfrac{4}{5} \div \dfrac{1}{4}$

(j) $\dfrac{2}{3} \div \dfrac{3}{4}$

10. Simplify by cancellation. (Show cancellations.)

Examples:

(1) $\dfrac{(\overset{1}{\cancel{5}})(\overset{2}{\cancel{12}})}{(\underset{1}{\cancel{6}})(\underset{3}{\cancel{15}})} = \dfrac{2}{3}$

(2) $\dfrac{(\overset{1}{\cancel{15}})(\overset{3}{\cancel{12}})(\overset{1}{\cancel{4}})}{(\underset{1}{\cancel{4}})(\underset{2}{\cancel{30}})(\underset{2}{\cancel{8}})} = \dfrac{3}{4}$

(a) $\dfrac{(6)(10)}{(5)(3)}$

(b) $\dfrac{(6)(15)(8)}{(3)(4)(5)}$

(c) $\dfrac{(9)(12)(15)}{(3)(8)(5)}$

(d) $\dfrac{(21)(32)(45)}{(7)(8)(15)}$

(e) $\dfrac{(14)(15)(24)}{(30)(28)(8)}$

(f) $\dfrac{(27)(45)(64)}{(15)(9)(16)}$

(g) $\dfrac{(15)(7)(8)(9)}{(21)(15)(5)}$

(h) $\dfrac{(72)(24)(15)(18)}{(8)(12)(5)(9)}$

11. Write in words.

Examples:

(1) 476.232: Four hundred seventy-six and two hundred thirty-two thousandths.

(2) 0.0521: Five hundred twenty-one ten-thousandths.

(a) 46.32

(b) 132.036

(c) 1.0066

(d) 0.400

(e) 0.20200

(f) 1.0001

(g) 2037.566

(h) 46,388.07

(i) 2,000.55

(j) 10.001

12. Find the products or quotients as indicated.

Examples:

(1) $537.26 \times 100 = 53,726$

(2) $\dfrac{6666}{0.02} = \dfrac{3333}{0.01} = 333,300$

(a) 12×10 (b) 23.16×100

(c) $\dfrac{0.3695}{100}$ (d) $\dfrac{863,914}{0.002}$

(e) $\dfrac{9,823}{0.01}$ (f) $\dfrac{0.3695}{0.001}$

(g) $\dfrac{274.767}{0.00001}$ (h) $\dfrac{6.37}{10,000}$

(i) $\dfrac{44.444}{0.04}$ (j) $\dfrac{0.032}{0.002}$

(k) $\dfrac{808.0}{0.004}$ (l) $\dfrac{0.003}{0.03}$

(m) 0.0063×0.2 (n) 0.272×0.002

(o) 32.42×0.002 (p) 0.0058×1000

(q) 11.11×0.01 (r) 0.000412×10

(s) $22.2 \div 0.0002$ (t) $0.000033 \div 0.03$

13. Change to a percent.

Examples: (1) $0.06 = 6\%$ (2) $5.45 = 545\%$

(a) 0.04   (b) 0.15   (c) 0.004
(d) 1.37   (e) 0.0002   (f) 0.275
(g) 2 1/2   (h) 1 1/4   (i) 1 1/10

14. Change from a percent to a decimal fraction.

Example: $0.6\% = 0.006$

(a) 13%   (b) 0.007%   (c) 125%
(d) 33 1/3%   (e) 250%   (f) 0.5%

15. Change from a common fraction to a percent.

Example: $1/2 = 50\%$

(a) 1/3   (b) 1 1/2   (c) 2 2/3
(d) 1 1/4   (e) 3/4   (f) 2/3
(g) 1 3/8   (h) 2 1/8   (i) 3 1/10

16. Find the percent.

Examples: (1) 50 is 50% of 100   (2) 6 is 25% of 24

PROFESSIONAL PUBLICATIONS, INC. ● P.O. Box 199, San Carlos, CA   94070

(a) 6 is x% of 12

(b) 20 is x% of 80

(c) 60 is x% of 90

(d) 8 is x% of 32

(e) 16 is x% of 64

(f) 5 is x% of 40

(g) 44 is x% of 22

(h) 36 is x% of 48

(i) 50 is x% of 20

(j) 60 is x% of 40

17. Find the number.

Example: 70 is 35% of 200 ($70 \div 0.35 = 200$)

(a) 8 is 50% of x

(b) 15 is 25% of x

(c) 60 is 66 2/3% of x

(d) 30 is 75% of x

(e) 8 is 12 1/2% of x

(f) 12 is 2% of x

(g) 2/3 is 75% of x

(h) 1/2 is 66 2/3% of x

(i) 1/8 is 12 1/2% of x

(j) 3/4 is 25% of x

18. Round-off each number to the nearest tenth.

Examples: (1) 4.62 <u>4.6</u>   (2) 2.56 <u>2.6</u>   (3) 6.55 <u>6.6</u>

(a) 2.48   (b) 3.429   (c) 5.552
(d) 6.77   (e) 8.009   (f) 3.651
(g) 2.09   (h) 8.039   (i) 4.272

19. Round-off each number to the nearest hundredth.

(a) 5.455   (b) 5.486   (c) 1.455
(d) 7.365   (e) 2.295   (f) 5.494
(g) 4.009   (h) 3.001   (i) 2.599

20. Round-off each number to the nearest hundred.

(a) 22,009   (b) 45,249   (c) 15,966
(d) 30,499   (e) 11,990   (f) 21,500
(g) 9,950   (h) 8,951   (i) 11,449
(j) 5,962   (k) 50,499   (l) 21,009

21. Write the number of significant digits in each of the following numbers.

Examples: (1) 2.506 <u>4</u>   (2) 0.0032 <u>2</u>   (3) 2.00 <u>3</u>

(a) 5651   (b) 0.002   (c) 348.4
(d) 2.407   (e) 7000   (f) 50.00
(g) 3.00   (h) 245.0   (i) 0.046

22. Round-off each number to three significant digits.

(a) 12.448   (b) 24,100   (c) 0.03606
(d) 323,202   (e) 447,501   (f) 3.585

23. Round-off each number to one significant digit.

(a) 47.2   (b) 1389   (c) 0.076
(d) 0.451   (e) 200,501   (f) 1.449

24. Express each of the following numbers in scientific notation.

(a) 848   (b) 132,000   (c) 0.011
(d) 0.0045   (e) 0.00015   (f) 10,800
(g) 0.549   (h) 0.00004   (i) 30,376

25. Find the square root or the approximate square root to the nearest tenth.

Examples: (1)

$$
(1) \quad \begin{array}{r} 6.3 \\ \sqrt{39.69} \\ \hline 36.00 \\ 123)\overline{3.69} \\ 3.69 \\ \hline 0.00 \end{array} \quad (2) \quad \begin{array}{r} 23.5 \\ \sqrt{552.25} \\ \hline 400.00 \\ 43)\overline{152.00} \\ 129.00 \\ 465)\overline{23.25} \\ 23.25 \\ \hline 0.00 \end{array}
$$

(a) 361   (b) 2916
(c) 1269.03   (d) 3177.75
(e) 17661.48   (f) 246932.58

26. Find the missing side of the right triangle.

Example: Given $a = 6$, $c = 10$, find b.

Solution: $b = \sqrt{c^2 - a^2} = \sqrt{10^2 - 6^2} = 8$

(a) Given: $a = 3$, $b = 4$, find c.

(b) Given: $a = 15$, $c = 17$, find b.

(c) Given: $b = 9$, $c = 15$, find a.

(d) Given: $a = 12$, $c = 13$, find b.

(e) Given: $b = 7$, $c = 25$, find a.

(f) A guy wire is attached to a TV-antenna tower 120 feet above ground. It is anchored 50 feet from the tower on level ground. What is the length of the guy wire?

(g) The foundation for a building is to be 80′ by 120′. The builder, in staking the building, wants to check to see if it is square. Compute the length of the diagonal.

(h) Crossroads is 14 miles due east of Pumpkin Hollow; Ghostown is 27 miles due south of Crossroads. How far is Pumpkin Hollow from Ghostown? (nearest mile)

27. Find the value of x which makes the ratio correct.

Examples:

(1) $x : 4 = 3 : 6$, $6x = 12$, $x = 2$

(2) $\dfrac{x}{3} = \dfrac{2}{6}$, $6x = 6$, $x = 1$

(a) $x : 4 = 12 : 144$

(b) $x : 3 = 18 : 6$

(c) $x : 6 = 12 : 72$

(d) $9 : x = 27 : 3$

(e) $3 : x = 4 : 24$

(f) $9 : 6 = 15 : x$

(g) $12 : 144 = 3 : x$

(h) $x : y = a : b$

(i) $x : 0.5 = 0.8 : 0.4$

(j) $\dfrac{12}{x} = \dfrac{16}{4}$

(k) $\dfrac{x}{8} = \dfrac{12}{6}$

(l) $\dfrac{2.5}{7.5} = \dfrac{3}{x}$

2 GEOMETRY

1 DEFINITION

Geometry is a branch of mathematics that deals with the measurement, properties, and relationships of points, lines, angles, surfaces, and solids. The word geometry is derived from the words geo, meaning earth, and metro, meaning measure.

2 HISTORY

The Egyptians first used geometry about 2500 B.C. because the seasonal overflowing of the Nile made it necessary to reestablish boundaries so that taxes could be levied and collected. About 500 B.C. the Greeks began to develope information received from the Egyptians into the branch of mathematics we now know as geometry. By the 4th century A.D., they had developed arithmetic and geometry into seperate branches of mathematical science.

3 POINTS AND LINES

Points and lines are undefined elements of geometry, yet everyone has some understanding of these terms. A point is understood to have no length, width or thickness but indicates a location. A point is usually shown on paper as a small dot and is named with a capital letter such as A.

A line is considered to have length but not width or thickness. A line connecting two points is said to be a *straight* line if it does not curve. A straight line is usually designated by two points which it connects, such as AB. A curved line is a line no part of which is straight.

4 PARALLEL LINES

Two lines which lie in the same plane and do not intersect are called parallel lines.

5 ANGLE

There are many definitions of an angle. In geometry, an angle may be defined as the space between two lines diverging from a common point; the point is called the *vertex*. In trigonometry, an angle may be defined as the amount of rotation required to bring one line into coincidence with another. In surveying, an angle may be defined as the difference in direction of two intersecting lines.

6 MEASURE OF ANGLES

The most common measure of an angle is the degree. It is defined as $\frac{1}{360}$ of a complete angle or turn. We usually think of a circle as divided into 360 equal arcs. If radii connect each end of these small arcs, the angle formed by the two radii measures one degree.

For closer measurement, the degree is divided into 60 equal parts, with one part measuring one minute. And for even closer measurement, the one minute angle is divided into 60 equal parts with each part measuring one second.

The symbols used for degrees, minutes, and seconds are: degrees ($°$), minutes ($'$), seconds ($''$). As an example, 36 degrees, 24 minutes, and 52 seconds is written $36°24'52''$.

The protractor can be used to measure angles on paper; the transit and theodolite measure angles in the field.

7 ACUTE ANGLE

An acute angle is an angle of less than $90°$.

8 RIGHT ANGLE

A right angle is an angle of $90°$.

PROFESSIONAL PUBLICATIONS, INC. • P.O. Box 199, San Carlos, CA 94070

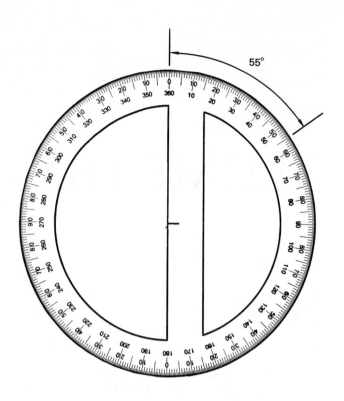

Figure 2.1

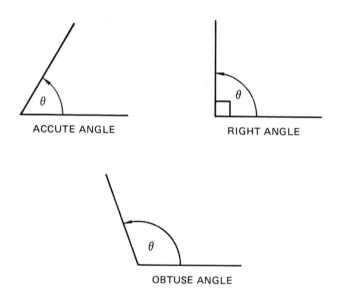

ACCUTE ANGLE

RIGHT ANGLE

OBTUSE ANGLE

Figure 2.2

9 OBTUSE ANGLE

An obtuse angle is an angle of more than 90° and less than 180°.

10 STRAIGHT ANGLE

A straight angle is an angle of 180°.

11 COMPLEMENTARY ANGLES

Two angles are said to be complementary if their sum is 90°.

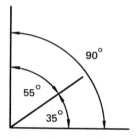

Figure 2.3

12 SUPPLEMENTARY ANGLES

Two angles are said to be supplementary if their sum is 180°.

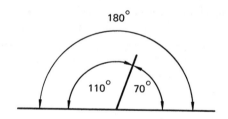

Figure 2.4

13 TRANSVERSAL

A line which cuts two or more lines is called a transversal.

14 ALTERNATE INTERIOR ANGLES

When two parallel lines are cut by a transversal, the alternate interior angles are equal.

15 ALTERNATE EXTERIOR ANGLES

When two parallel lines are cut by a transversal, the alternate exterior angles are equal.

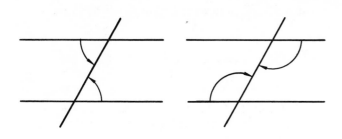

Figure 2.5 Alternate interior angles

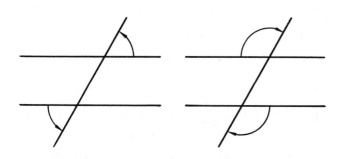

Figure 2.6 Alternate Exterior angles

16 ADDING AND SUBTRACTING ANGLES

The surveying technician is often called on to add the measurements of two angles or to find the difference in the measurements of two angles. The procedures often involve "borrowing" one degree and converting it to 60 minutes and "borrowing" one minute and converting it to 60 seconds.

Example 2.1

Add $24°40'$ and $16°30'$.

Solution

$$
\begin{aligned}
24°40' \\
+\,16°30' \\
\hline
40°70' = 41°10'
\end{aligned}
$$

Explanation:

When the number of minutes in the sum is sixty or more, sixty minutes are subtracted from the minutes column and one degree is added to the degree column. Likewise, when the number of seconds in the sum is sixty or more, sixty seconds are subtracted from the seconds column and one minute is added to the minutes column.

Example 2.2

Add $32°46'32''$ and $14°22'44''$.

Solution

$$
\begin{aligned}
32°46'32'' \\
+\,14°22'44'' \\
\hline
46°68'76'' = 46°69'16'' = 47°09'16''
\end{aligned}
$$

Example 2.3

Find the difference between $60°12'$ and $40°32'$.

Solution

$$
\begin{array}{cc}
60°12' & 59°72' \\
-\,40°32' & -\,40°32' \\
\hline
& 19°40'
\end{array}
$$

Explanation:

Before finding the difference, we have converted $60°12'$ to its equivalent $59°72'$. We have "borrowed" one degree from $60°$ leaving $59°$ and added its equivalent, 60 minutes, to 12 minutes, making 72 minutes.

Example 2.4

Find the difference between $96°08'14''$ and $52°33'50''$.

Solution

$$
\begin{array}{cc}
96°08'14'' & 95°67'74'' \\
-\,52°33'50'' & -\,52°33'50'' \\
\hline
& 43°34'24''
\end{array}
$$

Explanation:

One degree was borrowed from the degree column leaving $95°$ and making the minutes column $68'$. One minute was borrowed from the minutes column, leaving $67'$, and making the seconds column $74''$.

When a number of angle measurements are added, as is common in surveying, each column (degrees, minutes and seconds) is added separately and recorded. If the sum of either the minutes column or the seconds column, or both, is sixty or more, the same procedure is followed.

Example 2.5

Add the angle measurements

$$
\begin{aligned}
93° \; 18' \; 22'' \\
65° \; 13' \; 8'' \\
218° \; 19' \; 30'' \\
67° \; 05' \; 20'' \\
96° \; 04' \; 50''
\end{aligned}
$$

Solution

$$
\begin{array}{rrr}
93° & 18' & 22'' \\
65° & 13' & 08'' \\
218° & 19' & 30'' \\
67° & 05' & 20'' \\
96° & 04' & 50'' \\
\hline
539° & 59' & 130'' \\
+ & 02' & 120'' \\
\hline
539° & 61' & 10'' \\
+1° & -60' & \\
\hline
540° & 01' & 10''
\end{array}
$$

17 AVERAGE OF SEVERAL MEASURE-MENTS OF AN ANGLE

Surveyors often measure angles by repetition. An angle of $36°30'30''$ might have been read on the first reading as $36°30'$, but after turning the angle six times, the accumulated angle may have read $216°03'00''$. Dividing this by six gives $36°30'30''$ which is closer to the true measurement.

Example 2.6

An angle was doubled, and the accumulated reading was $84°26'$. What was the average?

Solution

$$84°26' \div 2 = 42°13'$$

Example 2.7

An angle which was doubled read $314°13'$. What was the closest value for the single angle?

Solution

$$314°13' \div 2 = 157°06'30''$$

Example 2.8

An angle reads $318°03'$ after having been turned six times. What was the average for the single angle?

Solution

$$318°03' \div 6 = 318°00'180'' \div 6 = 53°00'30''$$

18 CHANGING DEGREES AND MINUTES TO DEGREES AND DECIMALS OF A DEGREE

In some situations and in some tables of trigonometric functions, angles are expressed in degrees and decimals of a degree. To change degrees and minutes to degrees and decimals of a degree, first express the minutes as a common fraction with a denominator of 60, and then convert the common fraction to a decimal fraction. Add this fraction to the degrees.

Example 2.9

Change $73°15'$ to degrees and decimals of a degree.

Solution

$$73°15' = 73\frac{15}{60}° = 73.25°$$

19 CHANGING DEGREES, MINUTES, AND SECONDS TO DEGREES AND DECIMALS OF A DEGREE

To change degrees, minutes, and seconds to degrees and decimals of a degree, first convert degrees, minutes, and seconds to degrees, minutes, and decimals of a minute, then convert degrees, minutes, and decimals of a minute to degrees and decimals of a degree.

Example 2.10

Change $46°24'36''$ to degrees and decimals of a degree.

Solution

$$46°24'36'' = 46°24\frac{36}{60}' = 46°24.6' = 46\frac{24.6}{60}° = 46.41°$$

20 CHANGING DEGREES AND DECIMALS OF A DEGREE TO DEGREES, MINUTES AND SECONDS

To change degrees and decimals of a degree to degrees, minutes and seconds, multiply the decimal fraction by sixty and add the product (in minutes and decimals of a minute) to the degrees. Then multiply the decimal fraction in minutes by sixty and add the product (in seconds) to the degrees and minutes. The decimal fraction in seconds will be left as such.

Example 2.11

Change $36.12345°$ to degrees, minutes, and seconds.

Solution

$$0.12345° \times 60 = 7.407'$$
$$0.407' \times 60 = 24.42''$$
$$36.12345° = 36°07'24.42''$$

21 POLYGON

A closed figure bounded by straight lines lying in the same plane is known as a polygon.

The sum of the interior angles of a closed polygon is equal to:

$$(n-2)180°$$

where n is the number of sides. Thus, the sum of the interior angles of a triangle is $180°$; rectangle, $360°$; five-sided figure, $540°$, etc.

PROFESSIONAL PUBLICATIONS, INC. • P.O. Box 199, San Carlos, CA   94070

22 TRIANGLE

A polygon of three sides is known as a triangle.

23 RIGHT TRIANGLE

A right triangle is a triangle which has one right angle (90°).

24 ISOSCELES TRIANGLE

An isosceles triangle is a triangle which has two equal sides and two equal angles.

25 EQUILATERAL TRIANGLE

An equilateral triangle is a triangle which has three equal sides and three equal angles.

26 OBLIQUE TRIANGLE

An oblique triangle is a triangle which has no right angle and no two sides equal. It is also known as a scalene triangle.

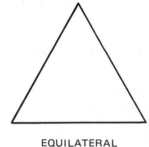

ISOSCELES   EQUILATERAL

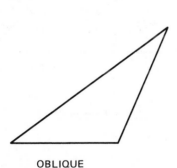

OBLIQUE

Figure 2.7

27 CONGRUENT TRIANGLES

Two triangles are congruent if their corresponding sides and corresponding angles are equal.

28 SIMILAR TRIANGLES

Two triangles are similar if their corresponding angles are equal and their corresponding sides are proportional.

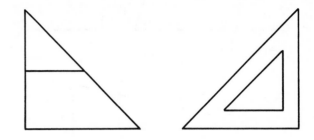

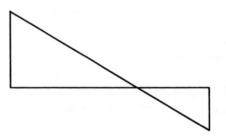

Figure 2.8

29 RECTANGLE

A rectangle is a four-sided polygon whose angles are right angles. A square is a rectangle whose sides are equal.

Figure 2.9

30 TRAPEZOID

A trapezoid is a four sided polygon which has two parallel sides and two non-parallel sides.

31 CIRCLE

A circle is a closed plane curve, all points on which are equidistant from a point within called the center.

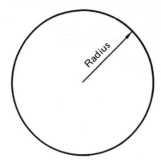

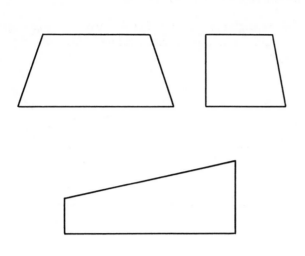

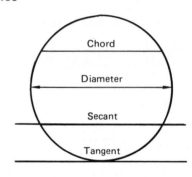

CIRCLE AND RADIUS

Figure 2.10 Trapezoids

32 RADIUS

The distance from the center of the circle to any point on the circle is called the radius of the circle.

33 DIAMETER

The distance across the circle through the center is called the diameter. One diameter is two radii.

34 CHORD

A straight line between two points on a circle is called a chord.

35 SECANT

A secant of a circle is a line that intersects the circle at two points.

36 TANGENT

A tangent of a circle is a line that touches the circle at only one point.

37 ARC

Any part of a circle is called an arc.

CHORD, DIAMETER
SECANT, TANGENT

Figure 2.11 Elements of a circle

38 SEMICIRCLE

An arc equal to one-half the circumference of a circle is a semicircle.

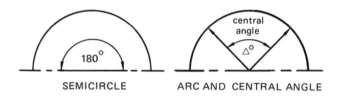

SEMICIRCLE   ARC AND CENTRAL ANGLE

Figure 2.12

39 CENTRAL ANGLE

A central angle is an angle formed by two radii. The Greek letter Δ (delta) is often used to denote a central angle. A central angle has the same number of degrees as the arc it intercepts. A 60° central angle intercepts a 60° arc, etc. Thus a central angle is measured by its intercepted arc.

40 SECTOR

A figure bounded by an arc of a circle and two radii of the circle is called a sector of the circle.

41 SEGMENT

A figure bounded by a chord and an arc of a circle is called a segment of a circle.

42 CONCENTRIC CIRCLES

Two circles of different radius but having the same center are called concentric circles.

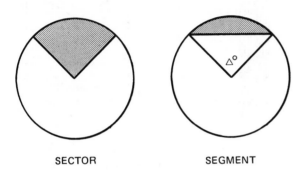

SECTOR   SEGMENT

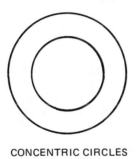

CONCENTRIC CIRCLES

Figure 2.13

43 RADIUS PERPENDICULAR TO TANGENT

The radius of a circle is perpendicular to a tangent to the circle at the point of tangency.

44 RADIUS AS PERPENDICULAR BISECTOR OF A CHORD

The perpendicular bisector of a chord passes through the center of the circle.

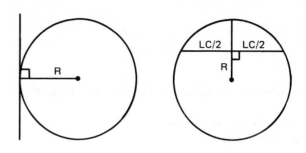

Figure 2.14

45 TANGENTS TO CIRCLE FROM OUTSIDE POINT

Tangents to a circle from an outside point are equal.

46 LINE FROM CENTER OF CIRCLE TO OUTSIDE POINT

A line from the center of a circle to an outside point bisects the angle between the tangents from the point to the circle.

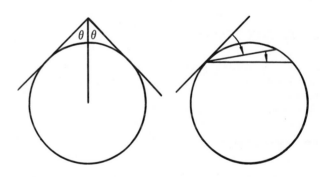

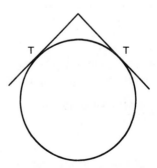

Figure 2.15

47 ANGLE FORMED BY TANGENT AND CHORD

The angle formed by a tangent and a chord is equal to one-half its intercepted arc.

48 ANGLE FORMED BY TWO CHORDS

The angle formed by two chords is equal to one-half its intercepted arc.

49 SOLID GEOMETRY

Figures shown thus far are plane figures and are included in the study of Plane Geometry. Solid Geometry is the study of figures of three dimensions such as cubes, cones, pyramids, and spheres.

50 POLYHEDRON

A polyhedron is any solid formed by plane surfaces.

51 PRISM

A prism is a polyhedron with parallel edges and parallel bases.

52 RIGHT PRISM

A prism with edges perpendicular to the bases is known as a right prism. A cylinder is a right prism with circular bases.

53 PYRAMID

A pyramid is a polyhedron having for its base a polygon and for faces, triangles with a common vertex.

The altitude of a pyramid is the perpendicular distance from the vertex to the base.

A right pyramid is a pyramid in which the base is a regular polygon and a line from the vertex to the center of the polygon is perpendicular to the polygon.

The slant height of a right pyramid is the altitude of one of the lateral faces.

54 FRUSTUM OF A PYRAMID

A frustum of a pyramid is the part left after cutting off the top part with a plane parallel to the base.

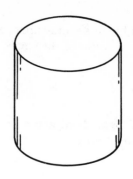
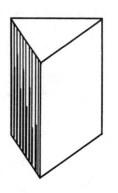
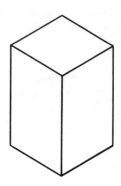

RIGHT PRISMS

Figure 2.16

55 CONE

A cone is a polyhedron having a circular base and with sides which taper evenly up to a vertex.

The altitude of a cone is the perpendicular distance from the vertex to the base.

A right circular cone is a cone in which a line from the vertex to the center of the base is perpendicular to the base.

The slant height of a cone is the distance from the vertex to the base measured along the surface.

56 FRUSTUM OF A CONE

The frustum of a right circular cone is the part left after cutting off the top part with a plane parallel to the base.

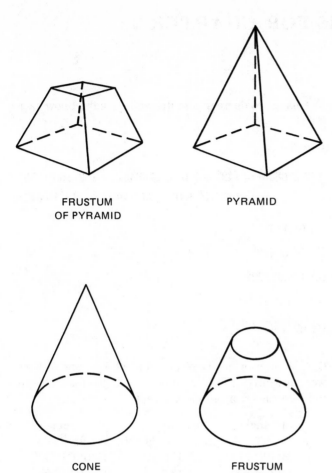

FRUSTUM
OF PYRAMID

PYRAMID

CONE

FRUSTUM
OF CONE

Figure 2.17

PROFESSIONAL PUBLICATIONS, INC. ● P.O. Box 199, San Carlos, CA 94070

PRACTICE PROBLEMS FOR CHAPTER 2

1. Add the angles.

Example:

$$21°41'12''$$
$$\underline{11°32'54''}$$
$$32°73'66'' = 33°14'06''$$

(a) $46°27'$
$\underline{+22°24'}$

(b) $56°24'$
$\underline{+33°26'}$

(c) $35°52'$
$\underline{+47°39'}$

(d) $21°46'52''$
$\underline{+40°25'26''}$

(e) $46°19'22''$
$\underline{+35°51'40''}$

(f) $13°49'58''$
$\underline{+12°21'32''}$

2. Find the average of the angles which were doubled in the field with the accumulated value shown.

Example:

$$2)311°17'20'' = 2)310°76'80''$$
$$155°38'40''$$

(a) $237°27'17''$

(b) $329°47'16''$

3. Find the average of the angles which were repeated six times in the field with the accumulated value shown.

$$390°13'24''$$

4. Change to degrees and decimals of a degree.

Example:

$$36°14'52'' = 36°14\frac{52'}{60} = 36°14.8667' = 36\frac{14.8667°}{60}$$
$$= 36.2478°$$

(a) $24°30'$

(b) $36°45'$

(c) $69°11'$

(d) $16°24'30''$

(e) $173°32'56''$

(f) $127°17'23''$

(g) $68°44'05''$

(h) $223°37'48''$

(i) $186°08'34''$

(j) $118°55'11''$

5. Change to degrees, minutes and seconds. Show each step.

Example:

$$142.276843° = 142° + (60)(0.276843)' = 142°16.61058'$$
$$= 142°16' + (60)(0.61058)'' = 142°16'37''$$

(a) $68.176°$

(b) $96.564722°$

(c) $145.882222°$

(d) $221.347778°$

(e) $303.107778°$

6. The interior angles of polygons of 5, 6, and 7 sides were measured. Find the sum of the angles for each and indicate the error of measurement.

	pt	angle		pt	angle		pt	angle
(a)	A	$83°23'$	(b)	A	$96°34'$	(c)	A	$98°08'05''$
	B	$105°27'$		B	$111°42'$		B	$149°16'12''$
	C	$158°31'$		C	$183°12'$		C	$134°12'55''$
	D	$53°19'$		D	$88°57'$		D	$93°20'10''$
	E	$\underline{139°18'}$		E	$139°21'$		E	$152°39'47''$
				F	$\underline{100°18'}$		F	$174°32'50''$
							G	$\underline{97°51'11''}$
	error			error			error	

7. Write the missing word in each sentence which is represented by the three dots.

(a) Two lines which lie in the same plane and do not intersect are called ... lines.

(b) An ... angle is an angle of less than 90°.

(c) A ... angle is an angle of 90°.

(d) An ... angle is an angle of more than 90° and less than 180°.

(e) A ... angle is an angle of 180°.

(f) Two angles are said to be ... if their sum is 90°.

(g) Two angles are said to be ... if their sum is 180°.

(h) A line which cuts two or more lines is called a ....

(i) When two parallel lines are cut by a transversal, the alternate ... angles are equal.

(j) When two parallel lines are cut by a transversal, the alternate ... angles are equal.

(k) A polygon of three sides is known as a ....

(l) A ... triangle is a triangle which has one right angle.

(m) An ... triangle is a triangle which has two equal sides and two equal angles.

(n) An ... triangle is a triangle which has three equal sides and three equal angles.

(o) An ... triangle is a triangle which has no right angle and no two sides equal.

(p) Two triangles are ... if their corresponding sides and corresponding angles are equal.

(q) Two triangles are ... if their corresponding angles are equal and their corresponding sides are proportional.

(r) A ... is a four-sided polygon which has two parallel sides and two non-parallel sides.

(s) A ... is a closed plane curve, all points on which are equidistant from a point within called the center.

(t) The distance from the center of the circle to any point on the circle is called the ....

(u) The distance across a circle through the center is called the ....

(v) A straight line between two points on a circle is called a ....

(w) A ... of a circle is a line that intersects the circle at two points.

(x) A ... is an angle formed by two radii.

(y) A ... of a circle is a line that touches the circle at only one point.

(z) A figure bounded by an arc of a circle and two radii of a circle is called a ... of a circle.

(aa) A figure bounded by a chord and an arc of a circle is called a ... of a circle.

(bb) Two circles of different radii but having the same center are called ... circles.

(cc) A ... is a polyhedron with parallel edges and parallel bases.

(dd) A prism with edges perpendicular to the base is known as a ... prism.

PROFESSIONAL PUBLICATIONS, INC. • P.O. Box 199, San Carlos, CA   94070

3
SYSTEMS OF WEIGHTS AND MEASURES

1 THE ENGLISH SYSTEM

American colonists brought with them from England the system of weights and measures in use at the time in England. When the American colonies separated from the mother country they adopted the English system which, at the time, was the best standardized system in the world.

The English system was originally based on standards determined by parts of the body, such as the foot, the hand, and the thumb. Several hundred years ago an English king proclaimed the length of the English inch to be the length of three barley corn grains laid end to end. Later, of course, more sophisticated methods were used for standardization, but the system has no uniform conversion factors. There are 12 inches per foot, 3 feet per yard, $5\frac{1}{2}$ yards per rod, and 16 ounces per pound.

During the Colonial Period, the English system was well standardized, but such was not the case in the rest of Europe. There was such a wide variety of weights and measures in use that commerce was difficult.

2 THE METRIC SYSTEM

This situation prompted the National Assembly of France to enact a decree in 1790 which directed the French Academy of Sciences to find standards for all weights and measures. The French Academy was to work with the Royal Society of London, but the English did not participate in the undertaking so the French proceeded alone. The result was the metric system which used the base ten in converting units of measure.

The system spread rapidly in the 19th century. In 1872, France called an international meeting which was attended by 26 nations including the United States to further refine the system.

The meeting resulted in the establishment of the International Bureau of Weights and Measures, and in 1960, an extensive revision and simplification resulted in the International System of Units which is in use in most countries today.

3 THE SI SYSTEM

The International System of Units—officially abbreviated SI—uses the base ten in expressing multiples and submultiples just as the metric system has always done. The six base units of measurement are:

Length	- Meter -	m
Time	- Second -	s
Mass	- Kilogram -	kg
Temperature	- Kelvin -	K
Electric current	- Ampere -	A
Luminous intensity	- Candela -	cd

The meter is defined as 1,650,763.73 wave lengths in vacuum of the orange-red line of the spectrum of krypton-86. The SI unit of area is the square meter (m^2). Land is measured by the hectare (10,000 square meters). The SI unit of volume is the cubic meter (m^3). Fluid volume is measured by the liter (0.001 cubic meter).

The second is defined as the duration of 9,192,631,770 cycles of the radiation associated with a specified transition of the atom.

The standard for the kilogram is a cylinder of platinum-iridium alloy kept by the International Bureau of Weights and Measures at Paris. A duplicate is in the custody of the National Bureau of Standards, Washington, D.C.

To make a conversion in the Metric System (SI System) the decimal is moved to the right or left just as we do in working with decimal fractions. To facilitate use of the system, names are given to the various powers of 10.

milli means	- one thousandth of -	.001
centi means	- one hundredth of -	.01
deci means	- one tenth of -	.1
deka means	- ten times -	10
hecto means	- one hundred times -	100
kilo means	- one thousand times -	1,000

Thus, centimeter means one hundredth of a meter, and kilometer means one thousand meters.

Example 3.1

To express meters in centimeters move decimal 2 places right.

To express meters in kilometers move decimal 3 places left.

Prefixes and symbols for all SI units are as shown:

Multiples and Submultiples	Prefixes	Symbols
$1,000,000,000,000 = 10^{12}$	tera	T
$1,000,000,000 = 10^9$	giga	G
$1,000,000 = 10^6$	mega	M *
$1,000 = 10^3$	kilo	k *
$100 = 10^2$	hecto	h
$10 = 10$	deka	da
$0.1 = 10^{-1}$	deci	d
$0.01 = 10^{-2}$	centi	c *
$0.001 = 10^{-3}$	milli	m *
$0.000\,001 = 10^{-6}$	micro	μ *
$0.000\,000\,001 = 10^{-9}$	nano	n
$0.000\,000\,000\,001 = 10^{-12}$	pico	p
$0.000\,000\,000\,000\,001 = 10^{-15}$	femto	f
$0.000\,000\,000\,000\,000\,001 = 10^{-18}$	atto	a

*Most commonly used

THE ENGLISH SYSTEM OF WEIGHTS AND MEASURES

LINEAR MEASURE

1 foot = 12 inches
1 yard = 3 feet
$16\frac{1}{2}$ feet = 1 rod
33 in. in California = 1 vara
$33\frac{1}{3}$ in. in Texas = 1 vara
66 feet = 1 Gunter's Chain
100 feet = 1 Station
5,280 feet = 1 mile

SQUARE MEASURE

1 sq ft = 144 sq in.
1 sq yd = 9 sq ft
1 acre = 43,560 sq ft
1 mi sq = 640 acres

CUBIC MEASURE

1728 cu in. = 1 cu ft
27 cu ft = 1 cu yd
231 cu in. = 1 gal
1 cu ft = 7.5 gal

WEIGHT

1 gal of water = 8.33 lb
1 cu ft of water = 62.5 lb
1 kip = 1,000 lb
1 ton = 2,000 lb

CONVERSION OF ENGLISH SYSTEM LINEAR AND SQUARE MEASURE TO VARAS AND SQUARE VARAS

LINEAR

1 inch = 0.03 vara
1 foot = 12 inches = 0.36 vrs
1 yard = 3 feet = 36 in. = 1.08 vrs
1 vara = $33\frac{1}{3}$ in. = 2.777 + ft
1 mile = 5,280 ft = 80 chains = 1,900.8 vrs

SQUARE

1 acre = 43,560 sq ft
= 5645.3757 sq vrs = 10 sq chains
1 sq mile = 27,878,400 sq ft
= 3,613,040.64 sq vrs = 640 ac
1 labor = 1000 vrs sq
= 2,777.77 + ft sq = 177.14 ac
1 league = 5000 vrs sq
= 13,888.88 ft sq = 4428.40 ac = 6.919 sq mi

4 CONVERSION OF ENGLISH UNIT TO METRIC UNIT

When the metric (SI) system is adopted in the United States, conversion factors will be very necessary. Until such time, little is to be gained by memorizing these factors.

5 CONVERSION OF INCHES TO DECIMALS OF A FOOT

Engineering plans usually show dimensions of structures in feet and inches, while elevations are established in feet and decimals of a foot. It is the surveying technician's lot to make the conversions necessary to establish finish evaluations. Construction stakes are usually set to the nearest hundredth of a foot for concrete, asphalt, pipe flow-lines, etc. For earthwork, stakes are set to the nearest tenth of a foot.

The key to conversion is shown below. An important part of the key is the value of 1 inch and $\frac{1}{8}$ inch.

1 in. = 0.08 ft
    because 1 in. = $\frac{1}{12}$ ft = 1 ÷ 12 = 0.083... ft

$\frac{1}{8}$ in. = 0.01 ft
    because $\frac{1}{8}$ in. = 1 in. ÷ 8 = 0.01 ft

2 in. = 0.17 ft
    because 1 in. + 1 in. = 0.166... ft

3 in. = 0.25 ft
    because $\frac{3}{12}$ ft = $\frac{1}{4}$ ft = 0.250 ft

4 in. = 0.33 ft
    because $\frac{4}{12}$ ft = $\frac{1}{3}$ ft = 0.333... ft

5 in. = 0.42 ft
    because 4 in. + 1 in. = 0.333 + 0.083 = 0.416 ft

6 in. = 0.50 ft
    because $\frac{6}{12}$ ft = $\frac{1}{2}$ ft = 0.500 ft

7 in. = 0.58 ft
    because 6 in. + 1 in. = 0.500 + 0.083 = 0.583 ft

8 in. = 0.67 ft
    because $\frac{8}{12}$ ft = $\frac{2}{3}$ ft = 0.666... ft

9 in. = 0.75 ft
    because $\frac{9}{12}$ ft = $\frac{3}{4}$ ft = 0.750 ft

10 in. = 0.83 ft
    because 9 in. + 1 in. = 0.750 + 0.083 = 0.833 ft

11 in. = 0.92 ft
    because 10 in. + 1 in. = 0.833 + 0.083 = 0.916 ft

12 in. = 1.00 ft

Conversions can be made mentally by following these steps:

1. First memorize:

    a. 6 in. = 0.50 ft

    b. 3 in. = 0.25 ft

    c. 9 in. = 0.75 ft

2. Next memorize:

    a. 4 in. = 0.33 ft

    b. 8 in. = 0.67 ft

3. Next memorize:

    a. 1 in. = 0.08 ft

    b. $\frac{1}{8}$ in. = 0.01 ft

4. In converting measurements expressed in feet, inches, and fractions of an inch to feet and decimals of a foot, convert the inches and fractions of an inch separately to decimals of a foot, then add the three parts. (In some cases subtraction can be used.)

Example 3.2

Convert the measurements to feet and decimals of a foot:

  (a) 1 ft 4 in.

  (b) 11 ft 9$\frac{1}{8}$ in.

  (c) 7 ft 5$\frac{3}{4}$ in.

  (d) 2 ft 8$\frac{7}{8}$ in.

  (e) 5 ft 11$\frac{1}{2}$ in.

Solutions

(a)
1 ft	= 1.00 ft
4 in.	= 0.33
1 ft 4 in.	= 1.33 ft

(b)
11 ft	= 11.00 ft
9 in.	= 0.75
$\frac{1}{8}$ in.	= 0.01
11 ft 9$\frac{1}{8}$ in.	= 11.76 ft

(c)
7 ft	= 7.00 ft
5 in.	= 0.42
$\frac{3}{4}$ in.	= 0.06
7 ft 5$\frac{3}{4}$ in.	= 7.48 ft

(d)
2 ft	= 2.00 ft
9 in. − $\frac{1}{8}$ in.	= 0.74
2 ft 8$\frac{7}{8}$ in.	= 2.74 ft

(e)
6 ft 0 in.	= 6.00 ft
−0 ft $\frac{1}{2}$ in.	= −0.04
5 ft 11$\frac{1}{2}$ in.	= 5.96 ft

6 CONVERSION OF DECIMALS OF A FOOT TO INCHES

In converting measurements expressed in feet and decimals of a foot to feet, inches and fractions of an inch, mentally recall the "decimal of foot value" for the full inch which is nearest and less than the given measurement. Then convert the remainder, which will be in hundredths of a foot, to a fraction which is expressed in eighths of an inch. Or, recall the "decimal of a foot value" for the full inch which is nearest and more than the given measurement and subtract the given measurement from it. It should be remembered that conversions are made only to the nearest $\frac{1}{8}$ inch in this procedure.

Example 3.3

  (a) 3.79 ft

  (b) 6.34 ft

  (c) 5.65 ft

  (d) 3.72 ft

Solutions

(a) $\begin{aligned} 3.00 &= 3' - 0'' \\ 0.79 = 0.75 + 0.04 &= 0' - 9\tfrac{1}{2}'' \\ \hline 3.79 &= 3' - 9\tfrac{1}{2}'' \end{aligned}$

(b) $\begin{aligned} 6.00 &= 6' - 0'' \\ 0.34 = 0.33 + 0.01 &= 0' - 4\tfrac{1}{8}'' \\ \hline 6.34 &= 6' - 4\tfrac{1}{8}'' \end{aligned}$

(c) $\begin{aligned} 5.00 &= 5' - 0'' \\ 0.65 = 0.67 - 0.02 &= 0' - 7\tfrac{3}{4}'' \\ \hline 5.65 &= 5' - 7\tfrac{3}{4}'' \end{aligned}$

(d) $\begin{aligned} 3.00 &= 3' - 0'' \\ 0.72 = 0.75 - 0.03 &= 0' - 8\tfrac{5}{8}'' \\ \hline 3.72 &= 3' - 8\tfrac{5}{8}'' \end{aligned}$

PRACTICE PROBLEMS FOR CHAPTER 3

1. At the end of the sentence in the space provided on the right, write the missing word or number in each sentence, which is represented by the three dots.

(a) 4 yards = ... inches

(b) 288 sq. in. = ... sq. ft.

(c) 54 cu. ft. = ... cu. yd.

(d) 3 acres = ... sq. ft.

(e) 0.5 sq. ft. = ... sq. in.

(f) 2 gal. = ... cu. in.

(g) 15 gal. = ... cu. ft

(h) 3 gal. = ... lb. (water)

(i) 250 lb. = ... cu. ft. (water)

(j) 3,000 lb. = ... ton.

(k) Kilo means ... times.

(l) Centi means ... of.

(m) Deci means ... of.

(n) Milli means ... of.

(o) Hecto means ... times.

(p) Deka means ... times.

(a) 3.52′	(k) 6.25′
(b) 4.76′	(l) 7.81′
(c) 9.23′	(m) 2.94′
(d) 6.16′	(n) 5.06′
(e) 8.72′	(o) 6.67′
(f) 2.69′	(p) 3.87′
(g) 4.79′	(q) 4.83′
(h) 8.21′	(r) 0.36′
(i) 6.08′	(s) 9.27′
(j) 5.60′	(t) 1.35′

2. Convert the following feet and inches to feet and decimals of a foot (to two decimals only).

Example: 3′-9 3/8″ = 3.78′

(a) 2′-6 3/4″	(k) 4′-6 1/8″
(b) 1′-10 1/8″	(l) 7′-2 7/8″
(c) 3′-7 5/8″	(m) 5′-4 3/4″
(d) 6′-3 1/2″	(n) 8′-7 1/8″
(e) 9′-8 1/2″	(o) 10′-5 1/4″
(f) 7′-2 1/2″	(p) 5′-0 1/4″
(g) 4′-9 3/4″	(q) 8′-8 5/8″
(h) 5′-10 5/8″	(r) 9′-4 1/8″
(i) 6′-4 7/8″	(s) 2′-6 1/2″
(j) 4′-3 3/8″	(t) 10′-11 1/4″

3. Convert the following feet and decimals to feet and inches.

PROFESSIONAL PUBLICATIONS, INC. • P.O. Box 199, San Carlos, CA   94070

4 PERIMETER AND CIRCUMFERENCE

1 DEFINITION

The sum of the lengths of the sides of a polygon is called the perimeter of the polygon.

Example 4.1

Find the perimeter of the right triangle shown.

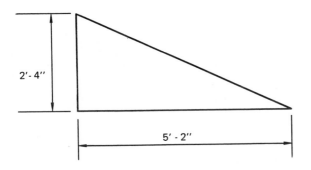

Figure 4.1

Solution

$$\text{Side} = \sqrt{5.17^2 + 2.33^2} = 5.67' = 5'8''$$
$$\text{Perimeter} = 5' - 2'' + 2' - 4'' + 5' - 8'' = 13'2''$$

Example 4.2

Find the perimeter of the isosceles trapezoid shown.

Solutions

$$\text{Side} = \sqrt{4^2 + 2^2} = 4.5''$$
$$\text{Perimeter} = 12'' + 4.5'' + 16'' + 4.5'' = 37''$$

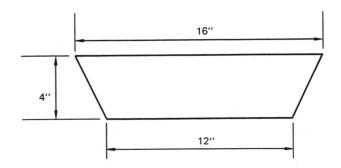

Figure 4.2   Example 4.2

2 CIRCUMFERENCE OF A CIRCLE

The circumference of a circle is the distance around the circle. It contains 360°. Regardless of the size, the circumference of any circle is always approximately 3.14 times the length of the diameter. The exact ratio of the circumference to the diameter is the number π. It is an irrational number, but its value is usually considered to be 3.1416 or 3.14, depending on the accuracy desired, based on the accuracy of the measurement of the diameter. Thus we can say that the circumference of any circle is

$$C = \pi D$$

where C is the circumference and D is the diameter of the circle. Because the diameter is twice the radius, we can also say

$$C = 2\pi R$$

where C is the circumference and R is the radius of the circle. Either of the two formulas is appropriate.

Example 4.3

Find the circumference of a 10'0'' circle.

Solution

A "10'0″ circle" implies 10'0″ diameter.

$$C = \pi D = (3.1416)(10) = 31.42' = 31'5''$$

Example 4.4

Find the circumference of a circle which has a radius of 21'9″.

Solution

$$C = 2\pi R = (2)(3.1416)(21.75) = 136.66' = 136'8''$$

Example 4.5

Find the outside circumference of a concrete pipe with inside diameter of 36″ and wall thickness of 3″.

Solution

$$C = \pi D = (3.1416)(42) = 132'' = 11'0''$$

Example 4.6

Find the diameter of a tank which measures 62'10″ around.

Solution

$$C = \pi D, \quad D = \frac{C}{\pi} = \frac{62.83}{3.1416} = 20'0''$$

3 LENGTH OF ARC OF A CIRCLE

The length of an arc of a circle is proportional to its central angle. A central angle of 90° ($\frac{1}{4}$ of 360°) subtends an arc which is one-fourth the circumference in length. Thus, an arc whose central angle is 45° is $\frac{45}{360} \times C$ in length.

Example 4.7

Find the length of an arc of a 100' circle which has a central angle of 36°.

Solution

$$\text{Arc} = \frac{(36)(100)\pi}{360} = \frac{100\pi}{10} = 31'$$

PRACTICE PROBLEMS FOR CHAPTER 4

Example: Find the perimeter of a right triangle with base 9″ and altitude 12″.

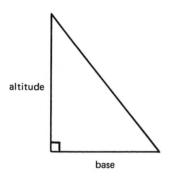

altitude

base

Figure

$$\text{Side} = \sqrt{9^2 + 12^2} = 15 \quad \text{Perimeter} = 36''$$

1. Find the perimeter of the floor of a room 18′ × 22′.

2. Find the perimeter of a right triangle with base 12″ and altitude 5″.

3. Find the perimeter of an isosceles triangle with base 12″ and altitude 8″. (Note: For an isosceles triangle, a line from the vertex perpendicular to the base bisects the base.)

4. Find the circumference of a 10″ circle (10″ diameter).

5. Find the circumference of a circle with radius of 7″.

6. Find the length of an arc of a circle of 24″ radius which has a central angle of 60°.

7. What is the diameter of a cylindrical tank which measures 47.10 ft. around the outside?

8. Find the length of an arc of a circle of 16″ diameter which has a central angle of 45°.

9. Find the diameter of a tree which measures 3′-1 3/4″ around.

10. Find the perimeter of each figure. Show computations for length of any side needed which is not dimensioned.

(a)

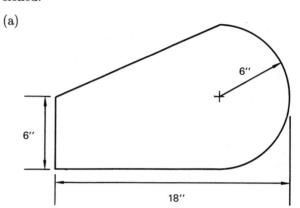

(b)

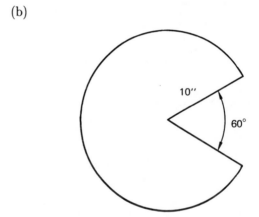

11. Find the length of the bars.

(a)

(b)

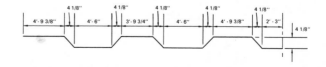

PROFESSIONAL PUBLICATIONS, INC. • P.O. Box 199, San Carlos, CA 94070

5 AREA

1 DEFINITION

Area is defined as the surface within a set of lines. Thus, the area of a triangle is the surface within the three sides; the area of a circle is the surface within the circumference.

Area is measured in square units: square inches, square feet, square miles, etc.

A square inch is a square, each side of which is 1 inch in length. The rectangle shown below has an area of 6 square inches. It could be exactly covered by six squares, each 1 inch on a side.

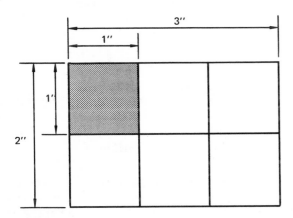

Figure 5.1

2 AREA OF A RECTANGLE

The area of a rectangle is equal to the product of the length and the width.

Example 5.1

Find the area of the floor of a room which is 20.25′ long and 16.33′ wide.

Solution

$$\text{Area} = \text{length} \times \text{width}$$
$$= 20.25' \times 16.33' = 330.7 \,\text{sq ft}$$

Example 5.2

Find the area of the walls of a room 8.0′ high if the length of the room is 20.0′ and the width is 15.0′.

Solution

$$\text{Area of 2 walls} = 2 \times 8' \times 20' = 320 \,\text{sq ft}$$
$$\text{Area of 2 walls} = 2 \times 8' \times 15' = \underline{240 \,\text{sq ft}}$$
$$\text{Total area} \qquad\qquad\quad = \overline{560 \,\text{sq ft}}$$

3 AREA OF A TRIANGLE

The area of a triangle is expressed in terms of its base and its altitude. Any side of a triangle can be called the base. The vertex of a triangle is the vertex opposite the base. The altitude of a triangle is the perpendicular distance from the vertex to the base.

The area of any triangle is equal to one-half the product of the base and the altitude.

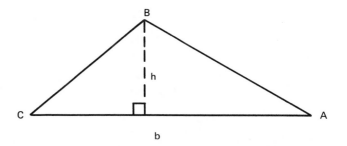

Figure 5.2

PROFESSIONAL PUBLICATIONS, INC. • P.O. Box 199, San Carlos, CA 94070

In the triangle ABC

$$A = \frac{1}{2}bh$$

where A = area, b = base and h = altitude.

Example 5.3

Find the area of a triangle with base $12''$ and altitude $4''$.

Solution

$$A = \frac{bh}{2} = \frac{(12)(4)}{2} = 24\,\text{sq in.}$$

4 AREA OF A RIGHT TRIANGLE

The area of a right triangle is equal to one-half the product of the base and the altitude.

In Fig. 5.3, a rectangle has a side b and a side a. The area of the rectangle is $a \times b$. If we cut the rectangle into two equal triangles as shown by the dashed line, the area of each of the triangles formed is $\frac{1}{2} \times a \times b$. If a represents the altitude of a right triangle and b represents the base of the right triangle, then the area of the triangle is $\frac{1}{2}(ab)$.

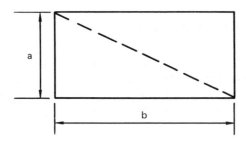

Figure 5.3

Example 5.4

Find the area of the right triangle shown.

Solution

$$\text{Base} = 5' - 2'' = 5.16'$$
$$\text{Altitude} = 2' - 4'' = 2.33'$$
$$\text{Area} = \frac{ab}{2}$$
$$= \frac{2.33 \times 5.16}{2} = 6.01\,\text{sq ft}$$

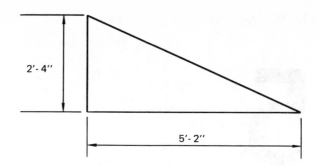

Figure 5.4 Example 5.4

5 AREA OF A TRIANGLE WITH KNOWN SIDES

If the length of the three sides of a triangle are known, the area of the triangle can be found from

$$A = \sqrt{s(s-a)(s-b)(s-c)}$$

where A = area,

$$s = \frac{1}{2}\,\text{the perimeter}$$

a, b, c = lengths of each of the sides

Example 5.5

Find the area of the triangle with sides $32'$, $46'$, and $68'$.

Solution

$$A = \sqrt{s(s-a)(s-b)(s-c)}$$
$$= \sqrt{73(73-32)(73-46)(73-68)} = 636\,\text{sq ft}$$

6 AREA OF A TRAPEZOID

The area of a trapezoid is equal to the average width times the altitude, or expressed in another way, the area of a trapezoid is equal to one-half the sum of the bases times the altitude.

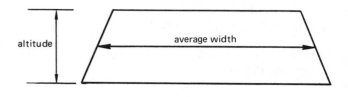

Figure 5.5

Example 5.6

Find the area of the trapezoid shown.

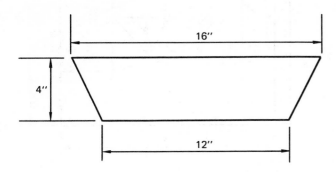

Figure 5.6

Solution

$$\text{Area} = \frac{1}{2}\,\text{sum of bases} \times \text{altitude}$$

$$= \frac{16'' + 12''}{2} \times 4'' = 56\,\text{sq in.}$$

Example 5.7

A swimming pool is 4' deep at one end, 8' deep at the other end, and 100' long. Find the area of the trapezoidal section through the long axis.

Solution

$$A = \frac{(4+8)(100)}{2} = 600\,\text{sq ft}$$

Example 5.8

A drainage ditch has a trapezoidal cross-section with bottom width 6', top width 24' and depth 4'. Find the area of the cross-section.

Solution

$$A = \frac{(6+24)(4)}{2} = 60\,\text{sq ft}$$

7 AREA OF A CIRCLE

We learned that there is a relation between the circumference of a circle and its radius so that the circumference is always 2π times the radius. There is also a relation between the area of a circle and its radius.

The area of a circle is always π times the square of its radius, or

$$A = \pi R^2$$

where A = the area of any circle and R is the radius of that circle.

Since the radius of a circle is equal to half the diameter, we can express the area of a circle as

$$A = \pi \left|\frac{D}{2}\right|^2 = \frac{\pi D^2}{4}$$

where D is the diameter.

Since $\frac{\pi}{4} = 0.785$, approximately, we can also say

$$A = 0.785 D^2$$

Example 5.9

Find the area of a 12.0 in. circle.

Solution

$$A = \pi R^2 = \pi 6^2 = 36\pi = (36)(3.14) = 113\,\text{sq in.}$$

Example 5.10

Find the area of a 10.0 in. circle.

Solution

$$A = \frac{\pi D^2}{4} = \frac{\pi(10.0)^2}{4} = \frac{100\pi}{4} = 78.5\,\text{sq in.}$$

Example 5.11

Find the area of an 11.0 in. circle.

Solution

$$A = 0.785 D^2 = 0.785(11.0)^2 = 95.0\,\text{sq in.}$$

8 AREA OF A SECTOR OF A CIRCLE

The area of a sector of a circle is a fractional part of the area of the circle. The central angle of the sector is a measure of the fraction. A sector whose central angle is 90° is one-fourth of a circle because 90° is one-fourth of 360°.

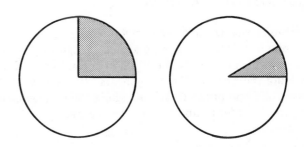

Figure 5.7

Example 5.12

Find the area of a 30° sector of a 6'' circle.

PROFESSIONAL PUBLICATIONS, INC. • P.O. Box 199, San Carlos, CA   94070

Solution

$$A = \frac{30}{360} \times 0.785 D^2$$
$$= \frac{30}{360} \times 0.785 \times 36 = 2.4 \,\text{sq in.}$$

9 AREA OF A SEGMENT OF A CIRCLE

A segment of a circle is bounded by an arc and a straight line which connects the ends of the arc. The area of a segment is found by subtracting the area of the triangle formed by the chord and the two radii to its end points from the area of the sector formed by the two radii and the arc.

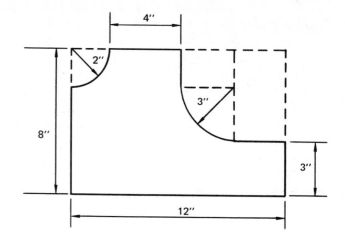

Figure 5.9

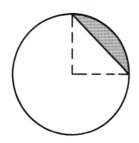

Figure 5.8

Example 5.13

Find the area of the segment whose arc subtends an angle of 90° in a circle of 8″ radius.

Solution

$$A = \frac{90}{360} \times 3.14 \times 8^2 - \frac{8 \times 8}{2}$$
$$= 50 - 32 = 18 \,\text{sq in.}$$

10 COMPOSITE AREAS

Irregular shaped areas can sometimes be divided into components which consist of geometric figures, the areas of which can be found. Total area can be found by adding the areas of the components. In some cases it may be appropriate to subtract the areas of geometric figures in order to find the net area desired.

Example 5.14

Find the area of the figure in Fig. 5.9.

Solution

$$\text{Area} = 12 \times 8 - \frac{\pi 2^2}{4} - 2 \times 3 - \frac{\pi 3^2}{4} - 3 \times 5 = 65 \,\text{sq in.}$$

PRACTICE PROBLEMS FOR CHAPTER 5

1. Find the area of a right triangle with base 12 in. and altitude 8 in.

2. Find the number of square feet of wall board needed to cover the walls and ceiling of a room 24 ft long, 16 ft wide and 8 ft high. Find the number of 4 ft × 8 ft sheets needed.

3. Find the cross-section area of a ditch of trapezoidal cross section with top width of 28 ft, bottom width of 4 ft and depth of 6 ft.

4. Find the cross-section area of a highway fill of trapezoidal cross-section with top width of 44 ft, base width of 92 ft and height of 8 ft.

5. Find the area of a circle of 20 ft radius.

6. Find the area of a 10 ft diameter circle.

7. Find the area of a 60° sector of a 6″ circle.

8. Find the area of the segment whose arc subtends an angle of 90° in a 12 ft circle (12 ft dia).

9. Find the area of a triangle with sides 18 ft, 12 ft, and 10 ft.

10. Divide the figures into component parts, then find the total area by either adding the areas of the component parts or by subtracting areas from a larger area which includes the area shown.

Example:

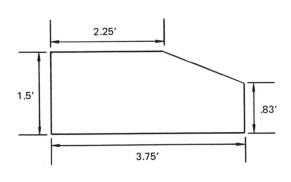

Figure

$$A = (3.75)(1.5) - 1/2(1.50)(0.67)$$
$$= 5.1\,\text{ft}^2$$

(a)

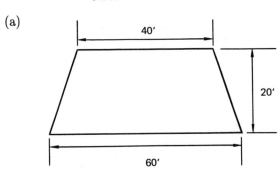

(b)

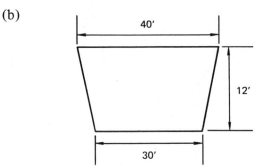

(c)

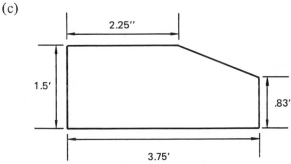

(d)

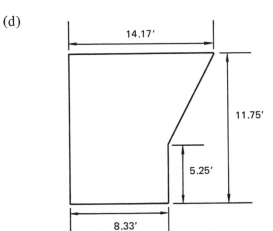

PROFESSIONAL PUBLICATIONS, INC. ● P.O. Box 199, San Carlos, CA 94070

(e)
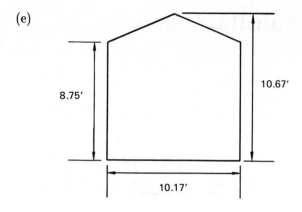

(i)
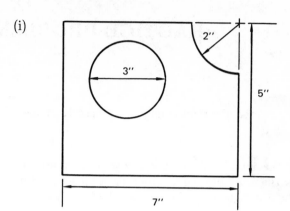

(f)
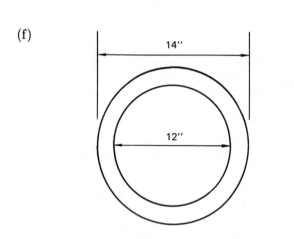

(j)
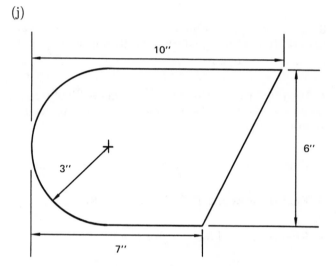

(g)
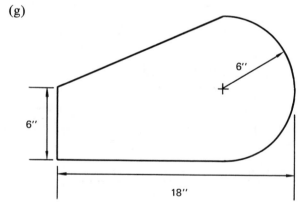

(h)
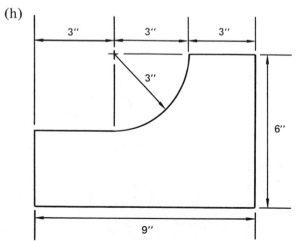

6

VOLUME

1 DEFINITION

Volume is defined as the amount of substance occupying a certain space. It is measured in cubic units. The block shown in figure 6.1 has a volume of 6 cubic inches. One cubic inch is a cube which measures 1 inch on each edge.

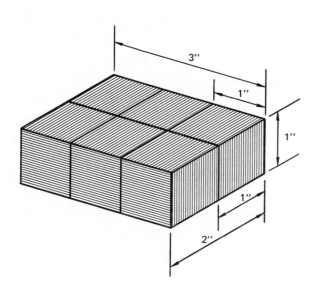

Figure 6.1

2 VOLUME OF RIGHT PRISMS AND CYLINDERS

The volume of a right prism or cylinder is the product of the area of the base and the altitude. Expressed as a formula

$$V = Ah$$

V is volume in cubic units, A is area in square units and h is altitude in linear units.

Example 6.1

Find the volume of a rectangular prism with base $8'' \times 6''$ and altitude $10''$.

Solution

$$V = Ah = 8'' \times 6'' \times 10'' = 480 \,\text{cu in.}$$

Example 6.2

Find the volume of a triangular prism with triangular base which has sides $3''$, $4''$, and $5''$, and with $8''$ altitude.

Solution

$$V = Ah = \frac{1}{2} \times 3 \times 4 \times 8 = 48 \,\text{cu in.}$$

Example 6.3

Find the number of cubic yards of dirt in $500'$ of a highway fill of trapezoidal cross-section with bottom base $112'$, top base $40'$, and height $12'$.

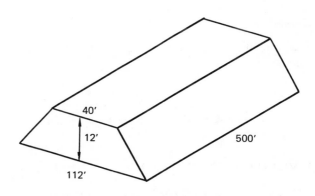

Figure 6.2

PROFESSIONAL PUBLICATIONS, INC. • P.O. Box 199, San Carlos, CA 94070

Solution

$$V = \frac{(112 + 40)(12)(500)}{(2)(27)} = 16{,}889 \,\text{cu yd}$$

Example 6.4

Find the volume of the shell of the hollow cylinder which has an outside diameter of 8″, inside diameter of 6″, and height of 5″.

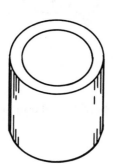

Figure 6.3

Solution

$$V = Ah$$
$$= (\text{outside area} - \text{inside area})h$$
$$= 0.785(D_o^2 - D_i^2)h$$
$$= 0.785(8^2 - 6^2)(5) = 110 \,\text{cu in.}$$

3 VOLUME OF CONE

The volume of a right circular cone is equal to one-third the product of the area of its base and its altitude, or

$$V = \frac{1}{3}\pi r^2 h$$

Example 6.5

Find the volume of a cone 6″ high with base 4″ in diameter.

Solution

$$V = \frac{1}{3}\pi r^2 h = \frac{1}{3}(\pi)(2)^2(6) = 8\pi = 25 \,\text{cu in.}$$

4 VOLUME OF PYRAMID

The volume of a pyramid is equal to one-third the product of the area of its base and its altitude, or

$$V = \frac{1}{3}Ah$$

5 VOLUME OF SPHERE

The volume of a sphere is equal to $\frac{4}{3}\pi r^3$, or

$$V = \frac{4}{3}\pi r^3$$

PRACTICE PROBLEMS FOR CHAPTER 6

(Note: When some dimensions are predominantly in feet but with one dimension in inches, convert inches to feet by using a common fraction: $3'' = 1/4'$, $4'' = 1/3'$, $6'' = 1/2'$. The denominators can be used for cancellation.)

1. Find the volume of a rectangular right prism with base $3' \times 4'$ and altitude $6'$.

2. Find the volume of a triangular right prism with sides of base $9''$, $12''$, and $15''$ and altitude $10''$.

3. Find the volume of a cylinder with base having a diameter of $10''$ and altitude of $8''$.

4. Find the number of cubic feet of concrete (to the nearest tenth) in a pipe of $8''$ inside diameter, wall thickness of $2''$ and length of $30''$.

5. Find the necessary height of a cylindrical tank of $6'$ diameter if its volume is to be 226 cu. ft. (to nearest tenth)

6. Find the volume of a right prism with base an isosceles triangle with base $8''$ and altitude $3''$ and altitude of prism $10''$.

7. Find the number of cubic yards of dirt in 810 ft of a highway fill of trapezoidal cross-section with base at bottom of 120 ft, base at top of 80 ft and height of fill 8 ft.

8. How many cubic yards of concrete are needed to pour a parking area $30'$ long, $27'$ wide and $4''$ thick?

PROFESSIONAL PUBLICATIONS, INC. • P.O. Box 199, San Carlos, CA 94070

PROFESSIONAL PUBLICATIONS, INC. • P.O. Box 199, San Carlos, CA 94070

7 DIMENSIONAL EQUATIONS

1 MEASUREMENT

A measurement consists of a number which expresses quantity and a unit of measure. The surveyor and the surveying technician are intricately involved in measurements and in converting measurements expressed in one unit of measure to an equivalent in another unit of measure.

In converting values from one unit of measure to another, it is just as important to find the correct unit of measure as it is to find the correct quantity.

2 DEFINITION OF DIMENSIONAL EQUATION

A dimensional equation is one which contains units of measure but does not contain the corresponding numerical values. Suppose we wish to express in cubic yards the volume of a dump truck bed with dimensions $6' \times 8' \times 4'$. Arithmetically we multiply $6 \times 8 \times 4$ and find the volume to be 192 cu ft. Since there are 27 cubic feet in a cubic yard, we divide 192 by 27 and find the bed has a volume of 7 cubic yards. This operation is

$$\frac{6 \times 8 \times 4}{27} = 7$$

The dimensional equation which corresponds to this is

$$(\text{ft} \times \text{ft} \times \text{ft}) \div (\text{cu ft per cu yd}) = \frac{\text{cu ft}}{1} \div \frac{\text{cu ft}}{\text{cu yd}}$$
$$= \frac{\text{cu ft}}{1} \times \frac{\text{cu yd}}{\text{cu ft}}$$
$$= \text{cu yd}$$

Including numbers in the equation

$$\frac{6}{1}\,\text{ft} \times \frac{8}{1}\,\text{ft} \times \frac{4}{1}\,\text{ft} \div \frac{27\,\text{cu ft}}{1\,\text{cu yd}} = \frac{192\,\text{cu ft}}{1} \times \frac{1}{27}\frac{\text{cu yd}}{\text{cu ft}}$$
$$= 7\,\text{cu yd}$$

Example 7.1

Write a dimensional equation for finding the area in acres of a rectangular tract of land 300 ft by 200 ft. Include the measured quantities in the equation.

Solution

$$\frac{300}{1}\,\text{ft} \times \frac{200}{1}\,\text{ft} \div \frac{43,560\,\text{sq ft}}{1\,\text{ac}}$$
$$= \frac{60,000}{1}\,\text{sq ft} \times \frac{1}{43,560}\frac{\text{ac}}{\text{sq ft}}$$
$$= 1.4\,\text{ac}$$

Example 7.2

Write a dimensional equation for finding the velocity in feet per second of a vehicle travelling 36 miles per hour. Include the measured quantity in the equation.

Solution

$$V = \frac{36\,\text{mi}}{1\,\text{hr}} \times \frac{5280\,\text{ft}}{1\,\text{mi}} \div \frac{3600\,\text{sec}}{1\,\text{hr}}$$
$$= \frac{36\,\text{mi}}{1\,\text{hr}} \times \frac{5280\,\text{ft}}{1\,\text{mi}} \times \frac{1}{3600}\frac{\text{hr}}{\text{sec}} = 53\,\frac{\text{ft}}{\text{sec}}$$

Example 7.3

Write a dimensional equation for finding the weight of water, in tons, in a full rectangular tank which is 10 ft long, 8 ft wide, and 6 ft deep.

Solution

$$W = \frac{10}{1}\,\text{ft} \times \frac{8}{1}\,\text{ft} \times \frac{6}{1}\,\text{ft} \times \frac{62.5}{1}\frac{\text{lb}}{\text{cu ft}} \div \frac{2000}{1}\frac{\text{lb}}{\text{ton}}$$
$$= \frac{480}{1}\,\text{cu ft} \times \frac{62.5}{1}\frac{\text{lb}}{\text{cu ft}} \div \frac{2000}{1}\frac{\text{lb}}{\text{ton}}$$
$$= \frac{480}{1}\,\text{cu ft} \times \frac{62.5}{1}\frac{\text{lb}}{\text{cu ft}} \times \frac{1}{2000}\frac{\text{ton}}{\text{lb}} = 15\,\text{tons}$$

3 FORM FOR PROBLEM SOLVING

Familiarity with conversion factors makes it unnecessary to set up a dimensional equation for solving problems involving several measured quantities, but setting up a single equation which includes the numbers expressing quantity but without units of measure, similar to the dimensional equation, is advantageous. It allows

cancellation and is easily followed by someone whose
task is to check its accuracy. For the benefit of the
checker, each number which expresses quantity should
be shown in the equation. A problem to find the area
of a 10″ circle should not be solved in this manner:
$A = 0.785D^2 = 78.5\,\text{sq in}$. It should be solved in this
manner: $A = 0.785(10)^2 = 78.5\,\text{sq in}$. For solutions
which involve unfamiliar formulas, it is good practice
to write the formula and then substitute the measured
quantities, but it is not necessary to write such well-
known formulas as that for the area of a circle.

Example 7.4

What is the cost of the concrete, delivered to the site
at \$36 per cubic yard, for a parking lot 100′ long, 54′
wide, and 4″ thick?

Solution

(Note: Where length and width are measured in feet
and thickness in inches, it is convenient to use the thickness measurement as a common fraction of a foot. Cancellation is often enhanced, and accuracy is improved.
Four inches is *exactly* one-third of a foot but *approximately* 0.33 of a foot.

$$\text{Cost} = \frac{(100)(54)(\$36)}{(27)(3)} = \$2,400$$

Example 7.5

What is the cost of filling a rectangular tank, 100′ long,
40′ wide, and 10′ deep, with water at \$0.60 per M gal-
lons?

Solution

$$\text{Cost} = \frac{(100)(40)(10)(7.5)(\$0.60)}{1000} = \$180.00$$

PRACTICE PROBLEMS FOR CHAPTER 7

1. Write a dimensional equation to convert the given quantities to an equivalent quantity in the unit of measure indicated.

Examples:

(1) Express 48 inches in feet.

$$48\,\text{in} \div \frac{12\,\text{in.}}{\text{ft}} = 48\,\text{in} \times \frac{1\,\text{ft}}{12\,\text{in.}} = 4\,\text{ft}$$

(2) Express 3 feet in inches.

$$3\,\text{ft} \times \frac{12\,\text{in.}}{\text{ft}} = 36\,\text{in.}$$

(3) Express 72 square feet in square yards.

$$72\,\text{sq ft} \div 9\frac{\text{sq ft}}{\text{sq yd}} = 72\,\text{sq ft} \times \frac{1\,\text{sq yd}}{9\,\text{sq ft}} = 8\,\text{sq yd}$$

(a) 588 ft to yards

(b) 121 yards to ft

(c) 2 miles to ft

(d) 4 sq ft to sq in.

(e) 432 sq in. to sq ft

(f) 5 sq yd to sq ft

(g) 81 sq ft to sq yd

(h) 2 acres to sq ft

(i) 21,780 sq ft to ac

(j) 3 cu ft to cu in.

(k) 3456 cu in. to cu ft

(l) 5 cu yd to cu ft

(m) 135 cu ft to cu yd

(n) 3 gal to cu in.

(o) 693 cu in. to gal

(p) 4 gal water to lb

(q) 25 lb water to gal

(r) 87,120 sq ft to ac

(s) 1320 ft to miles

(t) 7 sq yd to sq ft

2. Find the required quantities by including the given quantities within a dimensional equation.

Examples:

(1) Find the number of square yards in a driveway 20 ft wide and 54 ft long.

$$\text{Area} = 20\,\text{ft} \times 54\,\text{ft} \div \frac{9\,\text{sq ft}}{\text{sq yd}}$$
$$= 20\,\text{ft} \times 54\,\text{ft} \times \frac{1\,\text{sq yd}}{9\,\text{sq ft}} = 120\,\text{sq yd}$$

(2) Find the cost of a concrete sidewalk 4 ft wide, 81 ft long and 4 in. thick at $30 per cu yd.

$$\text{Cost} = 4\,\text{ft} \times 81\,\text{ft} \times 4\,\text{in} \div \frac{12\,\text{in.}}{1\,\text{ft}}$$
$$\div \frac{27\,\text{cu ft}}{\text{cu yd}} \times \frac{30\,\text{dollars}}{\text{cu yd}}$$
$$= 4\,\text{ft} \times 81\,\text{ft} \times 4\,\text{in.} \times \frac{1\,\text{ft}}{12\,\text{in.}}$$
$$\times \frac{1\,\text{cu yd}}{27\,\text{cu ft}} \times \frac{30\,\text{dollars}}{\text{cu yd}} = \$120$$

(a) How many acres are in a rectangular plot 545 ft long and 400 ft wide?

(b) What is the weight, in tons, of the water in a tank containing 2,000 gallons?

(c) Find the velocity, in feet per second, of a vehicle traveling 72 miles per hour.

(d) What is the weight of water, in tons, in a full cylindrical tank of 10 ft diameter and 10 ft height?

(e) What is the cost of excavation of a ditch of rectangular cross-section 3 ft wide, 4 ft deep and 324 ft long at 30¢ per cu yd?

3. Solve the problems by writing an equation in the form of a dimensional equation but include units of measure only if preferred.

Example: Find the weight of water in a rectangular tank, full, which is 8 ft long, 5 ft wide, and 5 ft deep.

$$\text{Wt} = \frac{8 \times 5 \times 5 \times 62.5}{2000} = \frac{200 \times 62.5}{2000}$$
$$= 6.25\,\text{tons}$$

(a) A 6′ × 3′ concrete box culvert, 54 ft long, is to be constructed. Walls, footing, and deck are 6″ thick. How many cubic yards of concrete are required? (Disregard wing walls.) Note: *Culvert dimensions refer to waterway openings.* Horizontal dimension is 6′, vertical dimension is 3′.

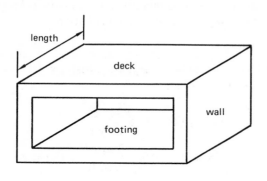

Figure

(b) The cross-section view of concrete curb and gutter to be used in street paving is shown. How many lineal feet of curb and gutter can be poured with one cubic yds of concrete?

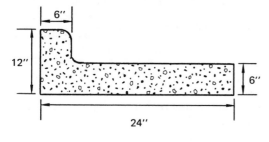

Figure

(c) A contractor is to be paid for sprinkling in units of M gallons. The empty weight of his water truck is 11,808 pounds. Loaded with water the truck weighs 28,468 pounds. How many M gallons of water does the truck hold? (M = 1000)

(d) A canal is to be excavated to a trapezoidal cross-section, 30 ft at the top and 6 ft at the bottom with a 5 ft depth. What will be the cost of excavation at $0.50 per cubic yard if the length is 540 ft?

(e) A drainage ditch has a 4 ft flat bottom, 12 ft top width and average depth of 2 ft through 162 ft of level ground. How many cubic yards of earth were excavated?

(f) How high must a cylindrical tank, 10 ft in diameter, be in order to have a capacity of 3000 gallons? (To the nearest tenth of a ft)

(g) A parking space is 100 ft × 81 ft. What is the cost of paving this area at $9.00 per square yard?

(h) A building lot has an area of 3,840 ft². How deep is the lot if it is 32 ft wide?

(i) An electric power line is to be built from one city to another. One city is 16 miles due north and 12 miles due west of the other. (a) What length of wire is needed to connect the two cities? (b) If the wire weighs 50 pounds per 100 ft, what weight of wire is needed?

(j) A piece of land that is triangular in shape has one side 320 yards long running north and south and another 1/4 mile long at right angles. A second piece of land, rectangular in shape and 250 yards on one side, has the same acreage as the triangular piece. Which piece of land would require the most fence to enclose?

(k) A 24 inch shaft was drilled 54 feet deep and filled with concrete as part of a bridge pier. How many cubic yards of concrete were poured?

(l) A swimming pool 100 ft long, 50 ft wide, 2 ft deep at the shallow end and 10 ft deep at the deep end is to be filled with water. What is the cost of the water at 20¢ per M gallons?

(m) A cylindrical piece of cheese 16 inches in diameter and 8 in. high weighs 24 pounds. If a 30° sector is cut from it, (a) what is the cost of the sector at $1.00 per pound and (b) how many cubic inches of cheese are in the sector?

(n) A rectangular concrete tank, 11 ft long, 6 ft wide, and 4′-6″ high (outside), is 3/4 full of water. Walls and floor of the tank are 6″ thick. How many gallons of water are in the tank?

(o) A piece of property to be purchased for highway right-of-way is bounded by an arc of a circle and a chord of that circle. The radius of the circle is 500 ft and the central angle formed by radii to the ends of the chord is 90°. Find the area of the segment.

(p) A cylindrical water tank contains 60,000 gallons of water when the water is 5 ft deep. What is the diameter of the tank?

(q) A lot 150 ft in depth and 100 ft wide is to be leveled for building construction. The fill at the front is 1.4 ft and at the rear is 2.2 ft. How many cubic yards of dirt will be required to make the fill? (disregard shrinkage of soil.)

PROFESSIONAL PUBLICATIONS, INC. • P.O. Box 199, San Carlos, CA 94070

8 SIGNED NUMBERS

1 POSITIVE AND NEGATIVE NUMBERS

In arithmetic the symbol for plus (+) indicates that something is to be added; the symbol for minus (−) indicates that something is to be subtracted. These same symbols are also used to show the values of numbers. The numbers are called *signed numbers*.

In algebra the sign + before a number indicates that the number is a positive number; the sign − before a number indicates that the number is a negative number. If there is no sign before a number, it is considered to be a positive number.

Positive numbers are greater than zero; negative numbers are less than zero. Zero is neither positive nor negative.

The relative value of numbers can be shown by a graduated horizontal line. (Fig. 8.1).

Figure 8.1

Numbers to the right of zero are positive; numbers to the left of zero are negative. Values of numbers increase from left to right. For instance, in Fig. 8.1, −4 is less than −3, and −3 is less than −2. This statement can be simplified by use of symbols: $-4 < -3 < -2$. The symbol < means "less than," the symbol > means "more than." $5 > 3 > 1$ means "5 is greater than 3, and 3 is greater than 1."

2 ABSOLUTE VALUE OF A NUMBER

The absolute value of a number is the numerical value of the number, the value of the number without regard to its sign. The absolute value of −5 is 5; the absolute value of +5 is 5. The symbol for the absolute value of −5 is $|-5|$.

3 ADDITION OF SIGNED NUMBERS

To add signed numbers, three rules should be remembered.

> Rule 1. To add two or more numbers with like signs, find the sum of their absolute values and prefix the common sign to the sum.

Example 8.1

(a) $\begin{array}{r} +9 \\ \underline{+6} \\ +15 \end{array}$

(b) $\begin{array}{r} -8 \\ \underline{-5} \\ -13 \end{array}$

(c) $(+9) + (+6) = +15$

(d) $(-8) + (-5) = -13$

> Rule 2. To add two numbers with unlike signs, find the difference between their absolute values and prefix the sign of the number having the greater absolute value.

Example 8.2

(a) $\begin{array}{r} +9 \\ \underline{-6} \\ +3 \end{array}$

(b) -8
     $\underline{+5}$
     -3

(c) $(+9) + (-6) = +3$

(d) $(-8) + (+5) = -3$

Rule 3. To add more than two numbers with unlike signs, add the positive numbers and negative numbers separately and use the two sums as in Rule 2.

Example 8.3

(a) -3
     $+6$
     $+4$
     $\underline{-2}$
     $+5$

(b) $+5$
     -8
     -3
     $\underline{-2}$
     -8

(c) -7
     $+4$
     $+5$
     $\underline{-2}$
     0

(a) $(-3) + (+6) + (+4) + (-2) = +5$

(b) $(+5) + (-8) + (-3) + (-2) = -8$

(c) $(-7) + (+4) + (+5) + (-2) = 0$

4 SUBTRACTION OF SIGNED NUMBERS

Subtracting $+3$ from $+10$ is the same as adding -3 to $+10$.

(a) $(+10) - (+3) = 7$ (c) $(+10) + (-3) = 7$
(b) $(+10) - (-3) = 13$ (d) $(+10) + (+3) = 13$

Rule: To subtract two signed numbers, change the sign of the subtrahend and add it to the minuend.

Example 8.4

Subtract the bottom number from the top number.

(a) $+12$
     $\underline{+3}$

(b) $+10$
     $\underline{-4}$

(c) -18
     $\underline{+6}$

(d) -16
     $\underline{-7}$

Solutions

(a) $+12$
     $(-)\underline{+3}$
          $+9$

(b) $+10$
     $(+)\underline{-4}$
          $+14$

(c) -18
     $(-)\underline{+6}$
          -24

(d) -16
     $(+)\underline{-7}$
          -9

(a) $+12 - (+3) = +9$

(b) $+10 - (-4) = +14$

(c) $-18 - (+6) = -24$

(d) $-16 - (-7) = -9$

5 HORIZONTAL ADDITION AND SUBTRACTION

Rules established for addition and subtraction of signed numbers apply to horizontal addition and subtraction.

Example 8.5

Combine into a single number.

(a) $(-8) + (-3) - (+2)$

(b) $(+12) - (-4) + (-3)$

(c) $(+16) + (-2) - (-3) - (+5)$

(d) $(+8) - (+1) + (-2) - (-5)$

Solutions

(a) $(-8) + (-3) - (+2) = -8 - 3 - 2 = -13$

(b) $(+12) - (-4) + (-3) = 12 + 4 - 3 = 13$

(c) $(+16) + (-2) - (-3) - (+5) = 16 - 2 + 3 - 5 = 12$

(d) $(+8) - (+1) + (-2) - (-5) = 8 - 1 - 2 + 5 = 10$

It should be noted that when parentheses are preceded by a positive sign, they can be removed and the sign of the term enclosed will remain unchanged, but when parentheses are preceded by a negative sign, they can be removed if the sign of the term enclosed is changed. This is in accordance with the rule for subtraction of signed numbers.

Example 8.6

(a) $(+5) + (-2) - (+4) - (-6) = 5 - 2 - 4 + 6 = 5$

(b) $(-7) - (+3) + (-5) - (-2) = -7 - 3 - 5 + 2 = -13$

(c) $(8) - (3) + (-7) - (6) = 8 - 3 - 7 - 6 = -8$

(d) $(-15) - (-12) + (-8) - (5) =$
$\quad -15 + 12 - 8 - 5 = -16$

(e) $(27) - (36) - (-45) + (-15) =$
$\quad 27 - 36 + 45 - 15 = 21$

(f) $(18) + (12) - (-21) - (16) = 18 + 12 + 21 - 16 = 35$

6 MULTIPLICATION OF SIGNED NUMBERS

The product of signed numbers is found by multiplying the absolute value of the numbers, as in arithmetic, and prefixing the sign of the product according to the following rule.

> Rule: The product of two numbers with like signs is a positive number. The product of two numbers with unlike signs is a negative number.

Example 8.7

(a) $(-3)(+4) = -12$

(b) $(-3)(-4) = 12$

(c) $(6)(-2)(3) = -36$

(d) $(4)(-5)(-3) = 60$

(e) $(5)(-4)(-3) = 60$

(f) $(-3)(4)(-5)(-6) = -360$

It should be noted that the product of two or more signed numbers is positive if there is an even number of negative factors and is negative if there is an odd number of negative factors.

7 DIVISION OF SIGNED NUMBERS

Division is the inverse of multiplication. The quotient of signed numbers is found by dividing the absolute value of the numbers as in arithmetic and prefixing the sign of the quotient according to the following rule:

> Rule: The quotient of two numbers with like signs is a positive number. The quotient of two numbers with unlike signs is a negative number.

Example 8.8

(a) $\dfrac{(-6)(-2)}{(3)(4)} = 1$

(b) $\dfrac{(-5)(-2)(21)}{(4)(15)(-7)} = -\dfrac{1}{2}$

(c) $\dfrac{(21)(-36)(-48)}{(-7)(12)(24)} = -18$

(d) $\dfrac{(16)(-24)(0)}{(12)(48)(18)} = 0$

(e) $\dfrac{(-15)(12)(-18)}{(9)(-5)(4)(36)} = -\dfrac{1}{2}$

(f) $\dfrac{(-30)(28)(-64)(72)}{(-45)(14)(-16)(-12)} = -32$

PRACTICE PROBLEMS FOR CHAPTER 8

1. Add algebraically.

Examples:

$$
\begin{array}{llll}
(1) & +8 & (3) & -12 & (5) & -4 & (6) & -12 \\
    & \underline{+7} & & \underline{+8} & & +5 & & -14 \\
    & +15 & & -4 & & +3 & & +16 \\
    & & & & & -2 & & -11 \\
(2) & -9 & (4) & +5 & & \underline{-7} & & +19 \\
    & \underline{-7} & & \underline{-12} & & -5 & & \underline{-2} \\
    & -16 & & -7 & & & &
\end{array}
$$

$$(7)(+7) + (-4) + (-3) + (-6) + (+9)$$
$$= 7 - 4 - 3 - 6 + 9 = 3$$

(a)
$$
\begin{array}{cccccc}
+12 & -22 & +35 & -19 & -51 & -76 \\
\underline{+27} & \underline{+13} & \underline{-18} & \underline{-26} & \underline{+51} & \underline{0} \\
\end{array}
$$

(b)
$$
\begin{array}{cccccc}
-18 & 22 & 14 & -21 & 13 & 90 \\
20 & -12 & 16 & -32 & 11 & 85 \\
12 & -13 & -10 & 18 & -44 & -25 \\
14 & 31 & -12 & -47 & 27 & -75 \\
-10 & -14 & 20 & 36 & -18 & -30 \\
\underline{-17} & \underline{25} & \underline{-15} & \underline{12} & \underline{-14} & \underline{45} \\
\end{array}
$$

(c) $(-8) + (-9) + (+5) + (-2)$

(d) $(+9) + (+5) + (-12) + (-10)$

(e) $(+21) + (-12) + (-13) + (+30)$

(f) $(-17) + (-13) + (+11) + (-10)$

(g) $(+13) + (+12) + (+20) + (-16)$

(h) $(+18) + (-16) + (-11) + (-10)$

(i) $(-15) + (+20) + (-12) + (+12)$

(j) $(+13) + (+11) + (-44) + (+27)$

2. Subtract the bottom number from the top number.

Examples:

$$
\begin{array}{llll}
 & +8 & & +9 & & -12 & & -10 \\
(-) & \underline{+5} & (+) & \underline{-3} & (+) & \underline{-8} & (-) & \underline{+3} \\
 & +3 & & +12 & & -4 & & -13 \\
\end{array}
$$

(a)
$$
\begin{array}{cccccc}
+8 & -12 & -15 & +21 & +32 & -28 \\
\underline{-3} & \underline{+6} & \underline{-5} & \underline{-9} & \underline{+18} & \underline{-32} \\
\end{array}
$$

(b)
$$
\begin{array}{cccccc}
47 & 65 & -38 & 18 & -62 & -27 \\
\underline{56} & \underline{-35} & \underline{24} & \underline{18} & \underline{0} & \underline{-27} \\
\end{array}
$$

Combine each of the following into a single number.

Example:

$$(+6) - (+3) - (-4) + (-5) + (+8)$$
$$= 6 - 3 + 4 - 5 + 8 = 10$$

(c) $(-6) + (-3) - (+4) - (-5)$

(d) $(12) - (+3) - (-4) + (-8)$

(e) $(9) - (4) + (-3) - (11)$

(f) $(-8) + (12) - (-8) - (10)$

(g) $(15) - (13) - (17) - (-14)$

(h) $(-28) - (36) + (32) + (-12)$

(i) $(57) - (43) - (68) + (22)$

(j) $(125) - (100) + (55) - (40)$

3. Multiply.

Examples:

(1) $(-6)(+3) = -18$ (2) $(-4)(-5) = 20$

(3) $(5)(-8)(-2) = 80$ (4) $(-6)(-5)(-3) = -90$

(a) $(-4)(5)(3)(2)$

(b) $(6)(-4)(-2)(3)$

(c) $(3)(2)(4)(-5)(6)$

(d) $(2)(3)(-5)(-4)(6)$

4. Perform the indicated operations. (Note: An odd number of negative factors will produce a negative result. First determine the sign of the result, then perform cancellation without regard to sign.)

Examples:

(1) $\dfrac{(-6)(2)}{(-3)(4)} = 1$ (2) $\dfrac{(4)(-3)}{(-2)(-6)} = -1$

(3) $\dfrac{(8)(15)(-6)}{(5)(-4)(2)} = 18$ (4) $\dfrac{(16)(-18)(24)}{(27)(-4)(-8)} = -8$

(a) $\dfrac{(-36)(21)(-48)}{(12)(-7)(24)}$ (d) $\dfrac{(28)(-3)(-48)}{(12)(-18)(14)}$

(b) $\dfrac{(24)(-16)(0)}{(38)(-14)(-12)}$ (e) $\dfrac{(45)(-9)(-20)}{(-27)(-15)(4)}$

(c) $\dfrac{(-28)(45)(21)}{(-15)(14)(-7)}$ (f) $\dfrac{(11)(-16)(15)}{(-5)(2)(3)}$

9 ALGEBRA

1 LITERAL NUMBERS

Letters of the alphabet used to represent numbers are called literal numbers. The letters a, b, c, x, y, and z are used more often than the others to represent a number, but any letter of the alphabet can be used. Literal numbers are called general numbers because they do not represent a specific number. The area of a certain room which is 12 feet long and 10 feet wide is equal to the product 12×10. If we want to express the area of any rectangle, we can say that $A = L \times W$, where A is area in square measure, L is length in linear measure, and W is width in linear measure.

2 USING LITERAL NUMBERS

Addition, subtraction, multiplication, and division, using literal numbers, are performed in the same manner as in arithmetic. If a and b represent any two numbers, their sum is $a + b$; their difference is $a - b$; their product is $a \times b$, $a \cdot b$ or ab; their quotient is $a \div b$ or $\frac{a}{b}$. Expressing the product as $a \cdot b$ or ab prevents confusing the letter x with the multiplication sign $\times$.

3 DEFINITIONS

A *term* is an algebraic expression not separated within itself by a plus or minus sign, such as $4xy$, $5x^2y$ or $2ab^2c$.

A *product* implies multiplication. The term $5xy$ means $5 \cdot x \cdot y$.

A *monomial* consists of one term, such as $4a^2b$; a binomial consists of two terms, such as $3x^2 - 5xy$; a trinomial consists of three terms, and a polynomial consists of any number of terms more than one.

The quantities multiplied together to form a product are called the *factors* of the product. The factors of $3xy$ are 3, x and y.

The numerical factor in a monomial is known as the numerical coefficient or simply the *coefficient*. The coefficient of $3x^2y$ is 3, and the coefficient of $-5ab$ is -5.

Like or similar terms are terms that have the same literal factors. Terms which do not have the same literal factors are called unlike, or dissimilar, terms.

$3x^2y$ and $5x^2y$ are like terms. $5x^2y$ and $5xy^2$ are unlike terms.

4 HORIZONTAL ADDITION AND SUBTRACTION OF MONOMIALS

Horizontal addition and subtraction of monomials is carried out according to the rules of addition and subtraction of signed numbers. Like terms can be combined; unlike terms cannot be combined.

Example 9.1

(a) $11ab + 4ab - 10ab - 8ab = -3ab$

(b) $5x^3 + 2x^2y - 8x^2y + 2x^3 = 7x^3 - 6x^2y$

(c) $2ab^2 - 3ab + 2 - 4ab^2 + 3ab = -2ab^2 + 2$

5 EXPONENTS

We have learned that 3 squared equals 9, meaning that 3 multiplied by itself equals 9. This can be written $3 \cdot 3 = 9$ or $3^2 = 9$. It is also true that 2 cubed equals 8, meaning that $2 \cdot 2 \cdot 2 = 8$. This can be written $2^3 = 8$. The small number 3 is known as the exponent; the number 2 is known as the base. If we write 4^6 we mean that six 4's are to be multiplied. The exponent is the power to which a number is to be raised. If a number does not have an exponent, it is of the first power; the exponent is considered to be 1. If we wish to raise a literal number to a power, we use an exponent to indicate this. Thus, x^5 means $x \cdot x \cdot x \cdot x \cdot x$ and is read "x raised to the fifth power" or "x to the fifth."

6 EXPONENTS USED IN MULTIPLICATION

Multiplication in algebra follows the same rules, or laws, as multiplication in arithmetic. Exponents simplify the process. If we want to multiply $x^3 \cdot x^4$ we can say

$$x^3 \cdot x^4 = (x \cdot x \cdot x) \cdot (x \cdot x \cdot x \cdot x) = x^7$$

But it is much easier to say that

$$x^3 \cdot x^4 = x^{3+4} = x^7$$

This can be expressed in the form of a rule.

> Rule: The exponent of the product of powers with the same base is the sum of the exponents of the factors.

The rule holds for either positive or negative exponents, and the base may be an arithmetic number. Thus,

$$2^5 \cdot 2^{-2} = 2^3 = 8$$

It must be remembered that the rule for multiplication holds only for the product of powers of the same base; however, multiplication involving powers of more than one base may be simplified. As an example,

$$(x^3 y^4 z^2)(x^2 y^{-2} z) = x^5 y^2 z^3$$

or

$$2^4 \cdot 3^3 \cdot 4^2 = 16 \cdot 27 \cdot 16 = 6{,}912$$

7 EXPONENTS USED IN DIVISION

Division is the inverse of multiplication, so it is logical to conclude that in division the exponent of the divisor is subtracted from the exponent of the dividend.

> Rule: The exponent of the quotient of two powers with the same base is the difference of the exponent of the dividend and the exponent of the divisor.

Example 9.2

(a) $\dfrac{x^5}{x^2} = \dfrac{x \cdot x \cdot x \cdot x \cdot x}{x \cdot x} = x \cdot x \cdot x = x^3$

(b) $3^5 \div 3^2 = 3^{5-2} = 3^3 = 27$

(c) $a^2 \div a^{-3} = a^{2+3} = a^5$

(d) $\dfrac{x^3}{x^5} = x^{3-5} = x^{-2}$

8 EXPONENT OF THE POWER OF A POWER

The exponent of a power is the product of the exponents of the powers.

Example 9.3

(a) $(x^3)^2 = (x \cdot x \cdot x) \cdot (x \cdot x \cdot x) = x^{3 \cdot 2} = x^6$

(b) $(-5x^2 y^3)^2 = (-5)^2 x^{2 \cdot 2} y^{3 \cdot 2} = 25 x^4 y^6$

(c) $\dfrac{a^3 (b^2)^4}{(ab^3)^2} = \dfrac{a^3 b^8}{a^2 b^6} = ab^2$

9 ZERO POWER

Any number, not zero, raised to the zero power is equal to 1. This applies to arithmetic numbers and to literal numbers as well as to a polynomial enclosed in parentheses.

Example 9.4

(a) $\dfrac{2^3}{2^3} = \dfrac{2 \cdot 2 \cdot 2}{2 \cdot 2 \cdot 2} = 2^{3-3} = 2^0 = 1$

(b) $\dfrac{x^3}{x^3} = x^{3-3} = x^0 = 1$

10 NEGATIVE EXPONENTS

A number with a negative exponent is equal to 1 divided by the number with the sign of the exponent changed to positive.

Example 9.5

(a) $\dfrac{2^2}{2^4} = \dfrac{4}{16} = \dfrac{1}{4}$

(b) $\dfrac{10^3}{10^4} = 10^{-1} = \dfrac{1}{10}$

(c) $\dfrac{x^{-2} y^3}{x^3 y^{-2}} = \dfrac{y^3 \cdot y^2}{x^2 \cdot x^3} = \dfrac{y^5}{x^5}$

(d) $\dfrac{a^{-3}}{b^{-2}} = \dfrac{b^2}{a^3}$

Example 9.6

(a) $5 \cdot 5 \cdot 5 = 125$

(b) $a^3 \cdot a^5 = a^8$

(c) $4^3 \cdot 4^2 = 4^5 = 1024$

(d) $(3x)(4x) = 12x^2$

(e) $x^3 \div x^{-2} = x^5$

(f) $2^2 \cdot 3^2 \cdot 4^2 = 4 \cdot 9 \cdot 16 = 576$

(g) $(ab^2)^3 = a^3 b^6$

(h) $(-3x)^5 = -243x^5$

(i) $(-2xy^2)^2 = 4x^2y^4$

(j) $\dfrac{x^3 \cdot x^4}{x^2} = x^5$

(k) $\dfrac{3^7 \cdot 3^2 \cdot 3}{3^5} = 3^5$

(l) $\dfrac{64a^3b^3}{4ab^2} = 16a^2b$

(m) $\dfrac{5x^5y^5}{45x^6y^6} = \dfrac{1}{9xy}$

(n) $\dfrac{(-2x^2y^2)^3}{(2xy^2)^2} = -2x^4y^2$

(o) $\dfrac{(-3ab^2)^3}{(12a^2b)^2} = \dfrac{-27b^4}{144a}$

(p) $5^{-2} \cdot 5^3 = 5$

(q) $a^3 \cdot a^{-1} = a^2$

(r) $10^0 = 1$

(s) $3a^0 = 3$

(t) $x^0y^2 = y^2$

(u) $(4-2)^0 = 1$

(v) $\dfrac{x^{-5}y^3}{xy} = \dfrac{y^2}{x^6}$

(w) $\dfrac{a^{-4}b^5c^{-6}}{ab^{-2}c^3} = \dfrac{b^7}{a^5c^9}$

(x) $\dfrac{x^3y^{-2}}{xy^3} = \dfrac{x^2}{y^5}$

11 MULTIPLYING A MONOMIAL AND A POLYNOMIAL

When a polynomial is multiplied by a monomial, each term of the polynomial must be multiplied by the monomial.

Example 9.7

$$-4x(2x^2 - 3x - 4) = (-4x)(2x^2)$$
$$- (-4x)(3x) - (-4x)(4)$$
$$= -8x^3 + 12x^2 + 16x$$

12 MULTIPLYING BINOMIALS OR TRINOMIALS

In multiplying two binomials, each term of the multiplicand must be multiplied by each term of the multiplier.

Example 9.8

Multiply

(a) $2x - 3y$
     $3x + 2y$

(b) $x^2 + 4x - 3$
     $3x - 2$

Solutions

(a) $2x - 3y$
     $3x + 2y$
     $6x^2 - 9xy$
     $\quad + 4xy - 6y^2$
     $6x^2 - 5xy - 6y^2$

(b) $x^2 + 4x - 3$
     $3x - 2$
     $3x^3 + 12x^2 - 9x$
     $\quad - 2x^2 - 8x + 6$
     $3x^3 + 10x^2 - 17x + 6$

In both solutions, the factors are arranged in a manner similar to multiplication in arithmetic with the multiplier placed under the multiplicand. Multiplication is performed from left to right.

Using example 9.8, steps in the operation are as follows:

1. The left term in the multiplier is multiplied by the left term in the multiplicand and the product is placed under the first column: $(3x)(2x) = 6x^2$.

2. The left term in the multiplier is multiplied by the next term (from left to right) in the multiplicand and the product is placed under the second column: $(3x)(-3y) = -9xy$.

3. The second term in the multiplier is multiplied by the first term in the multiplicand and the product is placed under $-9xy$ because it is a similar term: $(2y)(2x) = +4xy$.

4. The second term in the multiplier is multiplied by the second term in the multiplicand and the product is placed in a third column: $(2y)(-3y) = -6y^2$.

5. Similar terms are combined to give the product: $6x^2 - 5xy - 6y^2$.

Example 9.9

(a) $3xy(x^2 - xz + z^2) = 3x^3y - 3x^2yz + 3xyz^2$

(b) $2ab^2(a^2 + 2ab - 3b^2) = 2a^3b^2 + 4a^2b^3 - 6ab^4$

(c) $-3x^2y(xy - 2xy^2 - 3y^3) = -3x^3y^2 + 6x^3y^3 + 9x^2y^4$

(d) $(x+3)(x+4) = x^2 + 7x + 12$

(e) $(x-2)(x^2 + 4x - 8) = x^3 + 2x^2 - 16x + 16$

13 DIVISION OF A POLYNOMIAL BY A MONOMIAL

In dividing a polynomial by a monomial, each term in the dividend must be divided by the divisor.

Example 9.10

Divide:

$$\frac{6ax^3y^3 - 8a^2x^2y^2 + 4ax^3y}{2axy}$$

Solution

$$\frac{6ax^3y^3 - 8a^2x^2y^2 + 4ax^3y}{2axy}$$
$$= \frac{6ax^3y^3}{2axy} - \frac{8a^2x^2y^2}{2axy} + \frac{4ax^3y}{2axy}$$
$$= 3x^2y^2 - 4axy + 2x^2$$

14 DIVISION OF A POLYNOMIAL BY A POLYNOMIAL

In performing division of a polynomial by a polynomial the terms in the dividend and the divisor should be arranged in order of decreasing power. It is performed much like division in arithmetic.

Example 9.11

Divide: $(15x^2 - x - 6) \div (5x + 3)$

Solution

$$\begin{array}{r} 3x - 2 \\ 5x+3\overline{)15x^2 - x - 6} \\ \underline{15x^2 + 9x } \\ -10x - 6 \\ \underline{-10x - 6} \end{array}$$

Steps in the procedure:

1. Divide the first term in the dividend by the first term in the divisor, $15x^2 \div 5x = 3x$, and place $3x$ over the dividend.

2. Multiply the divisor by the first term in the quotient: $3x(5x + 3) = 15x^2 + 9x$. Place these two terms under similar terms in the dividend and subtract: $(15x^2 - x) - (15x^2 + 9x) = -10x$.

3. Bring down the next term in the dividend to form the new dividend: $-10x - 6$.

4. Divide the first term in the new dividend by the first term in the divisor: $-10x \div 5x = -2$ and place -2 as the second term in the quotient.

5. Multiply the divisor by the second term in the quotient: $-2(5x + 3) = -10x - 6$. Place these

two terms under similar terms in the dividend and subtract: $(-10x - 6) - (-10x - 6) = 0$.

If there had been a remainder of lower order than the first term in the divisor it would be written over the divisor, as in arithmetic, to represent the remainder. As there was no remainder, we can say that $(5x+3)(3x-2)$ are factors of $15x^2 - x - 6$.

Example 9.12

Perform the indicated divisions:

(a) $(36a^3 - 24a^2 - 12a) \div 2a = 18a^2 - 12a - 6$

(b) $\dfrac{21a^3b^2 - 14a^2b^3 + 7ab}{7ab} = 3a^2b - 2ab^2 + 1$

(c) $\dfrac{15a^4b^3c^2 + 20a^3b^2c^3 - 25a^2b^2c^3}{5a^2b^2c}$
$= 3a^2bc + 4ac^2 - 5c^2$

(d) $\dfrac{9x^5y^4z^3 - 18x^3y^2z^2 + 3x^4y^3z^2}{3x^3y^2z}$
$= 3x^2y^2z^2 - 6z + xy$

15 FACTORING

Factoring is the reverse of multiplication. If we multiply $8 \cdot 9$, we find the product to be 72. If we factor 72, we find the factors to be 8 and 9. We also find the factors of 72 to be 12 and 6. The prime factors of 72 are $2 \cdot 2 \cdot 2 \cdot 3 \cdot 3$.

16 FACTORING A POLYNOMIAL CONTAINING A COMMON MONOMIAL

The first step in factoring a polynomial is to factor a common monomial, if one exists. The other factor is found by dividing each term of the polynomial by the common factor and writing the result as the product of the common factor and the quotient.

Example 9.13

Factor: $20x^2 + 12x$

Solution

The greatest monomial factor in the binomial is $4x$.

$$20x^2 + 12x = 4x(5x + 3)$$

Example 9.14

Factor: $xy^2z^3 - x^2y^3z^4$

Solution

$$xy^2z^3 - x^2y^3z^4 = xy^2z^3(1 - xyz)$$

17 FACTORING A TRINOMIAL WHICH IS A PERFECT SQUARE

We have learned that 4 is a perfect square because both factors of 4 are the same. Likewise, 9, 16, and 25 are perfect squares.

A trinomial is also a perfect square if both factors are the same. If we multiply $(x + 2)(x + 2)$ the product is $x^2 + 4x + 4$ which is a perfect square. We can say then that

$$x^2 + 4x + 4 = (x + 2)^2$$

In order to factor a trinomial which is a perfect square we must be able to recognize a trinomial which is a perfect square. Let us examine

$$x^2 + 4x + 4 = (x + 2)(x + 2) = (x + 2)^2$$

In the factor $(x + 2)$:

1. x is the square root of the first term in the trinomial.

2. 2 is the square root of the third term in the trinomial.

3. The product of these two terms is $2x$, which is one-half the middle term of the trinomial.

4. The sign between the two terms is the same as the sign of the middle term of the trinomial.

In order for a trinomial to be a perfect square, the following must be true:

1. The first and third terms must be perfect squares.

2. The middle term must be twice the product of the square roots of the first and third terms.

3. The sign of the middle term can be plus or minus, but the sign of the second term of the factors must be the same as the sign of the middle term of the trinomial.

Let us examine the trinomial and its factors:

$$x^2 - 14x + 49 = (x - 7)^2$$

1. Are the first and third terms perfect squares? Yes.

2. Is the square root of the first term of the trinomial equal to the first term of the factors? Yes. $\sqrt{x^2} = x$.

3. Is the square root of the third term of the trinomial equal to the second term of the factors? Yes. $\sqrt{49} = 7$.

4. Is the sign of the second term in the factors the same as the sign of the middle term of the trinomial? Yes.

5. Is the middle term of the trinomial twice the product of the terms of the factors? Yes. $14x = (2)(7x)$.

The trinomial $x^2 - 14x + 49$ is a perfect square.

Example 9.15

(a) $25a^2 - 20a + 4 = (5a - 2)^2$

(b) $9 - 24x + 16x^2 = (3 - 4x)^2$

(c) $4a^2 - 4a + 1 = (2a - 1)^2$

(d) $x^2 - 6x + 9 = (x - 3)^2$

18 FACTORING THE DIFFERENCE BETWEEN TWO SQUARES

If we multiply $(a + b)(a - b)$ we find the product to be $a^2 - b^2$.

$$
\begin{array}{r}
a + b \\
a - b \\
\hline
a^2 + ab \\
- ab - b^2 \\
\hline
a^2 - b^2
\end{array}
$$

Notice that the factors are the same except for the algebraic sign. Also notice that the product $(a^2 - b^2)$ is equal to the square of the first term of the factors minus the square of the second term. There is no middle term in the product because the sum of $+ab$ and $-ab$ is zero.

We can see that the factors of the difference between two squares is the product of the square root of the first term plus the square root of the second term times the square root of the first term minus the square root of the second term.

Example 9.16

(a) $x^2 - 25 = (x + 5)(x - 5)$

(b) $9x^2 - 16y^2 = (3x + 4y)(3x - 4y)$

(c) $36x^2 - 81y^2 + 9(4x^2 - 9y^2) = 9(2x + 3y)(2x - 3y)$

(d) $1 - 64a^2 = (1 + 8a)(1 - 8a)$

PROFESSIONAL PUBLICATIONS, INC. • P.O. Box 199, San Carlos, CA 94070

19 FACTORING A TRINOMIAL OF THE FORM $Ax^2 + Bx + C$

The factors of a perfect square, such as $a^2 + 2ab + b^2$, or the difference between two squares, such as $a^2 - b^2$, are apparent as soon as the type of polynomial is identified, but if we try to type the polynomial $2x^2 - 5x - 12$, we find that it does not conform to either type.

Because factoring is the opposite of multiplication, let us start with the assumption that the factors of $(2x^2 - 5x - 12) = (2x + 3)(x - 4)$ and multiply the two factors.

$$
\begin{array}{r}
2x + 3 \\
\underline{x - 4} \\
2x^2 + 3x \\
\underline{-8x - 12} \\
2x^2 - 5x - 12
\end{array}
$$

Now examining this operation, we can see that the first term in the trinomial $2x^2 - 5x - 12$ is found by multiplying the first terms of the factors: $(x)(2x) = 2x^2$, and that the third term of the trinomial is found by multiplying the second terms of the factors: $(-4)(3) = -12$.

The middle term of the trinomial is the algebraic sum of the cross products: $(x)(3) + (-4)(2x) = -5x$.

Now let us use this information in factoring

$$2x^2 - x - 15$$

The factors are not immediately apparent, so let us set up two pairs of parentheses to contain the factors and insert trial numbers which must produce the trinomial.

$$2x^2 - x - 15 = (\qquad)(\qquad)$$

The product of the first terms of the factors must equal $2x^2$. The only possibility is $(x)(2x)$. But which of these two terms will be placed in the first set of parentheses and which in the second? Let us write

$$2x^2 - x - 15 = (x \quad)(2x \quad)$$

Because the middle and third terms of the trinomial are negative, one of the signs in the parentheses must be positive and one negative. (If both signs were negative, the sign of the third term in the trinomial would be positive.) But which will be positive and which negative? Let us try

$$2x^2 - x - 15 = (x - \quad)(2x + \quad)$$

The product of the second terms of the factors must be -15, so we insert 5 in the first parentheses and 3 in the second. The product of these two terms is -15, but the algebraic sum of the cross products gives $-10x$ so we try

$$2x^2 - x - 15 = (x - 3)(2x + 5)$$

The product of these two factors is $(2x^2 - x - 15)$, so the trinomial is factored.

With practice, various combinations of numbers within the parentheses can be tried by performing mental multiplication and addition.

Example 9.17

Factor:

   (a) $x^2 + 6x + 5$

   (b) $x^2 - 4x + 3$

   (c) $a^2 - 9a + 14$

   (d) $5a^2 + 11a + 6$

   (e) $9x^2 - 13x + 4$

Solutions

   (a) $x^2 + 6x + 5 = (x + 5)(x + 1)$

   (b) $x^2 - 4x + 3 = (x - 3)(x - 1)$

   (c) $a^2 - 9a + 14 = (a - 7)(a - 2)$

   (d) $5a^2 + 11a + 6 = (5a + 6)(a + 1)$

   (e) $9x^2 - 13x + 4 = (9x - 4)(x - 1)$

Example 9.18

   (a) $8x^4 - 4x^3 = 4x^3(2x - 1)$

   (b) $a^2 - 16 = (a + 4)(a - 4)$

   (c) $x^2 + 6x + 9 = (x + 3)^2$

   (d) $a^4 - 16 = (a^2 + 4)(a + 2)(a - 2)$

   (e) $5a^2 - 30a + 45 = 5(a - 3)^2$

   (f) $16x^2 - 24x + 9 = (4x - 3)^2$

   (g) $36a^2 - 81b^2 = 9(2a + 3b)(2a - 3b)$

   (h) $x^2 - 5x + 6 = (x - 3)(x - 2)$

   (i) $a^2 + 4a - 12 = (a + 6)(a - 2)$

   (j) $x^3 - 10x^2 + 9x = x(x - 9)(x - 1)$

20 EQUATIONS

Probably the most important concept in algebra is the equation. It is a means of solving problems in science, engineering, and everyday life.

An equation is a statement that two quantities are equal. The equal sign ($=$) separates the two quantities. The terms on the left of the equal sign are known as the left member; the terms on the right of the equal sign are known as the right member.

21 CONDITIONAL EQUATIONS

If we say "$2x + 4 = 10$", we can see that this statement is true if, and only if, x is equal to 3. The equation is true on the condition that x is equal to 3. We can say that 3 satisfies the equation.

Being able to determine an unknown quantity in an equation is what makes the equation important.

22 ROOT OF AN EQUATION

In the equation $2x + 4 = 10$, we can say that 3 is the solution of the equation or that 3 is the root of the equation, and 3 is the only root of the equation.

23 SOLVING AN EQUATION

Solving an equation simply means finding the value of the literal number in the equation; finding the root of the equation. The root of the equation $2x + 4 = 10$ can be found by inspection, but some equations are not so easily solved, so that we need to be familiar with certain principles of algebra in order to solve more difficult equations. These principles, or truths, are known as axioms.

24 AXIOMS

1. The same quantity may be added to both sides of an equation without altering the truth of the statement.

2. The same quantity may be subtracted from both sides of an equation without altering the truth of the statement.

3. Both sides of an equation may be multiplied by the same quantity, not zero, without altering the truth of the statement.

4. Both sides of an equation may be divided by the same quantity, not zero, without altering the truth of the statement.

5. The square root (or any root) of each side of an equation may be taken without altering the truth of the statement.

Example 9.19

(Axiom 1): Solve: $x - 3 = 12$

Solution

$$x - 3 = 12$$
Adding: $x - 3 + 3 = 12 + 3$
$$x = 15$$

Example 9.20

(Axiom 2): Solve: $x + 6 = 12$

Solution

$$x + 6 = 12$$
Subtracting: $x + 6 - 6 = 12 - 6$
$$x = 6$$

Example 9.21

(Axioms 3 & 4): Solve: $4 = \dfrac{12}{x}$

Solution

$$4 = \frac{12}{x}$$
Multiplying: $(x)(4) = \dfrac{(x)(12)}{x}$
$$4x = 12$$
Dividing: $\dfrac{4x}{4} = \dfrac{12}{4}$
$$x = 3$$

Example 9.22

(Axiom 5): Solve: $x^2 = 16$

Solution

$$x^2 = 16$$
Taking square root: $\sqrt{x^2} = \sqrt{16}$
$$x = \pm 4$$

25 TRANSPOSING

It should be noted in Example 9.19 (Axiom 1) that when 3 was added to both sides of the equation, -3 disappeared from the left side, leaving only x on that side. Also, in Example 9.20 (Axiom 2) when 6 was subtracted from both sides of the equation, $+6$ disappeared from the left side, leaving only x. Putting the unknown quantity x on the left side with no other quantity can be accomplished by a short-cut method known as transposing. It is not an axiom but gives the same results as were obtained in using axioms 1 and 2.

Transposing means that any term may be moved from one side of an equation to the other if its sign is changed.

Example 9.23

Solve: $3x - 2 = x + 6$

Solution

$$3x - 2 = x + 6$$
Transposing: $3x - x = 6 + 2$
$$2x = 8$$
Dividing: $x = 4$

Example 9.24

Solve: $4x + 2 - 3x - 6 = 5x + 4$

Solution

$$4x + 2 - 3x - 6 = 5x + 4$$

Transposing: $\quad 4x - 3x - 5x = 4 - 2 + 6$

$$-4x = 8$$

Dividing by -4: $\quad\quad x = -2$

26 PARENTHESES

If a quantity within parentheses, brackets, braces is preceded by a plus sign (+), the parentheses may be removed without changing the sign of the terms within the parentheses.

If a quantity within parentheses is preceded by a minus sign (−), the parentheses may be removed if the sign of each term within the parentheses is changed.

Example 9.25

Solve: $7x - (2x - 4) - 2(4x - 2) + (3x + 3)$
$\quad\quad\quad = 8 - (x - 5) - 3(x + 4)$

Solution

Removing parentheses: $\quad 7x - 2x + 4 - 8x + 4 + 3x + 3$
$$= 8 - x + 5 - 3x - 12$$

Transposing: $\quad 7x - 2x - 8x + 3x + x + 3x$
$$= 8 + 5 - 12 - 4 - 4 + 3$$

Combining: $\quad\quad 4x = -4$

Dividing: $\quad\quad\quad x = -1$

27 FRACTIONAL EQUATIONS

In solving fractional equations, the first step is to get rid of the denominators. This is done by multiplying both sides of the equation by the lowest common denominator. Then, after the equation is cleared of fractions, the equation is solved using the axioms.

Example 9.26

Solve: $\dfrac{3x}{4} + 5 = 2x - \dfrac{1}{3}$

Solution

In solving equations with fractions, it is convenient to write any term without a denominator with a denominator of 1.

$$\frac{3x}{4} + \frac{5}{1} = \frac{2x}{1} - \frac{1}{3}$$

The lowest common denominator (LCD) is 12.

Multiplying by LCD: $\dfrac{12}{1} \cdot \dfrac{3x}{4} + \dfrac{12}{1} \cdot \dfrac{5}{1} = \dfrac{12}{1} \cdot \dfrac{2x}{1} - \dfrac{12}{1} \cdot \dfrac{1}{3}$

$$9x + 60 = 24x - 4$$

Transposing: $\quad\quad -15x = -64$

$$x = \frac{64}{15}$$

Example 9.27

Solve the equations.

(a) $3x + 5 = 14$

(b) $5x - 8 = x + 16$

(c) $4x + 3(2x - 5) = 7 + 5(3x - 7)$

(d) $(2x - 3)(x + 4) - 23 = (x + 7)(2x)$

(e) $2(x + 3)(x - 4) = 6 + 2x(x - 5)$

(f) $\dfrac{x - 5}{x - 3} - \dfrac{x + 6}{x^2 + x - 12} = \dfrac{x - 2}{x + 4}$

Solution

(a)
$$3x + 5 = 14$$
$$3x = 14 - 5$$
$$3x = 9$$
$$x = 3$$

(b)
$$5x - 8 = x + 16$$
$$5x - x = 16 + 8$$
$$4x = 24$$
$$x = 6$$

(c)
$$4x + 3(2x - 5) = 7 + 5(3x - 7)$$
$$4x + 6x - 15 = 7 + 15x - 35$$
$$4x + 6x - 15x = 7 - 35 + 15$$
$$-5x = -13$$
$$x = \frac{13}{5}$$

(d)
$$(2x - 3)(x + 4) - 23 = (x + 7)(2x)$$
$$2x^2 + 5x - 12 - 23 = 2x^2 + 14x$$
$$5x - 14x = 35$$
$$-9x = 35$$
$$x = -\frac{35}{9}$$

(e)
$$2(x + 3)(x - 4) = 6 + 2x(x - 5)$$
$$2(x^2 - x - 12) = 6 + 2x^2 - 10x$$
$$2x^2 - 2x - 24 = 6 + 2x^2 - 10x$$
$$-2x + 10x = 6 + 24$$
$$8x = 30$$
$$x = \frac{30}{8} = \frac{15}{4}$$

(f)
$$\frac{x - 5}{x - 3} - \frac{x + 6}{x^2 + x - 12} = \frac{x - 2}{x + 4}$$
$$(x + 4)(x - 5) - (x + 6) = (x - 3)(x - 2)$$
$$x^2 - x - 20 - x - 6 = x^2 - 5x + 6$$
$$-x - x + 5x = 6 + 20 + 6$$
$$3x = 32$$
$$x = \frac{32}{3}$$

28 LITERAL EQUATIONS AND FORMULAS

Literal equations are equations in which some, or all, of the quantities are literal numbers.

Mentioned previously is the equation, or formula, for the area of a rectangle

$$A = L \cdot W$$

in which A represents area, L represents length and W represents width. Other formulas are: $C = 2\pi r$, the formula for the circumference of a circle and $V = \pi r^2 h$, the formula for the volume of a cylinder.

Literal equations are solved in the same way as any other equation by using the axioms mentioned previously.

If the length and width of a room are known and the area is needed, the formula $A = L \cdot W$ is in proper form, but if the length and area are known and the width is needed, the form can be written $W = A \div L$.

Example 9.28

From the formula for the area of a rectangle, $A = L \cdot W$, find the formula for the width of a rectangle.

Solution

$$A = L \cdot W$$

Transposing: $L \cdot W = A$

Dividing by L: $W = \dfrac{A}{L}$

Example 9.29

Solve the formula $F = \frac{9}{5}C + 32$ for C.

Solution

$$\frac{F}{1} = \frac{9C}{5} + \frac{32}{1}$$

Multiplying by LCD (5): $5F = 9C + (5)(32)$

Transposing: $9C = 5F - (5)(32)$

$= 5(F - 32)$

Dividing by 9: $C = \dfrac{5(F-32)}{9}$

Or, $C = \frac{5}{9}(F - 32)$

29 QUADRATIC EQUATIONS

Equations which have been solved to this point have been first degree equations known as linear equations.

A first degree equation is an equation which contains only the first power of x. A linear equation is a first degree equation of one or more variables such as

$$2x + 3 = 9$$
$$3x - 2y = 8$$

The general form of the linear equation is

$$Ax + By + C = 0$$

where A represents the coefficient of x, B represents the coefficient of y, and C represents any number which does not contain x; it is called the constant. In the equation $2x + 3 = 9$, the coefficient of y is zero (0).

A quadratic equation contains a term in the second degree of x but no higher, such as

$$2x^2 - 3x + 4 = 0$$

The general form of the quadratic equation is

$$Ax^2 + Bx + C = 0$$

When B is zero, the equation contains no term in x and is known as a pure quadratic, such as

$$3x^2 - 12 = 0$$

The root of a quadratic equation is any number that satisfies the equation. There are two roots for a quadratic equation.

30 SOLVING A PURE QUADRATIC EQUATION

In solving a pure quadratic equation, the term including x^2 is isolated on the left side and the other term is on the right side of the equation. If the coefficient of x^2 is not one (1), both sides of the equation are divided by this coefficient according to Axiom 4, Sec. 24. The value of x is found by taking the square root of each side of the equation, according to Axiom 5, Sec. 24.

Example 9.30

Solve $3x^2 - 12 = 0$

Solution

$$3x^2 - 12 = 0$$
$$3x^2 = 12$$
$$x^2 = 4$$
$$\sqrt{x^2} = \sqrt{4}$$
$$x = +2 \text{ or } -2$$

31 SOLVING A QUADRATIC EQUATION BY FACTORING

If we factor the left side of the equation $x^2 - 4x + 3 = 0$, we have

$$(x-3)(x-1) = 0$$

For this statement to be true—the product of two factors to be zero—one of the two factors must be zero or both must be zero. Consider two numbers, a and b, whose product is equal to zero:

$$ab = 0$$

For this to be true, either a must be zero, or b must be zero, or both must be zero. In the equation $(x-3)(x-1) = 0$, if we let $x = 3$, we have

$$(3-3)(3-1) = 0$$

which is a true statement. We can say then that the roots of the equation $x^2 - 4x + 3 = 0$ are 3 and 1. For proof, we substitute these numbers in the equation and have

$$(3)^2 - (4)(3) + 3 = 0$$
$$9 - 12 + 3 = 0$$
$$\text{and} \quad (1)^2 - (4)(1) + 3 = 0$$
$$1 - 4 + 3 = 0$$

To simplify the operation, we let each one of the factors equal zero and solve for x,

$$(x-3) = 0$$
$$x = 3$$
$$(x-1) = 0$$
$$x = 1$$

Example 9.31

Solve $2x^2 - x - 15 = 0$

Solution

$$2x^2 - x - 15 = 0$$
$$(x-3)(2x+5) = 0$$
$$x - 3 = 0$$
$$x = 3$$
$$2x + 5 = 0$$
$$x = -\frac{5}{2}$$

Example 9.32

Solve $2x^2 = 3x$

Solution

$$2x^2 = 3x$$
$$2x^2 - 3x = 0$$
$$x(2x-3) = 0$$
$$x = 0$$
$$2x - 3 = 0$$
$$2x = 3$$
$$x = \frac{3}{2}$$

32 SOLVING A QUADRATIC EQUATION BY COMPLETING THE SQUARE

The left side of the equation $x^2 + 4x + 4 = 16$ is a perfect square because the factors are $(x+2)(x+2) = (x+2)^2$. The right side of the equation is also a perfect square: $\sqrt{16} = \pm 4$. The values of x can be found by taking the square root of each side of the equation.

$$\sqrt{(x+2)^2} = \sqrt{16}$$
$$x + 2 = \pm 4$$
$$x = +2 \text{ or } -6$$

Quadratic equations which are not perfect squares can be solved by a method known as "completing the square." The method involves using the five axioms found in Sec. 24 to make both sides of the equation perfect squares and then solving as is shown above. The method is lengthy and will not be explained here. It is important because it has been used to derive a formula for the solution of quadratic equations.

33 SOLVING A QUADRATIC EQUATION BY FORMULA

In the general form of the quadratic equation, $Ax^2 + Bx + C = 0$, A is the coefficient of x^2, B is the coefficient of x, and C is the constant. As mentioned, if $B = 0$, there will be no middle term Bx. In the equation $3x^2 - 5x - 8 = 0$,

$$A = 3$$
$$B = -5$$
$$C = -8$$

The quadratic formula is

$$x = \frac{-B + \sqrt{B^2 - 4AC}}{2A} \text{ or } x = \frac{-B - \sqrt{B^2 - 4AC}}{2A}$$

It is usually written

$$x = \frac{-B \pm \sqrt{B^2 - 4AC}}{2A}$$

It is obvious that there are two roots for a quadratic equation.

Example 9.33

Solve the equation by formula

$$4x^2 - 5x - 6 = 0$$

Solution

$4x^2 - 5x - 6 = 0$: $A = 4$, $B = -5$, $C = -6$

$$x = \frac{-(-5) + \sqrt{(-5)^2 - (4)(4)(-6)}}{(2)(4)}$$

$$= \frac{5 + \sqrt{25 + 96}}{8}$$

$$= \frac{5 + \sqrt{121}}{8}$$

$$x = 2$$

or

$$x = \frac{-(-5) - \sqrt{(-5)^2 - (4)(4)(-6)}}{(2)(4)}$$

$$= \frac{5 - \sqrt{25 + 96}}{8}$$

$$= \frac{5 - \sqrt{121}}{8}$$

$$x = -\frac{3}{4}$$

In using the formula, the equation must be arranged in the form $Ax^2 + Bx + C = 0$ with A positive. If the equation is not in this form, it must be arranged.

Example 9.34

Arrange the equations in the form $Ax^2 + Bx + C = 0$ for solution by formula and write the values of A, B, C.

(a) $3x^2 = 4x + 2$

(b) $-4x^2 + 2x = 6$

(c) $x^2 = 4 - x$

(d) $x^2 - 7 = 0$

Solution

(a) $3x^2 - 4x - 2 = 0$: $A = 3$, $B = -4$, $C = -2$

(b) $4x^2 - 2x + 6 = 0$: $A = 4$, $B = -2$, $C = 6$

(c) $x^2 + x - 4 = 0$: $A = 1$, $B = 1$, $C = -4$

(d) $x^2 - 7 = 0$: $A = 1$, $B = 0$, $C = -7$

PRACTICE PROBLEMS FOR CHAPTER 9

1. Identify each of the following as a monomial, binomial, trinomial, or polynomial. (Note: *Mono* means one, *bi* means two, *tri* means three, and *poly* means many.)

(a) $x^2 + 2xy + y^2$

(b) $4a^2b^2c^2$

(c) $3x^2 + 5xy^2$

(d) $2x^3 + 3x^2 + 3x + 8$

2. Write the numerical coefficient of each term.

(a) $3x^2y$

(b) $6x^3$

(c) $4abc$

(d) $x^2y^2z^2$

3. Combine like terms into one algebraic expression.

Example:

$$3x^3 + 5xy + 2x^3 + x^2 + 2xy + 3x^2$$
$$= 5x^3 + 4x^2 + 7xy$$

(a) $2x^2 + xy + x^2 + 3xy + 4x^2$

(b) $x^2 + x^2y + xy^2 + 2x^2 + x^2y^2$

(c) $3ab^2 + 2ab - 4ab^2 - 2ab + 5$

(d) $x^2 + y^2 + 2 - y^2 + 4x^2 - 4$

(e) $x^3 + 2x^2 + 4 - 2x^3 - 5x^2 - x^3$

(f) $3x^3 + x^2y - 6x^2y + 2x + 4 - 6x^3$

(g) $x^2 + 3xy - x^3 + 3x^2 - 5xy + x^3$

(h) $x^2y^2 - 2xy^2 + 2x^2y^2 + 3xy^2 - 4x^2y + x^2y^2$

(i) $a^2b^2c - ab^2c + 2a^2b^2c - 2ab^2c + 4ab^2c$

(j) $2a - 3 + 11a + 5 - 4a - 9$

4. Find the indicated products and quotients.

(a) 2^4

(b) $x^3 \cdot x^4$

(c) $(-3)^4$

(d) $-(-4)^2$

(e) $(2a)(3a)$

(f) $-4x^2(2x)$

(g) $(x^2y^3)(xy)$

(h) $(3xy^2)(-4x^2y)$

(i) $2^3 \cdot 3^2 \cdot 4^2$

(j) $(-4x^3)(-2x^2)$

(k) $x^8 \div x^5$

(l) $a^2 \div a^{-3}$

(m) $\dfrac{x^3 \cdot x^4}{x^5}$

(n) $\dfrac{2^2 \cdot 2^3 \cdot 2}{2^5}$

(o) $\dfrac{15x^4y^3}{5x^3y^2}$

(p) $\dfrac{-64a^3b^2}{4ab^2}$

(q) $(x^2y)^3$

(r) $(-2ab^2c^3)^2$

(s) $\dfrac{(4xy^2)^4}{(-8x^2y)^2}$

(t) $\dfrac{a^3(b^2)^4}{(ab^3)^2}$

(u) $\dfrac{3^4}{3^4}$

(v) $\dfrac{x^4}{x^4}$

(w) $\dfrac{2^2}{2^5}$

(x) $\dfrac{a^{-2}b^3}{a^3b^{-2}}$

(y) $(4-2)^{-2}$

(z) $(a+b)^0$

5. Multiply.

(a) $2xy(x^2 - yz + z^2)$

(b) $3a^2b(a^2 - 2ab + 2b^2)$

(c) $5x^2(6x^2 - 3x + 4)$

(d) $-3xy(2x^2 - xy + 2y^2)$

(e) $\begin{array}{r} x+3 \\ x+2 \\ \hline \end{array}$

(f) $\begin{array}{r} a-4 \\ a+2 \\ \hline \end{array}$

(g) $\begin{array}{r} a^2 + 2a + 6 \\ a \quad\ -3 \\ \hline \end{array}$

(h) $\begin{array}{r} x^2 + 3x - 2 \\ 3x - 2 \\ \hline \end{array}$

(i) $\begin{array}{r} a^3 - 3a^2 + 2a - 8 \\ 5a + 4 \\ \hline \end{array}$

(j) $\begin{array}{r} x^2 - 2x - 3 \\ x^2 + x + 4 \\ \hline \end{array}$

6. Divide.

(a) $(8x^4 + 6x^3 + 4x^2 + 2x + 2) \div 2$

(b) $(24a^4 - 16a^3 - 12a) \div 4a$

(c) $\dfrac{14a^2b^3 - 21a^2b^2 - 28ab}{7ab}$

(d) $\dfrac{18x^4y^3z^2 - 6x^3yz^2 + 3x^2y^2z}{3x^2yz}$

PROFESSIONAL PUBLICATIONS, INC. • P.O. Box 199, San Carlos, CA   94070

(e) $\dfrac{24a^5bc^3 - 12a^3bc^2 - 18a^4bc^4}{6a^3bc^2}$

(f) $\dfrac{25xy^3z^5 - 20x^3y^2z^2 - 15x^4y^3z^3}{5xy^2z^2}$

(g) $x+3\overline{)x^2 + 8x + 15}$

(h) $x+2\overline{)2x^2 + 7x + 6}$

(i) $x+5\overline{)x^2 + 8x + 15}$

(j) $x+4\overline{)3x^2 + 10x - 6}$

7. Factor.

(a) $12x^3 - 3x^2$

(b) $x^3y - 3x^2y$

(c) $a^2 + 2ab + b^2$

(d) $a^2 - b^2$

(e) $4a^2 - 8a + 4$

(f) $x^2 - 16$

(g) $x^3 + 2x^2y + xy^2$

(h) $25a^2 - 49$

(i) $3a^2 - 18a + 27$

(j) $a^4 - 81$

(k) $x^2 - 4x - 12$

(l) $a^2 - 4a - 5$

(m) $y^2 - 5y + 6$

(n) $x^2 + 7x + 12$

(o) $2x^2 - 5x - 12$

8. Solve the equations.

Example:

$10x - 3(x+3)(2x-5) = 3 - 2(4x-7)(x+1) - x(3-2x)$

$10x - 6x^2 - 3x + 45 = 3 - 8x^2 + 6x + 14 - 3x + 2x^2$

$10x - 6x^2 - 3x + 8x^2 - 2x^2 - 6x + 3x = 3 + 14 - 45;$

$4x = -28; \quad x = -7$

(a) $3x + 4 = 6$

(b) $5a + 6 = a - 10$

(c) $15 - 2x = 1 - 5x$

(d) $9x - 12 = 7x - 11$

(e) $7x + 7 - x = 2x - 8$

(f) $5a - 7 - 4a - 8 + 8a - 15 = 0$

(g) $6 + 2x - 3 - 5x = x - 5 - 2x + 7$

(h) $12x - (4x - 6) = 3x - (9x - 27)$

(i) $5x - (x+3)(x-4) + 3 = 7 - x(x-7)$

9. Solve the equations.

Examples:

(1) $\quad \dfrac{x}{3} + \dfrac{x}{4} = \dfrac{7}{2}$

$\text{LCD} = 12. \quad \dfrac{(12)x}{3} + \dfrac{(12)x}{4} = \dfrac{(12)7}{2}$

$4x + 3x = 42; \quad 7x = 42; \quad x = 6$

(2) $\quad \dfrac{2x+2}{3} - \dfrac{3x-1}{4} = 1$

$\text{LCD} = 12. \quad \dfrac{12(2x+2)}{3} - \dfrac{12(3x-1)}{4} = 12(1)$

$8x + 8 - 9x + 3 = 12; \quad -x = 1; \quad x = -1$

(3) $\quad \dfrac{x-4}{x-2} - \dfrac{x}{x+2} = \dfrac{8-x}{x^2-4}$

$\text{LCD} = (x^2-4). \quad (x+2)(x-4) - x(x-2) = 8-x$

$x^2 - 2x - 8 - x^2 + 2x = 8 - x; \quad x = 16$

(a) $x + 3 = \dfrac{3x}{4} - \dfrac{x}{2}$

(b) $\dfrac{3x}{2} - x = \dfrac{x}{3} + 12$

(c) $\dfrac{3}{x+4} - \dfrac{4}{x-4} = \dfrac{x}{x^2-16}$

(d) $\dfrac{2}{x-1} - \dfrac{4}{x-3} = \dfrac{6}{x^2 - 4x + 3}$

10. Solve each formula for the letter indicated at the right.

(a) $C = \pi D$ $\quad\quad (D)$

(b) $C = 2\pi r$ $\quad\quad (r)$

(c) $A = \pi r^2$ $\quad\quad (r)$

(d) $A = \dfrac{\pi D^2}{4}$ $\quad\quad (D)$

(e) $V = \pi r^2 h$ $\quad\quad (h)$

(f) $V = \pi r^2 h$ $\quad\quad (r)$

(g) $A = 0.785 D^2$ $\quad\quad (D)$

(h) $A - P = Prt$ $\quad\quad (P)$

(i) $I = \dfrac{E}{R+r}$ $\quad\quad (R)$

(j) $\dfrac{a}{a+b} = \dfrac{c}{c+d}$ $\quad\quad (b)$

11. Write each equation in the form $Ax^2 + Bx + C = 0$.

(a) $7x^2 = 4x - 3$

(b) $x^2 = 4 - x$

12. Write the value of A, B, and C in each of the equations.

(a) $2x^2 + 2x + 5 = 0$

(b) $5x^2 - 9x = 0$

13. Solve by factoring.

(a) $3x^2 + 7x + 2 = 0$

(b) $4x^2 = 5x + 6$

(c) $3x^2 = 5x$

14. Solve by formula.

(a) $3x^2 + 7x + 2 = 0$

10 THE RECTANGULAR COORDINATE SYSTEM

1 DIRECTED LINE

Suppose a surveyor establishes a west-east line through a point on a monument which we will call the *origin*. Moving east from the origin, he marks off points one foot apart, labeling them +1, +2, +3, +4, +5,.... Then, moving west from the origin, the surveyor marks off points one foot apart, labeling them −1, −2, −3, −4, −5,....

Figure 10.1 is reproduced from his notebook, where he has established a *directed line*. From left to right (west to east), the direction is positive and numbers increase in value. From right to left (east to west), the direction is negative and numbers decrease in value.

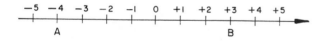

Figure 10.1 A directed line

A directed line, on paper, can be called an *axis*. A horizontal (west-east) line is called the *x-axis*. A point on the axis associated with a particular number is called the graph of the number; the number is called the *coordinate* of the point. This coordinate is the directed distance from the origin to the point. The point **A**, figure 10.1, has the coordinate −4, which is the distance from 0 to **A**, not the distance from **A** to 0. The point **B** has the coordinate +3, which is the distance from 0 to **B**, not the distance from **B** to 0.

$$\text{distance } \mathbf{AB} = 3 - (-4) = 7$$
$$\text{distance } \mathbf{BA} = -4 - (+3) = -7$$

The word *distance* is used loosely in this text. A more appropriate term for distance on a directed line would be the measure of travel in a specified direction. The actual length from **A** to **B**, or **B** to **A**, is 7. This actual length which disregards the negative sign is known as the *absolute value* of **AB** or of **BA**. Symbolically, $|-7| = 7$. The enclosure indicates absolute value.

In general, we can say that if $\mathbf{P}_1$ and $\mathbf{P}_2$ are any two points on the x-axis with coordinates x_1 and x_2, then

$$\mathbf{P}_1\mathbf{P}_2 = x_2 - x_1 \qquad 10.1$$
$$\mathbf{P}_2\mathbf{P}_1 = x_1 - x_2 \qquad 10.2$$

The surveyor can also establish a south-north line through the same point on the monument mentioned previously and mark points on the line at one foot intervals from the origin in a northerly direction. Points at one foot intervals on the line in a southerly direction are labeled −1, −2, −3, −4, −5,....

Figure 10.2 from his notebook shows a vertical line through the origin representing this south-north line. It is a directed line which is positive from south to north and is called the *y-axis*.

If $\mathbf{P}_1$ and $\mathbf{P}_2$ are any two points on the y-axis with coordinates y_1 and y_2, then

$$\mathbf{P}_1\mathbf{P}_2 = y_2 - y_1 \qquad 10.3$$
$$\mathbf{P}_2\mathbf{P}_1 = y_1 - y_2 \qquad 10.4$$

2 THE RECTANGULAR COORDINATE SYSTEM

It was the French mathematician Rene Descartes (17th century) who devised the *rectangular coordinate system*, sometimes called the *Cartesian plane*. The system uses an ordered pair of coordinates to locate a point. The ordered pair of coordinates are the x-coordinate and y-coordinate of a point, enclosed in parentheses with the x-coordinate always written first, followed by a comma and the y-coordinate.

The Cartesian plane consists of an x- (horizontal) and a y- (vertical) axis which are directed lines as shown in figure 10.3.

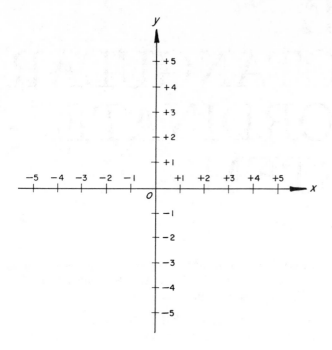

Figure 10.2　The x-y axis

10.4. Signs of the coordinates of points in each quadrant are also shown in figure 10.4. In quadrant I, x is positive, y is positive; in quadrant II, x is negative, y is positive; in quadrant III, x is negative, y is negative; in quadrant IV, x is positive, y is negative.

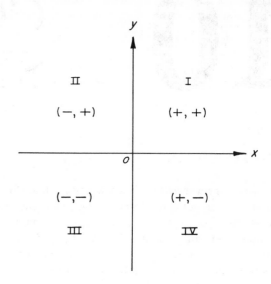

Figure 10.4　Signs of the quadrants

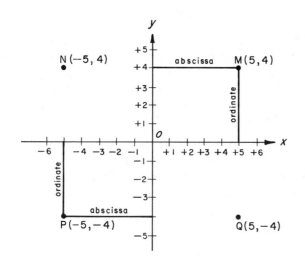

Figure 10.3　Abscissas and ordinates

The point **M** has the coordinates $(5, 4)$. The horizontal distance 5, from the y-axis to **M** is known as the *abscissa*. The vertical distance 4, from the x-axis to **M** is known as the *ordinate*. The abscissa and ordinate are measured from the axis to the point and not from the point to the axis, which is in accordance with distances on a directed line.

The point **N** has an abscissa of -5, an ordinate of $+4$.

The point **P** has an abscissa of -5, an ordinate of -4.

The point **Q** has an abscissa of $+5$, an ordinate of -4.

The x and y axes divide the plane into four parts, numbered in a counterclockwise direction as shown in figure

Example 10.1

Determine the coordinates of the points shown in figure 10.5.

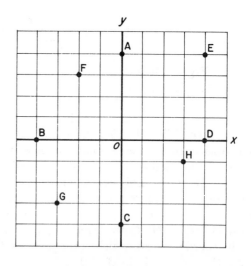

Figure 10.5　Example 10.1

Solution

A: $(0, 4)$; **B**: $(-4, 0)$; **C**: $(0, -4)$; **D**: $(4, 0)$; **E**: $(4, 4)$;

F: $(-2, 3)$; **G**: $(-3, -3)$; **H**: $(3, -1)$.

3 DISTANCE FORMULA

A formula for finding the distance between any two points in a rectangular coordinate system can be derived from the *Pythagorean theorem*: in a right triangle, the square of the length of the hypotenuse equals the sum of the squares of the lengths of the other two sides.

If we let c represent the length of the hypotenuse and a and b are the lengths of the other two sides,

$$c^2 = a^2 + b^2 \qquad 10.5$$

If we take the square root of both sides of the equation,

$$c = \sqrt{a^2 + b^2} \qquad 10.6$$

In figure 10.6, $\mathbf{P_1}$ and $\mathbf{P_2}$ represent any two points in a rectangular coordinate system with the coordinates (x_1, y_1) and (x_2, y_2). If we pass a horizontal line through $\mathbf{P_1}$ and a vertical line through $\mathbf{P_2}$ they will intersect at $\mathbf{Q}$, forming a right triangle with the line $\mathbf{P_1P_2}$ being the hypotenuse. The x-coordinate of $\mathbf{Q}$ will be the same as the x-coordinate of $\mathbf{P_2}$: x_2; the y-coordinate of $\mathbf{Q}$ will be the same as the y-coordinate of $\mathbf{P_1}$: y_1.

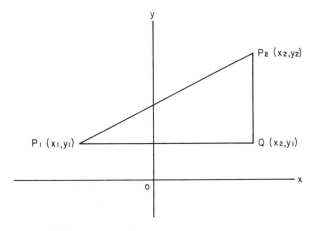

Figure 10.6   Distance between points

By the Pythagorean theorem,

$$(\mathbf{P_1P_2})^2 = (\mathbf{P_1Q})^2 + (\mathbf{QP_2})^2 \qquad 10.7$$

$$\mathbf{P_1P_2} = \sqrt{(\mathbf{P_1Q})^2 + (\mathbf{QP_2})^2}$$

$$= \sqrt{(x_2 - x_1)^2 + (y_2 - y_1)^2} \qquad 10.8$$

It is also true that

$$\mathbf{P_1P_2} = \sqrt{(x_1 - x_2)^2 + (y_1 - y_2)^2} \qquad 10.9$$

Example 10.2

Find the distance between $\mathbf{P_1}(-3, -1)$ and $\mathbf{P_2}(5, 5)$.

Solution

$$\mathbf{P_1P_2} = \sqrt{(x_2 - x_1)^2 + (y_2 - y_1)^2}$$
$$= \sqrt{[5 - (-3)]^2 + [5 - (-1)]^2}$$
$$= \sqrt{(5 + 3)^2 + (5 + 1)^2}$$
$$= 10$$

The distance r from the origin to any point $\mathbf{P}(x, y)$ is called the *radius vector*. This distance is always positive.

From the Pythagorean theorem,

$$r = \sqrt{x^2 + y^2}$$

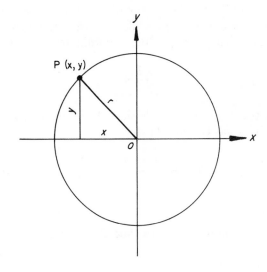

Figure 10.7   The radius vector

Example 10.3

Find the distance from the origin to the point $\mathbf{P}(-3, 4)$.

Solution

$$r = \sqrt{(-3)^2 + 4^2} = 5$$

4 MIDPOINT OF A LINE

In figure 10.8, the point $\mathbf{M}$ is the midpoint of the line $\mathbf{PQ}$.

The x-coordinate of $\mathbf{M}$ is the distance from the y-axis to $\mathbf{M}$. This is equal to the distance from y-axis to $\mathbf{L}$, which is the average of the x-coordinates of $\mathbf{P}$ and $\mathbf{Q}$.

$$\frac{6 + (-2)}{2} = 2$$

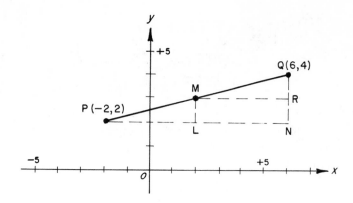

Figure 10.8 Midpoint of a line

The y-coordinate of **M** is the distance from the x-axis to **M**. This is equal to the distance from the x-axis to **R**, which is the average of the y coordinates of **P** and **Q**.

$$\frac{4+2}{2} = 3$$

From this we can conclude that the midpoint **M** of a line $\mathbf{P_1P_2}$ has the coordinates

$$\mathbf{M}\left(\frac{x_1+x_2}{2}, \frac{y_1+y_2}{2}\right) \qquad 10.10$$

Example 10.4

Find the midpoint **M** of the line $\mathbf{P_1}(-2,3)\mathbf{P_2}(-8,-3)$.

Solution

$$\mathbf{M}\left(\frac{x_1+x_2}{2}, \frac{y_1+y_2}{2}\right) = \mathbf{M}\left(\frac{-2-8}{2}, \frac{3-3}{2}\right)$$
$$= \mathbf{M}(-5,0)$$

PRACTICE PROBLEMS FOR CHAPTER 10

1. (a) In the rectangular coordinate system, plot the points and connect with lines in the order $PQRSTP$.

P(12,16)

Q(−14,18)

R(−12,1)

S(−14, −17)

T(14,−14)

(b) Find the length of each line segment (i.e., PQ, QR, etc.) and the perimeter.

PQ

QR

RS

ST

TP

2. Find the length of the radius vectors (r) for circles centered at the origin (0,0) and with the points on the circles given.

(a) (3,4)

(b) (−3, 3)

(c) (−5, −12)

(d) (5,−5)

11 TRIGONOMETRY FOR SURVEYORS

1 DEFINITION OF AN ANGLE

In trigonometry, an *angle* is considered to be the measure of the rotation of a *ray* (line) from one position to another in a counterclockwise direction.

2 STANDARD POSITION OF AN ANGLE

An angle is in *standard position* when its *vertex* is at the origin and the initial side coincides with the positive x-axis of a rectangular coordinate system.

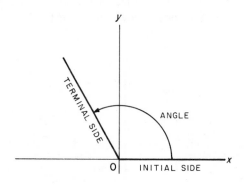

Figure 11.1　Angle nomenclature

3 QUADRANTS

The coordinate axes divide the plane into four *quadrants* designated I, II, III, and IV, as shown in figure 11.2. An angle is in one of the quadrants when its *terminal side* is in that quadrant.

4 TRIGONOMETRIC FUNCTIONS OF ANY ANGLE

For any angle in standard position, the six trigonometric functions are given by equations 11.1 through 11.6.

$$\sin\theta = \frac{y}{r} \qquad 11.1$$

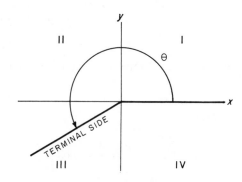

Figure 11.2　Angle quadrants

$$\cos\theta = \frac{x}{r} \qquad 11.2$$

$$\tan\theta = \frac{y}{x} \qquad 11.3$$

$$\csc\theta = \frac{r}{y} \qquad 11.4$$

$$\sec\theta = \frac{r}{x} \qquad 11.5$$

$$\cot\theta = \frac{x}{y} \qquad 11.6$$

The abbreviations of these functions are *sin, cos, tan, csc, sec,* and *cot.*

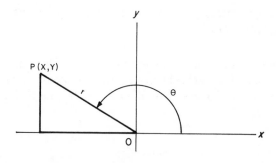

Figure 11.3　Functions of an angle

Because x, y, and r represent measured lengths, the

PROFESSIONAL PUBLICATIONS, INC. • P.O. Box 199, San Carlos, CA　94070

six trigonometric functions are actually ratios of two numbers. That is, they are ratios of the length of one side of a triangle to the length of another side. A *ratio* has been defined as a comparison of two numbers. The ratio 2 to 3 can be written 2/3 which is a fraction. Therefore, a trigonometric function is a number.

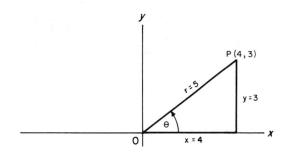

Figure 11.4   Functions related to sides

In figure 11.4,

$$\sin \theta = \frac{y}{r} = \frac{3}{5} = 0.60$$

$$\cos \theta = \frac{x}{r} = \frac{4}{5} = 0.80$$

$$\tan \theta = \frac{y}{x} = \frac{3}{4} = 0.75$$

The value of a trigonometric function of an angle θ depends on the value of θ. As θ changes, $\sin \theta$ changes, so that $\sin \theta$ is a *function* of θ.

5 RECIPROCAL OF A NUMBER

The *reciprocal* of a number is one (1) divided by the number. Thus, the reciprocal of 3 is 1/3 and the reciprocal of 2/3 is 3/2.

6 RECIPROCAL OF A TRIGONOMETRIC FUNCTION

Trigonometric functions are ratios of numbers and can be treated as such. Therefore, the reciprocal of $\sin \theta = 1/\sin \theta$. It follows that, if $\sin \theta = y/r$, the reciprocal is r/y.

$$\sin \theta = \frac{1}{\csc \theta} \qquad 11.7$$

$$\cos \theta = \frac{1}{\sec \theta} \qquad 11.8$$

$$\tan \theta = \frac{1}{\cot \theta} \qquad 11.9$$

$$\csc \theta = \frac{1}{\sin \theta} \qquad 11.10$$

$$\sec \theta = \frac{1}{\cos \theta} \qquad 11.11$$

$$\cot \theta = \frac{1}{\tan \theta} \qquad 11.12$$

Therefore, $\sin \theta$ and $\csc \theta$ are reciprocals, $\cos \theta$ and $\sec \theta$ are reciprocals, and $\tan \theta$ and $\cot \theta$ are reciprocals.

From the relation between a function and its reciprocal,

$$(\sin \theta)(\csc \theta) = \frac{y}{r} \times \frac{r}{y} = 1 \qquad 11.13$$

$$(\cos \theta)(\sec \theta) = \frac{x}{r} \times \frac{r}{x} = 1 \qquad 11.14$$

$$(\tan \theta)(\cot \theta) = \frac{y}{x} \times \frac{x}{y} = 1 \qquad 11.15$$

7 ALGEBRAIC SIGN OF TRIGONOMETRIC FUNCTIONS

An angle in standard position is considered to be in the quadrant in which its terminal side lies; therefore, the values of x and y have algebraic signs. The value of r is always considered to be positive. It follows then, that functions expressed as ratios of positive and negative numbers will have positive and negative values. The algebraic sign of any function can be determined by memorizing the terms shown in figure 11.5.

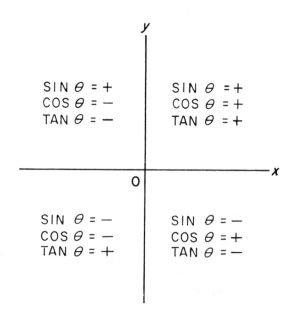

Figure 11.5   Signs of the natural functions

8 VALUES OF TRIGONOMETRIC FUNCTIONS OF QUADRANTAL ANGLES

The quadrantal angles are the angles which are common to two quadrants. They are: 0°, 90°, 180°, 270°, and

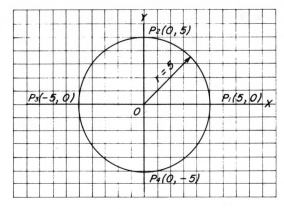

Figure 11.6   Quadrantal angles

$360°$. In figure 11.6, when $r = 5$, points $\mathbf{P}_1$, $\mathbf{P}_2$, $\mathbf{P}_3$, and $\mathbf{P}_4$ will have the coordinates shown.

If we consider the terminal side of $\theta_1\,(0°$ or $360°)$ passing through point $\mathbf{P}_1(5,0)$, $\theta_2(90°)$, passing through point $\mathbf{P}_2(0,5)$, $\theta_3(180°)$, passing through point $\mathbf{P}_3(-5,0)$, and $\theta_4(270°)$, passing through point $\mathbf{P}_4(0,-5)$, then

$$\sin 0° = \frac{y}{r} = \frac{0}{5} = 0$$

$$\sin 90° = \frac{y}{r} = \frac{5}{5} = +1$$

$$\sin 180° = \frac{y}{r} = \frac{0}{5} = 0$$

$$\sin 270° = \frac{y}{r} = \frac{-5}{5} = -1$$

$$\sin 360° = \frac{y}{r} = \frac{0}{5} = 0$$

As θ increases from $0°$ to $90°$, $\sin\theta$ increases from 0 to $+1$; as θ increases from $90°$ to $180°$, $\sin\theta$ decreases from $+1$ to 0; as θ increases from $180°$ to $270°$, $\sin\theta$ decreases from 0 to -1; as θ increases from $270°$ to $360°$, $\sin\theta$ increases from -1 to 0.

An important fact has been revealed which can be considered in the solution of right triangles: Except for the quadrantal angles, $\sin\theta$ always has a numerical value of less than one.

Considering the cosine function for the angles in figure 11.6,

$$\cos 0° = \frac{x}{r} = \frac{5}{5} = +1$$

$$\cos 90° = \frac{x}{r} = \frac{0}{5} = 0$$

$$\cos 180° = \frac{x}{r} = \frac{-5}{5} = -1$$

$$\cos 270° = \frac{x}{r} = \frac{0}{5} = 0$$

$$\cos 360° = \frac{x}{r} = \frac{5}{5} = +1$$

Except for the quadrantal angles, $\cos\theta$ also has a numerical value less than one.

The tangent function varies as follows:

$$\tan 0° = \frac{y}{x} = \frac{0}{5} = 0$$

$$\tan 90° = \frac{y}{x} = \frac{5}{0} = \infty$$

$$\tan 180° = \frac{y}{x} = \frac{0}{-5} = 0$$

$$\tan 270° = \frac{y}{x} = \frac{-5}{0} = \infty$$

$$\tan 360° = \frac{y}{x} = \frac{0}{5} = 0$$

The symbol ∞ is sometimes interpreted to mean infinity, but it must be remembered that $\tan 90°$ does not exist because a number cannot be divided by 0.

Table 11.1
Summary of functions of quadrantal angles

Angle	sin	cos	tan	cot	sec	csc
0°	0	1	0	∞	1	∞
90°	1	0	∞	0	∞	1
180°	0	−1	0	∞	−1	∞
270°	−1	0	∞	0	∞	−1
360°	0	1	0	∞	1	∞

9 TRIGONOMETRIC FUNCTIONS OF AN ACUTE ANGLE

All angles in quadrant I are *acute angles* and, therefore, positive. In dealing with acute angles only, it is more convenient to consider them as part of a right triangle and express the trigonometric functions in terms of the sides of a triangle which are given the names *opposite*, *adjacent* and *hypotenuse* as they are shown in figure 11.7.

$$\sin\theta = \frac{\text{opposite}}{\text{hypotenuse}} \qquad 11.16$$

$$\cos\theta = \frac{\text{adjacent}}{\text{hypotenuse}} \qquad 11.17$$

$$\tan\theta = \frac{\text{opposite}}{\text{adjacent}} \qquad 11.18$$

10 COFUNCTIONS

Any function of an acute angle is equal to the *cofunction* of its *complementary angle*. The sine and cosine are cofunctions, as are the tangent and the cotangent.

$$\sin 30° = \cos 60°$$

$$\tan 20° = \cot 70°$$

PROFESSIONAL PUBLICATIONS, INC. • P.O. Box 199, San Carlos, CA 94070

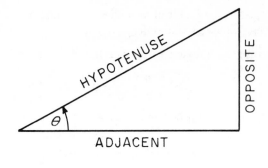
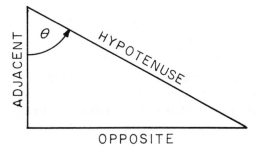

Figure 11.7　Sides of a triangle

11 TRIGONOMETRIC FUNCTIONS OF 30°, 45°, and 60°

Consider an equilateral triangle with sides of length 2. The bisector of any of the 60° angles will bisect the opposite side and form a right triangle with acute angles of 30° and 60° as shown in figure 11.8.

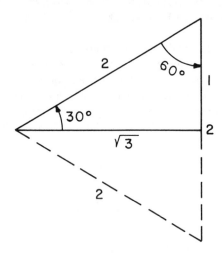

Figure 11.8　A 30-60-90 triangle

In figure 11.8,

$$\sin 30° = \frac{1}{2} = \cos 60° = 0.500$$

$$\sin 60° = \frac{\sqrt{3}}{2} = \cos 30° = 0.866\ldots$$

$$\tan 30° = \frac{1}{\sqrt{3}} = \cot 60° = 0.577\ldots$$

Consider an isosceles right triangle with two of the sides equal to one (1) as shown in figure 11.9. The angles opposite the sides will be 45°. Then

$$\sin 45° = \frac{1}{\sqrt{2}} = \frac{\sqrt{2}}{2} = 0.707\ldots$$

$$\cos 45° = \frac{1}{\sqrt{2}} = \frac{\sqrt{2}}{2} = 0.707\ldots$$

$$\tan 45° = \frac{1}{1} = 1 = 1.000$$

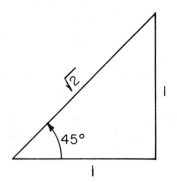

Figure 11.9　A 45-45-90 triangle

12 TABLE OF VALUES OF TRIGONOMETRIC FUNCTIONS

In sections 7 and 10, we computed certain trigonometric functions of angles of 0°, 30°, 45°, 60°, 90°, 180°, 270°, and 360°. Many years ago, mathematicians compiled tables showing values of trigonometric functions expressed in decimal form. Some tables show values for each degree of angle, some for each degree and minute of angle, and some for each degree, minute and second of angle. Values of functions are computed to four decimal places in some tables, up to ten decimal places in others.

The use of hand-held calculators has essentially eliminated the need for tables of trigonometric functions. However, the technique of interpolation is still often required in surveying work.

13 INTERPOLATION

Interpolation has been defined as finding an intermediate term in a sequence. Tables often give the sine and cosine functions for each minute of angle, but to determine the sine and cosine functions of an angle measured in degrees, minutes and seconds, interpolation is necessary. The process can be best explained with examples.

Example 11.1

Find $\sin 52°15'24''$ if $\sin 52°15' = 0.7906896$ and $\sin 52°16' = 0.7908676$.

Solution

The angle $52°15'24''$ is $24/60 = 0.4$ of the way from $52°15'00''$ to $52°16'00''$. The difference in the sine function of these two angles is $0.7908676 - 0.7906896 = 0.0001780$, remembering that as the angle increases, the sine increases. Because the angle $52°15'24''$ is 0.4 of the way between the other two angles, the sine of $52°15'24''$ is also 0.4 of the way between the sines of the other two angles.

$$\sin 52°15'24'' = 0.7906896 + (0.4)(0.0001780)$$
$$= 0.7907608.$$

Example 11.2

Find $\cos 37°25'48''$ given that $\cos 37°25' = 0.7942379$ and $\cos 37°26' = 0.7940611$.

Solution

As θ increases $\cos \theta$ decreases, so the cosine of $37°25'48''$ will be less than the cosine of $37°25'00''$.

$$48/60 = .8$$
$$0.7942379 - 0.7940611 = 0.0001768$$
$$\cos 37°25'48'' = 0.7942379 - (0.8)(0.0001768)$$
$$= 0.7940965$$

Example 11.3

Find θ if $\cos \theta = 0.8047643$.

Solution

The number 0.8047643 is not found in the cosine column in the tables. The cosine of $36°24'00''$, 0.8048938, and the cosine of $36°25'00''$, 0.8047211, are just greater than and just less than 0.8047643. So, θ is more than $36°24'00''$ and less than $36°25'00''$. The cosine of θ is $1295/1727 = 0.75$ of the way between 0.8048938 and 0.8047211, so θ is 0.75 of the way between $36°24'00''$ and $36°25'00''$.

$$\text{arccos } 0.8047643 = 36°24'00'' + (0.75)(60'') = 36°24'45''$$

Example 11.4

Find $\tan 44°17'06''$.

Solution

$$\tan 44°17'00'' = 0.9752914$$
$$\tan 44°18'00'' = 0.9758591$$
$$\tan 44°17'06'' = 0.9752914$$
$$+ (6/60)(0.9758591 - 0.9752914)$$
$$= 0.9753482$$

Example 11.5

Find θ if $\cot \theta = 1.0967405$.

Solution

$$\cot 42°21' = 1.0970609$$
$$1.0970609 - 1.0967405 = 0.0003204$$
$$\cot 42°22' = 1.0964201$$
$$1.0970609 - 1.0964201 = 0.0006408$$
$$\text{arccot } 1.0967405 = 42°21'00''$$
$$+ (0.0003204/0.0006408)(60'')$$
$$= 42°21'30''$$

14 BEARING OF A LINE

The *bearing of a line* is the acute horizontal angle between the meridian (north line) and the line.

15 ANGLE OF ELEVATION AND ANGLE OF DEPRESSION

The *angle of elevation* is the vertical angle between the horizontal and a line rotated upward from the horizontal. It is positive. The *angle of depression* is the vertical angle between the horizontal and a line rotated downward. It is negative.

16 SOLUTION OF RIGHT TRIANGLES

Solving a right triangle means finding the value of its three angles and length of each of its sides. In order to solve a right triangle, two of these values must be known. Either two sides must be known or one side and an acute angle must be known.

If an acute angle of a right triangle is known, the other acute angle is the complement of it, since the sum of the interior angles of a triangle equals $180°$.

Example 11.6

Solve the right triangle ABC having angle $A = 23°30'$ and side $a = 400$. (Note: It is customary to use capital letters to name the vertices of a triangle and the corresponding lower case letter to name the side opposite each of the vertices.)

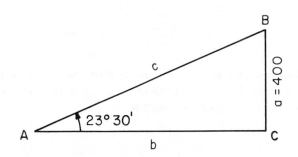

Figure 11.10   Example 11.6

Solution

$$B = 90° - 23°30' = 66°30'$$

In solving for c, the side opposite the known angle A is the known side a and the side c is the hypotenuse. So, select the sine function (opp/hyp) for use in the solution.

$$\sin A = \frac{\text{opp}}{\text{hyp}}$$
$$\sin 23°30' = \frac{400}{c}$$

To remove c from the denominator, multiply both sides of the equation by c.

$$c \sin 23°30' = \frac{400c}{c}$$
$$c \sin 23°30' = 400$$

To isolate c, divide both sides of the equation by $\sin 23°30'$.

$$\frac{c \sin 23°30'}{\sin 23°30'} = \frac{400}{\sin 23°30'}$$
$$c = \frac{400}{\sin 23°30'}$$

From trigonometric tables or calculator,

$$\sin 23°30' = 0.3987$$
$$c = \frac{400}{0.3987} = 1000$$

b is the side adjacent to the known angle A, so we select the tangent function (opp/adj).

$$\tan A = \frac{\text{opp}}{\text{adj}}$$
$$\tan 23°30' = \frac{400}{b}$$

Multiplying both sides of the equation by b and dividing both sides by $\tan 23°30'$,

$$b = \frac{400}{\tan 23°30'}$$
$$b = \frac{400}{0.4348} = 920$$

We have avoided use of c and B after they have been computed, because an error in the computations will be carried into the computation for b.

Example 11.7

Solve the right triangle EFG having angle $E = 40°$ and the hypotenuse, side $g = 200$.

Solution

$$F = 90° - 40° = 50°$$

To solve for e, the sine function is selected (opp/hyp).

$$\sin E = \frac{\text{opp}}{\text{hyp}}$$
$$\sin 40° = \frac{e}{200}$$

To isolate e, multiply both sides of the equation by 200.

$$200 \sin 40° = \frac{200}{200} e$$
$$e = 200 \sin 40°$$

From trigonometric tables or calculator,

$$\sin 40° = 0.6428$$
$$e = 200(0.6428) = 130$$

To solve for f, select the cosine function (adj/hyp).

$$\cos E = \frac{\text{adj}}{\text{hyp}}$$
$$\cos 40° = \frac{f}{200}$$

Multiplying both sides by 200,

$$f = 200 \cos 40°$$
$$= 200(0.7660) = 150$$

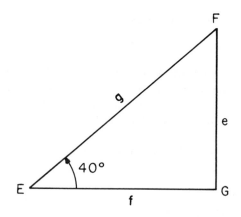

Figure 11.11　Example 11.7

17 ALTERNATE SOLUTION METHODS FOR RIGHT TRIANGLES

Since there are many different relationships between the sides and angles of right triangles, it is not surprising that triangle problems may have more than a single correct solution method. Some solutions are simpler than others. Practice is required to be able to select the simplest solution procedure.

Additional examples are provided to illustrate various solution procedures.

Example 11.8

Solve the right triangle ABC having angle $A = 23°30'$ and side $a = 400$. (Refer again to figure 11.10).

Solution

$$B = 90° - 23°30' = 66°30'$$

The known side is opposite the known angle. To find the hypotenuse, the sine function is selected.

$$c = \frac{400}{\sin 23°30'} = 1000$$

To solve for b, the tangent function is selected (opp/adj).

$$b = \frac{400}{\tan 23°30'} = 920$$

Example 11.9

Solve the right triangle EFG in figure 11.11 having angle $E = 40°$ and side $g = 200$.

Solution

$$F = 90° - 40° = 50°$$

The hypotenuse is known and is the longest side; therefore, to find the other sides, the hypotenuse must be multiplied by the sine or cosine function.

To solve for e, the sine function is used (opp/hyp).

$$e = 200 \sin 40° = 130$$

To solve for f, the cosine function is used (adj/hyp).

$$f = 200 \cos 40° = 150$$

Example 11.10

Solve the right triangle ABC shown in figure 11.12.

Solution

The solution is found from the inverse sine function on a calculator.

$$\sin A = \frac{235.62}{328.72} = 0.7167802$$

$$A = \arcsin 0.7167802 = 45°47'21''$$

$$B = 90°00'00'' - 45°47'21'' = 44°12'39''$$

$$b = \sqrt{328.72^2 - 235.62^2} = 229.22$$

Example 11.11

In order to measure the width of a river, a surveyor establishes points A and B on the west bank and finds the distance between them to be 90.0 feet. He sets up a transit on point B and establishes point C on the east bank so that BC is at right angles to AB. He then measures the angle at A between AB and AC and finds it to be $68°20'$. How wide is the river?

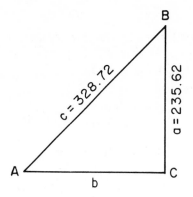

Figure 11.12   Example 11.10

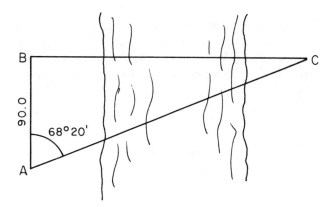

Figure 11.13   Example 11.11

Solution

The tangent function (opp/adj) is selected.

$$BC = 90.0 \tan 68°20'$$
$$= 90.0(2.157) = 226.5 \,\text{ft}$$

Example 11.12

San Angelo is due west of Waco and Arlington is 100 miles due north of Waco. The bearing of Arlington from San Angelo is N 66°30' E. (a) How far is San Angelo from Arlington? (b) How far is San Angelo from Waco?

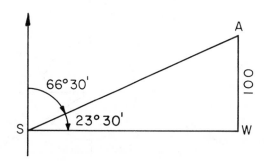

Figure 11.14   Example 11.12

Solution

The triangle formed by the three cities is a right triangle. The angle at S is the complement of the bearing angle. In solving for SA, the sine function (opp/hyp) is selected.

$$SA = \frac{100}{\sin 23°30'} = 250 \text{ miles}$$

In solving for SW, select the tangent function (opp/adj).

$$SW = \frac{100}{\tan 23°30'} = 230 \text{ miles}$$

18 RELATED ANGLES

Trigonometric tables give values for acute angles only. So, it is necessary to express larger angles in terms of an acute angle which has the same values for the functions. These angles are known as *related angles* or *reference angles.*

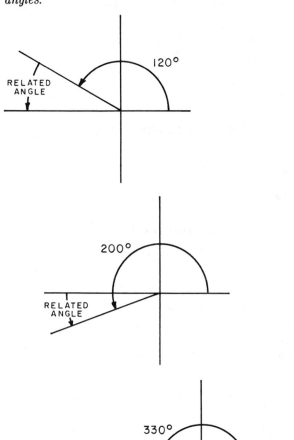

Figure 11.15　Related angles

The related angle of θ is the positive acute angle between the x-axis and the terminal side of the angle. For example, in figure 11.15,

- The related angle of 120° is 60°.
- The related angle of 200° is 20°.
- The related angle of 330° is 30°.

Regardless of which quadrant an angle lies in, the numerical value of any one of the six trigonometric functions is the same as that of the related angle. But the algebraic sign of any function depends on the quadrant in which the angle lies.

Figure 11.2 and section 3 illustrated that an angle is a third quadrant angle if its terminal side lies in the third quadrant. Likewise, an angle is in one of the other quadrants if its terminal side lies in that quadrant. Figure 11.5 gives the signs of the natural functions for each quadrant.

As solving oblique triangles involves first and second quadrant angles only, it is important to remember that the sine function will always be positive and that the cosine function will be negative for second quadrant angles (90° to 180°).

Example 11.13

Find the sine, cosine, and tangent of 150°.

Solution

The related angle is $180° - 150° = 30°$. The angle is less than 180° and more than 90°; therefore, it is a second quadrant angle. The sign of the sine function is positive; the sign of the cosine function is negative; and the sign of the tangent function is negative.

$$\sin 150° = +0.500$$
$$\cos 150° = -0.866\ldots$$
$$\tan 150° = -0.577\ldots$$

Example 11.14

Find the sine, cosine, and tangent of 315°.

Solution

The related angle is $360° - 315° = 45°$.

$$\sin 315° = -0.707\ldots$$
$$\cos 315° = +0.707\ldots$$
$$\tan 315° = -1.000$$

19 THE SINE CURVE

The variations in the value of $\sin \theta$ can be shown by plotting θ as the abscissa and $\sin \theta$ as the ordinate on a system of coordinate axes. This is done in figure 11.16.

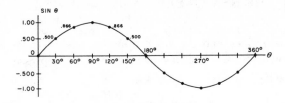

Figure 11.16   Sine curve

20 THE COSINE CURVE

The cosine curve has the same shape as the sine curve, but is offset 90°.

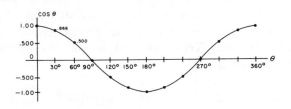

Figure 11.17   Cosine curve

21 OBLIQUE TRIANGLES

An *oblique triangle* is a triangle which does not contain a right angle. All of the angles in an oblique triangle may be acute, or there may be one obtuse angle and two acute angles.

As with right triangles, the three angles in an oblique triangle are identified with capital letters, and most often, the letters A, B, and C are used. The sides of the triangle are often identified with small (lower case) letters, with side a opposite angle A, side b opposite angle B and side c opposite angle C. However, in surveys, angles can also be identified with letters other than A, B, and C and sometimes are identified with numbers. Sides can be identified with two capital letters.

The three angles and three sides of any triangle make up the six parts of the triange. An oblique triangle can be solved if three of its parts, at least one of which is a side, are known. However, the solution is not as simple as the solution of a right triangle. Oblique triangles can be solved by forming two right triangles within the oblique triangle, but the task is made easier by the use of formulas. The most important are the law of sines and the law of cosines. The choice of which of the two laws to use to solve a particular triangle depends on which three parts of the triangle are known.

If one side and two angles of a triangle are known, the triangle can be solved by the law of sines. This case is represented by the symbol **SAA** (side, angle, angle).

Example 11.15

Given: $a = 32.16$, $B = 64°20'$, $C = 50°20'$

If two sides and the angle opposite one of them are known, the triangle can also be solved by the law of sines. This case is represented by the symbol **SSA** (side, side, angle).

Example 11.16

Given: $a = 251.5$, $b = 647.3$, $A = 22°20'$

If two sides and the angle included between the two sides are known, the triangle cannot be solved by the law of sines alone. However, it can be solved by the law of cosines and the law of sines together. This case is represented by the symbol **SAS** (side, angle, side). After the third side is found by the law of cosines, the second angle can be found by the law of sines.

If three sides of a triangle are known, the triangle can be solved by the law of cosines. This case is represented by the symbol **SSS** (side, side, side).

22 THE LAW OF SINES

In any triangle, the sides are proportional to the sines of the opposite angles. This is the *the law of sines.*

$$\frac{a}{\sin A} = \frac{b}{\sin B} = \frac{c}{\sin C} \qquad 11.19$$

$$\frac{\sin A}{a} = \frac{\sin B}{b} = \frac{\sin C}{c} \qquad 11.20$$

23 THE SAA CASE

In solving the **SAA** case (side, angle, angle), the law of sines can be expressed as equations 11.21 through 11.23.

$$a = \sin A \left(\frac{b}{\sin B}\right)$$
$$= \sin A \left(\frac{c}{\sin C}\right) \qquad 11.21$$
$$b = \sin B \left(\frac{a}{\sin A}\right)$$
$$= \sin B \left(\frac{c}{\sin C}\right) \qquad 11.22$$
$$c = \sin C \left(\frac{a}{\sin A}\right)$$
$$= \sin C \left(\frac{b}{\sin B}\right) \qquad 11.23$$

In the case with two angles known, the third angle can be readily found, so there will always be a known side opposite a known angle.

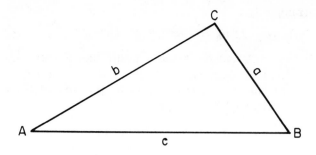

Figure 11.18   Example 11.17

Example 11.17

Solve the triangle ABC in figure 11.18 with $C = 83°$, $B = 61°$, and $c = 150$.

Solution

$$A = 180° - (83° + 61°) = 36°$$
$$a = \sin A \left(\frac{c}{\sin C}\right)$$
$$= \sin 36° \left(\frac{150}{\sin 83°}\right)$$
$$= 89$$
$$b = \sin B \left(\frac{c}{\sin C}\right)$$
$$= \sin 61° \left(\frac{150}{\sin 83°}\right) = 132$$

Example 11.18

Solve the triangle EFG, in figure 11.19.

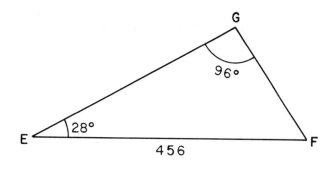

Figure 11.19   Example 11.18

Solution

$$F = 180° - (28° + 96°) = 56°$$
$$FG = \sin 28° \left(\frac{456}{\sin 96°}\right) = 215$$
$$GE = \sin 56° \left(\frac{456}{\sin 96°}\right) = 380$$

24 THE SSA CASE

If two sides and an angle opposite one of them are known (**SSA**), the law of sines can be expressed as equation 11.24.

$$\sin A = \frac{a \sin B}{b} \qquad 11.24$$

Example 11.19

Solve the triangle **ABC** in figure 11.20 with $A = 36°$, $a = 50$, and $b = 70$.

In using the law of sines to solve triangles in which two sides and the angle opposite one of them are given (**SSA**), it is possible to construct two different triangles from the given information. For example, triangles **ABC** and **AB'C** can both be constructed from the given information.

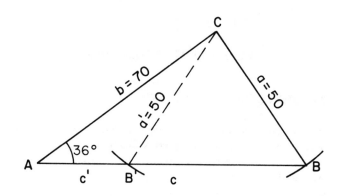

Figure 11.20   Example 11.19

Solution of triangle **ABC**

$$\sin B = \frac{70 \sin 36°}{50} = 0.8229$$
$$B = 55°$$
$$C = 180° - (36° + 55°) = 89°$$
$$c = \sin 89° \left(\frac{50}{\sin 36°}\right) = 85$$

Solution of triangle **AB'C**

Angle B could be the related angle to $55°$. Then,

$$B' = 125°$$
$$C = 180° - (36° + 125°) = 19°$$
$$c' = \sin 19° \left(\frac{50}{\sin 36°}\right) = 28$$

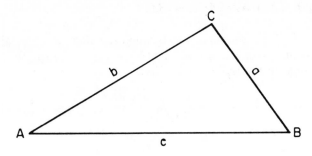

Figure 11.21 A general triangle

25 THE LAW OF COSINES

Equations 11.25 through 11.26 are alternate forms of the law of cosines. Referring to figure 11.21,

$$a^2 = b^2 + c^2 - 2bc \cos A \qquad 11.25$$
$$b^2 = c^2 + a^2 - 2ca \cos B \qquad 11.26$$
$$c^2 = a^2 + b^2 - 2ab \cos C \qquad 11.27$$

It must be remembered that the cosine of an angle greater than 90° and less than 180° has a negative algebraic sign. In substituting the cosines of angles in that range, the negative sign must be included, meaning that the value of $2bc \cos A$ in equation 11.25 will be added to the value of $b^2 + c^2$.

$$\cos A = \frac{b^2 + c^2 - a^2}{2bc} \qquad 11.28$$
$$\cos B = \frac{a^2 + c^2 - b^2}{2ac} \qquad 11.29$$
$$\cos C = \frac{a^2 + b^2 - c^2}{2ab} \qquad 11.30$$

26 THE SAS CASE

The law of cosines can be used when two sides and the included angle (**SAS**) of a triangle are known. In this case, the law of cosines can be expressed as equations 11.31 through 11.33.

$$a = \sqrt{b^2 + c^2 - 2bc \cos A} \qquad 11.31$$
$$b = \sqrt{a^2 + c^2 - 2ac \cos B} \qquad 11.32$$
$$c = \sqrt{a^2 + b^2 - 2ab \cos C} \qquad 11.33$$

Example 11.20

Solve the triangle shown in figure 11.22.

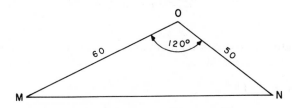

Figure 11.22 Example 11.20

Solution

$$MN = \sqrt{60^2 + 50^2 - 2(60)(50) \cos 120°}$$
$$= \sqrt{3600 + 2500 - 6000(-0.500)} = 95$$
$$\sin M = \frac{50 \sin 120°}{95} = 0.45580 \quad \text{(law of sines)}$$
$$M = 27°$$
$$\sin N = \frac{60 \sin 120°}{95} = 0.54696 \quad \text{(law of sines)}$$
$$N = 33°$$

27 THE SSS CASE

When three sides of a triangle are known (**SSS**), the law of cosines can be expressed as

$$\cos A = \frac{b^2 + c^2 - a^2}{2bc} \qquad 11.34$$
$$\cos B = \frac{a^2 + c^2 - b^2}{2ac} \qquad 11.35$$
$$\cos C = \frac{a^2 + b^2 - c^2}{2ab} \qquad 11.36$$

Example 11.21

Solve the triangle ABC with sides $a = 3.0$, $b = 5.0$, and $c = 6.0$.

Solution

$$\cos A = \frac{5.0^2 + 6.0^2 - 3.0^2}{2(5.0)(6.0)}$$
$$= \frac{25 + 36 - 9}{60}$$
$$A = 30°$$
$$\cos B = \frac{3.0^2 + 6.0^2 - 5.0^2}{2(3.0)(6.0)}$$
$$= \frac{9 + 36 - 25}{36}$$
$$B = 56°$$
$$\cos C = \frac{3.0^2 + 5.0^2 - 6.0^2}{2(3.0)(5.0)}$$
$$= \frac{9 + 25 - 36}{30} = -0.067$$
$$C = 94°$$

The negative value of $\cos C$ indicates that the angle is greater than 90° and the related angle is 86°.

28 OBLIQUE TRIANGLES USED IN SURVEYING

Example 11.22

Austin is 100 miles S 23° W from Waco. The bearing of Houston from Waco is S 40° E, and the bearing of Austin from Houston is N 76° W. What is the distance from Waco to Houston? From Houston to Austin?

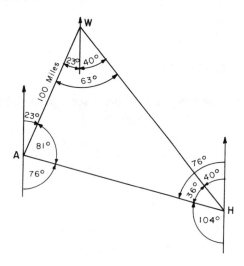

Figure 11.23   Example 11.22

Solution

Draw a meridian (north line) and place point W (for Waco) on it. From W draw a line which makes an angle of about 23° with the meridian in a southwesterly direction. Using any scale, place point A (for Austin) on this line. Also from W draw a line in a southeasterly direction making an angle of about 40° with the meridian. From the point A, draw a line in a southeasterly direction making an angle of about 76° with a meridian through A. This line intersects the southeasterly line from W at Point H.

$$WH = \frac{100 \sin 81°}{\sin 36°} = 170 \text{ miles}$$
$$HA = \frac{100 \sin 63°}{\sin 36°} = 150 \text{ miles}$$

29 SELECTION OF LAW TO BE USED

In selecting the proper law for solution of triangles, the following is recommended:

Law of sines: One side and two angles (**SAA**).

Two sides and the angle opposite one (**SSA**).

Law of cosines: Two sides and the included angle (**SAS**).

Three sides (**SSS**).

30 RADIAN MEASURE

A *radian* is an angle which, when situated as a central angle of a circle, is subtended by an arc whose length is equal to the radius of the circle.

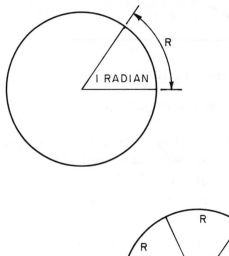

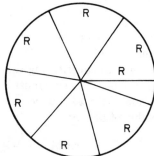

Figure 11.24   Radians

The circumference of a circle is 2π times the length of the radius, R. Therefore, the number of arcs of length R which can be applied to the circumference of a circle is 2π.

$$2\pi \text{ radians} = 360° \qquad\qquad 11.37$$
$$\pi \text{ radians} = 180° \qquad\qquad 11.38$$

Dividing both sides of equation 11.38 by π,

$$1 \text{ radian} = \frac{180°}{\pi} \qquad\qquad 11.39$$

Dividing both sides of equation 11.38 by 180°,

$$1° = \frac{\pi}{180} \text{ radians} \qquad\qquad 11.40$$

Rule: To convert degrees to radians, multiply the number of degrees by $\pi/180$.

For example,

$$30° = \frac{30\pi}{180} = \frac{\pi}{6} \text{ radians}$$

Rule: To convert radians to degrees multiply the number of radians by $180/\pi$.

For example,

$$2\,\text{radians} = \frac{(2)(180)}{\pi} = \frac{360°}{\pi}$$

31 LENGTH OF ARC OF CIRCLE

If we let S equal the length of any arc of a circle, the relationship between the arc length and radius is

$$S = R\theta \qquad\qquad 11.41$$

θ is expressed in radians in equation 11.41.

Example 11.23

What is the length of the arc subtended by a central angle of 30° in a circle of 10 inch radius?

Solution

$$S = R\theta = \frac{(10)(30)\pi}{180} = 5.2\,\text{in.}$$

Example 11.24

What is the radius of the circle on which an arc of 100 feet subtends an angle of 1°?

Solution

$$R = \frac{S}{\theta} = \frac{100}{\frac{(1)\pi}{180}} = \frac{18000}{\pi} = 5729.58\,\text{ft.}$$

PRACTICE PROBLEMS FOR CHAPTER 11

1. Each of the following points lies on the terminal side of an angle θ in standard position. Plot the point, draw the terminal side through the point, and measure the angle with a protractor.

(a) (12,6):

(b) (−12,4):

(c) (−10,−10):

(d) (7,−9):

2. Write the sin, cos, tan, cot, sec, and csc of angles θ as a common fraction. Show algebraic sign.

(a)

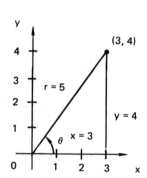

(b)

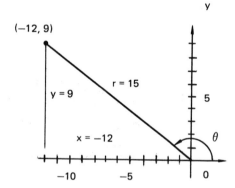

(c)

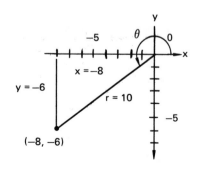

(d)

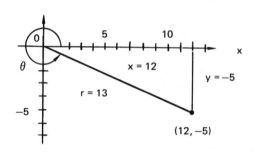

3. Write the reciprocal of each number.

Example: $\dfrac{2}{3}$ $\quad$ $\dfrac{1}{\frac{2}{3}} = \dfrac{3}{2}$

(a) 3 $\qquad$ (f) $\dfrac{1}{x}$

(b) $\dfrac{3}{4}$ $\qquad$ (g) $\dfrac{1}{y}$

(c) x $\qquad$ (h) $\dfrac{x}{r}$

(d) y $\qquad$ (i) $\dfrac{y}{r}$

(e) r $\qquad$ (j) $\dfrac{y}{x}$

4. Show that $\sin\theta$ and $\csc\theta$, $\cos\theta$ and $\sec\theta$, and $\tan\theta$ and $\cot\theta$ are reciprocals.

Example: $\sin\theta$ $\quad$ $\dfrac{1}{\sin\theta} = \dfrac{1}{\frac{y}{r}} = \dfrac{r}{y} = \csc\theta$

(a) $\cos\theta$

(b) $\tan\theta$

(c) $\cot\theta$

5. Using a calculator and the reciprocal, find

(a) $\sec 60°$

(b) $\csc 30°$

(c) $\cot 45°$

6. Using the graph below, find the sin, cos, and tan functions of the angles indicated.

Example:

$\theta = 150°$

$\sin 150° = 5.0/10 = 0.50$

$\cos 150° = -8.7/10 = -0.87$

$\tan 150° = 5.0/-8.7 = -0.57$

(a) $\theta = 30°$

(b) $\theta = 135°$

(c) $\theta = 300°$

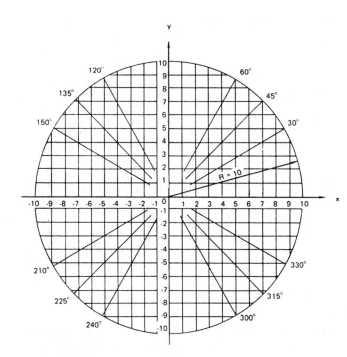

7. Indicate the algebraic sign of the sin, cos, and tan functions of each of the angles.

Example: $30°$ $\sin + \cos + \tan$

(a) $185°$ (h) $175°$
(b) $225°$ (i) $\ \ 95°$
(c) $350°$ (j) $275°$
(d) $190°$ (k) $300°$
(e) $265°$ (l) $110°$
(f) $100°$ (m) $\ \ 85°$
(g) $\ \ 89°$ (n) $290°$

8. Using the information provided, identify the quadrant in which the terminal side of the angle θ lies.

Example: $\sin \theta = +, \cos \theta = -:$ II

(a) $\sin \theta = -, \cos \theta = +:$

(b) $\tan \theta = -, \sin \theta = -:$

(c) $\tan \theta = +, \sin \theta = -:$

(d) $\sin \theta = +, \cos \theta = +:$

(e) $\tan \theta = +, \cos \theta = -:$

(f) $\sin \theta = +, \tan \theta = +:$

(g) $\sin \theta = -, \tan \theta = -:$

9. Indicate the value of the sin, cos, and tan functions of the quadrantal angles.

Example: $0°$ $\sin = 0$ $\cos = +1$ $\tan = 0$

(a) $90°$

(b) $180°$

(c) $270°$

(d) $360°$

10. Using the figures, write the sin, cos, and tan function of each acute angle. Express as a common fraction and as a decimal fraction to two significant digits.

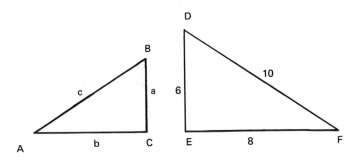

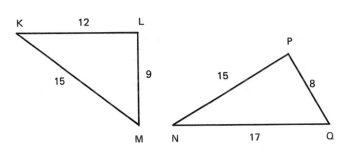

Examples: (1) $\sin A = a/c$

(2) $\sin K = 9/15 = 0.60$

(a) $\cos A$ (l) $\cos D$
(b) $\tan A$ (m) $\tan D$
(c) $\sin B$ (n) $\sin F$
(d) $\cos B$ (o) $\cos F$
(e) $\tan B$ (p) $\tan F$
(f) $\cos K$ (q) $\sin N$
(g) $\tan K$ (r) $\cos N$
(h) $\sin M$ (s) $\tan N$
(i) $\cos M$ (t) $\sin Q$
(j) $\tan M$ (u) $\cos Q$
(k) $\sin D$ (v) $\tan Q$

11. Find the trigonometric function for the angle indicated or find the angle for the function indicated.

Example: (1) $\sin 53°13'36'' = \underline{0.8010101}$

(a) $\sin 37°29'16''$

(b) $\cos 52°42'51''$

(c) $\cos^{-1} 0.7918605$

(d) $\tan 43°17'28''$

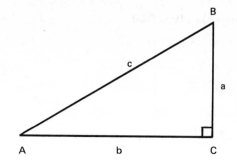

12. For the triangle ABC, select the trigonometric function to be used to solve the part of the triangle indicated.

Examples: (1) $A = 36°52'$, $a = 600$.

$b : \tan$ $c : \sin$

(2) $a = 300$, $b = 400$

$A : \tan$ $B : \tan$

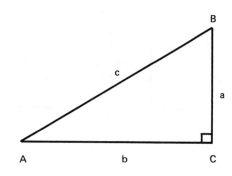

(a) $B = 51°40'$, $a = 650$.   (b) $A = 46°44'$, $b = 156$.

$b : \underline{\quad}$ $c : \underline{\quad}$ $a : \underline{\quad}$ $c : \underline{\quad}$

(c) $B = 53°21'$, $c = 300$.   (d) $A = 38°19'$, $c = 700$

$a : \underline{\quad}$ $b : \underline{\quad}$ $a : \underline{\quad}$ $b : \underline{\quad}$

(e) $a = 600$, $c = 1000$ (f) $b = 400$, $c = 500$

$A : \underline{\quad}$ $B : \underline{\quad}$ $A : \underline{\quad}$ $B : \underline{\quad}$

(g) $B = 55°10'$, $b = 378$ (h) $A = 33°40'$, $a = 250$.

$a : \underline{\quad}$ $c : \underline{\quad}$ $b : \underline{\quad}$ $c : \underline{\quad}$

13. Completely solve the right triangle ABC.

Example: $A = 36°52'$, $a = 600$

$B = 90° - 36°52' = 53°08'$

$b = \dfrac{600}{\tan 36°52'} = 800$

$c = \dfrac{600}{\sin 36°52'} = 1000$

(a) $A = 28°41'$, $b = 540$

(b) $B = 55°13'$, $a = 371$

(c) $B = 61°29'$, $b = 466$

(d) $A = 33°15'$, $c = 263$

(e) $B = 58°55'$, $c = 562$

(f) $a = 300$, $b = 400$

14. Find the height to the nearest inch of a man who casts a shadow 10.0 ft long when the angle of elevation of the sun is $31°18'$.

15. From a point 160 ft from the foot of a flagpole, the angle of elevation to the top of the flagpole is $50°40'$. Find the height of the flagpole.

16. From the top of a tower 500 ft high, the angle of depression to a road intersection is $30°$. How far from the tower is the road intersection?

17. Temple is due south of Fort Worth and 115 miles due east of Brady. The bearing of Brady from Fort Worth is $S45°W$. How far is Brady from Fort Worth?

18. A 20 ft ladder reaches from the ground to the roof of a building. The angle of elevation of the ladder is $60°$. How high is the roof?

19. Completely solve the right triangle ABC, as in problem 13.

(a) $B = 41°12'38''$, $a = 625.18'$

(b) $A = 66°22'37''$, $a = 492.72'$

(c) $B = 38°04'48''$, $c = 585.20'$

(d) $A = 22°13'50''$, $a = 376.26'$

(e) $B = 75°35'41''$, $b = 237.68'$

(f) $a = 427.82'$, $b = 396.95'$

(g) $b = 445.64'$, $c = 616.38'$

20. Write the related (or reference) angle of each angle.

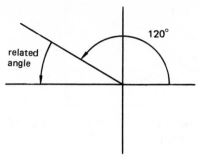

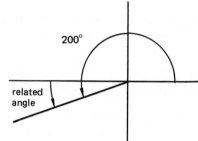

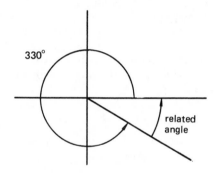

Example: 150°   30°

(a) 260°   (l) 155°
(b) 210°   (m) 110°
(c) 160°   (n) 140°
(d) 220°   (o) 135°
(e) 190°   (p) 100°
(f) 350°   (q) 120°
(g) 330°   (r) 175°
(h) 170°   (s) 95°
(i) 200°   (t) 185°
(j) 300°   (u) 280°
(k) 125°

21. Solve for the missing sides and angles for the oblique triangles. (*SAA*)

Example: Triangle MNO: $M = 38°48'45''$, $0 = 82°23'56''$, $MN = 298.34'$.

$$N = 180° - (38°48'45'' + 82°23'56'') = 58°47'19''$$

$$NO = \frac{\sin 38°48'45''(298.34)}{\sin 82°23'56''} = 188.65'$$

$$OM = \frac{\sin 58°47'19''(298.34)}{\sin 82°23'56''} = 257.42'$$

(a) Triangle 1–2–3: $1 = 34°18'24''$, $3 = 62°12'55''$, $1–2 = 1347.77'$.

(b) Triangle PQR: $P = 118°34'24''$, $Q = 23°06'54''$, $QR = 526.30'$.

(c) Triangle 1–2–3: $1 = 82°46'58''$, $2 = 58°54'20''$, $2–3 = 345.43'$

22.   Solve for the missing sides and angles for the oblique triangles. (*SSA*)

Example: Triangle ABC: $A = 76°00'09''$, $a = 256.07$, $b = 172.28$

$$B: \sin B = \frac{172.28 \sin 76°00'09''}{256.07} \quad B = 40°45'13''$$

$$C = 180° - (76°00'09'' + 40°45'13'') = 63°14'38''$$

$$c = \frac{\sin 63°14'38''(256.07)}{\sin 76°00'09''} = 235.65$$

(a) Triangle EFG: $E = 25°40'55''$, $FG = 646.13$, $EF = 1296.20$.

(b) Triangle ABC: $A = 51°20'14''$. $a = 445.23$, $b = 526.17$. (2 solutions.)

23.   Solve for the missing sides and angles for the oblique triangles. (*SAS*)

Example: Triangle ABC: $A = 58°33'47''$, $b = 204.38$, $c = 152.15$

$$a = $$
$$\sqrt{204.38^2 + 152.15^2 - (2)(204.38)(152.15)\cos 58°33'47''}$$
$$= 180.23$$

$$B: \sin B = \frac{204.38 \sin 58°33'47''}{180.23} \quad B = 75°21'43''$$

$$C = 180° - (75°21'43'' + 58°33'47'') = 46°04'30''$$

(a) Triangle EFG: $G = 95°12'50''$, $FG = 146.25$, $GE = 122.31$

(b) Triangle KLM: $L = 35°19'16''$, $KL = 595.45$, $LM = 851.78$ (Hint: Largest angle must be opposite longest side. Sine function may represent related angle.)

(c) Triangle NOP: $N = 46°07'01''$, $NO = 138.38$, $PN = 165.12$

24.   Solve for the missing sides and angles for the oblique triangles. (*SSS*)

Example: Triangle ABC: $a = 48.79'$, $b = 62.45'$, $c = 30.13'$.

A: $\cos A = \dfrac{62.45^2 + 30.13^2 - 48.79^2}{(2)(62.45)(30.13)}$ $A = 49°49'59''$

B: $\cos B = \dfrac{48.79^2 + 30.13^2 - 62.45^2}{(2)(48.79)(30.13)}$ $B = 102°00'32''$

C: $\cos C = \dfrac{48.79^2 + 62.45^2 - 30.13^2}{(2)(48.79)(62.45)}$ $C = 28°09'29''$

$\qquad\qquad\qquad$ check $\qquad$ $180°00'00''$

(a) Triangle EFG: $EF = 125.83'$, $FG = 171.25'$, $GE = 155.13'$.

(b) Triangle MNO: $MN = 298.34'$, $NO = 188.65'$, $OM = 257.42'$.

25. Given the parts of an oblique triangle ABC indicated, select the law applicable for the solution.

Example: a, c, A   law of <u>sines</u>

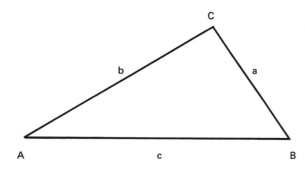

(a) a, b, A

(b) b, c, A

(c) b, A, C

(d) A, B, a

(e) a, c, B

(f) b, c, B

(g) A, a, b

(h) a, b, c

(i) A, C, a

(j) c, b, a

26. Express the angles in radians using π in each answer.

Example: $30° = \pi/6$ radians

(a) 45°   (b) 120°   (c) 90°   (d) 15°   (e) 270°

27. Express the angles in degrees.

Example: π radians $= 180°$

(a) $\dfrac{\pi}{3}$ radians   (b) $\dfrac{\pi}{2}$ radians   (c) $\dfrac{\pi}{4}$ radians

(d) $\dfrac{\pi}{12}$ radians   (e) $\dfrac{3\pi}{2}$ radians

28. What is the length of the arc subtended by a central angle of 60° in a circle with 300' radius?

29. What is the radius of the circle on which an arc of 300' subtends an angle of 12°?

30. Find the radius of a circle on which an arc of 25 inches has a central angle of 2.4 radians.

31. Find the length along the equator of an arc subtended by a central angle of 1° if the diameter of the earth at the equator is 7927 miles.

32. The end of a 25 inch pendulum swings through a 3.4 inch arc. What is the size of the angle through which the pendulum swings?

33. Using a calculator, find the value of the sine function for each angle indicated, plot on the coordinate system marked sine curve, and connect the points with a curved line. Repeat the procedure for the cosine curve.

(a)

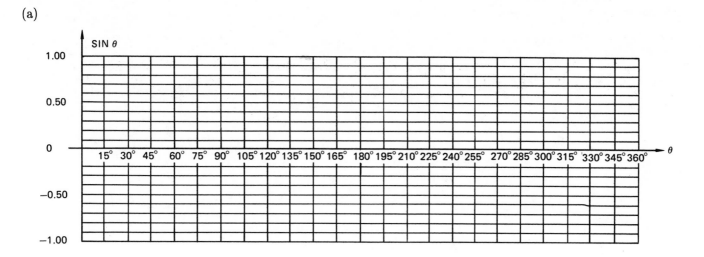

(b)

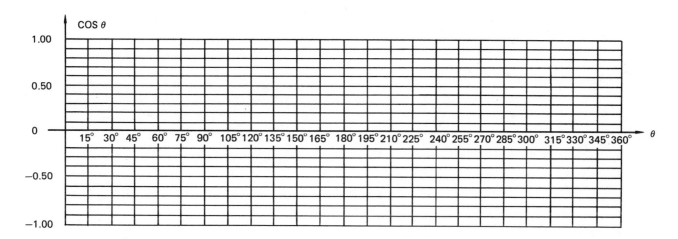

PROFESSIONAL PUBLICATIONS, INC. ● P.O. Box 199, San Carlos, CA 94070

12 ANALYTIC GEOMETRY FOR SURVEYORS

1 FIRST DEGREE EQUATIONS

If we say "4 times a number minus three is equal to 5," we can express this statement in algebraic terms as $4x - 3 = 5$. The letter x represents the unknown number. The number 2 satisfies the equation, and we say that 2 is the *root* of the equation.

If we say "the sum of two numbers is 8," we write this in algebraic terms as $x + y = 8$. In this equation we have two unknowns, represented by x and y. And, more than one pair of numbers will make the statement true. (If $x = 1$, $y = 7$; if $x = 2$, $y = 6$; if $x = 3$, $y = 5$; etc.) If we always express these pairs of numbers in the order of x first and y second, we say that they are *ordered pairs* and write them symbolically as $(1, 7)$; $(2, 6)$; $(3, 5)$ etc. Because x and y are of the first power, we say that $x + y = 8$ is an *equation of the first degree*. We say that $x + y = 8$ is an equation in two *unknowns*, or two *variables*. The equation $4x - 3 = 5$, mentioned above, is an equation of the first degree in one unknown.

2 GRAPHS OF FIRST DEGREE EQUATIONS IN TWO VARIABLES

If we consider each ordered pair which satisfies a first degree equation in two variables (the roots of the equation, or the solution set of the equation) to be the coordinates of a point, and if we plot several of these points on a rectangular coordinate system and connect them with a line, we have what is called the *graph of the equation*.

Consider the graph of the equation

$$2x - 3y = 12$$

In order to solve the equation, that is, to find ordered pairs which satisfy it, first rearrange the equation.

$$-3y = -2x + 12$$
$$y = \frac{2}{3}x - 4$$

Next, give various values to x and solve for the corresponding values of y (for instance, when $x = 0$, $y = -4$),

$$x = -3, \quad 0, \quad 3, 6, 9$$
$$y = -6, \ -4, \ -2, 0, 2$$

Plotting these ordered pairs as coordinates of a point, we obtain the graph of the equation $2x - 3y = 12$ shown in figure 12.1. When the points are connected, they lie in a straight line. Also, it is obvious that we could give an infinite number of values to x and find an infinite number of values for y.

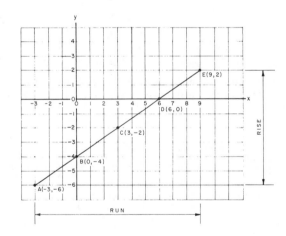

Figure 12.1 A straight line

Example 12.1

Find three roots of each equation. Give your answers as (x, y).

(a) $2x + y = 10$ (c) $x + y = 5$

(b) $3x + 2y = 0$ (d) $x - y = 0$

Solution

(a) $(3, 4)$; $(2, 6)$; $(5, 0)$ (c) $(1, 4)$; $(2, 3)$; $(0, 5)$

(b) $(2, -3)$; $(4, -6)$; $(0, 0)$ (d) $(-4, -4)$; $(0, 0)$; $(2, 2)$

Example 12.2

Find the coordinates of the points at which each of the following equations intersect the x and y axes.

(a) $x - y + 5 = 0$ (c) $2x - 5y = 10$

(b) $3x - y + 6 = 0$ (d) $2x + 3y + 6 = 0$

Solution

Let $y = 0$, and solve for x. Then, let $x = 0$, and solve for y.

(a) $(0, 5); (-5, 0)$ (c) $(0, -2); (5, 0)$

(b) $(0, 6); (-2, 0)$ (d) $(0, -2); (-3, 0)$

3 SLOPE OF A LINE

Slope is the ratio of the **change** in the vertical distance to the change in the **horizontal** distance. If we let m represent slope for the line connecting points (x_1, y_1) and (x_2, y_2), then

$$m = \frac{y_2 - y_1}{x_2 - x_1} \qquad 12.1$$

To describe the steepness or *slope* of a line (such as the graph of the equation $2x - 3y = 12$ shown in figure 12.1), choose two points on it (such as $A(-3, -6)$ and $E(9, 2)$).

$$\text{slope} = \frac{\text{rise}}{\text{run}} = \frac{\text{ordinate of } E - \text{ordinate of } A}{\text{abscissa of } E - \text{abscissa of } A}$$

$$= \frac{2 - (-6)}{9 - (-3)} = \frac{8}{12} = \frac{2}{3}$$

Alternately,

$$\text{slope} = \frac{\text{ordinate of } A - \text{ordinate of } E}{\text{abscissa of } A - \text{abscissa of } E}$$

$$= \frac{-6 - 2}{-3 - 9} = \frac{8}{12} = \frac{2}{3}$$

Example 12.3

Determine the slope of the line through each pair of points.

(a) $(1, 2); (3, 6)$ (c) $(-2, -3); (2, 5)$

(b) $(-4, 3); (3, -4)$ (d) $(-4, 3); (4, 3)$

Solution

(a) $m = \frac{6 - 2}{3 - 1} = +2$

(b) $m = \frac{-4 - 3}{3 - (-4)} = -1$

(c) $m = \frac{5 - (-3)}{2 - (-2)} = +2$

(d) $m = \frac{3 - 3}{4 - (-4)} = 0$

In plotting the lines it can be seen that when a line rises from left to right, its slope is positive; when a line falls from left to right, its slope is negative.

4 LINEAR EQUATIONS

We have determined that the slope of the line represented by the linear equation $2x - 3y = 12$ is $+2/3$. The graph of the equation, shown in figure 12.1, rises from left to right. Writing the equation in the form

$$y = \frac{2}{3}x - \frac{12}{3}$$

we notice that the coefficient of x is $+2/3$, which is the slope of the line. Notice also that the numerator 2 is the coefficient of x, and the denominator 3 is the coefficient of y when the equation is written in the form

$$2x - 3y = 12$$

Also notice that when the coefficient of x is $+2$ and the coefficient of y is -3, the slope is positive.

If we let the equation

$$Ax + By + C = 0 \qquad 12.2$$

represent the *general form* of a linear equation where A is the positive coefficient of x, B is the coefficient of y, and C is a constant, then the slope is

$$m = -\frac{A}{B} \qquad 12.3$$

Example 12.4

Rearrange the equations to the form $Ax + By + C = 0$ with A positive. Determine the slope m of each.

(a) $3x + 4y = 8$

(b) $2x = 6y$

(c) $2y + 3x - 6 = 0$

(d) $-3x + 4y + 10 = 0$

Solution

(a) $3x + 4y - 8 = 0$

$$m = -\frac{A}{B} = -\frac{+3}{+4} = -\frac{3}{4}$$

(b) $2x - 6y = 0$

$$m = -\frac{+2}{-6} = \frac{1}{3}$$

(c) $3x + 2y - 6 = 0$

$$m = -\frac{+3}{+2} = -\frac{3}{2}$$

(d) $3x - 4y - 10 = 0$

$$m = -\frac{+3}{-4} = \frac{3}{4}$$

5 EQUATIONS OF HORIZONTAL AND VERTICAL LINES

Consider the line which contains the points $(-4, 5)$ and $(4, 5)$. The slope of the line is

$$m = \frac{5 - 5}{4 + 4} = \frac{0}{8} = 0$$

This is a horizontal line which has the equation $y = 5$ (figure 12.2). In considering the general equation of a line, $Ax + By + C = 0$, the coefficient of x for a horizontal line is 0. So, the equation of a horizontal line is $0x + By + C = 0$, or $By = C$.

The line which contains the points $(4, 5)$ and $(4, -5)$ has the slope

$$m = \frac{5 - (-5)}{4 - 4} = \frac{10}{0}$$

which is infinite, so we say that the line has an infinite slope. The equation of the line is $x = 4$, and the general equation of a vertical line is $Ax + 0y + C = 0$.

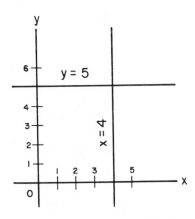

Figure 12.2 Horizontal and vertical lines

6 X AND Y INTERCEPTS

If we let $x = 0$ in the equation $2x - 3y = 12$, we find that $y = -4$. The point $(0, -4)$ is the point where the graph of the equation crosses the y-axis. (Figure 3-1). The distance from the x-axis to this point is known as the *y-intercept* and is given the symbol b.

If we let $y = 0$ in the equation $2x - 3y = 12$, we find that $x = 6$. The point $(6, 0)$ is where the graph of the equation crosses the x-axis. The distance from the y-axis to this point is known as the *x-intercept* and is given the symbol a.

We can say, then, that the coordinates of the point where any line crosses the y-axis are $(0, b)$, and the coordinates of the point where any line crosses the x-axis are $(a, 0)$. In figure 12.1, these points are $(0, -4)$ and $(6, 0)$.

If the equation $2x - 3y = 12$ is written

$$y = \frac{2}{3}x - 4$$

the constant term -4 is the y-coordinate of the point where the line crosses the y-axis. It is, in fact, the y-intercept. In finding this equivalent equation of $2x - 3y = 12$, both sides of the equation were divided by -3, which is the coefficint of y, so that the constant 12 was divided by -3 to obtain the quotient -4.

For any equation $Ax + By + C = 0$, the y-intercept is

$$b = -\frac{C}{B} \qquad\qquad 12.4$$

If the equation $2x - 3y = 12$ is written

$$x = \frac{3}{2}y + 6$$

the constant term 6 is the x-coordinate of the point where the line crosses the x-axis. It is the x-intercept. For any equation $Ax + By + C = 0$, the x-intercept is

$$a = -\frac{C}{A} \qquad\qquad 12.5$$

In summary, for any linear equation $Ax + By + C = 0$,

$$m = \text{slope} \ = -\frac{A}{B} \qquad\qquad 12.6$$

$$a = x\text{-intercept} \ = -\frac{C}{A} \qquad\qquad 12.7$$

$$b = y\text{-intercept} \ = -\frac{C}{B} \qquad\qquad 12.8$$

Example 12.5

Find the slope, x-intercept, and y-intercept of the line $3x + 2y - 6 = 0$.

PROFESSIONAL PUBLICATIONS, INC. ● P.O. Box 199, San Carlos, CA 94070

Solution

$$m = -\frac{A}{B} = -\frac{3}{2}$$

$$a = -\frac{C}{A} = -\frac{-6}{3} = 2$$

$$b = -\frac{C}{B} = -\frac{-6}{2} = 3$$

7 PARALLEL LINES

Two different lines having the same slope are *parallel lines.*

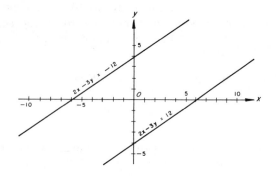

Figure 12.3 Parallel lines

In figure 12.3, the lines $2x - 3y = -12$ and $2x - 3y = 12$ have the same slope, $+2/3$. Writing the equations in equivalent form shows that the slope for each line is the same, only the y-intercepts differ.

$$y = \frac{2}{3}x + 4$$

$$y = \frac{2}{3}x - 4$$

8 PERPENDICULAR LINES

Two lines intersecting at right angles are called *perpendicular lines.* For two lines to be perpendicular, the slope of one must be the negative reciprocal of the other, or

$$m_1 = -\frac{1}{m_2} \qquad 12.9$$

In figure 12.4, the lines $2x - 3y = 12$ and $3x + 2y = 8$ are perpendicular lines, as can be seen when written in the form $y = mx + b$.

$$y = \frac{2}{3}x - 4$$

$$y = -\frac{3}{2}x + 4$$

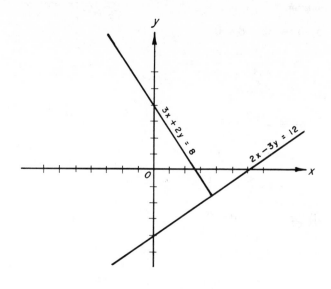

Figure 12.4 Perpendicular lines

9 PERPENDICULAR DISTANCE FROM A POINT TO A LINE

A formula for finding the perpendicular distance from a point of known coordinates (x, y) to a line of known equation can be found from equation 12.10.

$$D = \frac{|Ax + By + C|}{\sqrt{A^2 + B^2}} \qquad 12.10$$

Example 12.6

Find the perpendicular distance D from the point $P(-2, 4)$ to the line $4x - 3y - 16 = 0$.

Solution

$$D = \frac{|(4)(-2) - (3)(4) - 16|}{\sqrt{(4)^2 + (-3)^2}} = \frac{|(-36)|}{5} = \frac{36}{5} = 7.2$$

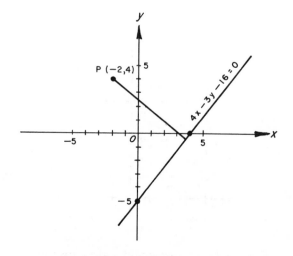

Figure 12.5 Example 12.6

10 WRITING THE EQUATION OF A LINE

An equation may be written for a straight line if sufficient information is known. For example, if

- one point on the line and the slope are known, or
- two points on the line are known, or
- the x intercept and y intercept are known, or
- the slope of the line and the y intercept are known.

11 POINT-SLOPE FORM OF THE EQUATION OF A LINE

If $\mathbf{P}(x, y)$ is any point on a line with slope m through point $\mathbf{P}_1(x_1, y_1)$, then the *point-slope form* of the line is

$$\frac{y - y_1}{x - x_1} = m \qquad 12.11$$

$$y - y_1 = m(x - x_1) \qquad 12.12$$

Example 12.7

Write the equation of the line through the given point with the given slope m in the form $Ax + By + C = 0$.

(a) $(4, -2)$; $m = 2$

(b) $(3, -2)$; $m = -\dfrac{3}{2}$

(c) $(-4, 5)$; $m = 0$

Solution

(a)
$$y - y_1 = m(x - x_1)$$
$$y - (-2) = 2(x - 4)$$
$$y + 2 = 2x - 8$$
$$-2x + y + 10 = 0$$
$$2x - y - 10 = 0$$

(b) The slope m can be written $-(3/2)$, $(-3)/2$, or $3/(-2)$. For ease in performing the algebraic operation, the numerator should carry the negative sign.

$$y - y_1 = m(x - x_1)$$
$$y + 2 = \frac{-3(x - 3)}{2}$$
$$2y + 4 = -3x + 9$$
$$3x + 2y - 5 = 0$$

(c)
$$y - y_1 = m(x - x_1)$$
$$y - 5 = 0(x + 4)$$
$$y = 5$$

This is a linear equation. The graph is a line parallel to the x-axis, 5 units above.

12 THE TWO-POINT FORM OF THE EQUATION OF A LINE

If $\mathbf{P}(x, y)$ is any point on a line which passes through the points $\mathbf{P}_1(x_1, y_1)$ and $\mathbf{P}_2(x_2, y_2)$, then the *two-point form* of the line is

$$y - y_1 = m(x - x_1) \qquad 12.13$$

$$y - y_1 = \frac{y_2 - y_1}{x_2 - x_1}(x - x_1) \qquad 12.14$$

$$\frac{y - y_1}{x - x_1} = \frac{y_2 - y_1}{x_2 - x_1} \qquad 12.15$$

In writing the two-point form, either point may be designated point 1.

Example 12.8

Write the equation of the line through the two points.

(a) $(1, 4)$; $(3, -2)$ (c) $(1, -3)$; $(-2, 1)$

(b) $(-2, 2)$; $(1, -3)$ (d) $(3, 4)$; $(1, 4)$

Solution

(a)
$$\frac{y - y_1}{x - x_1} = \frac{y_2 - y_1}{x_2 - x_1}$$
$$\frac{y - 4}{x - 1} = \frac{-2 - 4}{3 - 1} = \frac{-6}{2} = -3$$
$$\frac{y - 4}{x - 1} = -3$$
$$y - 4 = -3(x - 1)$$
$$y - 4 = -3x + 3$$
$$3x + y - 7 = 0$$

(b)
$$\frac{y - y_1}{x - x_1} = \frac{y_2 - y_1}{x_2 - x_1}$$
$$\frac{y - 2}{x + 2} = \frac{-3 - 2}{1 + 2} = \frac{-5}{3}$$
$$\frac{y - 2}{x + 2} = \frac{-5}{3}$$
$$-5(x + 2) = 3(y - 2)$$
$$-5x - 10 = 3y - 6$$
$$-5x - 3y - 4 = 0$$
$$5x + 3y + 4 = 0$$

(c)
$$\frac{y - y_1}{x - x_1} = \frac{y_2 - y_1}{x_2 - x_1}$$
$$\frac{y + 3}{x - 1} = \frac{1 + 3}{-2 - 1} = \frac{4}{-3}$$
$$\frac{y + 3}{x - 1} = \frac{4}{-3}$$
$$4(x - 1) = -3(y + 3)$$
$$4x - 4 = -3y - 9$$
$$4x + 3y + 5 = 0$$

PROFESSIONAL PUBLICATIONS, INC. • P.O. Box 199, San Carlos, CA 94070

(d)
$$\frac{y - y_1}{x - x_1} = \frac{y_2 - y_1}{x_2 - x_1}$$
$$\frac{y - 4}{x - 3} = \frac{4 - 4}{1 - 3} = \frac{0}{-2}$$
$$\frac{y - 4}{x - 3} = \frac{0}{-2}$$
$$-2(y - 4) = 0(x - 3)$$
$$-2y + 8 = 0$$
$$2y - 8 = 0$$
$$y = 4$$

13 INTERCEPT FORM OF THE EQUATION OF A LINE

A line with x-intercept a and y-intercept b, (where both a and b are not zero), has the equation

$$\frac{x}{a} + \frac{y}{b} = 1 \qquad 12.16$$

Example 12.9

Write the equation of the line with x-intercept 3 and y-intercept -4.

Solution

$$\frac{x}{3} + \frac{y}{-4} = 1$$
$$-4x + 3y = -12$$
$$4x - 3y = 12$$

14 SLOPE-INTERCEPT FORM OF THE EQUATION

If the slope of a line and its y-intercept are known, the *slope-intercept form* of the line is

$$y = mx + b \qquad 12.17$$

Example 12.10

Write the equation of the line of slope 2 and y-intercept -3.

Solution

$$y = 2x - 3$$
$$2x - y = 3$$

Example 12.11

Write the equation of the line through $(3, -1)$ perpendicular to the line $2x + 3y = 6$.

Solution

The slope of $2x + 3y = 6$ is $m_1 = -\dfrac{2}{3}$

The slope of the perpendicular line is $m_2 = \dfrac{3}{2}$

The equation of the perpendicular line is:

$$y - y_1 = m_2(x - x_1)$$
$$y - (-1) = \frac{3}{2}(x - 3)$$
$$3x - 2y = 11$$

15 SYSTEMS OF LINEAR EQUATIONS

If the graphs of two linear equations lie in the same xy plane, then one of three conditions must be true:

- The two lines are parallel and will never intersect.
- The two lines coincide.
- The two lines will intersect at a point.

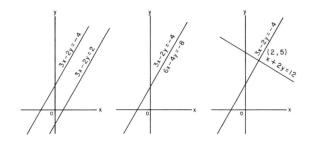

Figure 12.6   System of linear equations

When a point must be located on two straight lines, we say that the line equations form a system of simultaneous equations. Any ordered pair which satisfies both equations is called a solution or a *root* of the system.

When two lines intersect at a point, there can be only one root, and this root can be found by solving the two equations simultaneously. The ordered pair found to be the root of a system will be the coordinates of the point of intersection of the two graphs of the equations.

In figure 12.6(a), the two lines are parallel and will not intersect. Therefore, the two equations cannot be solved simultaneously. The slopes of the two lines will indicate whether or not they are parallel.

In figure 12.6(b), the two lines intersect everywhere because they have the same solution set. It can be seen that the equation $6x - 4y = -8$ is equivalent to the equation $3x - 2y = -4$. If both sides of the equation are divided by 2, the result will be $3x - y = -4$. Thus, both equations have the same graph.

PROFESSIONAL PUBLICATIONS, INC. • P.O. Box 199, San Carlos, CA   94070

In figure 12.6(c), the two lines will intersect at a point. This point can be found by solving the equations simultaneously.

16 SOLVING SYSTEMS OF SIMULTANEOUS EQUATIONS

Several methods can be used to solve a system of equations. One method is known as the *method of reduction*.

Consider the equations $x + y = 9$ and $x - y = 3$. If we add the two equations, we get

$$\begin{array}{rl} x +y = & 9 \\ x -y = & 3 \\ \hline 2x = & 12 \\ x = & 6 \end{array}$$

We have reduced the set of equations to an equation of one variable. Substituting the value of x in either equation and solving for y,

$$y = 3$$

We can get the same results by subtracting one equation from the other,

$$\begin{array}{rl} x + y = 9 \\ x - y = 3 \\ \hline 2y = 6 \\ y = 3 \end{array}$$

Substituting,

$$x = 6$$

In the above example, the coefficient of x and the coefficient of y are the same, but this will not always be the case. If we consider the equations $3x + 2y = 5$ and $2x - 3y = 7$, we can see that adding or subtracting the two equations will not eliminate one of the unknowns as it did in the first example. However, one or both of the two equations can be converted into an equivalent equation which will make it possible to do so.

$$3x + 2y = 4$$
$$2x - 3y = 7$$

Multiplying the first equation by 3 and the second equation by 2 and adding will reduce the system of equations to a single variable equation.

$$\begin{array}{rl} 9x +6y = & 12 \\ 4x -6y = & 14 \\ \hline 13x = & 26 \\ x = & 2 \\ y = & -1 \text{ (obtained by substitution)} \end{array}$$

If the graphs of the two equations are plotted, the two lines will intersect at $(2, -1)$.

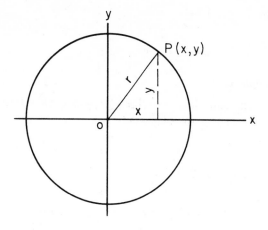

Figure 12.7   A circle centered at $(0,0)$

17 EQUATION OF A CIRCLE

A *circle* is a curve, all points on which are equidistant from a point called the *center*. The distance of all points from the center is known as the *radius*.

If the center of the circle is at the origin as in figure 12.7, the equation of the circle is

$$x^2 + y^2 = r^2 \qquad 12.18$$

If **P** is any point on the circle, its coordinates must satisfy the equation $x^2 + y^2 = r^2$.

If the center of the circle is at point $\mathbf{Q}(h, k)$, the equation becomes,

$$(x - h)^2 + (y - k)^2 = r^2 \qquad 12.19$$

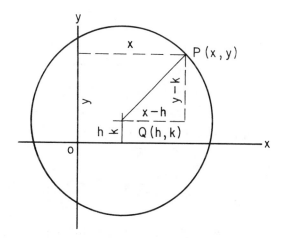

Figure 12.8   A circle centered at (h, k)

The *general form* for this equation is,

$$x^2 + y^2 + Dx + Ey + F = 0 \qquad 12.20$$

Example 12.12

Find the equation of the circle with center at $(2, -1)$ and radius 3.

PROFESSIONAL PUBLICATIONS, INC. • P.O. Box 199, San Carlos, CA 94070

Solution

$$(x-2)^2 + (y+1)^2 = 3^2$$
$$x^2 + y^2 - 4x + 2y - 4 = 0$$

Example 12.13

Write the equation $x^2 + 10x + y^2 - 6y + 18 = 0$ in the form $(x-h)^2 + (y-k)^2 = r^2$.

Solution

Completing the square,

$$x^2 + 10x + y^2 - 6y = -18$$
$$x^2 + 10x + 25 + y^2 - 6y + 9 = -18 + 25 + 9$$
$$(x+5)^2 + (y-3)^2 = 16$$

The center of the circle is at $(-5, 3)$ and the radius is 4.

18 LINEAR–QUADRATIC SYSTEMS

The intersections of a circle and a straight line can be found by solving the system of the linear equation and the quadratic equation. This is illustrated in example 12.14.

Example 12.14

Find the intersections of the graphs of the system,

$$x^2 + y^2 = 100$$
$$2x - y = 8$$

Solution

Transform the linear equation, by isolating one of the variables.

$$y = 2x - 8$$

Substitute this value of y into the quadratic equation,

$$x^2 + (2x-8)^2 = 100$$
$$x^2 + 4x^2 - 32x + 64 = 100$$
$$5x^2 - 32x - 36 = 0$$

Use the quadratic formulas to solve for x.

$$x = \frac{-B \pm \sqrt{B^2 - 4AC}}{2A}$$
$$x = \frac{-(-32) + \sqrt{-32^2 - (4)(5)(-36)}}{(2)(5)},$$
$$= \frac{-(-32) - \sqrt{-32^2 - (4)(5)(-36)}}{(2)(5)}$$
$$x = 7.38 \text{ and } x = -0.98$$

Substituting the values of x in the equation $y = 2x - 8$,

$$y = (2)(7.38) - 8 = 6.76$$

and

$$y = (2)(-0.98) - 8 = -9.96$$

Intersections: $(7.38, 6.76)$; $(-0.98, -9.96)$

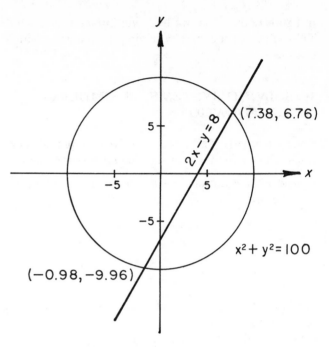

Figure 12.9　Example 12.14

19 INCLINATION OF A LINE

The *inclination of a line* not parallel to the x-axis is the angle measured counterclockwise from the positive direction of the x-axis. (The inclination of a line parallel to the x-axis is zero.) The symbol α denotes inclination.

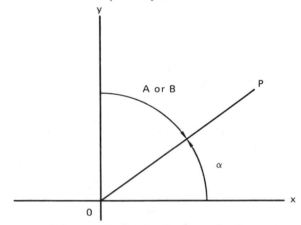

Figure 12.10　Inclination of a line

In figure 12.10, the inclination of the line **OP** is α. Considering the trigonometric ratios,

$$m = \frac{y}{x} = \tan \alpha \qquad 12.21$$

The *azimuth from north A (bearing angle B)* of a line calculated from the slope is the complement of inclination, and it can be calculated from the slope.

$$m = \cot A$$
$$= \cot B \qquad 12.22$$

20 THE ACUTE ANGLE BETWEEN TWO LINES

If the equations of two intersecting lines are known, the acute angle between them can be found by using the *law of tangents*.

$$\tan \theta = \tan(\alpha_2 - \alpha_i)$$

$$= \frac{\tan \alpha_2 - \tan \alpha_1}{1 + \tan \alpha_1 \tan \alpha_2}$$

$$= \left| \frac{m_2 - m_1}{1 + m_1 m_2} \right| \text{ for } m_1 m_2 \neq -1 \quad 12.23$$

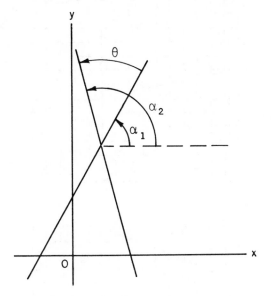

Figure 12.11   Angle between two lines

Example 12.15

Find the acute angle θ between the two lines

$$2x + 3y - 12 = 0$$
$$3x - 4y - 12 = 0$$

Solution

$$m_1 = -\frac{A}{B} = -\frac{2}{3}$$

$$m_2 = -\frac{A}{B} = +\frac{3}{4}$$

$$\tan \theta = \frac{\frac{3}{4} + \frac{2}{3}}{1 - \frac{2}{3} \times \frac{3}{4}} = 2.833$$

$$\theta = 70°30'$$

21 TRANSLATION OF AXES

Solving simultaneous equations in which the coefficients of x and y are large numbers can be simplified by reducing the value of the coefficients. This can be done without changing the values of the equations by translating the axes.

In figure 12.12, let **P** be any point with coordinates (x, y) with respect to the axes **OX** and **OY**. Establish the new axes, **O'X'** and **O'Y'** respectively, parallel to the old axes, so that the new origin **O'** has the coordinates (h, k) with respect to the old axes. The coordinates of the point **P** will then be (x', y') with respect to the new axes.

$$x = x' + h \qquad 12.24$$
$$x' = x - h \qquad 12.25$$
$$y = y' + k \qquad 12.26$$
$$y' = y - k \qquad 12.27$$

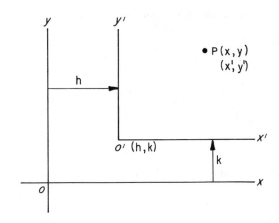

Figure 12.12   Transformation of axes

Example 12.16

The point **P** has the coordinates $(5, 3)$. Find the coordinates of **P** from the origin **O'**$(3, 1)$.

Solution

$$x' = x - h = 5 - 3 = 2$$
$$y' = y - k = 3 - 1 = 2$$

coordinates of **P** : $(2, 2)$

PROFESSIONAL PUBLICATIONS, INC. ● P.O. Box 199, San Carlos, CA 94070

PRACTICE PROBLEMS FOR CHAPTER 12

1. Find two sets of roots of each equation.

Example: $2x + y = 7$ $(2,3)$ $(3,1)$

(a) $2x - 3y = 5$

(b) $x - 2y = 4$

(c) $3x + 2y = 6$

2. Find the coordinates of the points at which the graph of each equation intersects the two axes.

Example: $3x - 2y = 12$ $(0,-6)$ $(4,0)$

(a) $x - y = 0$

(b) $2x + 3y = 18$

(c) $x + y = 4$

3. Determine the slope of the line through each pair of points.

Example: $(-2,4)$; $(4,-3)$: $m = \dfrac{-3-4}{4+2} = -\dfrac{7}{6}$

(a) $(3,2)$; $(6,8)$:

(b) $(1,3)$; $(4,5)$:

(c) $(0,-2)$; $(-3,5)$:

(d) $(6,-4)$; $(2,-3)$:

(e) $(-3,4)$; $(3,4)$:

(f) $(1,-5)$; $(-1,3)$:

4. Rearrange the equations in the form $Ax + By + C = 0$ with A positive and determine the slope m of each.

Example: $3y - 2x - 4 = 0$ $2x - 3y + 4 = 0$

$m = -\dfrac{A}{B} = -\dfrac{+2}{-3} = \dfrac{2}{3}$

(a) $3x + 4y = 6$

(b) $y = -2x + 5$

(c) $-4x + 2y + 8 = 0$

(d) $y = -5x$

5. Find the slope of each line. Express as a common fraction showing algebraic sign.

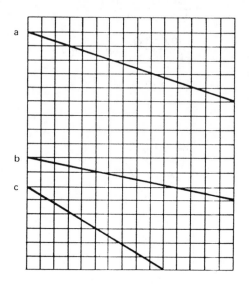

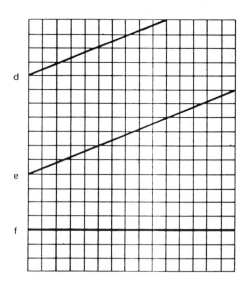

6. Graph each equation, plotting at least 3 points, and write the equation along the line.

(a) $3x - 2y = 0$

(b) $2x - 3y + 30 = 0$

(c) $3x + 2y + 6 = 0$

(d) $x + y = 0$

(e) $y + 11 = 0$

PROFESSIONAL PUBLICATIONS, INC. • P.O. Box 199, San Carlos, CA   94070

7. Write each equation in the form $y = mx + b$.

Example: $3x + 2y + 6 = 0$ $y = -\dfrac{3}{2}x - 3$

(a) $4x + 5y + 10 = 0$

(b) $2x - y - 10 = 0$

(c) $3x + 4y + 8 = 0$

(d) $x - y = 0$

(e) $2x - 3y - 12 = 0$

8. Write each equation in the form $Ax + By + C = 0$

Example: $y = -\dfrac{2}{3}x - 2$ $\underline{2x + 3y + 6 = 0}$

(a) $y = \dfrac{2}{5}x - 2$

(b) $y = 2x - 10$

(c) $y = -\dfrac{3}{2}x + \dfrac{5}{2}$

(d) $y = -x + 5$

(e) $y = -\dfrac{2}{3}x + 4$

9. Write each equation in the form $y = mx + b$ and plot the graph. Write the equation in the form $Ax + By + C = 0$ along each graph.

Example: $2x - 3y + 30 = 0$ $y = \dfrac{2}{3}x + 10$

(a) $x + 5y - 60 = 0$

(b) $x - y = 0$

(c) $3x + 4y + 24 = 0$

(d) $x + y + 4 = 0$

(e) $2x - 5y - 50 = 0$

10. Write each equation in the form $Ax + By + C = 0$ and indicate the slope, the x-intercept, and the y-intercept of each. (Hint: $m = -A/B$, x-intercept $= -C/A$ and y-intercept $= -C/B$.)

Example: $4y - 3x = 10$ $\underline{3x - 4y + 10 = 0}$

$m = 3/4$ $\underline{x_{y=0} = -10/3}$ $\underline{y_{x=0} = 5/2}$

(a) $x - y + 5 = 0$

(b) $-2x + 3y = 12$

(c) $x - y = 0$

(d) $6y = 4x + 2$

(e) $2x - 4y = 2$

11. Indicate which lines are parallel and which lines are perpendicular.

Example:

$3x + 4y = 8$

$m = -\dfrac{3}{4}$

PARALLEL TO EQUATION (h)

PERPENDICULAR TO EQUATION (f)

(a) $2x - 3y - 3 = 0$

(b) $10x - 6y = 18$

(c) $y = -\dfrac{3}{2}x + 5$

(d) $5x - 3y = 9$

(e) $6x + 8y = 12$

(f) $4x - 3y = 7$

(g) $3x + 5y = -10$

(h) $y = -\dfrac{3}{4}x - 3$

12. Find the perpendicular distance from the point **P** to the line indicated.

Example:

$$\mathbf{P}(-8, -1); \ 2x - 3y - 6 = 0$$

$$D = \frac{|(2)(-8) + (-3)(-1) + (-6)|}{\sqrt{2^2 + (-3)^2}}$$

$$= \frac{|-16 + 3 - 6|}{\sqrt{13}} = 5.3$$

(a) $\mathbf{P}(3, 3); \ 3x - 2y + 4 = 0$

(b) $\mathbf{P}(-10, 8); \ x - y + 5 = 0$

(c) $\mathbf{P}(9, 6); \ 5x - 2y + 10 = 0$

(d) $\mathbf{P}(-8, -6); \ 3x + 2y = 0$

(e) $\mathbf{P}(12, -6); \ 3x - 2y - 6 = 0$

13. Write the equation of the line through the given point with the given slope.

Example:

$(1, 4); \ m = -\dfrac{1}{2}$

$y - 4 = \dfrac{-1(x - 1)}{2}$ $2y - 8 = -x + 1$ $\underline{x + 2y - 9 = 0}$

(a) $(3, 1); \ m = -2$

(b) $(-4, 3); \ m = \dfrac{2}{3}$

(c) $(-2, 5); \ m = 0$

(d) $(2, -3); \ m = -\dfrac{2}{3}$

14. Write the equation of the line through the two given points in the form $Ax + By + C = 0$.

Example:

$$(-4, 3); (0, -2)$$
$$\frac{y - 3}{x + 4} = \frac{-2 - 3}{0 + 4}$$
$$\frac{y - 3}{x + 4} = \frac{-5}{4}$$
$$-5(x + 4) = 4(y - 3)$$
$$-5x - 20 = 4y - 12$$
$$5x + 4y + 8 = 0$$

(a) $(-3, 2); (1, 4)$

(b) $(2, -3); (5, -2)$

(c) $(3, 4); (-3, 4)$

(d) $(-2, -6); (3, -4)$

15. Write the equation of the lines with the given x and y intercepts in the form $Ax + By + C = 0$.

Example:

$$a = -3; \; b = 4$$
$$\frac{x}{-3} + \frac{y}{4} = 1$$
$$4x - 3y = -12$$
$$4x - 3y + 12 = 0$$

(a) $a = -4; \; b = 3$

(b) $a = 1; \; b = -4$

16. Write the equation of the line which has a slope of 3/4 and y-intercept -3.

17. Write the equation of the line whose y-intercept is 4 and which is perpendicular to the line $4x + 3y + 9 = 0$.

18. Write the equation of the line through the point $(0,8)$ and parallel to the line whose equation is $y = -3x + 4$.

19. Write the equations of two lines through the point $(5,5)$, one parallel and one perpendicular to the line $2x + y - 4 = 0$.

20. Write the equation of the line through the point $(-2, 1)$ and parallel to the line through the points $(1,4)$ and $(2, -3)$.

21. Solve the systems of simultaneous equations by addition or subtraction.

(a) $x + y = 8$
$\underline{x - y = 4}$

(b) $x + 2y = 6$
$\underline{x + 2y = 4}$

(c) $2x + 5y = -8$
$\underline{2x + 3y = 5}$

(d) $5x - 4y = -15$
$\underline{2x - 12y = 7}$

(e) $7x - 2y = 3$
$\underline{2x + 3y = 9}$

(f) $7x - 2y = -11$
$\underline{8x + 3y = -39}$

(g) $5x - 7y = 3$
$\underline{-3x + 6y = 4}$

(h) $5x - 4y = -17$
$\underline{2x - 12y = 14}$

(i) $9x + 10y = 9$
$\underline{6x - 25y = -13}$

22. Graph the systems of equations and find the intersection of the two lines in each system if the two lines do intersect.

(a) $4x + 3y = 24$
$4x - 3y = -48$

(b) $3x + 2y = 24$
$x - 2y = -11$

(c) $5x + 11y = -55$
$5x + 11y = -11$

(d) $2x + 4y = -48$
$x + 2y = -24$

23. Find the coordinates of the point of intersection of the diagonal lines CA and EB.

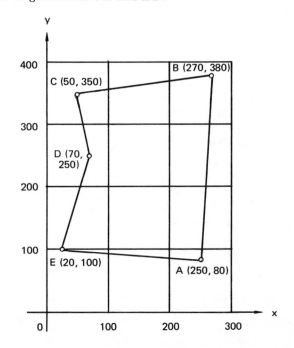

24. The figure $ABCDEA$ represents a tract of land which is to be subdivided by a line from D parallel to EA. Find the coordinates of the point of intersection of DH and AB. (Designate the point of intersection as H).

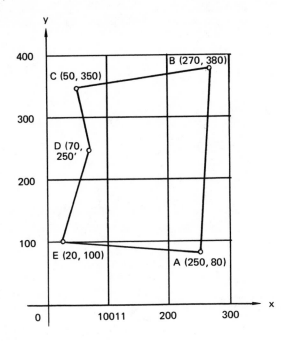

25. Find the acute angle θ between the two lines.

(a) $4x + 3y = \ \ \ 24$
$\ \ \ 4x - 3y = -48$

(b) $3x + 2y = \ \ \ 24$
$\ \ \ \ \ x - 2y = -11$

(c) $\ \ \ \ \ 2x - y = 5$
$\ \ \ \ \ 4x + y = 2$

(d) $2x + \ \ y = -\ 2$
$\ \ \ 6x - 5y = \ \ \ 18$

(e) $\ \ \ \ \ x - \ \ y = 5$
$\ \ \ \ \ x + 2y = 2$

26. Find the new coordinates if the axes are translated to the new origin located at (4,3).

(a) $(4,6)$

(b) $(-7,3)$

(c) $(-3,-2)$

(d) $(0,0)$

(e) $(8,-2)$

(f) $(7,7)$

13 THE TRAVERSE

1 INTRODUCTION

A *traverse* is a series of lines connecting successive instrument stations of a survey. The relative position of the stations is determined by the direction and length of the lines. In land surveys the lines are the boundaries of the land; in topographic surveys the lines are the control net to which physical features are tied. The traverse is also used as control for construction surveys.

2 OPEN TRAVERSE

An *open traverse* is a series of lines which do not return to the starting point. It is used in route surveying for the location of highways, pipelines, canals, etc. In order to check the accuracy of an open traverse, it must start and end at points of known position.

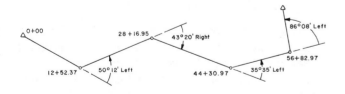

Figure 13.1   An open traverse

3 CLOSED TRAVERSE

The *closed traverse*, also called *loop traverse*, starts and ends at the same point. Because it is a closed polygon, the interior angles and the lengths of the sides may be checked for accuracy and mathematically adjusted.

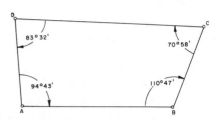

Figure 13.2   A closed traverse

4 HORIZONTAL ANGLES

Angles measured for open traverses are usually deflection angles as shown in figure 13.1. Angles measured for closed traverses are usually *interior angles*, as shown in figure 13.2, but they can be deflection angles as shown in figure 13.9.

Interior angles can be turned clockwise (right) or counterclockwise (left), but usually are turned by the method known as *angles to the right*.

5 DEFLECTION ANGLES

A *deflection* angle is an angle between a line and the extension of the preceding line, as shown in figure 13.1. It may be turned either right or left from the extension, but this direction of turning must be recorded with the angular measurement. In the open traverse, the straight lines between the points of change in direction are known as *tangents*, and the points of change in direction are known as *points of intersection*, abbreviated *PI*.

6 ANGLES TO THE RIGHT

The interior angles of the closed traverse in figure 13.2 are known as *angles to the right*. They are measured from the *backsight station* to the *foresight station* in a clockwise direction. With the instrument at station **A**, a backsight was made on station **D**, the telescope was turned in a clockwise direction to station **B**, and the angle 94°43′ was read and recorded. With the instrument at station **B**, the backsight was on station **A** and foresight was on station **C**. At each point, the angle was measured in a clockwise direction, or to the right.

The stations in the traverse **ABCDA** run in a counterclockwise direction, alphabetically, which is very appropriate for the angles to the right method. Most theodolites measure angles only to the right, which makes the counterclockwise lettering of the stations necessary. But it is not improper to letter traverse points in a clockwise direction.

7 DIRECTION OF SIDES

The direction of the sides of the traverse may be determined if the direction of one of the sides is known. If the direction of none of the sides is known, the direction of one side may be determined by measuring the angle at the intersection of this line and a line outside the traverse which has a known direction, or by making an observation on the sun or the stars. Otherwise, the direction of one of the lines must be assumed.

8 ANGLE CLOSURE

The sum of the interior angles of a polygon depends on the number of sides of the polygon. For a triangle, the sum is 180°; for a four-sided polygon, the sum is 360°. For any polygon with n sides, the sum is

$$(n-2)180° \qquad 13.1$$

After the interior angles of a traverse have been measured, they should be adjusted so that their sum agrees with equation 13.1. The error may be distributed evenly at each angle or it may be distributed arbitrarily in accordance with the surveyor's knowledge of the conditions of the survey. The error should be within the limits allowed in specifications for the survey. In land surveying, these specifications are usually based on the value of the land. Surveys of metropolitan areas are performed at much more rigid standards than surveys of arid ranch lands.

The interior angles of an arbitrary five-sided traverse are balanced here as an illustration.

angles as measured	
A	96°03′30″
B	95°19′30″
C	65°13′00″
D	216°19′30″
E	67°06′00″
	540°01′30″

balanced angles $(n-2)180° = 540°$	
A	96°03′
B	95°19′
C	65°13′
D	216°19′
E	67°06′
	540°00′

The angles in this traverse were measured with a one minute vernier, and the balanced angles reflect this accuracy. However, the total discrepancy of 90″ could be distributed evenly over the five angles by subtracting 18″ from each recorded angle.

A	96°03′12″
B	95°19′12″
C	65°12′42″
D	216°19′12″
E	67°05′42″
	540°00′00″

9 METHODS OF DESIGNATING DIRECTION

The *direction of a line* is expressed as the angle between a meridian and the line. The *meridian* may be a *true meridian* (which is a great circle of the earth passing through the poles), a *magnetic meridian* (the direction of which is defined by a compass needle), or a *grid meridian* which is established for a plane coordinate system.

The direction of a line may be expressed as its *bearing* or its *azimuth*. *True bearing* or *true azimuth* is measured from true north, referred to as *geodetic north*.

Old land surveys were usually referenced to the magnetic meridian. The direction of a line was determined by reading the angle between the line and the compass needle, but this method has long since been discarded for most surveys. Modern surveys are referred to geodetic north or grid north, but old surveys must be retraced. Therefore, an understanding of the magnetic meridian is essential to the land surveyor.

10 BEARING

The *bearing of a line* is the horizontal acute angle between the meridian and the line. Because the bearing of a line cannot exceed 90°, the full horizontal circle is divided into four *quadrants*: northeast, southeast, southwest, and northwest. An angle of 40°, measured

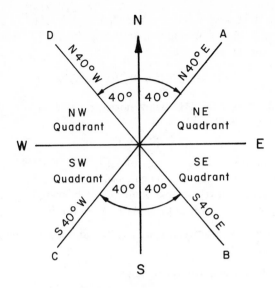

Figure 13.3 Bearing quadrants

between the meridian and a line in each of the four quadrants, is shown in figure 13.3.

Angles are measured from either the north or south, but are never measured from the east or west. The quadrants are designated in the bearing of a line by preceding the angle with North (N) or South (S) and following the angle with East (E) or West (W).

11 BACK BEARING

If the bearing of a line **AB** is N 65° E, the back bearing of **AB** is S 65° W. In other words, a man standing at **A** looking at **B** is looking northeast; if he stands at **B** and looks at **A**, he is looking southwest. The angle with the meridian is the same. The prefix is changed from N to S, and the suffix is changed from E to W. This is illustrated in figure 13.4

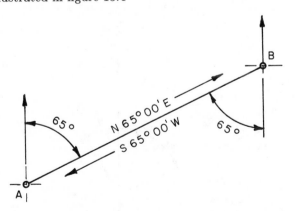

Figure 13.4 A back bearing

When two parallel lines are cut by a transversal, the alternate interior angles are equal. In figure 13.4, the

meridians through **A** and **B** are parallel lines. The line **AB** is the transversal, and the alternate interior angles are the bearing angles at **A** and **B**. The interior angle at **A** is the bearing angle of **AB**, and the interior angle at **B** is the bearing angle of **BA**. This also demonstrates that the bearing angle of **AB** is equal to the back bearing angle. Only the prefix and the suffix differ.

12 COMPUTATION OF BEARINGS OF A CLOSED TRAVERSE

Before the directions of the sides of a closed traverse are computed, it is essential that the interior angles of the traverse be adjusted so that their sum agrees with equation 13.1.

In figure 13.5, the bearing of **AB** is known, and the interior angles have been measured and adjusted. The bearings of the other sides are to be computed.

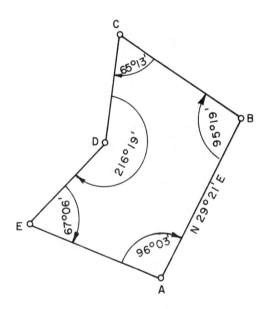

Figure 13.5 Interior angles of a closed traverse

The first step in computing the bearing of **BC** is to draw a sketch around point **B** showing:

- the meridian through **B**.

- the angle made by **BA** with the meridian. (The bearing of **BA** is the back bearing of **AB**, which is known.)

- the interior angle at **B** with its field measurement.

- the bearing angle of **BC**, which is the angle to be computed and by definition is the angle between the meridian and the line **BC**.

PROFESSIONAL PUBLICATIONS, INC. • P.O. Box 199, San Carlos, CA 94070

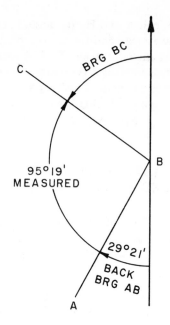

Figure 13.6 Angles about a point

A straight angle is an angle that equals 180°. The meridian makes an angle of 180° at point **B**. Therefore, the bearing angle of **BC** is

$$180° - (29°21' + 95°19') = 55°20'$$

From figure 13.6, **BC** bears northwest. Therefore, the bearing of **BC** is

N 55°20′ W

Computations for bearings of **CD**, **DE**, and **EA** are shown in figure 13.7. In each case, a meridian is drawn through the next traverse point after a bearing has been computed. At each point, three angles are identified:

- the angle between the meridian and the preceding side

- the measured interior angle

- the angle between the meridian and the succeeding side

Accuracy of the computations can be checked by calculating the bearing of **AB** (the known bearing), using the computed bearing of **EA**. The computed bearing of **AB** must be equal to the given bearing.

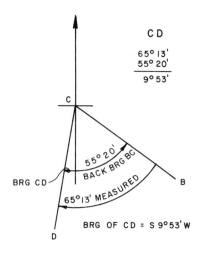

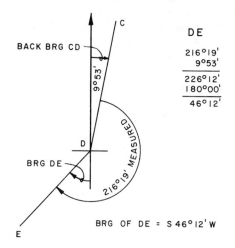

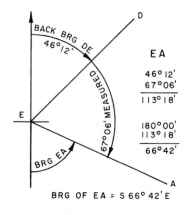

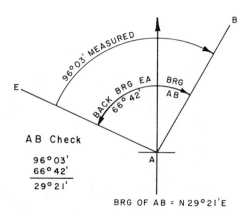

Figure 13.7 Bearing computations for traverse in figure 13.5

The importance of a sketch at each traverse point cannot be overemphasized. No set rule for computing bearings can be made, but a sketch which properly identifies the three angles mentioned makes a lengthy explanation unnecessary. The meridian through each traverse point always makes an angle of 180°, and the bearing angle is the angle between the meridian and the line (not between an east-west line and the line).

13 AZIMUTH

The *azimuth* of a line is the horizontal angle measured clockwise from the meridian. Azimuth is usually measured from the north, but the National Geodetic Survey (NGS) measures azimuth from the south. In this text, azimuth will be measured from the north.

Azimuths are not limited to 90°. Therefore, there is no need to divide the circle into quadrants. Azimuths vary from 0° to 360°. Figure 13.8 shows azimuths of 40°, 140°, 220°, and 320°.

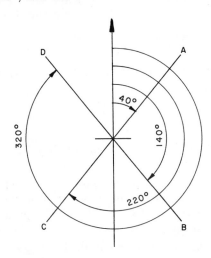

Figure 13.8 Representative azimuths

14 BACK AZIMUTH

As with bearing, the *back azimuth* of a line **AB** is the azimuth of line **BA**. The back azimuth of a line may be found by either adding 180° to the azimuth of the line or by subtracting 180° from the azimuth of the line. If the azimuth is less than 180°, add 180°; if the azimuth is more than 180°, subtract 180°.

Referring to figure 13.8,

- For azimuth 40°, back azimuth = 220°.
- For azimuth 140°, back azimuth = 320°.
- For azimuth 220°, back azimuth = 40°.
- For azimuth 320°, back azimuth = 140°.

15 CONVERTING BEARING TO AZIMUTH

Bearing is used to give the direction of a course in most land surveys. Azimuth is used in topographic surveys and some route surveys. Bearings may be converted to azimuth, and azimuths may be converted to bearings.

In converting bearing to azimuth in the northeast quadrant, the azimuth angle equals the bearing angle.

$$N\,76°30'\,E = 76°30'$$

In the southeast quadrant, the azimuth angle equals 180° minus the bearing angle.

$$S\,42°28'\,E = 180° - 42°28' = 137°32'$$

In the southwest quadrant, 180° is added to the bearing angle.

$$S\,36°47'\,W = 180° + 36°47' = 216°47'$$

In the northwest quadrant, the bearing angle is subtracted from 360°.

$$N\,62°56'\,W = 360° - 62°56' = 297°04'$$

16 CONVERTING AZIMUTH TO BEARING

In the northeast quadrant, the bearing angle equals the azimuth. The prefix N and the suffix E must be added.

In the southeast quadrant, the azimuth is subtracted from 180° and the prefix S and suffix E are added.

$$168°40' = 180° - 168°40' = S\,11°20'\,E$$

In the southwest quadrant, 180° is subtracted from the azimuth.

$$195°22' = 195°22' - 180° = S\,15°22'\,W$$

In the northwest quadrant, the azimuth is subtracted from 360°.

$$314°35' = 360° - 314°35' = N\,45°25'\,W$$

17 CLOSED DEFLECTION ANGLE TRAVERSE

In a closed deflection angle traverse (figure 13.9), the difference between the sum of the right deflection angles and the sum of the left deflection angles is 360°. Before bearings are computed, deflection angles must be adjusted.

If, during the adjustment of angles, it is found that the sum of the right deflection angles is greater than the sum of the left deflection angles, the sum of the right deflection angles must be reduced and the sum of the left deflection angles must be increased. The correction may be distributed arbitrarily or evenly.

Bearings of the sides of the traverse are computed in much the same manner as with the interior angle traverse, with a sketch drawn at each traverse point showing the angles involved in the computation.

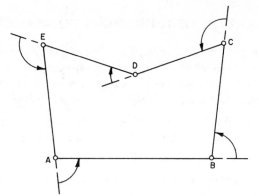

Figure 13.9   A closed deflection angle traverse

18 ANGLE TO THE RIGHT TRAVERSE

Open or closed traverses can be run by the *angle to the right method*. All angles are measured from the backsight to the foresight in a clockwise direction, as shown in figure 13.10.

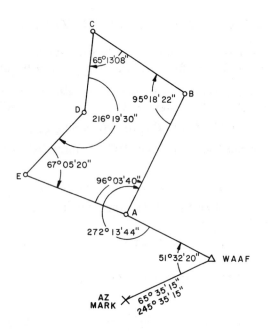

Figure 13.10   An angle to the right traverse

To determine the forward azimuth from a traverse point, the angle to the right is added to the back azimuth of the preceding line. In other words, the angle to the right is added to the azimuth of the preceding line ±180°. (It is sometimes necessary to subtract 360°.)

Traverse **ABCDEA** in figure 13.10 is tied to triangulation station **WAAF** for direction. National Geodetic Survey data shows the azimuth to the azimuth mark from station **WAAF** to be 65°35′15″ (from the south). Converted to azimuth from the north, this azimuth is

180°+65°35′15″ = 245°35′15″. Angles to the right, adjusted, are shown in figure 13.10. Computations for the azimuths of the traverse courses are shown in table 13.1.

Table 13.1
Computation of azimuth

245°35′15″ Az **WAAF-Mark**	124°39′41″ Az **CB**
51°32′20″ angle right	65°13′08″ angle right
297°07′35″	189°52′49″ Az **CD**
−180°	−180°
117°07′35″ Az **A-WAAF**	9°52′49″ Az **DC**
272°13′44″ angle right	216°19′30″ angle right
389°21′19″	226°12′19″ Az **DE**
−360°	−180°
29°21′19″ Az **AB**	46°12′19″ Az **ED**
+180°	67°05′20″ angle right
209°21′19″ Az **BA**	113°17′39″ Az **EA**
95°18′22″ angle right	+180°
304°39′41″ Az **BC**	293°17′39″ Az **AE**
−180°	96°03′40″ angle right
124°39′41″ Az **CB**	389°21′19″
	−360°

check   29°21′19″ Az **AB**

19 LATITUDES AND DEPARTURES

Latitudes and departures are similar in concept to the projections of a line. The projection of a line can be compared to the shadow of a line. The projection of a line can be compared to the shadow of a building. When the sun is nearly overhead, the shadow is short; when the sun is sinking in the west, the shadow becomes long. The height has not changed, but the length of its shadow has changed.

In figure 13.11, the line **AB** is projected on the *y*-axis of a rectangular coordinate system by dropping perpendiculars from **A** to the *y*-axis and from **B** to the *y*-axis. The interval between these two perpendiculars along the *y*-axis is the projection of **AB** on the *y*-axis. As **AB** changes its position relative to the *y*-axis, as shown in figure 13.11, the length of the projection becomes longer or shorter. As the position of **AB** nears the vertical, the projection nears the length of the line. As the projection of **AB** nears the horizontal, the projection of **AB** becomes very short.

In figure 13.12, **AB** is projected on the *x*-axis. As in figure 13.11, the length of the projection of **AB** on the *x*-axis changes as the position of **AB** changes relative to the *x*-axis.

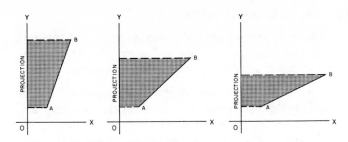

Figure 13.11   Projections on the y-axis

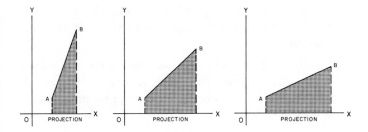

Figure 13.12   Projections on the x-axis

In surveying, the projection of a side of a traverse on the north-south (y) axis of a rectangular coordinate system is known as its *latitude*. The projection on the east-west (x) axis is known as its *departure*.

Figure 13.13 shows the projection of a line **AB** on the y- and x-axes. The latitude of line **AB** is the length of the right angle projection of **AB** on a meridian. The departure of line **AB** is the length of the right angle projection of line **AB** on a line perpendicular to the meridian, an east-west line.

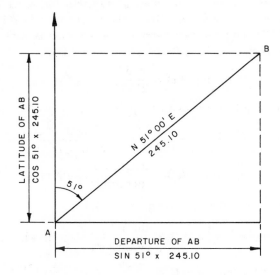

Figure 13.13   Latitude and departure

The *latitude of a course* is equal to the cosine of its bearing angle multiplied by its length. The *departure of a course* is equal to the sine of its bearing angle multiplied by its length.

$$\text{latitude} = \cos \text{bearing} \times \text{length} \qquad 13.2$$
$$\text{departure} = \sin \text{bearing} \times \text{length} \qquad 13.3$$

The latitude of the course **AB** in figure 13.13 is

$$\text{latitude} = \cos 51°00' \times 245.10 = 154.25$$

The departure of the course **AB** is

$$\text{departure} = \sin 51°00' \times 245.10 = 190.48$$

Figure 13.14 shows the latitude and departure for each of the courses in the traverse **ABCDEA**. Course **AB** has a north latitude and an east departure, **BC** has a south latitude and east departure, **CD** has a south latitude and west departure, **DE** has a south latitude and west departure, and **EA** has a north latitude and west departure.

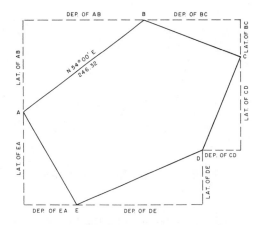

Figure 13.14   Latitudes and departures
of traverse legs

20   ERROR OF CLOSURE

If the traverse **ABCDEA** in figure 13.14 starts at point **A** and ends at point **A**, it is obvious that the distance traversed north equals the distance traversed south. Likewise, the distance traversed east equals the distance traversed west. In other words, the sum of the north latitudes must equal the sum of the south latitudes; the sum of the east departures must equal the sum of the west departures.

However, because measurements are not exact due to human and instrument errors, these sums will not be equal and their differences may be used to determine the *linear error of closure* due to errors caused by angular and linear measurements. The actual error for each course cannot be determined, but the total error can be distributed over the entire length of the traverse so that north latitudes equal south latitudes and east departures equal west departures. This is called *balancing the traverse* or *closing the traverse*.

The *error of closure* of a traverse is a measure of the precision of a survey.

Figure 13.15 shows a traverse which does not close. The point **A** is the *point of beginning*. Point **A′** is the *ending point*, found by plotting the latitude and departure of each course before the traverse is balanced. The distance from **A′** to **A** is the error of closure. For the traverse to close, **A′** would move in the direction of **A′A**. Balancing the traverse does just that—it makes **A′** coincide with **A**.

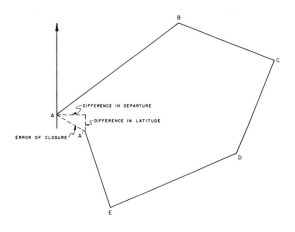

Figure 13.15   Error of closure

The error of closure is found by the Pythagorean theorem.

$$\text{error of closure} = \sqrt{(\text{dif in lat})^2 + (\text{dif in dep})^2} \quad 13.4$$

The direction of the line **A′A** is found by determining the angle it makes with the meridian. The tangent of this bearing angle is equal to the difference in departures divided by the difference in latitudes.

$$\tan \text{ bearing } \mathbf{A'A} = \frac{\text{dif in dep}}{\text{dif in lat}} \quad 13.5$$

21 BALANCING THE TRAVERSE

The traverse can be balanced by using one of several methods:

- least squares adjustment
- the compass rule
- the transit rule
- the Crandall method

The *least squares method* is adaptable to any traverse, whether angular accuracy is higher, equal to, or less than linear accuracy. It is difficult and seldom used.

The *compass rule*, also known as the *Bowditch rule*, is adaptable to traverses in which angular accuracy and linear accuracy are about the same. In this type of traverse, the compass rule will give very nearly the same results as the least squares adjustment. It is used in a great majority of traverse closures.

The *transit rule* is adaptable to traverses in which angular accuracy is much higher than linear accuracy. It is seldom used.

The *Crandall method* is also used where angular accuracy is much higher than linear accuracy. It is more accurate than the transit rule but requires more time. It, also, is seldom used.

22 THE COMPASS RULE

Using the compass rule, the difference in the sums of the north and south latitudes is distributed over the latitudes of the traverse. A correction is made in the latitude of each side to bring the north and south latitudes into balance. The difference in the sums of the east and west departures is distributed in the same way.

The correction to be applied to the latitude of each side is a fraction of the total difference in the north and south latitudes. The fraction is the ratio of the length of each side to the perimeter of the entire traverse.

As an example, if the difference in north and south latitudes of a traverse is 0.86, the length of side **AB** is 356.73, and the perimeter of the traverse is 2156.78, the correction in latitude for the side **AB** is

$$\text{correction } \mathbf{AB} = \frac{356.73}{2156.78} \times 0.86 = 0.14$$

This can be expressed as a proportion:

$$\frac{\text{correction for lat of } \mathbf{AB}}{\text{dif in N-S latitudes}} = \frac{\text{length of } \mathbf{AB}}{\text{traverse perimeter}} \quad 13.6$$

$$\text{correction for } \mathbf{AB} = \frac{\text{length } \mathbf{AB}}{\text{perimeter}} \times \text{dif in lat} \quad 13.7$$

This can be written as equation 13.8.

$$\text{correction for } \mathbf{AB} = \frac{\text{dif in lat}}{\text{perimeter}} \times \text{length } \mathbf{AB} \quad 13.8$$

$$\text{correction for } \mathbf{AB} = \frac{0.86}{2156.78} \times 356.73 = 0.14 \, \text{ft}$$

Equation 13.8 is more efficient because 0.86/2156.78 is a constant which can be applied to the other sides.

After the corrections are made in the latitude and departure for each side, the traverse is balanced.

23 RATIO OF ERROR

The *ratio of error*, or *precision*, of a traverse is the ratio of the error of closure to the perimeter. It is expressed with the numerator as one (1) and the denominator in round numbers. It is a measure of the precision of a traverse.

If the error of closure of a traverse is 0.76 and the perimeter is 5214.75, the ratio of error is

$$\frac{0.76}{5214.75}$$

Dividing numerator and denominator by 0.76 (in order to have 1 in the numerator), and rounding off,

$$\text{ratio of error} = \frac{1}{6900}$$

This indicates an error of 1 foot per 6900 feet in distance.

24 SUMMARY OF COMPUTATIONS FOR BALANCING A TRAVERSE

The steps in balancing a traverse are summarized here.

Step 1. Compute the angular error and adjust to make the sum of the angles agree with equation 13.1.

Step 2. Compute the bearings for each course.

Step 3. Compute the latitudes and departures.

Step 4. Compute the error of closure.

Step 5. Compute the ratio of error.

Step 6. Compute the latitude and departure corrections for each course.

Step 7. Adjust the latitudes and departures.

Example 13.1

Given the traverse **ABCDEA** shown in figure 13.16 with angles adjusted, balance the traverse using the compass rule, and compute the error of closure and ratio of error.

Solution

Table 13.2
Example 13.1

pt	bearing	distance	cosine sine	N	S	cor	E	W	cor	N	S	E	W
A			.597625										
	S53°18′E	660.27	.801776		334.83	−.10	449.21		−.06		334.73	449.16	
B			.743145										
	N42°00′E	484.18	.669131	359.81		+.08	323.98		−.05	359.89		323.93	
C			.295986										
	N72°47′W	375.42	.955192	111.12		+.06		358.60	+.04	111.18			358.64
D			.804376										
	N36°27′W	311.44	.594121	250.52		+.05		185.03	+.04	250.57			185.07
E			.860298										
	S30°39′W	449.83	.509792		386.99	−.08		229.32	+.05		386.91		229.37
A		2181.14		721.45	721.82		773.19	772.95		721.64	721.64	773.08	773.08
					721.45		772.95						
					.37		.24						

$$\text{error of closure} = \sqrt{.37^2 + .24^2} = .44$$
$$\text{ratio of error} = .44/2181 = 1/5000$$

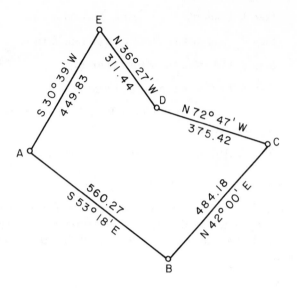

Figure 13.16   Example 13.1

For each course, the cosine of the bearing angle times the distance has been recorded in one of the two latitude columns. Latitudes of courses with a north prefix are recorded under N; latitudes of courses with a south prefix are recorded under S.

Likewise, the sine of the bearing angle times the distance has been recorded in one of the two departure columns. Departures of courses with an east suffix are recorded under E; departures of courses with a west suffix are recorded under W.

Next, the N, S, E, and W columns have been added and the totals recorded. The smaller of the two latitude totals has been subtracted from the larger total, and the smaller of the departure totals has been subtracted from the larger.

The error of closure is the square root of the sum of the squares of these two numbers.

Table 13.3
Example 13.1

course	latitude corrections	departure corrections
AB	$\dfrac{0.37}{2181} \times 560.27 = -0.10$	$\dfrac{0.24}{2181} \times 560.27 = -0.06$
BC	$\dfrac{0.37}{2181} \times 484.18 = +0.08$	$\dfrac{0.24}{2181} \times 484.18 = -0.05$
CD	$\dfrac{0.37}{2181} \times 375.42 = +0.06$	$\dfrac{0.24}{2181} \times 375.42 = +0.04$
DE	$\dfrac{0.37}{2181} \times 311.44 = +0.05$	$\dfrac{0.24}{2181} \times 311.44 = +0.04$
EA	$\dfrac{0.37}{2181} \times 449.83 = -0.08$	$\dfrac{0.24}{2181} \times 449.83 = +0.05$

The ratio of error, or precision, is the ratio of the error of closure to the perimeter. Ratio of error, in this case, indicates that for every 5,000 feet measured, an error of one foot was made.

To balance the traverse, the difference in the north and south latitude totals has been prorated to the latitudes of each course. A correction is made for each latitude in proportion to the ratio of the length of the course to the perimeter. Computations for corrections to latitudes and departures are shown in table 13.3.

In table 13.2, the sum of the north latitudes is smaller than the sum of the south latitudes. Therefore, corrections for north latitudes are positive and are added to the computed latitudes. Corrections for south latitudes are negative and are subtracted from the computed latitudes.

The sum of the west departures is smaller than the sum of the east departures. Therefore, corrections for west departures were added, and corrections for east departures were subtracted. After corrections were made, north latitudes equal south latitudes, and east departures equal west departures. The traverse is balanced.

It should be noted that the sum of the correction column equals the difference in the sums of the north and south latitudes. Likewise, the departure corrections equal the difference in east and west departures.

25 COORDINATES

After latitudes and departure have been computed, coordinates of the traverse points are easily computed. Coordinates for one of the points may be known, or they can be assumed. If coordinates are assumed, they should be large enough so that no negative coordinates will occur.

Coordinates of a point are computed by adding a north latitude to or subtracting a south latitude from the y-coordinate of the preceding point, and by adding an east departure to or by subtracting a west departure from the x-coordinate of the preceding point.

Example 13.2

Latitudes and departures for the traverse **ABCDEA** have been computed and balanced as shown in table 13.4. Assume the coordinates of point **A** to be $y = 1000.00$, $x = 1000.00$. Compute the coordinates of points **B**, **C**, **D**, and **E**. Check arithmetic by computing coordinates of point **A** from coordinates of point **E**.

Solution

Table 13.4
Example 13.2

point	latitude north	latitude south	departure east	departure west	coordinates y	coordinates x
A					1000.00	1000.00
		334.73	449.15		−334.73	+449.15
B					665.27	1449.15
	359.89		323.93		+359.89	+323.93
C					1025.16	1773.08
	111.18			358.64	+111.18	−358.64
D					1136.34	1414.44
	250.57			185.07	+250.57	−185.07
E					1386.91	1229.37
		386.91		229.37	−386.91	−229.37
A					1000.00	1000.00
	721.64	721.64	773.08	773.08		

In the solution, north latitudes are given a positive sign and south latitudes are given a negative sign; east departures are given a positive sign and west departures are given a negative sign. y-coordinates are associated with latitude; x-coordinates are associated with departure. Also, the cosine function is associated with latitude; the sine function is associated with departure.

26 FINDING BEARING AND LENGTH OF A LINE FROM COORDINATES

It is often necessary to find the bearing and length of a line between two points of known coordinates. Figure 13.17 illustrates that the tangent of the bearing angle can be determined from the latitude and departure.

$$\text{tangent of bearing angle} = \frac{\text{departure}}{\text{latitude}} \qquad 13.9$$

$$\text{tangent of bearing angle} = \frac{\text{dif in } x}{\text{dif in } y} \qquad 13.10$$

$$\begin{aligned} \text{length} &= \sqrt{\text{dep}^2 + \text{lat}^2} \\ &= \sqrt{(\text{dif in } x)^2 + (\text{dif in } y)^2} \\ &= \frac{\text{latitude}}{\cos \text{brg}} \\ &= \frac{\text{departure}}{\sin \text{brg}} \qquad 13.11 \end{aligned}$$

It is also true that

$$\text{cotangent of bearing angle} = \frac{\text{lat}}{\text{dep}} = \frac{\text{dif in } y}{\text{dif in } x} \qquad 13.12$$

For large angles, equation 13.12 is recommended.

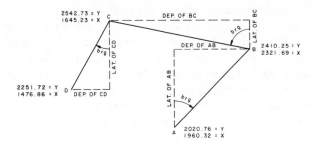

Figure 13.17 Bearing and length from coordinates

Example 13.3

Using the coordinates of points **A**, **B**, **C**, and **D** as shown in figure 13.17, find the bearing and lengths of **AB**, **BC**, and **CD**.

Solution

$$\tan \text{brg } \mathbf{AB} = \frac{2321.69 - 1960.32}{2410.25 - 2020.76}$$

$$\text{brg } \mathbf{AB} = \text{N } 42°51' \text{ E}$$

AB is northerly because the y-coordinate of **B** is greater than the y-coordinate of **A**. **AB** is easterly because the x-coordinate of **B** is greater than the y-coordinate of **A**.

$$\begin{aligned} \mathbf{AB} &= \sqrt{(2321.69 - 1960.32)^2 + (2410.25 - 2020.76)^2} \\ &= 531.31' \end{aligned}$$

$$\tan \text{brg } \mathbf{BC} = \frac{2321.69 - 1645.23}{2542.73 - 2410.25}$$

$$\text{brg } \mathbf{BC} = \text{N } 78°55' \text{ W}$$

$$\begin{aligned} \mathbf{BC} &= \sqrt{(2321.69 - 1645.23)^2 + (2542.73 - 2410.25)^2} \\ &= 689.31' \end{aligned}$$

$$\tan \text{brg } \mathbf{CD} = \frac{1645.23 - 1476.86}{2542.73 - 2251.72}$$

$$\text{brg } \mathbf{CD} = \text{S } 30°03' \text{ W}$$

$$\begin{aligned} \mathbf{CD} &= \sqrt{(1645.23 - 1476.86)^2 + (2542.73 - 2251.72)^2} \\ &= 336.21' \end{aligned}$$

27 COMPUTING TRAVERSES WHERE TRAVERSE POINTS ARE OBSTRUCTED

In land surveying, it is common to find boundary corners occupied by fence posts or other obstructions, making it impractical to retrace courses as they were orginally run. In such cases, a traverse can be run very near the original one, and ties can be made to the original corners. The adjacent traverse can be closed, and the original survey can be computed. By making the points on the adjacent traverse very close to the corresponding points on the original survey, error in measurement of the ties is lessened. However, error of closure cannot be computed for the original survey.

The direction and distance from an adjacent point to corresponding original corner can be determined by measurement, and from this information, latitude and departure for this tie can be computed. Knowing the coordinates of each point in the adjacent traverse, the coordinates of original corners can be determined by using these coordinates and the latitudes and departures of each of the ties to the original corners.

Knowing the coordinates of the original corners, the bearing and length of each course of the original survey can be computed.

$$\tan \mathrm{brg} = \frac{\mathrm{dif\ in\ } x}{\mathrm{dif\ in\ } y} \qquad 13.13$$

$$\mathrm{length} = \sqrt{(\mathrm{dif\ in\ } x)^2 + (\mathrm{dif\ in\ } y)^2} \qquad 13.14$$

Example 13.4

The traverse **ABCDEA** has been run inside the original survey represented by the traverse **MNOPQM** as shown in figure 13.18. The traverse **ABCDEA** has been closed, and balanced latitudes and departures are shown. Ties to the original survey have been made from corresponding points on the adjacent traverse, and azimuth and distance for each tie are shown. Coordinates of the point **A** are assumed to be $x = 1000.00$, $y = 1000.00$.

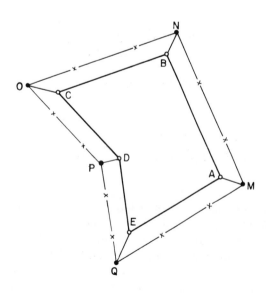

Figure 13.18   Example 13.4

Traverse **ABCDEA**

line	bearing	length	balanced north	balanced south	balanced east	west
AB	N25°45′W	560.27	504.60			243.45
BC	S 69°33′W	484.18		169.19		453.70
CD	S 45°14′E	375.42		264.40	266.51	
DE	S 08°54′E	311.44		307.70	48.16	
EA	N58°11′E	449.83	236.69		382.48	
			741.29	741.29	697.15	697.15

Corner ties

line	azimuth	length
AM	106°20′	58.30
BN	33°52′	64.65
CO	271°59′	71.22
DP	256°11′	72.30
EQ	203°47′	77.35

Solution

The coordinates of points **B**, **C**, **D**, and **E** are computed in table 13.5.

Table 13.5
Example 13.4

point	latitude north	south	departure east	west	coordinates y	x
A					1000.00	1000.00
	504.60			243.45		
B					1504.60	756.55
		169.19		453.70		
C					1335.41	302.85
		264.40	266.51			
D					1071.01	569.36
		307.70	48.16			
E					763.31	617.52
	236.69		382.48			
A					1000.00	1000.00

After the coordinates of the traverse points of **ABCDEA** are computed, latitudes and departures of the ties to each survey corner are determined. Then, the coordinates of each survey corner are computed. Directions of the ties have been recorded in azimuth, so these azimuths need to be converted to bearings. The tie from traverse point **A** to survey corner **M** is shown in figure

13.19. The coordinates of **M** are

$$\text{brg } \mathbf{AM} = 180° - 106°20' = \text{S } 73°40' \text{ E}$$
$$\text{latitude } \mathbf{AM} = 58.30 \cos 73°40' = 16.40 \text{ ft}$$
$$\text{departure } \mathbf{AM} = 58.30 \sin 73°40' = 55.95 \text{ ft}$$
$$y\text{-coordinate of } \mathbf{M} = 1000.00 - 16.40 = 983.60$$
$$x\text{-coordinate of } \mathbf{M} = 1000.00 + 55.95 = 1055.95$$

Computations of the coordinates of **N**, **O**, **P**, and **Q** are performed in a similar manner. Tabulations showing data used in these computations is shown in table 13.6.

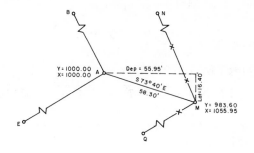

Figure 13.19   Example 13.4

Table 13.6
Computation of coordinates for Example 13.4

point	bearing	length	latitude north	latitude south	departure east	departure west	coordinates y	coordinates x
A							1000.00	1000.00
	S 73°40′ E	58.30		16.40	55.95			
M							983.60	1055.95
B							1504.60	756.55
	N 33°52′ E	64.65	53.68		36.03			
N							1558.28	792.58
C							1335.41	302.85
	N 88°01′ W	71.22	2.46			71.18		
O							1337.87	231.67
D							1071.01	569.36
	S 76°11′ W	72.30		17.27		70.21		
P							1053.74	499.15
E							763.31	617.52
	S 23°47′ W	77.35		70.78		31.19		
Q							692.53	586.33

Computation of traverse **MNOPQM**

point	bearing	length	latitude north	latitude south	departure east	departure west	coordinates y	coordinates x
M							983.60	1055.95
	N 24°37′ W	632.16	574.68			263.37		
N							1558.28	792.58
	S 68°33′ W	602.66		220.41		560.91		
O							1337.87	231.67
	S 43°16′ E	390.22		284.13	267.48			
P							1053.74	499.15
	S 13°34′ E	371.58		361.21	87.18			
Q							692.53	586.33
	N 58°12′ E	552.51	291.07		469.62			
M							983.60	1055.95
			865.75	865.75	824.28	824.28		

With the coordinates of **M** and **N** known, the bearing and length of **MN** are computed:

$$\tan \text{brg}\,\mathbf{MN} = \frac{1055.95 - 792.58}{1558.28 - 983.60}$$

$$\text{brg}\,\mathbf{MN} = \text{N}\,24°37'\,\text{W}$$

$$\text{length}\,\mathbf{MN} = \sqrt{(263.37)^2 + (574.68)^2} = 632.16\,\text{ft}$$

Tabulations of bearings and lengths of **NO**, **OP**, **PQ**, and **QM** are shown in table 13.6.

Example 13.5

The tract of land represented by the figure **MNOPQM** in figure 13.20 has been resurveyed. It is completely enclosed by a fence, so that corners could not be occupied. The traverse **ABCDEA** was run as close to the survey as possible. The point **A** is a monument of known position, and the azimuth from **A** to monument **X** is known to be 24°10′30″. The traverse was tied to this line for direction. All angles were turned to the right, beginning with the instrument at **A** with backsight on **X**, using the known azimuth. The coordinates of **A** are known to be $y = 1276.28$, $x = 1533.45$, and coordinates of the traverse points and the original survey corners were referred to the coordinates of **A**. The interior angle at each point of the traverse and the angle to the survey corner were turned from the same instrument setup in the order shown in the field notes. Horizontal distances were also measured in the sequence shown in the notes. Field notes are as shown.

Field notes

𝜋 sta	angle rt	measured angle	adjusted angle	distance from-to	dist
A	X-M	270°00′30″		A-M	85.20
A	X-B	342°15′30″		A-B	986.10
A	E-B	85°16′30″	85°16′00″		
B	A-C	104°42′00″	104°42′00″	B-N	72.67
B	A-N	232°10′00″		B-C	930.01
C	B-D	57°37′00″	57°37′00″	C-O	81.31
C	B-O	315°30′00″		C-D	690.70
D	C-P	98°30′00″		D-P	70.54
D	C-E	210°55′00″	210°55′00″	D-E	510.22
E	D-A	81°30′00″	81°30′00″	E-Q	77.80
E	D-Q	225°10′00″		E-A	809.00
			540°00′00″		

Solution

The interior angles were adjusted and recorded in the field notes. Using the azimuth of **A-X**, azimuths of the traverse sides and corner ties were computed and converted to bearings. Computations are shown in tables

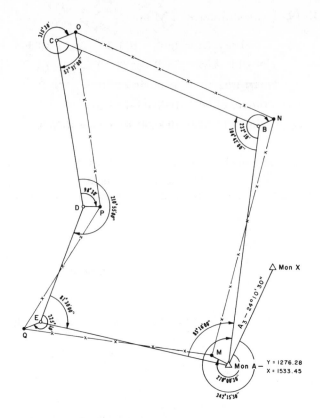

Figure 13.20   Example 13.5

13.7 and 13.8. Traverse closure and coordinate computations for the traverse **ABCDEA** are shown in table 13.9. Computations for coordinates of **M**, **N**, **O**, **P**, and **Q**, and for bearings and lengths of **MN**, **NO**, **OP**, and **PQ** are shown in table 13.10.

Table 13.7
Traverse bearing computations for Example 13.5

point	angle right	azimuth	bearing
X			
		24°10′30″	
A	342°15′30″		
		6°26′00″	N06°26′ E
B	104°42′00″		
		291°08′00″	N68°52′ W
C	57°37′00″		
		168°45′00″	S11°15′ E
D	210°55′00″		
		199°40′00″	S19°40′ W
E	81°30′00″		
		101°10′00″	S78°50′ E
A	85°16′00″		
		6°26′00″	Check
B			

Table 13.8
Corner tie bearing computation for Example 13.5

point	angle right	azimuth	bearing
X			
		24°10′30″	
A	270°00′30″		
		294°11′00″	N 65°49′ W
M			
A			
		6°26′00″	
B	232°10′00″		
		58°36′00″	N 58°36′ E
N			
B			
		291°08′00″	
C	315°30′00″		
		66°38′00″	N 66°38′ E
O			
C			
		168°45′00″	
D	98°30′00″		
		87°15′00″	N 87°15′ E
P			
D			
		199°40′00″	
E	225°10′00″		
		244°50′00″	S 64°50′ W
Q			

28 LATITUDES AND DEPARTURES USING AZIMUTH

In the preceding example, it was not necessary to convert azimuth to bearing; latitudes and departures can be found from azimuth with calculators which will give the algebraic sign of the latitudes and departures.

Example 13.6

Given the azimuths and lengths of the sides of the traverse **ABCDEA**, as shown tabulated, compute latitudes and departures, balance the traverse, and compute error of closure and precision.

line	azimuth	length
AB	29°21′23″	560.06
BC	304°39′45″	484.14
CD	189°52′53″	375.48
DE	226°12′23″	311.52
EA	113°17′43″	449.79

Table 13.9
Traverse computation for example 13.5

point	bearing	length	latitude N	latitude S	cor	departure E	departure W	cor	latitude N	latitude S	departure E	departure W	coordinates y	coordinates x
A													1276.28	1533.45
	N06°26′ E	986.10	979.89		−.16	110.49		+.06	979.73		110.55			
B													2256.01	1644.00
	N68°52′ W	930.01	335.81		−.15		867.46	−.06	335.16			867.40		
C													2591.17	776.60
	S11°15′ E	690.70		677.43	+.11	134.75		+.04		677.54	134.79			
D													1913.63	911.39
	S19°40′ W	510.22		480.46	+.08		171.71	−.03		480.54		171.68		
E													1433.09	739.71
	S78°50′ E	809.00		156.67	+.14	793.68		+.06		156.81	793.74			
A													1276.28	1533.45
		3926.03	1315.20	1314.56		1038.92	1039.17		1314.89	1314.89	1039.08	1039.08		
			1314.56				1038.92							
			0.64				0.25							

$$\text{error of closure} = \sqrt{0.64^2 + 0.25^2} = 0.69$$

$$\text{precision} = \frac{0.69}{3926.03} = \frac{1}{5700}$$

Solution

Table 13.11
Example 13.6

line	azimuth	length	latitude	departure	balanced latitude	balanced departure
AB	29°21′23″	560.06	+484.14	+274.56	+488.11	+274.52
BC	304°39′45″	484.14	+275.35	−398.21	+275.33	−398.25
CD	189°52′53″	375.48	−369.91	−64.44	−369.93	−64.47
DE	226°12′23″	311.52	−215.59	−224.87	−215.61	−224.89
EA	113°17′43″	449.79	−177.88	+413.12	−177.90	+413.09
		2180.99	+0.11	+0.16	0.00	0.00

$$\text{error of closure} = \sqrt{0.11^2 + 0.16^2} = 0.19$$
$$\text{precision} = 0.19/2180.99 = 1/11,500$$

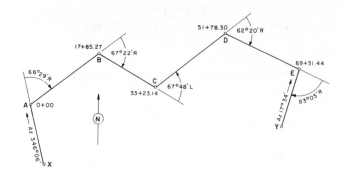

Figure 13.21   A deflection angle traverse

Table 13.10
Summary of Example 13.5

point	bearing	length	latitude north	latitude south	departure east	departure west	coordinates y	coordinates x
A							1276.28	1533.45
	N65°49′W	85.20	34.90			77.72		
M							1311.18	1455.73
B							2256.01	1644.00
	N58°36′E	72.67	37.86		62.03			
N							2293.87	1706.03
C							2591.17	776.60
	N66°38′E	81.31	32.25		74.64			
O							2623.42	851.24
D							1913.63	911.39
	N87°15′E	70.54	3.38		70.46			
P							1917.01	981.85
E							1433.09	739.71
	S64°50′W	77.80		33.08		70.41		
Q							1400.01	669.30
M							1311.18	1455.73
	N14°17′E	1014.07	982.69		250.30			
N							2293.87	1706.03
	N68°55′W	916.12	329.55			854.79		
O							2623.42	851.24
	S10°28′E	718.38		706.41	130.61			
P							1917.01	981.85
	S31°09′W	604.13		517.00		312.55		
Q							1400.01	669.30
	S83°33′E	791.43		88.83	786.43			
M							1311.18	1455.73

29 ROUTE LOCATION BY DEFLECTION ANGLE TRAVERSE

Deflection angle traverses are suitable for highway locations because the deflection angle at the point of intersection of two tangents along the centerline of a highway is equal to the central angle of the circular arc which is inserted to connect two tangents. Straight sections along the centerline are known as *tangents*, and *circular* arcs are known as *simple curves*. Curves are not always *circular* arcs, but normally they are. Curves will be discussed in detail in Chapter 16.

The deflection angle traverse shown in figure 13.21 begins at point **A** which is a point on the line **XA**, the azimuth of which is 346°06′, and ends at point **E** on the line **YE**, the azimuth of which is 17°34′. The azimuths of these two lines are used to check the angular closure of the traverse. The azimuth of **AB** is found by adding the deflection angle at **A** to the azimuth **XA**. Azimuths of the other lines of the traverse are found by adding right deflection angles to the forward azimuth of the preceding line and subtracting left deflection angles from the forward azimuth of the preceding line.

Following this procedure, the azimuth of **EY**, back azimuth of **YE**, fails to check by 2′00″. The deflection angles were measured to the nearest 30″, so in adjusting the five deflection angles an arbitrary correction of 30″ was made in the deflection angles at **A**, **B**, **C**, and **D**. The angle at **B** received a correction of 1′00″, at **C** the correction was 1′30″ and at **D** the correction was 2′00″. Computations for azimuths and bearings are shown in table 13.12.

Table 13.12
Calculations for Figure 13.21

line	azimuth	correction	adjusted azimuth	adjusted bearing
XA	346°06′	fixed	346°06′00″	
	+66°29′			
	412°35′			
	−360°00′			
AB	52°35′	−0′30″	52°34′30″	N52°34′30″E
	+67°22′			
BC	119°57′	−1′00″	119°56′00″	S60°04′00″E
	−67°48′			
CD	52°09′	−1′30″	52°07′30″	N52°07′30″E
	+62°20′			
DE	114°29′	−2′00″	114°27′00″	S65°33′00″E
	+83°07′			
EY	197°36′	fixed	197°34′00″	S17°34′00″W
	−180°00′			
YE	+17°36′			
	−17°34′			
	+02′ = closure error			

30 CONNECTING TRAVERSE

Figure 13.22 shows a connecting traverse between triangulation station **WAAF** and triangulation station **PRICE**. Traverse stations **A**, **B**, and **C** are established as part of the connecting traverse. x- and y-coordinates for these stations are to be computed.

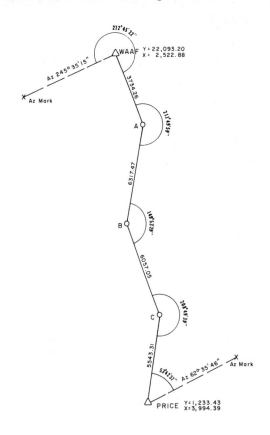

Figure 13.22 A connecting traverse

Coordinates for triangulation stations **WAAF** and **PRICE** are known and are shown in figure 13.22. Azimuth from each triangulation station to an azimuth mark is also known and shown in figure 13.22.

Computations for determining coordinates of stations **A**, **B**, and **C** are shown in tables 13.14, 13.15, and 13.16.

Table 13.13 shows computations for direction of traverse courses. The direction of **WAAF** azimuth mark is fixed, as is the direction of the **PRICE** azimuth mark. Angular closure was found to be 0′35″, which was distributed evenly to angles at stations **WAAF**, **A**, **B**, **C**, and **PRICE**.

Table 13.14 shows traverse computations with the error of closure and precision.

Table 13.15 shows latitudes and departures balanced by the compass rule and coordinates of stations **A**, **B**, and **C**.

Table 13.13
Computations of directions for Figure 13.22

azimuth		correction	adjusted azimuth
245°35′15″ **WAAF-MK**			245°35′15″
+272°45′22″ angle right		−07″	272°45′15″
$\overline{518°20′37″}$			$\overline{518°20′30″}$
−360°00′00″			360°00′00″
$\overline{158°20′37″}$ **WAAF-A**			$\overline{158°20′30″}$
+211°49′59″ angle right		−07″	211°49′52″
$\overline{370°10′36″}$			$\overline{370°10′22″}$
−180°00′00″			180°00′00″
$\overline{190°10′36″}$ **A-B**			$\overline{190°10′22″}$
+149°53′29″ angle right		−07″	149°53′22″
$\overline{340°04′05″}$			$\overline{340°03′44″}$
−180°00′00″			180°00′00″
$\overline{160°04′05″}$ **B-C**			$\overline{160°03′44″}$
+208°49′39″ angle right		−07″	208°49′32″
$\overline{368°53′44″}$			$\overline{368°53′16″}$
−180°00′00″			180°00′00″
$\overline{188°53′44″}$ **C-PRICE**			$\overline{188°53′16″}$
+53°42′37″ angle right		−07″	53°42′30″
$\overline{+242°36′21″}$			$\overline{242°35′46″}$
−180°00′00″			180°00′00″
$\overline{62°36′21″}$ **PRICE-Mk**		check	$\overline{62°35′46″}$
−62°35′46″ **PRICE-Mk**			
$\overline{\text{clos} = +35″}$			

Table 13.14
Traverse computations for Figure 13.22

line	azimuth	length	latitude	departure
WAAF-A	158°20′30″	3734.26	−3470.63	+1378.21
A-B	190°10′22″	6317.47	−6218.16	−1115.77
B-C	160°03′44″	6057.05	−5694.01	+2065.45
C-PRICE	188°53′16″	5543.31	−5476.75	−856.44
		$\overline{21652.09}$	$\overline{-20859.55}$	$\overline{+1471.45}$
			+20859.77	−1471.51
			$\overline{+0.22}$	$\overline{-0.06}$

error of closure $= \sqrt{0.22^2 + 0.06^2} = 0.23$

precision $= 1/94000$

Table 13.15
Coordinate computations for Figure 13.22

station	balanced latitude	balanced departure	coordinates y	x
WAAF			22,093.20	2,522.88
	−3470.67	+1378.22		
A			18,622.53	3,901.10
	−6218.22	−1115.75		
B			12,404.31	2,785.35
	−5694.07	+2065.47		
C			6,710.24	4,950.82
	−5476.81	−856.43		
PRICE			1,233.43	3,994.39
	$\overline{-20859.77}$	$\overline{+1471.51}$		

PRACTICE PROBLEMS FOR CHAPTER 13

1. Adjust the interior angles of the closed traverses arbitrarily.

(a)
pt	measured angle
A	67°06′30″
B	216°19′
C	65°12′30″
D	95°18′30″
E	96°02′

(b)
pt	measured angle
A	92°38′
B	117°21′30″
C	129°13′
D	261°44′30″
E	55°28′30″
F	207°10′
G	70°52′30″
H	145°34′

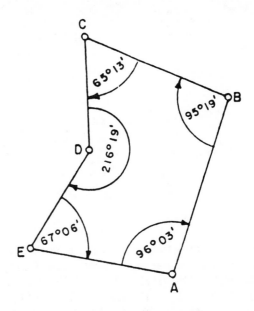

2. Adjust the interior angles of the closed traverse by applying the same correction to each angle.

pt	measured angle
A	90°19′59″
B	106°19′55″
C	318°51′26″
D	48°29′12″
E	150°55′29″
F	195°44′47″
G	87°11′56″
H	193°08′23″
I	212°18′51″
J	47°27′37″
K	288°30′28″
L	60°43′21″

4. Compute the bearings of the sides of the traverse *ABCDEA*. Interior angles are as shown. Draw a sketch for each traverse station showing the two known angles and designate the unknown bearing angle with a question mark. Show computations at each station. Bearing of *AB* is N 65°04′ W.

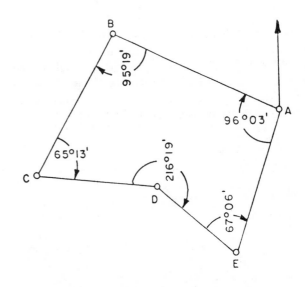

3. Compute the bearings of the sides of the traverse *ABCDEA*. Interior angles are as shown. Draw a sketch for each traverse station showing the two known angles and designate the unknown bearing angle with a question mark. Show computations at each station. Bearing of *AB* is N 15°22′ E.

5. Use the figure showing the traverse *ABCDEA* to compute the interior angles of the traverse. Draw a meridian (north line) through each traverse station, identify the angles at each station which are used in the computations, and show the computations in the space allocated. Then, record the interior angles on the figure.

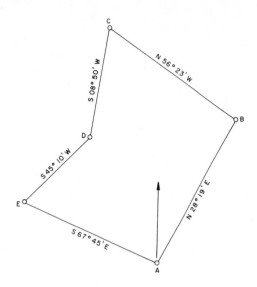

6. Using a protractor and a scale, plot the traverse *ABCDEFGHA* to a scale of $1'' = 100'$. *AB*: N 21°12′ E, 289.07; *BC*: S 78°41′ E, 366.63; *CD*: S 25°24′ W, 228.60; *DE*: S 35°23′ E, 232.44; *EF*: S 47°51′ W, 281.24; *FG*: N 78°30′ W, 316.68; *GH*: N 29°29′ E, 192.13; *HA*: N 25°57′ W, 174.36.

7. On a Cartesian *x-y* plane, draw lines with azimuths: *OA* = 30°, *OB* = 60°, *OC* = 105°, *OD* = 135°, *OE* = 180°, *OF* = 210°, *OG* = 265°, *OH* = 270°, *OI* = 315°. Show the back azimuth of each line.

8. Traverse *ABCDEA* is tied to station *WAAF* for direction. Azimuth to the azimuth mark from station *WAAF* is 14°37′19″ (from the north). Angle-to-the-right at station *WAAF* (*Az Mk* to A) is 21°14′46″. Angles-to-the-right at other stations are shown. Compute the azimuth of each side of the traverse.

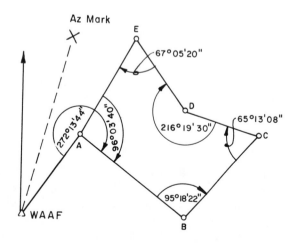

9. Convert bearing to azimuth.

(a) S 85°13′16″ W   (f) N 07°26′33″ W
(b) S 08°19′19″ E   (g) N 27°57′45″ E
(c) N 74°24′01″ W   (h) S 05°17′25″ W
(d) N 84°28′13″ E   (i) S 04°18′12″ E
(e) S 83°03′28″ E   (j) N 57°08′02″ W

10. Convert azimuth to bearing.

(a) 291°37′06″   (f) 65°11′37″
(b) 106°12′46″   (g) 337°15′11″
(c) 12°13′47″   (h) 267°25′51″
(d) 232°31′18″   (i) 102°27′38″
(e) 93°04′02″   (j) 317°40′53″

11. Determine the latitude and departure of each side of the traverse *ABCDA* and record in the proper column. Total the north and south latitudes and the east and west departures.

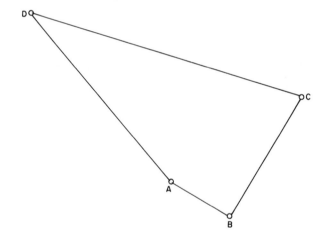

pt	bearing	length
A		
	S 60°00′ E	50.0
B		
	N 30°00′ E	100.0
C		
	N 75°00′ W	200.0
D		
	S 41°30′ E	151.0
A		

12. Compute the error of closure and ratio of error, and balance the traverse by the compass rule. Show computations for corrections for each side of the traverse.

pt	bearing	length
A		
	N 28°19′ E	560.27
B		
	N 56°23′ W	484.18
C		
	S 08°50′ W	375.42
D		
	S 45°10′ W	311.44
E		
	S 67°45′ E	449.83
A		

13. Balance the traverse by the compass rule.

pt	bearing	length
A		
	S 53°18′ E	560.27
B		
	N 42°00′ E	484.18
C		
	N 72°47′ W	375.42
D		
	N 36°27′ W	311.44
E		
	S 30°39′ W	449.83
A		

14. Balance the traverse.

pt	bearing	length
A		
	N 80°00′ W	1015.43
B		
	S 66°30′ W	545.52
C		
	S 12°00′ E	480.97
D		
	S 88°30′ E	750.26
E		
	N 69°00′ E	639.18
F		
	N 10°00′ E	306.28
A		

15. Given the balanced latitudes and departures of the traverse *ABCDEA* with the coordinates of station *A* being $Y = 9012.34$, $X = 1234.56$, determine the coordinates of stations *B*, *C*, *D*, and *E*.

pt	bearing	length	latitude N	S	departure E	W	coordinates Y	X
A							9012.34	1234.56
	N 28°19′ E	560.27	493.12		265.66			
B								
	N 56°23′ W	484.18	267.97			403.29		
C								
	S 08°50′ W	375.42		371.04		57.72		
D								
	S 45°10′ W	311.44		219.64		220.91		
E								
	S 67°45′ E	449.83		170.41	416.26			
A								

16. Given the open traverse *ABCD* with coordinates of *A*, *B*, *C*, and *D* as shown, compute the bearings and lengths of *AB*, *BC*, and *CD*.

pt	coordinates Y	X
A	9090.90	9090.90
B	9480.39	9452.27
C	9612.87	8890.70
D	9321.86	8775.81

17. The traverse *ABCDEA* has been run inside the tract *MNOPQM* because corners of the tract cannot be occupied. The traverse has been balanced, and coordinates of each traverse station have been computed. Ties to each corner of the tract have been made from the adjacent traverse station and are shown. Compute the bearing and length of each course of the trace *MNOPQM*.

pt	bearing	length	latitude N	S	departure E	W	coordinates Y	X
A							1000.00	1000.00
	N 28°19 E	560.27	493.12		265.66			
B							1493.12	1265.66
	N 56°23′ W	484.18	267.97			403.29		
C							1761.09	862.37
	S 08°50′ W	375.42		371.04		57.72		
D							1390.05	804.65
	S 45°10′ W	311.44		219.64		220.91		
E							1170.41	583.74
	S 67°45′ E	449.83		170.41	416.26			
A							1000.00	1000.00

	corner ties	
line	azimuth	length
AM	167°58′	59.20
BN	88°13′	65.13
CO	346°08′	70.85
DP	303°51′	72.45
EQ	258°31′	76.51

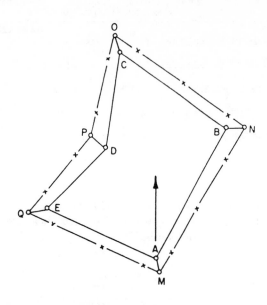

18. The traverse *ABCDEA* has been run and ties to the corners of the tract *MNOPQM*, which cannot be occupied, have been made. The traverse has been balanced and coordinates of each traverse station have been computed as shown. Compute the bearing and length for each course of the tract *MNOPQM*.

pt	bearing	length	latitude N	latitude S	departure E	departure W	coordinates Y	coordinates X
A							1000.00	1000.00
	N 28°19′ E	560.27	493.12		265.66			
B							1493.12	1265.66
	N 56°23′ W	484.18	267.97			403.29		
C							1761.09	862.37
	S 08°50′ W	375.42		371.04		57.72		
D							1390.05	804.65
	S 45°10′ W	311.44		219.64		220.91		
E							1170.41	583.74
	S 67°45′ E	449.83		170.41	416.26			
A							1000.00	1000.00

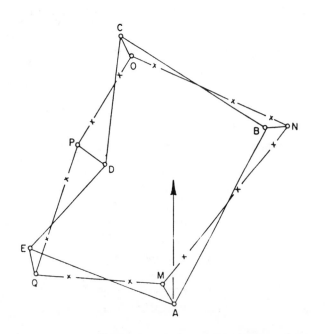

corner ties

line	azimuth	length
AM	331°12′	58.46
BN	87°33′	66.72
CO	152°13′	68.89
DP	304°31′	74.13
EQ	166°22′	66.52

19. Balance the traverse, and find coordinates of points *B, C, D, E, F, G* and *H*. Enter azimuth directly into calculator with north and east plus (+) and south and west minus (−).

line	azimuth	length
AB	21°12′	289.07
BC	101°19′	366.63
CD	205°24′	228.60
DE	144°37′	232.44
EF	227°51′	281.24
FG	281°30′	316.68
GH	29°29′	192.13
HA	334°03′	174.06

20. The open traverse *ABCDE* for highway location begins at point *A*, which lies on line *XA*, the azimuth of which is known to be 21°44′, and ends at point *E*, which lies on line *YE*, the azimuth of which is known to be 350°14′. Using the azimuth of *XA* and *YE* as fixed, find the angular error in the traverse, correct arbitrarily, and convert azimuths to bearings.

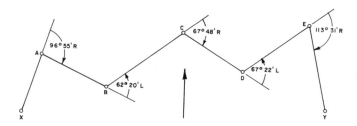

21. Traverse FOX-A-B-C-DOG connects triangulation stations FOX and DOG, coordinates of which are known and shown. The direction FOX-Az Mk and DOG-Az Mk are fixed. The traverse stations *A*, *B*, and *C* were set. Find the angular closure, correct azimuths, and compute the coordinates of *A*, *B*, and *C*. Az FOX-Az Mk = 246°45′46″. Az DOG-Az Mk = 69°45′15″.

line	length
FOX	
−A	5543.31
AB	6057.05
BC	6317.47
C−	3734.26
DOG	

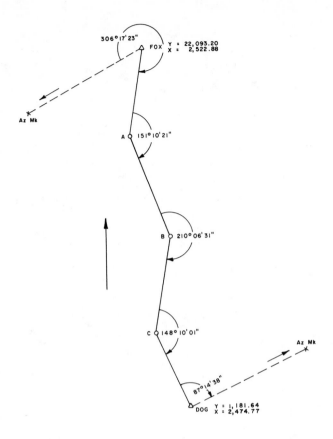

PROFESSIONAL PUBLICATIONS, INC. • P.O. Box 199, San Carlos, CA   94070

14

AREA

1 METHODS FOR COMPUTATION OF AREA

The area within a traverse can be computed by the double meridian distance (**DMD**) method, the coordinate method, and use of geometric or trigonometric formulas. The most frequently used method for land area calculations is the **DMD** method.

2 DOUBLE MERIDIAN DISTANCES

The **DMD** method makes use of balanced latitudes and departures. It is widely used because it can be computed quickly.

The **DMD** method sets up a series of trapezoids and triangles, both inside and outside of the traverse. It calculates each of these areas and determines the area of the traverse from them.

The double meridian distance is simply twice the meridian distance. (See Section 3.) The **DMD** is used instead of the **MD** to simplify the arithmetic. If the **MD** were used, division by two would be required several times. Using **DMD**, division by two is required only once.

The area of a trapezoid is one-half the sum of the bases times the altitude (i.e., the average of the bases times the altitude.) In the **DMD** method the meridian distance for each course of the traverse serves as the average of the bases of a trapezoid. The **DMD**'s of the courses are obtained from the departures of the courses. Thus, the only data needed are latitudes and departures.

3 MERIDIAN DISTANCES

The *meridian distance* of a course is the right angle distance from the midpoint of the course to a reference meridian. **MD**'s are illustrated in figure 14.1. Since east and west departures are used, algebraic signs must be considered. To simplify the use of plus and minus values for departure, the entire traverse should be placed in the north-east quadrant. This can be done by taking the reference meridian through the most westerly point in the traverse.

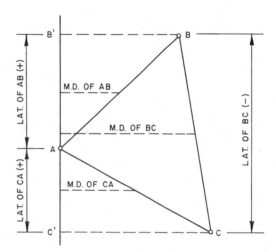

Figure 14.1 Meridian distances

4 DETERMINING MOST WESTERLY POINT

Determining the most westerly point can usually be accomplished by studying the east and west departures.

5 RULES FOR USE WITH DMD CALCULATIONS

In figure 14.2, the **MD** of $EA = \frac{1}{2}$ departure of EA. **DMD** of $EA = $ departure of EA.

The **MD** of $AB = $ **MD** of $EA + \frac{1}{2}$ departure $EA + \frac{1}{2}$ departure AB. **DMD** of $AB = $ **DMD** of $EA + $ departure $EA + $ departure AB.

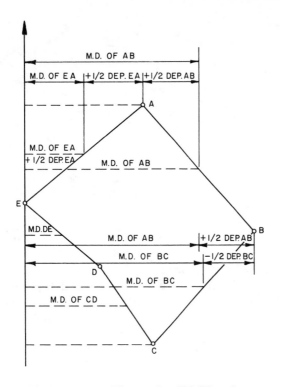

Figure 14.2   Illustrating **DMD** rules

The **MD** of $BC = $ **MD** of $AB + \frac{1}{2}$ departure $AB - \frac{1}{2}$ departure BC.   **DMD** of $BC = $ **DMD** of $AB +$ departure $AB -$ departure BC.

From these examples and from further examination, the following rules can be inferred for computing **DMD**'s.

rule 14.1   The **DMD** of the first course is equal to the departure of the first course.

rule 14.2   The **DMD** of any course is equal to the **DMD** of the preceding course plus the departure of the preceding course plus the departure of the course itself.

rule 14.3   The **DMD** of the last course is equal to the departure of the last course with opposite sign. ("Opposite" sign means that for a westerly course, the **DMD** will be positive.)

6 AREAS BY DMD

The area of a traverse can be found by multiplying the **DMD** of each course by the latitude of that course, with north latitudes producing positive areas and south latitudes producing negative areas, adding the areas algebraically, and dividing by two. The algebraic sign of **DMD**'s is always positive.

In figure 14.1,

area $ABC = $ area $B'BCC' - $ area $AB'B - $ area ACC'

If **MD**'s are positive, north latitudes are positive, and south latitudes are negative,

$$\text{area of } B'BCC' = \textbf{MD} \text{ of } BC \times \text{latitude of } BC$$
$$\text{(negative)}$$
$$\text{area of } AB'B = \textbf{MD} \text{ of } AB \times \text{latitude of } AB$$
$$\text{(positive)}$$
$$\text{area of } ACC' = \textbf{MD} \text{ of } CA \times \text{latitude of } CA$$
$$\text{(positive)}$$
$$\text{area of } ABC = \text{algebraic sum of } B'BCC', AB'B,$$
$$ACC'$$

The sign of the sum can be either plus or minus.

If **DMD**'s are used, the area will be double the area determined by using **MD**'s, and the area of the traverse can be determined by dividing the double area by two.

Example 14.1

Table 14.1 shows the tabulation of computations for the area of a traverse $ABCDEA$ by the **DMD** method. Latitudes and departures shown are balanced. (Traverse is not illustrated.)

The first step is to determine the most westerly traverse point. In lieu of a sketch showing the traverse, it is found as follows:

1. Because AB has a northwest direction, B is west of A. Looking at the departure column, it is 507.97 feet west of A.

Table 14.1
Example 14.1

line	bearing	distance	latitude		departure		DMD	double	area
			north	south	east	west		plus	minus
AB	N65°04'W	560.27	236.11			507.97	995.41	235,026	
BC	S30°14'W	484.14		418.39		243.72	243.72		101,970
CD	S84°33'E	375.42		35.71	373.77		373.77		13,347
DE	S48°13'E	311.44		207.56	232.27		979.81		203,369
EA	S18°53'E	449.83	425.55		145.65		1357.73	577,782	
								812,808	318,686

2. C is 243.72 feet west of B, so B is not the most westerly point.

3. Courses CD, DE, and EA have east departures; therefore, C is the most westerly point.

With C the most westerly point, the first **DMD** computed is for the course CD. Remembering that the **DMD** for the first course is the departure of the first course, and also remembering the definition of the **DMD** for any course,

departure of $CD = +373.77 = $ **DMD** of CD

departure of $CD = +373.77$

departure of $DE = +232.27$

$+979.81 = $ **DMD** of DE

departure of $DE = +232.27$

departure of $EA = +145.16$

$+1357.73 = $ **DMD** of EA

departure of $EA = +145.65$

departure of $AB = -507.97$

$+995.41 = $ **DMD** of AB

departure of $AB = -507.97$

departure of $BC = -243.72$

$+243.72 = $ **DMD** of BC

It should be noted that the **DMD** of BC is the same as its departure except that it has a positive sign.

The area of the traverse $ABCDEA$ is found from the double area sums in table 14.1.

$$\text{area} = \frac{1}{2}(812{,}808 - 318{,}686) = 247{,}061 \text{ square feet}$$

If necessary, this area can be converted to acres by dividing by 43,560.

$$\text{area} = \frac{247{,}061}{43{,}560} = 5.672 \text{ acres}$$

7 AREA BY COORDINATES

After the coordinates of the corners of a tract of land are determined, the area of the tract can be computed by the *coordinate method.*

The coordinate formula is derived by forming trapezoids and determining their areas just as is done in the **DMD** method. Meridian distances are not used; trapezoids are formed by the abscissas of the corners. Ordinates of the corners serve as the altitude of the trapezoids. Alternatively, the trapezoids can be formed by the ordinates of the corners, and the abscissas serve as the

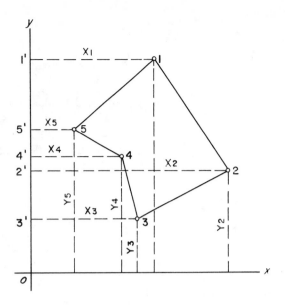

Figure 14.3 Coordinates of traverse points

altitudes. In figure 14.3, the abscissas are used to form the trapezoids.

$$\text{area } 123451 = \text{area } 1'122' + \text{area } 2'233' - \text{area } 1'155'$$
$$- \text{area } 5'544' - \text{area } 4'433'$$
$$= \frac{1}{2}[X_1(Y_5 - Y_2) + X_2(Y_1 - Y_3)$$
$$+ X_3(Y_2 - Y_4) + X_4(Y_3 - Y_5)$$
$$+ X_5(Y_4 - Y_1)]$$

The same results can be found by using the ordinates to form the trapezoids. The formula is,

$$\text{area } 123451 = \frac{1}{2}[Y_1(X_5 - X_2) + Y_2(X_1 - X_3)$$
$$+ Y_3(X_2 - X_4) + Y_4(X_3 - X_5)$$
$$+ Y_5(X_4 - X_1)]$$

Example 14.2

Given the traverse 123451 in figure 14.3 with coordinates as shown, find the area inside the traverse by the coordinate method.

point	coordinates	
	Y	X
1	1000.00	1000.00
2	1236.11	492.03
3	817.72	248.31
4	782.01	622.01
5	574.45	854.35

Solution

station	coordinates Y	X		double area plus	minus
1	1000.00	1000.00	1000.00(574.45−1236.11)		661,660
2	1236.11	492.03	492.03(1000.00−817.72)	89,687	
3	817.72	248.31	248.31(1236.11−782.01)	112,758	
4	782.01	622.08	622.08(817.72−574.45)	151,333	
5	574.45	854.35	854.35(782.01−1000.00)		186,240
				353,778	847,900
					353,778
					2)494,122
					247,061

$$\text{area} = \frac{847{,}900 - 353{,}778}{(2)(43{,}560)} = 5.672 \text{ acres}$$

8 AREA BY TRIANGLES

When small traverses do not warrant computations of latitudes and departures, their areas can be determined by using formulas for the area of a triangle.

$$\text{area} = 1/2 \text{ ab } \sin C \qquad 14.1$$

(*a* and *b* are any two sides and *C* is the angle included between them.)

In figure 14.4(a), a tract of land has been divided into two triangles. Two sides and an included angle have been measured in each triangle. The areas of the triangles can be computed by using equation 14.1. Their sum is the area of the tract.

In figure 14.4(b), the property line has grown up in brush so that four triangles have been formed from a central point. Angles at the central point have been measured for each triangle, and distances from the central point to each corner have also been measured. Areas can again be computed by using equation 14.1.

Also applicable in computing areas is equation 14.2.

$$A = \sqrt{(s)(s-a)(s-b)(s-c)} \qquad 14.2$$

s is one-half of the perimeter of the triangle, and *a*, *b*, and *c* are the sides of the triangle.

Example 14.3

Find the area of a triangle with sides 32 ft, 46 ft, and 68 ft.

Solution

$$s = \frac{1}{2}(32 + 46 + 68) = 73$$

$$\text{area} = \sqrt{(73)(73-32)(73-46)(73-68)} = 636 \,\text{ft}^2$$

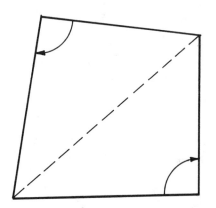

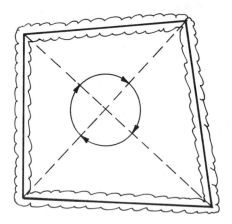

Figure 14.4 Area by triangles

9 AREA ALONG AN IRREGULAR BOUNDARY

When a tract of land is bounded on one side by an irregular boundary such as a stream or lake, the traverse can be composed of straight lines so that closure can be computed. Points along the irregular side can be tied to one of the sides of the traverse by right angle offset measurements. The area between the irregular side and the traverse line is approximated by dividing the area into trapezoids and triangles formed by the ties to the breaks in the irregular side. This irregular area is then added to the traverse area. The irregular area can be computed by applying the trapezoidal rule or Simpson's one-third rule.

10 THE TRAPEZOIDAL RULE

In using the *trapezoidal rule*, it is assumed that the irregular boundary is made up of a series of straight lines. When the ties are taken close enough, a curved line connecting the ends of any two ties is very nearly a straight line, and no significant error is introduced.

The trapezoidal rule applies only to the part of the area where the ties are at regular intervals and form trapezoids. Triangles and trapezoids which do not have altitudes of the regular interval are computed separately and added to the area found by applying the trapezoidal rule.

The rule is given by equation 14.3.

$$\text{area} = D \left(\frac{T_1}{2} + T_2 + T_3 + T_4 + \cdots + \frac{T_n}{2} \right) \quad 14.3$$

D is the regular interval, and T is the tie distance.

Example 14.4

The area in figure 14.5 between points B and H is

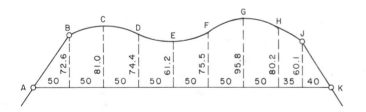

Figure 14.5   Example 14.4

$$\text{area} = 50 \left(\frac{72.6}{2} \right)$$
$$+ 50 \left(\frac{72.6}{2} + 81.0 + 74.4 + 61.2 \right.$$
$$+ 75.5 + 95.8 + \left. \frac{80.2}{2} \right)$$
$$+ 35 \left(\frac{80.2 + 60.1}{2} \right) + 40 \left(\frac{60.1}{2} \right)$$
$$= 28,687 \, \text{ft}^2$$

11 SIMPSON'S ONE-THIRD RULE

A more accurate method of determining areas bounded by irregular boundaries made up of curved lines is known as *Simpson's one-third rule*. This rule applies only to the portion lying between an odd number of ties. It states that the area is equal to one-third the regular interval times the sum of the end ties plus four times the sum of the even ties plus twice the sum of the odd ties.

$$\text{area} = \frac{D}{3} \left[T_1 + T_n + 4 \left(\sum T_{\text{even}} \right) + 2 \left(\sum T_{\text{odd}} \right) \right] \quad 14.4$$

Example 14.5

The area between points B and H in figure 14.5 is

$$\text{area} = 50 \left(\frac{72.6}{2} \right) + \frac{50}{3} [72.6 + 80.2$$
$$+ 4(81.0 + 61.2 + 95.8)$$
$$+ 2(74.4 + 75.5)] + 35 \left(\frac{80.2 + 60.1}{2} \right)$$
$$+ 40 \left(\frac{60.1}{2} \right) = 28,882 \, \text{ft}^2$$

12 AREA OF SEGMENT OF CIRCLE

Land along highways, streets, and railroads often has a circular arc for a boundary. A traverse of straight lines can be run by using the long chord of the circular arc as one of the sides of the traverse. The area of the tract can be found by adding the area of the segment formed by the chord and the arc to the area within the traverse. It is usually practical to measure the chord length and the middle ordinate length. Using these two lengths and formulas derived for computing circular curves, the area of the segment can be found.

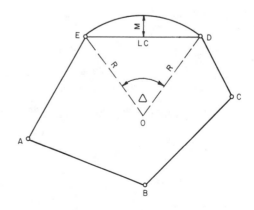

Figure 14.6   Circular segment areas

PROFESSIONAL PUBLICATIONS, INC. • P.O. Box 199, San Carlos, CA 94070

$$\text{Area of sector} = \frac{\Delta^\circ \pi R^2}{360^\circ} \qquad 14.5$$

$$\text{Area of triangle} = \frac{R \times R \sin \Delta}{2}$$

$$= \frac{R^2 \sin \Delta}{2} \qquad 14.6$$

The radius R is not known, but it can be determined if the long chord DE and the middle ordinate M are known. These lengths can be measured. Formulas used to compute R are,

$$\tan \Delta/4 = \frac{2M}{LC} \qquad 14.7$$

$$R = \frac{LC}{2 \sin \Delta/2} \qquad 14.8$$

Δ is the *central angle*, M is the *middle ordinate*, LC is the *long chord*, and R is the radius.

If A = area of the segment, A_s = area of the sector, and A_t = area of the triangle,

$$A = A_s - A_t = \frac{\Delta \pi R^2}{360} - \frac{R^2 \sin \Delta}{2}$$

$$= R^2 \left[\frac{\Delta \pi}{360} - \frac{\sin \Delta}{2} \right] \qquad 14.9$$

Example 14.6

The tract of land shown in figure 14.6 consists of the area within the traverse $ABCDEA$ plus the area in the segment bounded by side DE and arc DE. The area of the segment is equal to the area of the sector DOE minus the area of the triangle DOE.

Find the area of the segment bounded by side DE and arc DE in figure 14.6 if the long chord is 325.48 ft and the middle ordinate is 42.16 ft.

Solution

$$\tan \Delta/4 = \frac{(2)(42.16)}{325.48}$$

$$\Delta = 58.0958^\circ$$

$$R = \frac{LC}{2 \sin \Delta/2} = \frac{325.48}{2 \sin 58.0958^\circ/2} = 335.17 \, \text{ft}$$

$$A = [335.17]^2 \left[\frac{58.0958\pi}{360} - \frac{\sin 58.0958^\circ}{2} \right]$$

$$= 9270 \, \text{ft}^2$$

PRACTICE PROBLEMS FOR CHAPTER 14

1. Find the most westerly point in each traverse, and indicate which side would be the first for DMD computations.

(a)

line	departure east	west
AB	600	
BC	50	
CD		1000
DE		500
EF	100	
FA	750	

(b)

line	departure east	west
AB		500
BC	100	
CD	750	
DE	600	
EF	50	
FA		1000

(c)

line	departure east	west
AB	559.88	
BC	60.84	
CD		357.90
DE		294.87
EA	32.05	

(d)

line	departure east	west
AB		507.90
BC		243.75
CD	373.76	
DE	232.26	
EA	145.63	

(e)

line	departure
AB	+536.87
BC	+96.62
CD	+487.82
DE	−102.54
EF	−629.31
FG	−583.40
GA	+193.96

(f)

line	departure
AB	+629.31
BC	+102.54
CD	−487.82
DE	−96.62
EF	−536.87
FG	−193.94
GA	+583.40

2. Compute the area of the traverse *ABCDEFA* by the DMD method.

pt	bearing	length	latitude N	S	departure E	W
A						
	north	500.0	500		0	
B						
	N 45°00′ W	848.6	600			600
C						
	S 69°27′ W	854.4		300		800
D						
	S 11°19′ W	1019.8		1000		200
E						
	S 79°42′ E	1118.0		200	1100	
F						
	N 51°20′ E	640.3	400		500	
A						

3. Compute the area of the traverse *ABCDEA* by the DMD method.

pt	bearing	length	latitude N	S	departure E	W
A						
	N 28°19′ E	560.27	493.12		265.66	
B						
	N 56°23′ W	484.18	267.97			403.29
C						
	S 08°50′ W	375.42		371.04		57.72
D						
	S 45°10′ W	311.44		219.64		220.91
E						
	S 67°45′ E	449.83		170.41	416.26	
A						

4. Compute the area of the traverse *ABCDEFGHA* by the DMD method.

ln	bearing	length	latitude	departure
AB	S 25°57′ E	174.36	−156.78	+76.30
BC	S 29°29′ W	192.13	−167.25	−94.56
CD	S 78°30′ E	316.68	−63.14	+310.32
DE	N 47°51′ E	281.24	+188.73	+208.51
EF	N 35°23′ W	232.44	+189.51	−134.59
FG	N 25°24′ E	228.60	+206.50	+98.05
GH	N 78°41′ W	366.63	+71.94	−359.50
HA	S 21°12′ W	289.07	−269.51	−104.53

5. Compute the area of the traverse 1-2-3-4-5-1 by the coordinate method.

pt	coordinates	
	Y	X
1	1000.00	1000.00
2	1559.43	1029.10
3	1629.21	549.93
4	1269.27	656.63
5	976.31	550.83

6. Compute the area of the traverse $ABCDEFA$ by the coordinate method.

pt	coordinates	
	Y	X
A	1000.00	1000.00
B	524.31	1071.00
C	458.92	1818.41
D	651.13	2428.01
E	943.43	2499.63
F	1186.48	1512.25

7. Compute the area of the city lot shown by using the formula $A = 1/2\,ab\sin C$. $AB = 437.2$; $BC = 400.6$; $CD = 378.8$; $DA = 469.1$.

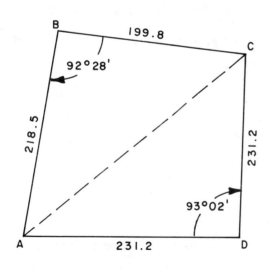

8. Compute the area of the city lot shown by using the formula $A = 1/2\,ab\sin C$. $OE = 282.3$; $OF = 264.7$; $OG = 288.1$; $OH = 342.8$.

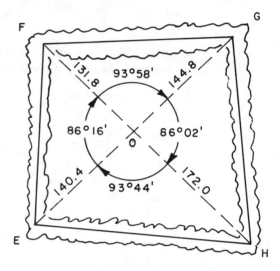

9. Find the area of the triangle with sides 12 feet, 14 feet, and 20 feet by using the formula

$$A = \sqrt{s(s-a)(s-b)(s-c)}.$$

10. Compute the area along the irregular boundary by both the trapezoidal rule and Simpson's one-third rule. Ties: $B = 48.1$; $C = 52.6$; $D = 46.8$; $E = 39.9$; $F = 43.7$; $G = 58.0$; $H = 51.6$; $J = 40.0$.

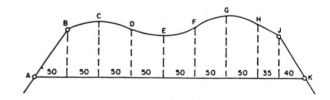

11. Compute the area of the segment with long chord $= 491.67$ and middle ordinate $= 98.23$.

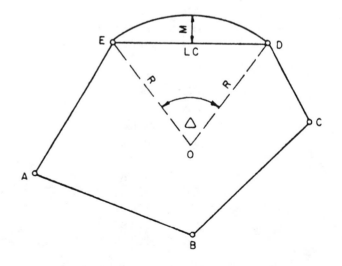

15 PARTITIONING OF LAND

1 INTRODUCTION

An important part of the surveyor's work is the subdivision of tracts of land into two or more parts. Each partition of land is a separate problem, but there are basic techniques which can be used. These techniques will be illustrated in this chapter through the use of numerous examples.

In solving the examples, latitudes and departures and **DMD**'s will be used where applicable. Unknown sides and angles of triangles will be solved by trigonometry. Formulas for area of triangles and trapezoids will be used to find lengths and bearings of sides.

2 LENGTH AND BEARING OF ONE SIDE UNKNOWN (THE CUTOFF LINE)

Example 15.1

The bearing and length of the side DE of the traverse $ABCDEA$ shown in figure 15.1 are missing and are to be computed.

line	bearing	length
AB	N28°19′ E	560.27
BC	N56°23′ W	484.18
CD	S08°50′ W	375.42
DE		
EA	S67°45′ E	449.83

Solution

With the bearing and length of a side missing, the error of closure and precision cannot be computed. However, the traverse can be mathematically closed by giving a value to the latitude and departure of the side with missing bearing and length. These assigned values should be chosen to balance north and south latitudes and east and west departures of the traverse. In effect,

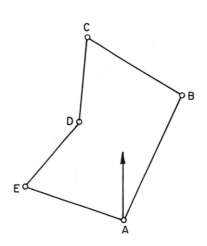

Figure 15.1 Example 15.1

this assumes that the latitudes and departures of the other sides are correct and any errors in them will be contained in the latitude and departure of the side with missing measurements.

line	bearing	length	latitude	departure
AB	N28°19′ E	560.27	+493.23	+265.76
BC	N56°23′ W	484.18	+268.06	−403.21
CD	S08°50′ W	375.42	−370.97	−57.65
DE				
EA	S67°45′ E	449.83	−170.33	+416.34
			+219.99	+221.24

$$\text{tangent of bearing } DE = \text{dep} \div \text{lat} = 221.24 \div 219.99$$
$$\text{bearing } DE = \text{S} 45°09′ \text{W}$$
$$\text{length } DE = \sqrt{\text{lat}^2 + \text{dep}^2}$$
$$= \sqrt{219.99^2 + 221.24^2}$$
$$= 312.00\,\text{ft}$$

In the solution, the sum of the latitude column is +219.99, which represents the latitude of side DE with opposite sign. The sum of the departure column is +221.24, which represents the departure of side DE with opposite sign.

3 LENGTHS OF TWO SIDES UNKNOWN

Example 15.2

The lengths for side DE and EA of the traverse $ABCDEA$ in figure 15.2 are unknown and are to be computed.

line	bearing	length
AB	N65°04'W	560.27
BC	S30°14'W	484.18
CD	S84°33'E	375.42
DE	S48°13'E	
EA	N18°53'E	

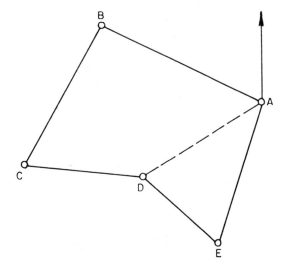

Figure 15.2   Example 15.2

Solution

With measurements of two sides missing, the method of the *cutoff line* can be applied. The traverse $ABCDA$ can be formed, excluding the sides with missing measurements. The side DA of this traverse will have an unknown bearing and length which can be computed by the method used in the preceding example. The side DA is the cutoff line. The cuoff line is usually used to isolate sides with missing measurements from sides with known measurements.

After the bearing and length of DA are computed, the side DA, together with sides DE and EA, can be considered to be the triangle ADE, which can be solved by the law of sines. When the triangle is solved, the solution is complete.

line	bearing	length	latitude	departure
AB	N65°04'W	560.27	+236.19	−508.05
BC	S30°14'W	484.18	−418.32	−243.80
CD	S84°33'E	375.42	−35.66	+373.72
DA				
			−217.79	−378.13

tangent of bearing $DA = 378.13 \div 217.79$

bearing $DA = \mathrm{N}\,60°03'\,\mathrm{E}$

length $DA = \sqrt{378.13^2 + 217.79^2}$

$= 436.37\,\mathrm{ft}$

In triangle DEA,

$$D = 180° - (60°03' + 48°13') = 71°44'$$
$$A = 60°03' - 18°53' = 41°10'$$
$$E = 180° - (71°44' + 41°10') = 67°06'$$
$$DE = \frac{\sin 41°10'\,(436.37)}{\sin 67°06'}$$
$$= 311.82\,\mathrm{ft}$$
$$EA = \frac{\sin 71°44'\,(436.37)}{\sin 67°06'}$$
$$= 449.83\,\mathrm{ft}$$

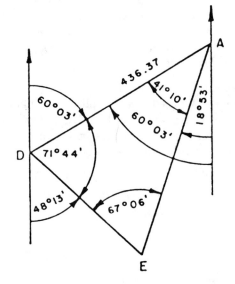

Figure 15.3   Example 15.2, continued

4 BEARING OF TWO SIDES UNKNOWN

Example 15.3

The bearings of the sides DE and EA of the traverse $ABCDEA$, figure 15.4, are unknown and are to be computed.

line	bearing	length
AB	N65°04'W	560.27
BC	S30°14'W	484.18
CD	S84°33'E	375.42
DE		311.82
EA		449.87

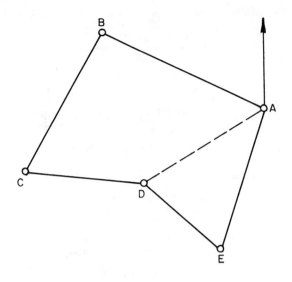

Figure 15.4 Example 15.3

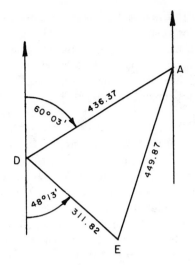

Figure 15.5 Example 15.3, continued

5 BEARING OF ONE SIDE AND LENGTH OF ANOTHER SIDE UNKNOWN

Solution

Bearings and lengths for sides AB, BC, and CD are the same as they are in the preceding example. Side DA is used as a cutoff line in traverse $ABCDA$ again. The sides DA, DE, and EA form the triangle ADE again, but the known information is different from that in the preceding example. In this example, the lengths of all three sides of the triangle are known and the interior angles are to be determined in order to find bearings of the sides of the triangle. The law of cosines is appropriate for the solution.

$$\text{bearing } DA = N\,60°03'\,E$$
$$\text{length } DA = 436.37\,\text{ft}$$

In triangle DEA,

$$\cos D = \frac{436.37^2 + 311.82^2 - 449.87^2}{(2)(436.37)(311.82)}$$
$$D = 71°44'$$
$$\cos E = \frac{311.82^2 + 449.87^2 - 436.37^2}{(2)(311.82)(449.87)}$$
$$E = 67°06'$$
$$A = 180° - (71°44' + 67°06')$$
$$= 41°10'$$
$$\text{bearing } DE = 180° - (60°03' + 71°44')$$
$$= S\,48°13'\,E$$
$$\text{bearing } EA = 60°03' - 41°10' = N\,18°53'\,E$$

Example 15.4

The length of CD and the bearing of EA of the traverse $ABCDEA$ in figure 15.6 are unknown and are to be computed.

line	bearing	length
AB	N65°04′W	560.27
BC	S30°14′W	484.18
CD	S84°33′E	
DE	S48°13′E	311.82
EA		449.87

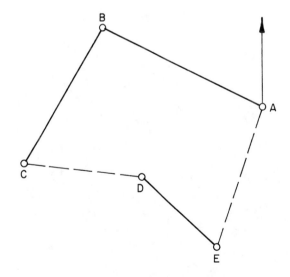

Figure 15.6 Example 15.4

Solution

The sides AB, BC, and DE, which have no missing measurements, can be connected in sequence to form a

PROFESSIONAL PUBLICATIONS, INC. • P.O. Box 199, San Carlos, CA 94070

traverse with the missing side X in figure 15.7. Arranging the sides in this order does not change the latitudes and departures of the sides. (Designating the sides in the manner shown is done to avoid confusion.) The bearing and length of the cutoff line X can be computed as in previous examples.

The line CD is given its correct bearing from the terminal of DE, but with an indefinite length. The length of EA is used as the radius of an arc, with center at the beginning point of AB, to intersect CD. In swinging the arc, it can be seen that it intersects CD at two points, giving two possible locations for EA, thus making two bearings. From the information given, the correct solution cannot be determined. Further information from the field would be necessary before determination of the correct solution. (Not all problems of this nature will have two solutions.)

line	bearing	length	latitude	departure
AB	N65°04'W	560.27	+236.19	−508.05
BC	S30°14'W	484.18	−418.32	−243.80
DE	S48°13'E	311.82	−207.77	+232.51
			−389.90	−519.34

$$\text{tangent bearing } X = 519.34 \div 389.90$$
$$\text{bearing } X = \text{N}\,53°06'\,\text{E}$$
$$\text{length } X = \sqrt{519.34^2 + 389.90^2} = 649.41\,\text{ft}$$

In the triangle bounded by sides X, CD_1, and EA_1:

$$\sin \alpha_1 = \frac{649.41(\sin 42°21')}{449.87}$$
$$\alpha_1 = 76°31'$$
$$\beta_1 = 180° - (76°31' + 42°21') = 61°08'$$
$$CD_1 = \frac{\sin 61°08'(449.87)}{\sin 42°21'} = 584.82\,\text{ft}$$
$$\text{bearing } EA_1 = 61°08' - 53°06' = \text{N}\,08°02'\,\text{W}$$

In the triangle bounded by sides X, CD_2, and EA_2:

$$\sin \alpha_2 = \frac{649.41(\sin 42°21')}{449.87}$$
$$\alpha_2 = 103°29'(\text{related angle to } 76°31')$$
$$\beta_2 = 180° - (103°29' + 42°21') = 34°10'$$
$$CD_2 = \frac{\sin 34°10'(449.87)}{\sin 42°21'} = 375.04\,\text{ft}$$
$$\text{bearing } EA_2 = 53°06' - 34°10'$$
$$= \text{N}\,18°56'\,\text{E}$$

6 AREAS CUT OFF BY A LINE BETWEEN TWO POINTS ON THE PERIMETER

Example 15.5

The tract of land represented by the traverse $ABCDEA$, Figure 15.8, is to be divided into two parts by a line from D to A.

line	bearing	length
AB	N65°04'W	560.27
BC	S30°14'W	484.18
CD	S84°33'E	375.42
DE	S48°13'E	311.44
EA	N18°53'E	449.83

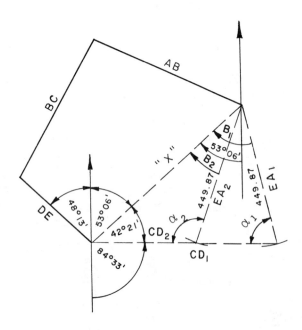

Figure 15.7　Example 15.4, continued

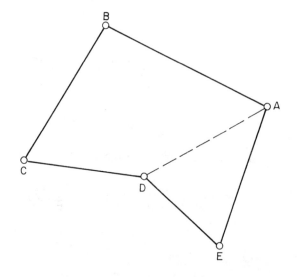

Figure 15.8　Example 15.5

Solution

The area of the entire tract can be computed by **DMD**. The bearing and length for *DA* can be computed as in previous examples, and the areas of the two parts can be computed by **DMD** and their sum checked against the area of the entire tract.

total area

line	bearing	length	latitude	departure	DMD	area
AB	N65°04'W	560.27	+236.08	−507.90	995.40	+234,994
BC	S30°14'W	484.18	−418.38	−243.75	243.75	−101,980
CD	S84°33'E	375.42	−35.71	+373.76	373.76	−13,347
DE	S48°13'E	311.44	−207.56	+232.26	979.78	−203,363
EA	N18°53'E	449.83	+425.57	+145.63	1357.67	+577,784
			0.00	0.00		494,088

$$\text{area} = 494{,}088 \div 2 = 247{,}044 \text{ ft}^2$$

cutoff line

line	bearing	length	latitude	departure
AB			+236.08	−507.90
BC			−418.38	−243.75
CD			−35.71	+373.76
DA				
			−218.01	−377.89

$$\text{tangent bearing } DA = 377.80 \div 218.01$$

$$\text{bearing } DA = \text{N}\,60°01'\,\text{E}$$

$$\text{length } DA = \sqrt{218.01^2 + 377.89^2}$$
$$= 436.27 \text{ ft}$$

area *ABCDA*

line	bearing	length	latitude	departure	DMD	area
AB			+236.08	−507.90	995.40	+234,994
BC			−418.38	−243.75	243.75	−101,980
CD			−35.71	+373.76	373.76	−13,347
DA	N60°01'E	436.27	+218.01	+377.89	1125.41	+245,351
			0.00	0.00		365,018

$$\text{area } ABCDA = 365{,}018 \div 2 = 182{,}509 \text{ ft}^2$$

area *ADEA*

line	latitude	departure	DMD	area
AD	−218.01	−377.89	377.89	−82,384
DE	−207.56	+232.26	232.26	−48,208
EA	+425.57	+145.63	610.15	+259,666
	0.00	0.00		129,074

$$\text{area } ADEA = 129{,}074 \div 2 = 64{,}537 \text{ ft}^2$$

$$182{,}509 + 64{,}537 = 247{,}046 \text{ ft}^2$$

7 AREAS CUT OFF BY A LINE IN A GIVEN DIRECTION FROM A POINT ON THE PERIMETER

Example 15.6

The tract of land represented by the traverse *ABCDEA* in figure 6.9 is to be divided into two parts by a line from point *D* parallel to *BC*.

line	bearing	length
AB	N65°04'W	560.27
BC	S30°14'W	484.18
CD	S84°33'E	375.42
DE	S48°13'E	311.44
EA	N18°53'E	449.83

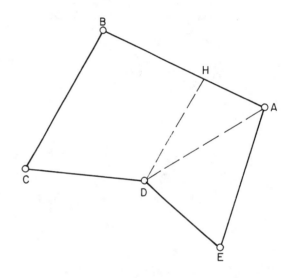

Figure 15.9 Example 15.6

Solution

The line *DH* is drawn parallel to *BC* to represent the dividing line. The line *DA* is used as a cutoff line for the traverse *ABCDA*. The bearing and length for *DA* are computed as in previous examples. The triangle *AHD* is solved by the law of sines for sides *AH* and *DH*. *HB* is found by subtracting length *AH* from *AB*. Areas of *HBCDH* and *AHDEA* are computed by **DMD** and checked against the total area of *ABCDEA*.

From the preceding example:

$$\text{bearing } DA = \text{N}\,60°01'\,\text{E}$$
$$\text{length } DA = 436.27 \text{ ft}$$
$$\text{bearing } DH = \text{bearing } CB = \text{N}\,30°14'\,\text{E}$$
$$\text{bearing } AH = \text{bearing } AB = \text{N}\,65°04'\,\text{W}$$

In triangle AHD,

$$AH = \frac{\sin 29°47'(436.27)}{\sin 95°18'}$$
$$= 217.64\,\text{ft}$$
$$DH = \frac{\sin 54°55'(436.27)}{\sin 95°18'}$$
$$= 358.53\,\text{ft}$$
$$HB = AB - AH = 560.27 - 217.64$$
$$= 342.63\,\text{ft}$$

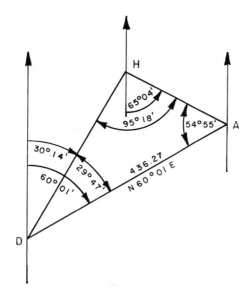

Figure 15.10   Example 15.6, continued

area $HBCDH$

line	bearing	length	latitude	departure	DMD	area
HB	N65°04′W	342.62	+144.44	−310.69	798.19	+115,291
BC	S30°14′W	484.18	−418.38	−243.75	243.75	−101,980
CD	S84°33′E	375.42	−35.71	+373.76	373.76	− 13,347
DH*	S30°14′E	358.54	+309.65	+180.68	928.20	+287,417
			0.00	0.00		287,381

* closure forced

area $HBCDH = 287,381 \div 2 = 143,690\,\text{ft}^2$

area $AHDEA$

line	bearing	length	latitude	departure	DMD	area
AH	N65°04′W	217.65	+91.75	−197.36	558.42	+ 51,235
HD	S30°14′W	358.53	−309.75	−180.53	180.53	− 55,921
DE	S48°13′E	311.44	−207.56	+232.26	232.26	− 48,208
EA	N18°43′E	449.83	+425.57	+145.63	610.15	+259,662
			0.00	0.00		+206,768

area $AHDEA = 206,768 \div 2 = 103,384\,\text{ft}^2$

$143,690 + 103,384 = 247,074\,\text{ft}^2$

8 DIVIDING TRACTS INTO TWO EQUAL PARTS BY A LINE FROM A POINT ON THE PERIMETER

Example 15.7

The tract represented by traverse $ABCDEA$, figure 15.11, is to be divided into two equal parts by a line from point D. The traverse $ABCDEA$ is the same as that in the preceding example.

line	bearing	length
AB	N65°04′W	560.27
BC	S30°14′W	484.18
CD	S84°33′E	375.42
DE	S48°13′E	311.44
EA	N18°53′E	449.83

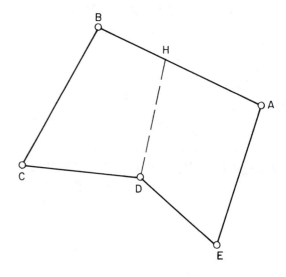

Figure 15.11   Example 15.7

Solution

The line DH is drawn by inspection to divide the tract into two equal parts. In making computations for the solution of the problem, the line is considered to be in the exact location. The area of the entire tract is computed by **DMD** and the area of traverse $HBCDH$ must be exactly one-half of the total area.

The area of the traverse $ABCDA$ can be computed by **DMD** after the bearing and length of DA have been computed as in previous examples. Then the area of the triangle AHD can be found by subtracting area $HBCDH$ from area $ABCDA$. Angle A of the triangle AHD can be found from bearings. Using equation 5.1, the formula for triangle AHD is written $A = 1/2(AH)(DA)\sin A$.

The triangle AHD is solved for DH by the law of cosines. Angle D is found by the law of sines. Bearing DH can now be found, as can length HB. Areas can be checked by **DMD**.

From example 15.5,

$$\text{bearing } DA = \text{N}\,60°01'\,\text{E}$$
$$\text{length } DA = 436.27\,\text{ft}$$
$$\text{area } ABCDEA = 247{,}044\,\text{ft}^2$$
$$\text{area } ABCDA = 182{,}509\,\text{ft}^2$$

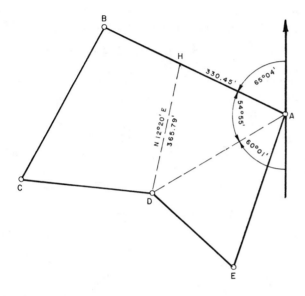

Figure 15.12   Examle 15.7, continued

$$\text{area } HBCDH = 1/2 \text{ area } ABCDEA = 123{,}522\,\text{ft}^2$$
$$\text{area } AHD = \text{area } ABCDA - \text{area } HBCDH$$
$$= 182{,}509 - 123{,}522 = 58{,}987\,\text{ft}^2$$

In triangle AHD,

$$\text{angle } A = 180° - (65°04' + 60°01')$$
$$= 54°55'$$
$$\text{area } AHD = \frac{(AH)(AD)\sin 54°55'}{2}$$
$$AH = \frac{2A}{AD\sin 54°55'}$$
$$= \frac{(2)(58{,}987)}{436.27\sin 54°55'} = 330.45\,\text{ft}$$

In triangle AHD, using the law of cosines,

$$DH =$$
$$\sqrt{330.45^2 + 436.27^2 - (2)(330.45)(436.27)(\cos 54°55')}$$
$$= 365.79\,\text{ft}$$

In triangle AHD, using the law of sines,

$$\sin D = \frac{330.45\sin 54°55'}{365.79}$$
$$D = 47°40'$$
$$\text{bearing } DH = 60°01' - 47°40' = \text{N}\,12°21'\,\text{E}$$

Also,

$$HB = AB - AH = 560.27 - 330.45 = 229.82\,\text{ft}$$

area $AHDEA$

line	bearing	length	latitude	departure	**DMD**	area
AH	N65°04'W	330.45	+139.31	−299.65	456.13	+63,543
HD	S12°21'W	365.79	−357.32	−78.24	78.24	−27,957
DE	S48°13'E	311.44	−207.56	+232.26	232.26	−48,106
EA	N18°53'E	449.83	+425.57	+145.63	610.15	+259,662
						247,142

$$\text{area } AHDEA = 247{,}142 \div 2 = 123{,}571\,\text{ft}^2$$

9 DIVIDING AN IRREGULAR TRACT INTO TWO EQUAL PARTS

Example 15.8

The tract of land represented by the traverse $ABCDE$-FA, figure 15.13, is to be divided into two equal parts by a line parallel to CD.

line	bearing	length
AB	N80°00'W	1015.43
BC	S66°30'W	545.22
CD	S12°00'E	480.97
DE	S88°30'E	750.26
EF	N69°00'E	639.18
FA	N10°00'E	306.78

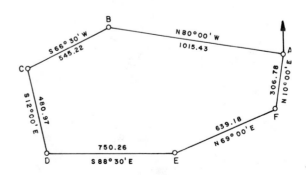

Figure 15.13   Example 15.8

Solution

The problem will be solved in a manner similar to that used in the preceding example. A line KL will be drawn parallel to CD at the approximate location of the dividing line, using a scale drawing of the traverse. It can be seen by inspection that K will fall on AB and L will fall on DE.

From the point nearest to K, point B, the line BJ is drawn parallel to CD. (Point E could be used in place of point B.)

There are now four traverses formed:

1. The original traverse $ABCDEFA$, for which the total area can be found by the **DMD** method.

2. The traverse $KBCDLK$, for which the area can be found by taking one-half of the total area.

3. The traverse $BCDJB$, for which the area can be found after first determining the bearing and length of BJ by use of the cutoff line DB as in previous examples.

4. The traverse $KBJLK$, a trapezoid, for which the area can be found by subtracting the area of $BCDJB$ from the area of $KBCDLK$.

The altitude X of a trapezoid can be found as the unknown quantity in the formula for the area of a trapezoid in which the area and the lengths of bases BJ and KL are known. KL can be expressed in terms of BJ and X to give a quadratic equation of the form $Ax^2 + Bx + C = 0$, which can be solved by the quadratic formula

$$x = \frac{-B \pm \sqrt{B^2 - 4AC}}{2A}$$

line	bearing	length	latitude	departure	DMD	area
				total area		
AB	N80°00′W	1015.43	+176.33	−1000.00	2000.00	+352,660
BC	S66°30′W	545.22	−217.41	−500.00	500.00	−108,705
CD	S12°00′E	480.97	−470.46	+100.00	100.00	−47,046
DE	S88°30′E	750.26	−19.64	+750.00	950.00	−18,658
EF	N69°00′E	639.18	+229.06	+596.73	2296.73	+526,089
FA	N10°00′E	306.78	+302.12	+53.27	2946.73	+890,266
			0.00	0.00		1,594,606

$$\text{total area} = 1{,}594{,}606 \div 2 = 797{,}303 \text{ ft}^2$$
$$\text{area } KBCDLK = 797{,}303 \div 2 = 398{,}652 \text{ ft}^2$$

To find the bearing and length of BJ, the cutoff line DB is found as the missing side of $BCDB$ and the triangle BDJ is solved.

line	latitude	departure
BC	−217.41	−500.00
CD	−470.46	+100.00
DB	−687.87	−400.00

$$\text{tangent bearing } DB = 400.00 \div 687.87$$
$$\text{bearing } DB = \text{N } 30°11′ \text{ E}$$
$$\text{length } DB = \sqrt{687.87^2 + 400.00^2}$$
$$= 795.72 \text{ ft}$$

In triangle DJB,

$$DJ = \frac{\sin 42°11′(795.72)}{\sin 76°30′} = 549.46 \text{ ft}$$
$$JB = \frac{\sin 61°19′(795.72)}{\sin 76°30′} = 717.94 \text{ ft}$$
$$\text{bearing } JB = \text{brg } DC = \text{N } 12°00′ \text{ W}$$

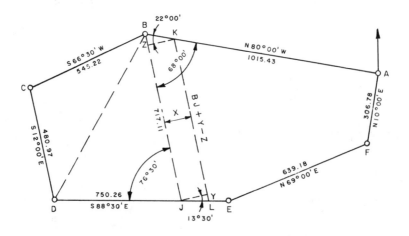

Figure 15.14　Example 15.8, continued

area *BCDJB*

line	bearing	length	latitude	departure	DMD	area
BC	S66°30′W	545.22	−217.41	−500.00	500.00	−108,705
CD	S12°00′E	480.97	−470.46	+100.00	100.00	−47,046
DJ	S88°30′E	549.46	−14.38	+549.27	749.27	−10,774
JB	N12°00′W	717.94	+702.25	−149.27	1149.27	+807,075
			0.00	0.00		640,550

area $BCDJB = 640{,}550 \div 2 = 320{,}275 \text{ ft}^2$

In the trapezoid *BJLK*,

$$Y = X \tan 13°30′$$
$$Z = X \tan 22°00′$$
$$LK = JB + Y - Z$$
$$= JB + X \tan 13°30′ - X \tan 22°00′$$

$$\text{area } KBCDLK = 398{,}652 \text{ ft}^2$$
$$\text{area } BCDJB = 320{,}275$$
$$\text{area } BJLKB = \overline{78{,}377}$$

$$\text{area } BJLKB = \frac{(JB + LK)X}{2}$$

$$78{,}377 = \frac{(JB + JB + X \tan 13°30′ - X \tan 22°00′)X}{2}$$
$$= \frac{(2JB + X \tan 13°30′ - X \tan 22°00′)X}{2}$$
$$= (JB)X - \frac{(\tan 22°00′ - \tan 13°30′)X^2}{2}$$
$$= (717.94)X - (0.0820)X^2$$
$$(0.0820)X^2 - 717.94X + 78377 = 0$$

Substituting in the quadratic formula,

$$X = \frac{717.94 \pm \sqrt{(-717.94)^2 - (4)(0.0820)(78{,}377)}}{(2)(0.0820)}$$
$$= 110.57 \text{ ft}$$

Then,

$$JL = 110.57 \div \cos 13°30′ = 113.71 \text{ ft}$$
$$KB = 110.57 \div \cos 22°00′ = 119.25 \text{ ft}$$
$$LE = DE - DJ - JL$$
$$= 750.26 - 549.46 - 113.71 = 87.09 \text{ ft}$$
$$AK = AB - KB = 1015.43 - 119.25 = 896.18 \text{ ft}$$
$$LK = JB + Y - Z$$
$$= 717.94 + 110.57 \tan 13°30′ - 110.57 \tan 22°00′$$
$$= 699.81 \text{ ft}$$
$$DL = DE - LE = 750.26 - 87.09 = 663.17 \text{ ft}$$

area *AKLEFA*

line	bearing	length	latitude	departure	DMD	area
AK	N80°00′W	896.14	+155.61	−882.53	882.53	+137,330
KL	S12°00′E	699.81	−684.51	+145.50	145.50	−99,596
LE	S88°30′E	87.09	−2.28	+87.06	378.06	−862
EF	N69°00′E	639.18	+229.06	+596.73	1061.85	+243,214
FA	N10°00′E	306.78	+302.12	+53.27	1711.85	+517,166
			0.00	0.00		797,252

area $AKLEFA = 797{,}252 \div 2 = 398{,}626 \text{ ft}^2$

Error in area $= 398{,}656 - 398{,}626 = 30 \text{ ft}^2$

10 CUTTING A GIVEN AREA FROM AN IRREGULAR TRACT

Example 15.9

Two acres are to be cut off of the westerly end of the tract represented by the traverse *ABCDA*, figure 15.15.

line	bearing	length
AB	N51°00′E	647.81
BC	S30°22′E	449.76
CD	S64°14′W	596.15
DA	N39°00′W	308.17

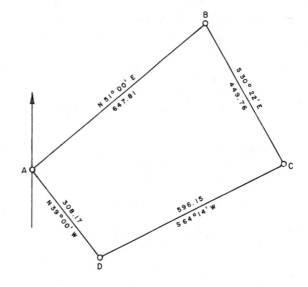

Figure 15.15   Example 15.9

Solution

After plotting the traverse, it can be seen that the interior angle at *A* is a right angle.

Cutting two acres off the westerly end of the tract implies that exactly two acres are to be cut off and that the

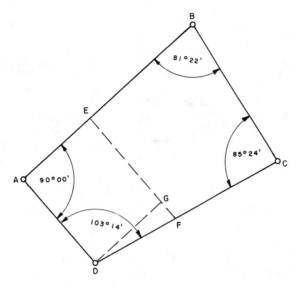

Figure 15.16   Example 15.9, continued

line cutting off the tract will be parallel to the westerly side DA.

Computing the area of the entire tract by **DMD** gives an area of 223,440 ft^2 = 5.359 acres. The area to be cut off is 2 acres = 87,120 ft^2.

First draw EF parallel to DA at the approximate location of the dividing line, forming the trapezoid $AEFD$. Then drop a perpendicular to EF from D, which intersects EF at G. DG is parallel to AE. Using bearings,

$$\text{angle } FDG = 39°00' + 64°14' - 90°00'$$
$$= 13°14'$$

Also,

$$GF = EF - 308.17$$

In triangle DGF,

$$DG = \frac{GF}{\tan 13°14'} = \frac{EF - 308.17}{\tan 13°14'}$$

In trapezoid AED,

$$\text{area} = \frac{(\text{base} + \text{base})(\text{altitude})}{2}$$

$$87,120 = \frac{(EF + 308.17)DG}{2}$$

$$87,120 = \frac{(EF + 308.17)(EF - 308.17)}{2\tan 13°14'}$$

$$87,120 = \frac{EF^2 - 308.17^2}{2\tan 13°14'}$$

$$EF^2 = 2\tan 13°14'(87,120) + 308.17^2$$

$$EF = 368.70\,\text{ft}$$

Solving for DG,

$$87,120 = \frac{(EF + 308.17)DG}{2}$$

$$87,120 = \frac{(368.70 + 308.17)DG}{2}$$

$$DG = \frac{(2)(87,120)}{(368.70 + 308.17)} = 257.42\,\text{ft}$$

$$DF = \frac{DG}{\cos 13°14'} = \frac{257.42}{\cos 13°14'} = 264.44\,\text{ft}$$

Check:

$$\frac{(368.70 + 308.17)(257.42)}{2} = 87,120\,\text{ft}^2$$

11 ANALYTIC GEOMETRY IN PARTING LAND

In many situations, analytic geometry is more applicable to the solution of problems encountered in surveying than is trigonometry. Both analytic geometry and surveying deal with the coordinates of points. The equations of analytic geometry can be used in surveying problems as written or with slight modification. (Analytic geometry for surveyors has been discussed in chapter 3, and can be reviewed at this time.)

Example 15.10

For the traverse $ABCDEA$ shown in figure 15.8, find the intersection of the line connecting points A and C and the line connecting points B and D.

point	coordinates	
	y	x
A	1000.00	1000.00
B	1236.08	492.10
C	817.70	248.35
D	782.01	622.11
E	574.45	854.37

Solution

Using the two-point form of the equation of a line and designating A as point 1 and C as point 2,

$$y - y_1 = \frac{y_2 - y_1}{x_2 - x_1}(x - x_1)$$

$$y - 1000.00 = \frac{817.70 - 1000.00}{248.35 - 1000.00}(x - 1000.00)$$

$$y - 1000.00 = 0.24253(x - 1000.00)$$

$$y - 1000.00 = 0.24253x - 242.53$$

$$0.24253x - y = -757.47$$

Designating B as point 1 and D as point 2,

$$y - y_1 = \frac{y_2 - y_1}{x_2 - x_1}(x - x_1)$$

$$y - y_1 = \frac{782.01 - 1236.08}{622.08 - 492.03}(x - x_1)$$

$$y - 1236.08 = -3.49150(x - 492.03)$$

$$y - 1236.08 = -3.49150x + 1717.92$$

$$3.49150x + y = 2954.00$$

Solving simultaneously,

$$
\begin{aligned}
0.24253x - y &= -757.47 \\
3.49150x + y &= 2954.00 \\
\hline
3.73403x &= 2196.53 \\
x &= 588.25
\end{aligned}
$$

Substituting,

$$y = 900.14$$

The two lines intersect at: $(588.25, 900.14)$

12 AREAS CUT OFF BY A LINE IN A GIVEN DIRECTION FROM A POINT ON THE PERIMETER USING ANALYTIC GEOMETRY

Example 15.11

The tract of land represented by the traverse $ABCDEA$, figure 15.9, is to be divided into two parts by a line from point D parallel to side BC. Coordinates of traverse points are shown.

point	coordinates	
	y	x
A	1000.00	1000.00
B	1236.08	492.03
C	817.72	248.31
D	782.01	622.08
E	574.45	854.35

Solution

$$\text{slope } DH = \text{slope } CB = \frac{1236.08 - 817.72}{492.03 - 248.31} = 1.716560$$

(The slope of line DH can also be found as the cotangent of the bearing angle.)

The equation of line DH is

$$
\begin{aligned}
y - y_1 &= m(x - x_1) \\
y - 782.01 &= 1.716560(x - 622.08) \\
y - 782.01 &= 1.716560x - 1067.84 \\
1.7167x - y &= 285.83
\end{aligned}
$$

The equation of line AB is

$$
\begin{aligned}
y - y_1 &= \frac{y_2 - y_1}{x_2 - x_1}(x - x_1) \\
y - 1000.00 &= \frac{1236.08 - 1000.00}{492.03 - 1000.00}(x - 1000.00) \\
y - 1000.00 &= -0.464752(x - 1000.00) \\
y - 1000.00 &= -0.4648x + 464.75 \\
0.4648x + y &= 1464.75
\end{aligned}
$$

Solving simultaneously,

$$
\begin{aligned}
1.7167x - y &= 285.83 \\
0.4648x + y &= 1464.75 \\
\hline
2.1815x &= 1750.58 \\
x &= 802.47 \\
y &= 1091.76
\end{aligned}
$$

coordinates of point H : $(802.47, 1091.76)$

length $DH =$
$$\sqrt{(802.47 - 622.08)^2 + (1091.76 - 782.01)^2}$$
$$= 358.45 \,\text{ft}$$

length $AH =$
$$\sqrt{(1000.00 - 802.47)^2 + (1000.00 - 1091.76)^2}$$
$$= 217.80 \,\text{ft}$$

length $HB = 560.27 - 217.80 = 342.47 \,\text{ft}$

PRACTICE PROBLEMS FOR CHAPTER 15

1. Compute the bearing and length of the side *DE* of the traverse *ABCDEA*.

line	bearing	length	latitude	departure
AB	N 28°19′ E	560.27	+493.23	+265.76
BC	N 56°23′ W	484.18	+268.06	−403.21
CD	S 08°50′ W	375.42	−370.97	−57.65
DE				
EA	S 67°45′ E	449.83	−170.33	+416.34

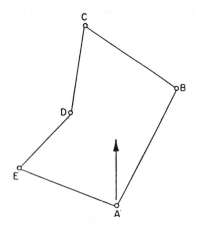

2. Compute the length of the sides *DE* and *EA* of the traverse *ABCDEA*. Use *DA* as a cut-off line. Draw a sketch of triangle *DEA* to a scale of 1″ = 200′. Show size of angles in sketch.

line	bearing	length
AB	N 28°19′ E	560.27
BC	N 56°23′ W	484.18
CD	S 08°50′ W	375.42
DE	S 45°10′ W	
EA	S 67°45′ E	

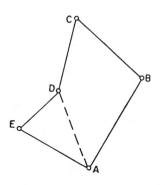

3. Compute the bearing of the sides *DE* and *EA* of the traverse *ABCDEA*. Use *DA* as a cut-off line. Draw a sketch of triangle *DEA* to a scale of 1″ = 200′. Show size of angles in the sketch.

line	bearing	length
AB	N 28°19′ E	560.27
BC	N 56°23′ W	484.18
CD	S 08°50′ W	375.42
DE		311.44
EA		449.83

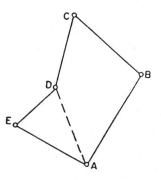

4. Compute the length of side *CD* and the bearing of side *EA* of the traverse *ABCDEA*. Use the following procedure: 1. Designate the cut-off line as *XA*. 2. From *X* draw *CD* in its given direction. 3. Using *A* as a compass point and length of side *EA* as the radius, draw an arc to interesect *CD*. A line from this intersection to *A* gives the direction of *EA*. There are two intersections so there will be two solutions. *AB*: N 28°19′ E, 560.27; *BC*: N 56°23′ W, 484.18. *CD*: S 08°50′ W; *DE*: S 45°10′ W, 311.44. *EA*: 449.83.

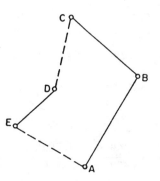

5. The tract of land represented by the traverse *ABCDEA* has been divided into two parts by a line from D to A. Balanced latitudes and departures are shown. Compute the area of the entire tract, the area of the tract $ABCDA$ and the area of the tract $DEAD$ by the DMD method.

line	bearing	length	latitude	departure
AB	N 28°19′ E	560.27	+493.12	+265.66
BC	N 56°23′ W	484.18	+267.97	−403.29
CD	S 08°50′ W	375.42	−371.04	−57.72
DE	S 45°10′ W	311.44	−219.64	−220.91
EA	S 67°45′ E	449.83	−170.41	+416.26

6. The tract of land represented by the traverse $ABCDEA$ is to be divided into two parts by a line from D parallel to the side BC, which intersects the side AB at point H. The length AH must be computed so that point H can be located by measurement from point A. To find AH, the triangle AHD is formed by using the cut-off line AD. The bearing of AD has been computed to be $S\,26°36'\,E$ and the length of AD has been computed to be 436.23′. Area of the tract has been computed as shown. Find the lengths AH, DH, and HB and compute the area of the two subdivided tracts by the DMD method and check the sum of the two areas against the total area.

ln	bearing	length	lat	dep	DMD	area
AB	N 28°19′ E	560.27	+493.12	+265.66	1098.18	+541,535
BC	N 56°23′ W	484.18	+267.97	−403.29	960.55	+257,399
CD	S 08°50′ W	375.42	−371.04	−57.72	499.54	−185,349
DE	S 45°10′ W	311.44	−219.64	−220.91	220.91	−48,521
EA	S 67°45′ E	449.83	−170.41	+416.26	416.26	−70,935
						494,129

AREA = 494,129 ÷ 2 = 247,065 sq ft = 5.672 ac

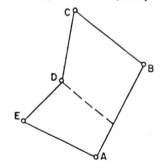

7. The tract of land represented by the traverse $ABCDEA$ in problem 6 is to be divided into two equal parts by a line from point D. The dividing line intersects side AB at the point H. The tract has been surveyed and the following information has been found from computations.

1. Area $ABCDEA$ = 247,065 sq ft

2. Area $ABCDA$ = 182,300 sq ft

3. Brg DA = S 26°36′ E

4. Lth DA = 436.23′

Compute the length of AH, HB, and DH and the bearing DH.

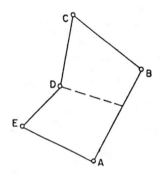

8. The tract of land represented by the traverse $ABCDEFA$ is to be divided into two equal parts by a line parallel to AB. A survey has been made and computations for the area of the tract have been made as shown. Designate dividing line as KL and line parallel to AB as FJ. Find the length of LC, BL, KF, EK, and KL and compute the area of $ABLKFA$ by the DMD method.

ln	bearing	length	lat	dep	DMD	area
AB	S 08°30′ E	480.97	−475.69	+71.09	71.09	−33,817
BC	S 85°00′ E	750.26	−65.39	+747.41	889.59	−58,170
CD	N 72°30′ E	639.18	+192.21	+609.60	2246.60	+431,819
DE	N 13°30′ E	306.78	+298.30	+71.62	2927.82	+873,369
EF	N 76°30′ W	1015.43	+237.05	−987.37	2012.07	+476,961
FA	S 70°00′ W	545.22	−186.48	−512.35	512.35	−95,543
						1,594,619

AREA = 1,594,619 ÷ 2 = 797,310 sq ft = 18.304 ac

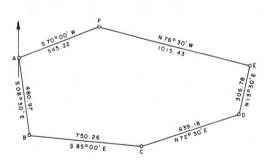

PROFESSIONAL PUBLICATIONS, INC. • P.O. Box 199, San Carlos, CA   94070

16 HORIZONTAL CURVES

1 SIMPLE CURVES

Highways consist of a series of straight sections joined by curved sections. The straight sections are known as *tangents*. The curves are most often circular arcs, known as *simple curves*, but may be spiral curves. Spiral curves are encountered more often on railroads.

Initial locations of highways usually consist of straight lines. Curves are later inserted to connect two intersecting tangents. Many curves of different radii, or degrees of curve, may be selected for any given intersection of tangents.

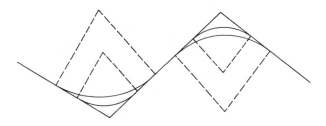

Figure 16.1 Curves connecting tangent lines

Figure 16.1 shows that a choice of circular arcs, or curves, can be made after the tangent locations have been made. The curves are usually classified as to their *degree of curve*. In selecting the degree of curve, consideration is given to design speed, topographic features, economy, and other variables.

2 GEOMETRY

Derivation of formulas for use in computations involving circular curves depends on certain principles of geometry and trigonometry.

3 INSCRIBED ANGLE

An inscribed angle is an angle which has its vertex on a circle and which has chords for its sides. (Figure 16.2a)

4 MEASURE OF AN INSCRIBED ANGLE

An inscribed angle is measured by one-half its intercepted arc. (Figure 16.2a)

5 MEASURE OF ANGLE FORMED BY A TANGENT AND A CHORD

An angle formed by a tangent and a chord is measured by one-half its intercepted arc. (Figure 16.2b)

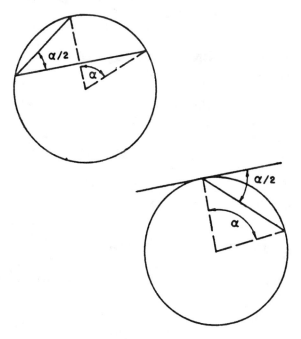

Figure 16.2 Tangent and arc geometry

6 RADIUS IS PERPENDICULAR TO TANGENT

The radius of a circle is perpendicular to a tangent at the point of tangency. (Figure 16.3a)

7 RADIUS IS PERPENDICULAR BISECTOR OF A CHORD

The perpendicular bisector of a chord passes through the center of the circle. (Figure 16.3b)

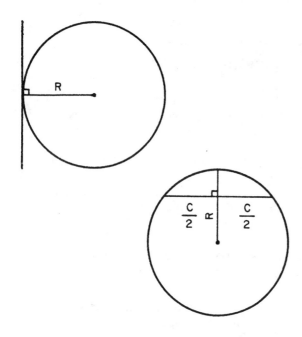

Figure 16.3　Chord geometry

8 DEFLECTION ANGLE EQUALS CENTRAL ANGLE

In Figure 16.4, the sum of the interior angles of the polygon 0-*PC-PI-PT* equals 360°. The angles at the *PC* and *PT* each equals 90°. The sum of the interior angle at the *PI* and the deflection angle equals 180°. The sum of the interior angle at the *PI* and the central angle equals 180°. Therefore, the deflection angle equals the central angle.

9 DEFINITION AND SYMBOLS

Definitions of symbols used in curve computations are given here.

PI-point of intersection. Point where two tangents intersect.

Δ-*deflection angle*. Central angle. Angle at *PI*, or angle at center.

D-degree of curve.	Central angle which subtends 100 feet arc.
L-length of curve.	Distance from *PC* to *PT* along the arc.
T-tangent distance.	Distance from *PI* to *PC*, or distance from *PI* to *PT*.
R-radius.	
LC-length of long chord.	Distance from *PC* to *PT*. Chord length for angle Δ.
C-length of chord.	Chord length for angle *D*.
PC-point of curvature.	Beginning of curve.
PT-point of tangency.	End of curve.
M-middle ordinate.	Length of ordinate from middle of long chord to middle of curve.
E-external distance.	Distance from *PI* to middle of curve.

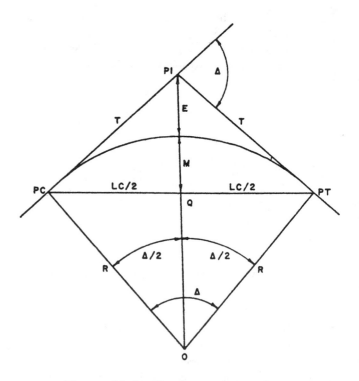

Figure 16.4　Circular arc nomenclature

10 FORMULAS

In computing various components of a circular curve, certain formulas derived from trigonometry are useful and necessary. In addition to the usual six trigonometric functions, two others are used in curve computations: *versed sine* (vers) and *external secant*. (exsec).

Most surveying textbooks include tables for these functions.

$$\text{vers}\,\phi = 1 - \cos\phi \qquad 16.1$$
$$\text{exsec}\,\phi = \sec - 1 \qquad 16.2$$

Other formulas commonly encountered are:

$$T = R\tan\Delta/2 \qquad 16.3$$
$$LC = 2R\sin\Delta/2 \qquad 16.4$$
$$C = 2R\sin D/2 \qquad 16.5$$
$$M = R(1 - \cos\Delta/2) = R\,\text{vers}\,\Delta/2 \qquad 16.6$$
$$E = R(\sec\Delta/2 - 1) = R\,\text{exsec}\,\Delta/2 \qquad 16.7$$

11 DEGREE OF CURVE

There are two definitions of degree of curve. The *arc definition* is used for highways and streets. The *chord definition* is used for railroads. By the arc definition, *degree of curve*, D, is the central angle which subtends a 100 foot arc. By the chord definition, degree of curve, D, is the central angle which subtends a 100 foot chord.

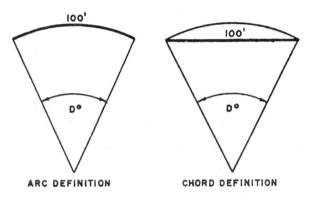

Figure 16.5 Arc and chord definitions

Using the arc definition,

$$D : 100' = 360° : 2\pi R$$
$$R = \frac{36{,}000}{2\pi D} = \frac{5729.58}{D} \qquad 16.8$$
For $D = 1°$, $R = 5729.58\,\text{ft}$

Using the chord definition,
$$R = \frac{50}{\sin D/2} \qquad 16.9$$
For $D = 1°$, $R = 5729.65\,\text{ft}$

Values of R for various values of D are given in appendix E.

When using the arc definition for curve computations with a 100 foot tape to lay out the curve in the field, measurements are actually chord lengths of 100 feet; the arc length is somewhat greater. For curves up to 4°, the difference in arc length and chord length is negligible. For instance, the chord length for a 100 foot arc on a 4° curve is 99.980 feet. For curves of greater degree of curve, the actual chord length can be found in the tables in the appendix. Chord lengths can be measured accordingly.

12 CURVE LAYOUT

Due to their long radii, most curves cannot be laid out by swinging an arc from the center of the circle. They must be laid out by a series of straight lines (chords). This is done by use of transit and tape.

13 DEFLECTION ANGLE METHOD

This method is based on the fact that the angle between a tangent and a chord, or between two chords which form an inscribed angle, is one-half the intercepted arc. (See figure 16.2a–b). In figure 16.6, the angle formed by the tangent at the PC and a chord from the PC to a point 100 feet along the arc is equal to one half the degree of curve. Likewise, the angle formed by this chord and a chord from the PC to a point 100 feet farther along the arc is also equal to one-half the degree of curve. These angles are known as *deflection angles*. The deflection angle from the PC to the PT is one-half the central angle Δ, which provides an important check in computing deflection angles.

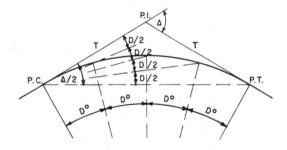

Figure 16.6 Laying out a curve

In laying out the curve, the transit is set up at the PC, PT, or other point on the curve, and deflection angles are turned for 100 foot arcs along the curve as 100 foot arcs are marked off by a 100 foot tape. For degree of curve up to 4°, the difference in length between chord and arc is slight. For sharper curves, this discrepancy can be corrected by laying out a chord slightly less than

---

[1] The arc definition is used exclusively in this book.

PROFESSIONAL PUBLICATIONS, INC. ● P.O. Box 199, San Carlos, CA 94070

100 feet. This length can be found in appendix D or from equation 16.5.

On route surveys, stationing is carried continuously along tangent and curve. Thus, the PC and PT will seldom fall on a full station. The first full station will not likely be 100 feet from the PC. And, the deflection angle to the first full station will not be $D/2$ but will be a fraction of $D/2$.

When stakes are required on closer intervals than full stations, such as 50 feet, the true length of the 50 foot chord (known as a *subchord*) will be less than 50 feet and this length can be found in the appendix or from equation 16.5. (The central angle which subtends a 50 foot arc is $D/2$).

14 LENGTH OF CURVE

The *length of curve*, L, (arc definition), is the distance along the arc from the PC to PT. As any two arcs are proportional to their central angles,

$$L = 100\Delta/D \qquad 16.10$$

15 FIELD PROCEDURE IN STAKING A SIMPLE CURVE

The following steps can be used when staking out a curve from the PC or PT:

Step 1: Measure the deflection angle.

Step 2: Select D by considering design criteria.

Step 3: Compute the tangent distance T from the PI to the PC.

Tangent distances for a 1° curve for various values of Δ can be found in appendix B. Tangent distance for any degree of curve can be found by dividing the tangent distance for a 1° curve by D (degree of curve).

Step 4: Measure the tangent distance T from the PI to the PT and set tacked hub. Measure T from the PI to the PC and set tacked hub.

Step 5: Compute PC station by subtracting T from PI station.

Step 6: Compute length of curve L. Compute PT station by adding L to PC station.

Step 7: Compute deflection angles at PC for each station, checking to see that deflection angle to PT is $\Delta/2$.

Step 8: (a) Set up transit at PC and take foresight on PI with telescope normal and with A

vernier set on 0°00'. Turn deflection angle for each station as chainpeople mark off corresponding arc length. Make check at PT for angle and distance.

(b) Set up transit at PT and take backsight on PC with telescope normal and with A vernier set on 0°00'. Turn deflection angle for each station as chainpeople mark off corresponding arc length, with chainpeople starting at PC. Deflection angles are the same as computed for staking curve from PC. Make check by sighting on PI with deflection angle to PT.

16 CIRCULAR CURVE COMPUTATIONS

Example 16.1

Parameters of a simple curve are to be computed.

$$PI = \text{Sta}\,25 + 05$$
$$\Delta = 20°$$
$$D = 2°$$

Solution

$$L = 100\frac{\Delta}{D} = \frac{(100)(20)}{2} = 1000\,\text{ft}$$

$$R = 2864.79 \qquad \text{(Appendix)}$$

$$R = 5729.58 \div 2 = 2864.79 \qquad \text{(See section 11)}$$

$$T = \frac{T\text{ for }1°\text{ curve}}{D} = \frac{1010.28}{2}$$
$$= 505\,\text{ft} \qquad \text{(Appendix)}$$

$$T = R\tan\Delta/2 = 2864.79\tan(20°/2)$$
$$= 505\,\text{ft} \qquad \text{(Equation 16.3)}$$

$$E = \frac{88.39}{2} = 44.19\,\text{ft} \qquad \text{(Appendix)}$$

$$E = R(\sec\Delta/2 - 1) = R\left(\frac{1}{\cos\Delta/2} - 1\right)$$
$$= 44.19\,\text{ft} \qquad \text{(Equation 16.7)}$$

$$M = R(1 - \cos\Delta/2) = 5729.58(1 - 0.9848)$$
$$= 43.52\,\text{ft} \qquad \text{(Equation 16.6)}$$

$$
\begin{array}{rl}
PI = \text{Sta} & 25 + 05 \\
T = & -5 + 05 \\
\hline
PC = \text{Sta} & 20 + 00 \\
L = & +10 + 00 \\
\hline
PT = \text{Sta} & 30 + 00 \\
\end{array}
$$

The deflection angles are:

point	station		deflection angle
PC	$20 + 00$	=	$0°00'$
	$21 + 00 = D/2$	=	$1°00'$
	$22 + 00 =$ deflection angle of station $21 + 00 + D/2 =$		$2°00'$
	$23 + 00 =$	$22 + 00 + D/2 =$	$3°00'$
	$24 + 00 =$	$23 + 00 + D/2 =$	$4°00'$
	$25 + 00 =$	$24 + 00 + D/2 =$	$5°00'$
	$26 + 00 =$	$25 + 00 + D/2 =$	$6°00'$
	$27 + 00 =$	$26 + 00 + D/2 =$	$7°00'$
	$28 + 00 =$	$27 + 00 + D/2 =$	$8°00'$
	$29 + 00 =$	$28 + 00 + D/2 =$	$9°00'$
PT	$30 + 00 =$	$29 + 00 + D/2 =$	$10°00'$

(Check: deflection angle to $PT = \Delta/2 = 10°00'$)

Example 16.2

Parameters of a simple curve are to be computed.

$$PI = \text{Sta}\, 45 + 78.39$$
$$\Delta = 43°39'$$
$$D = 4°15'$$

Solution

$$L = \frac{43°39'}{4°15'} \times 100 = \frac{43.65°}{4.25°} \times 100 = 1027.06\,\text{ft}$$
$$R = 1348.14 \qquad\qquad\text{(Appendix)}$$
$$T = \frac{T \text{ for } 1° \text{ curve}}{D}$$
$$= \frac{2294.57}{4.25} = 539.90\,\text{ft} \qquad\text{(Appendix)}$$
$$R = 1348.14\,\text{ft} \qquad\qquad\text{(Appendix)}$$

$$
\begin{aligned}
PI &= \text{Sta} & 45 + 78.39 \\
T &= & -5 + 39.90 \\
PC &= \text{Sta} & \overline{40 + 38.49} \\
L &= & +10 + 27.06 \\
PT &= \text{Sta} & \overline{50 + 65.55}
\end{aligned}
$$

The deflection angles are:

point	station		deflection angle
PC	$40 + 38.49$	=	$0°00'$
	$41 + 00 =$	$\frac{100-38.49}{100} \cdot \frac{D}{2} =$	$1°18'$
	$42 + 00 =$	$1°18' + D/2 =$	$3°26'$
	$43 + 00 =$	$3°26' + D/2 =$	$5°33'$
	$44 + 00 =$	$5°33' + D/2 =$	$7°41'$
	$45 + 00 =$	$7°41' + D/2 =$	$9°48'$
	$46 + 00 =$	$9°48' + D/2 =$	$11°56'$
	$47 + 00 =$	$11°56' + D/2 =$	$14°03'$
	$48 + 00 =$	$14°03' + D/2 =$	$16°11'$
	$49 + 00 =$	$16°11' + D/2 =$	$18°18'$
	$50 + 00 =$	$18°18' + D/2 =$	$20°26'$
PT	$50 + 65.55 =$	$20°26' + (0.6555)(2.125) =$	$21°49.5'$

(Check: Deflection angle to $PT = \dfrac{\Delta}{2} = \dfrac{43°39'}{2}$
$$= 21°49.5')$$

Field notes for this curve are shown in Figure 16.7.

17 TRANSIT AT POINT ON CURVE

On long highway curves, it is often impossible to locate the entire curve from one point. Obstructions along the arc or along the line of sight from the transit to a station may prevent location of the curve from one point. In these situations, part of the curve can be located from the *PC* and then the transit can be moved to a point on the curve which has been located from the *PC*. The balance of the curve can then be located. On extremely long curves, part of the curve can be located from the *PC* and part from the *PT*. Any error will be located where the two parts meet.

point	station	deflection calculated angle	bearing	curve data
⊙PT	50 + 65.55	21°49.5′	N 71°54′ E	
	50 + 00.00	20°26′		
	49 + 00.00	18°18′		
	48 + 00.00	16°11′		
	47 + 00.00	14°03′		
	46 + 00.00	11°56′		Δ = 43°39′ RT.
				D = 4°15′
⊙PI	45 + 78.39			T = 539.90′
	45 + 00.00	9°48′		L = 1027.06′
				R = 1348.14′
	44 + 00.00	7°41′		
	43 + 00.00	5°33′		
	42 + 00.00	3°26′		
	41 + 00.00	1°18′		
⊙PC	40 + 38.49	0°00′	N 28°15′ E	

Figure 16.7   Field notes for alignment of Reisel-Mart Highway

In all cases, when it is necessary to move up on a curve, the deflection angles are computed as if the curve were to be located from the *PC*. When it is decided to move the transit, tack points are set at the station which is to be occupied and at the station which is to be used as a backsight, unless the *PC* can be used as a backsight. In either case, to orient the transit at the new station, the deflection angle of the station which is to be used for the backsight is set on the horizontal circle with the upper clamp before setting on the backsight with the lower clamp. If the *PC* is to be used for the backsight, the vernier would be set on 0°00′. After orientation, deflection angles as originally computed are turned for the balance of the curve.

Example 16.3

Station 41 + 00, 42 + 00, and 43 + 00 for the curve tabulated in figure 16.7 have been set from the *PC*, but because of an obstruction on the line of sight, station 44 + 00 cannot be located. The transit is to be moved to station 43 + 00 for continuation of the curve location; the *PC* is to be used for the backsight. Locate station 44 + 00.

Figure 16.8   Example 16.3

Solution

Step 1: Set the transit on station 43 + 00.

Step 2: Set 0°00′ on the *A* vernier.

Step 3: Backsight on the *PC* with the telescope inverted.

Step 4: Plunge the telescope and continue the location using the deflection angles as originally computed. (Note: When the vernier reads 5°33′, the line of sight is tangent to the curve at the point occupied.)

This procedure can be used in locating *culverts* on curves. The culvert station is occupied, line of sight is made tangent at this point, and 90° is turned off this tangent. This puts the centerline of the culvert on a radial line.

Example 16.4

Stations 41 through 46 of the curve tabulated in figure 16.7 have been set from the *PC*. The balance of the curve is to be located from station 46 + 00, from which the *PC* is not visible. Station 43 + 00 is to be used as a backsight. Outline the procedure.

Figure 16.9   Example 16.4

Solution

Step 1: Set the transit on station 46 + 00.

Step 2: Set 5°33′ on the *A* vernier which is the deflection angle for station 43 + 00.

Step 3: Backsight on station 43 + 00 with the telescope inverted.

Step 4: Plunge the telescope and continue the location using deflection angles as originally computed. (The vernier is set on the deflection angle of the station sighted.)

18 COMPUTING TRANSIT STATIONS FOR HIGHWAY LOCATION

As has been mentioned previously, initial locations of highways usually consist of a series of tangents. Curves are inserted later to connect two intersecting tangents. The original tangents are chained from PI to PI, and deflection angles are measured. When curves are inserted, the original stations at PI's must be corrected. Stations of PC's and PT's must be computed by considering distances along the curves and not along the tangents. Obviously, the total length of the project will be less than the tangents' lengths.

Example 16.5

A preliminary highway location has been made. PI stations, deflection angles, and the end of the line are as shown in figure 16.5. (The beginning station is $0 + 00$.) Degree of curve for each curve has been selected. Stations for each PC and PT and the station for the end of the line are to be computed.

PI number	original station	Δ	D
1	$12 + 24.31$	$21°28'R$	$1°30'$
2	$31 + 12.48$	$40°56'L$	$2°30'$
3	$51 + 90.13$	$22°12'R$	$2°00'$
end	$65 + 56.03$	end of line	

Solution

A sketch is drawn showing curves connecting tangents. Distances from PI to PI, tangent distances and lengths of curves are then computed and tabulated. Using the tabulated distances and the sketch, stations are computed in sequence.

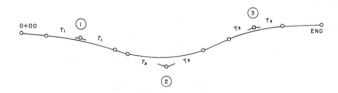

Figure 16.10 Example 16.5

distance $PI_1 - PI_2 = 3112.48 - 1224.31 = 1888.17$

distance $PI_2 - PI_3 = 5190.13 - 3112.48 = 2077.65$

distance $PI_3 - \text{End} = 6556.03 - 5190.13 = 1365.90$

$PC_1 = PI_1 - T_1 = 12 + 24.31 - 724.05 = 5 + 00.26$

$PT_1 = PC_1 + L_1 = 5 + 00.26 + 1431.11$
$\quad = 19 + 31.37$

$PC_2 = PT_1 + (PI_2 - PI_1 - T_1 - T_2)$
$\quad = 19 + 31.37 + (1888.17 - 724.05 - 855.36)$
$\quad = 22 + 40.13$

$PT_2 = PC_2 + L_2 = 22 + 40.13 + 1637.32$
$\quad = 38 + 77.46$

$PC_3 = PT_2 + (PI_3 - PI_2 - T_2 - T_3)$
$\quad = 38 + 77.46 + (2077.65 - 855.36 - 562.05)$
$\quad = 45 + 37.70$

$PT_3 = PC_3 + L_3 = 45 + 37.70 + 1110.00$
$\quad = 56 + 47.70$

$\text{End} = PT_3 + (\text{End} - PI_3 - T_3)$
$\quad = 56 + 47.70 + (1365.90 - 562.05) = 64 + 51.55$

The stations can be computed by successive additions.

$1224.31 - 724.05 + 1431.11 + 3112.48 - 1224.31$
$- 724.05 - 855.36 + 1637.33 + 5190.13 - 3112.48$
$- 855.36 - 562.05 + 1110.00 + 6556.03 - 5190.13$
$- 562.05 = 6451.11(\text{End}).$

The data can now be summarized.

PI	original station	T	L	PC station	PT station
1	$12 + 24.31$	724.05	1431.11	$5 + 00.26$	$19 + 31.37$
	1888.17				
2	$31 + 12.48$	855.36	1637.33	$22 + 40.13$	$38 + 77.46$
	2077.65				
3	$51 + 90.13$	562.05	1110.00	$45 + 37.70$	$56 + 47.70$
	1365.90				
4	$65 + 56.03$				end $= 64 + 51.55$

19 LOCATING CURVE WHEN PI IS INACCESSIBLE

It is often necessary to find the deflection angle Δ and locate the PC and PT of a curve when the PI is inaccessible as shown in figure 16.11.

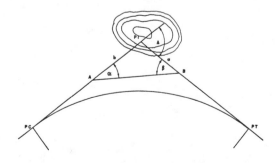

Figure 16.11 Inaccessible PI

To solve the problem, a *point on tangent* (POT) is established on the back tangent at any point **A** which is visible from a point **B** on the forward tangent. The station number of point **A** is established by chaining, and angles α and β, the sum of which equals Δ, and the distance AB, are measured.

Distances a and b in triangle PI-A-B are computed from the law of sines. The station number of the PI is found by adding distance b to the station number of point **A**. The tangent distance T is computed, and the PC and PT are then located by measuring from points **A** and **B**.

Example 16.6

Two tangents of a highway location intersect in a lake, which makes the PI inaccessible. A POT has been established at station 23+45.67 and designated as point **A**. A 4° curve is to be located on the ground.

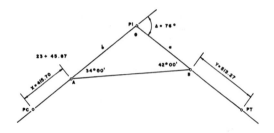

Figure 16.12   Example 16.6

Solution

Step 1: Point **B** is located on the forward tangent so that it is visible from point **A** on the back tangent.

Step 2: Angle A is measured and found to be 34°00′.

Step 3: Angle B is measured and found to be 42°00′.

Step 4: Distance AB is measured and found to be 1020.00 feet.

Step 5: The deflection angle Δ is the sum of angles A and B, which equals 76°00′.

Step 6: From the law of sines, in triangle PI-A-B,

$$a = \frac{1020.00\sin 34°}{\sin 104°} = 587.84\,\text{ft}$$

$$b = \frac{1020.00\sin 42°}{\sin 104°} = 703.41\,\text{ft}$$

Step 7: Using $\Delta = 76°00′$ and $D = 4°$,

$$T = 1119.10\,\text{ft}.$$

Step 8: Adding distance b to station of point **A**, PI is at station $30 + 49.08$. PC is at station $19 + 29.98$.

Step 9: PC is located by measuring 415.69 feet from point **A**.

Step 10: PT is located by measuring 531.26 feet from point **B**.

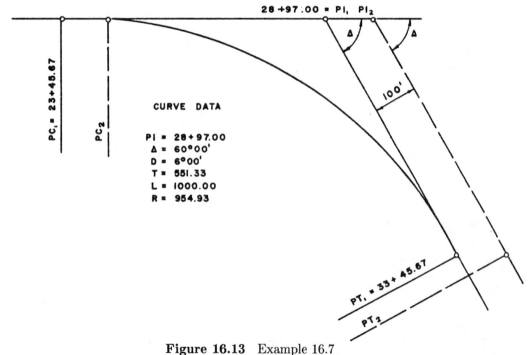

CURVE DATA

PI = 28+97.00
Δ = 60°00′
D = 6°00′
T = 551.33
L = 1000.00
R = 954.93

Figure 16.13   Example 16.7

20 SHIFTING FORWARD TANGENT

Route locations often require changes in curve locations. One such case involves shifting the forward tangent to a new location parallel to the original tangent and keeping the back tangent in its original location. This produces a change in both the *PC* and *PT* stations.

Example 16.7

The forward tangent of the highway curve shown in figure 16.13 is to be shifted outward so that it will be parallel to and 100 feet from the original tangent. Curve data for the original curve is shown in the figure. Degree of curve is to remain unchanged. *PC* and *PT* stations for the new curve are to be computed.

Solution

$$PI_1 - PI_2 = \frac{100}{\sin 60°} = 115.47\,\text{ft}$$
$$PI_2 = 28 + 97.00 + 115.47 = 30 + 12.47$$
$$PC_2 = 30 + 12.47 - 551.33 = 24 + 61.14$$
$$PT_2 = 24 + 61.14 + 1000.00 = 34 + 61.14$$

21 EASEMENT CURVES

Where simple curves are used, the tangent changes to a curved line at the *PC*. This means that a vehicle on a tangent arriving at the *PC* changes direction to a curved path instantaneously. At high speed this is impossible. What actually happens in an automobile is that the driver adjusts the steering wheel to make a gradual transition from a straight line to a curved line.

For railroad cars travelling at high speed, the problem is more acute than for automobiles. The rigidity and the length of a railroad car causes a sharp thrust on the rails by the wheel flanges. To alleviate this situation, curves which provide a gradual transition from tangent to circular curve (and back again) are inserted between the tangent and the circular curve. These transition curves are known as *easement curves*. These curves also provide a place to increase superelevation from zero to the maximum required for the circular curve. The spiral is a curve which fulfills the requirements for the transition.

22 SPIRALS

The simple spiral has certain features which make it useful as an easement curve.

- The degree of curve of the spiral increases from zero at the beginning to the degree of curve of the circular arc where they meet. Likewise, the radius

decreases from infinity at the beginning to the radius of the circular arc.

- Because the degree of curve changes uniformly, the central angle of the spiral equals the length of the spiral in stations times the average degree of curve.

$$\theta_s = \frac{L_s}{100} \times \frac{D}{2} = \frac{L_s D}{200} \qquad 16.11$$

- Spiral central angles are directly proportional to the squares of the lengths from the beginning of the spiral, as are the deflection angles.

23 LENGTH OF SPIRAL

The spiral provides a transition for superelevation. The length required to attain maximum superelevation is a function of the speed of the vehicle. Experiments have provided a formula for the length of the spiral:

$$L_s = \frac{1.6 V^3}{R_c} \qquad 16.12$$
$$L_s = \text{length of spiral (feet)}$$
$$V = \text{design speed (mph)}$$
$$R = \text{radius of circular arc (feet)}$$

24 COMPUTATIONS AND PROCEDURE FOR STAKING

Spirals can be computed by using spiral tables. Symbols used on spirals should be memorized to facilitate use of such tables in other texts. These symbols are illustrated in figure 16.14.

Figure 16.14 shows that the simple curve has been shifted inward in order to insert the easement curve, while the radius of the simple curve has been maintained.

Example 16.8

A circular curve with spiral transitions is to be computed and staked. Curve data is as follows:

$$PI = \text{Sta } 72 + 58.00$$
$$\Delta = 42°00'$$
$$D_c = 5°$$
$$V = 60\,\text{mph}$$

Solution

(This solution requires use of spiral curve tables.)

$$R_c = 1145.92$$
$$L_s = 1.6 \times \frac{V^3}{R_c} = 1.6 \times \frac{60^3}{1145.92} = 301, \text{ use } 300\,\text{ft}$$

Transit stations

$$\theta_s = \frac{L_s D_c}{200} = \frac{300 \times 5°}{200} = 7°30'$$

$$\Delta_c = \Delta - 2\theta_s = 42° - 2 \times 7°30' = 27°00'$$

$$L_c = \frac{\Delta_c}{D_c} \times 100 = \frac{27}{5} \times 100 = 540.00 \text{ ft}$$

$$p = 3.27 \text{ ft}; \quad k = 149.91 \text{ ft}$$

$$T_s = (R_c + p) \tan \frac{\Delta}{2} + k$$

$$= (1145.92 + 3.27)(0.38386) + 149.91$$

$$= 591.04$$

$$
\begin{aligned}
PI &= 72 + 58.00 \\
T_s &= -5 + 91.04 \\
TS &= \overline{66 + 66.96} \\
L_s &= +3 + 00.00 \\
SC &= \overline{69 + 66.96} \\
L_c &= +5 + 40.00 \\
CS &= \overline{75 + 06.96} \\
L_s &= +3 + 00.00 \\
ST &= \overline{78 + 06.06}
\end{aligned}
$$

$$L.T. = 200.18$$
$$S.T. = 100.16$$
$$LC = 299.77$$

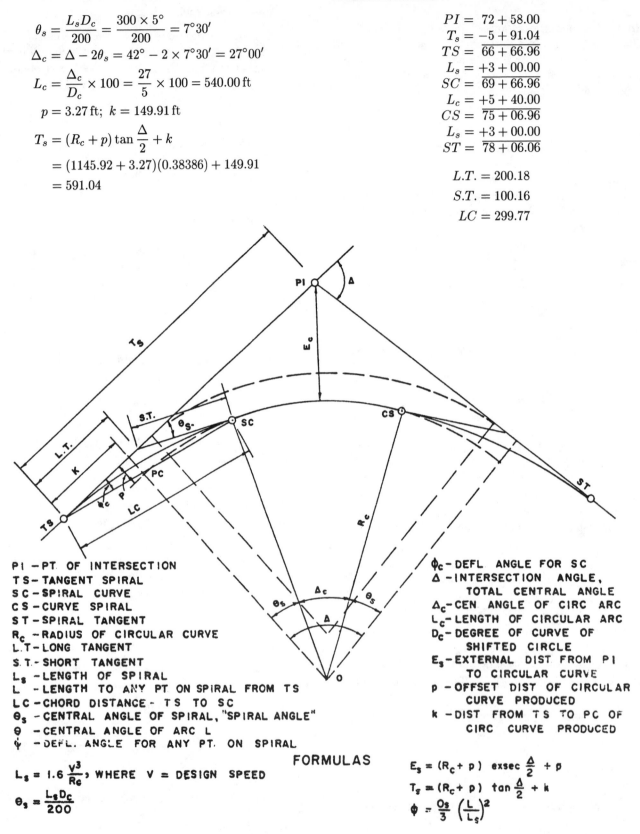

PI – PT. OF INTERSECTION
TS – TANGENT SPIRAL
SC – SPIRAL CURVE
CS – CURVE SPIRAL
ST – SPIRAL TANGENT
R_c – RADIUS OF CIRCULAR CURVE
L.T. – LONG TANGENT
S.T. – SHORT TANGENT
L_s – LENGTH OF SPIRAL
L – LENGTH TO ANY PT ON SPIRAL FROM TS
LC – CHORD DISTANCE – TS TO SC
θ_s – CENTRAL ANGLE OF SPIRAL, "SPIRAL ANGLE"
θ – CENTRAL ANGLE OF ARC L
ψ – DEFL. ANGLE FOR ANY PT. ON SPIRAL

ϕ_c – DEFL ANGLE FOR SC
Δ – INTERSECTION ANGLE,
    TOTAL CENTRAL ANGLE
Δ_c – CEN ANGLE OF CIRC ARC
L_c – LENGTH OF CIRCULAR ARC
D_c – DEGREE OF CURVE OF
    SHIFTED CIRCLE
E_s – EXTERNAL DIST FROM PI
    TO CIRCULAR CURVE
p – OFFSET DIST OF CIRCULAR
    CURVE PRODUCED
k – DIST FROM TS TO PC OF
    CIRC CURVE PRODUCED

FORMULAS

$$L_s = 1.6 \frac{V^3}{R_c}, \text{ WHERE } V = \text{DESIGN SPEED}$$

$$\theta_s = \frac{L_s D_c}{200}$$

$$E_s = (R_c + p) \text{ exsec} \frac{\Delta}{2} + p$$

$$T_s = (R_c + p) \tan \frac{\Delta}{2} + k$$

$$\phi = \frac{\theta_s}{3} \left(\frac{L}{L_s}\right)^2$$

Figure 16.14   Spiral curve nomenclature

PROFESSIONAL PUBLICATIONS, INC. ● P.O. Box 199, San Carlos, CA 94070

Deflection angles for spiral

$$\phi = \frac{\theta_s}{3} \times \frac{(L)^2}{(L_s)^2} = \frac{1}{3}\frac{(7.5)L^2}{(300)^2} = 0.0000278L^2$$

point	station	L	L^2	ϕ	ϕ
TS	66 + 66.96	0.00	0	0°	= 0°
	67 + 00	33.04	1,092	0.0303°	= 0°02′
	67 + 50	83.04	6,896	0.1915°	= 0°11.5′
	68 + 00	133.04	17,700	0.4916°	= 0°29.5′
	68 + 50	183.04	33,503	0.9306°	= 0°56′
	69 + 00	233.04	54,307	1.5082°	= 1°30.5′
	69 + 50	283.04	80,111	2.2253°	= 2°13.5′
SC	69 + 66.96	300.00	90,000	2.5000°	= 2°30′

Field procedure

Step 1: Locate TS by measuring from PI.

Step 2: Locate intersection of $L.T.$ and $S.T.$ (200.18′ from TS)

Step 3: Set up at intersection of $L.T.$ and $S.T.$; sight on PI. Locate SC by turning 7°30′ (θ_s) and measuring 100.16′. ($S.T.$)

Step 4: Locate CS as in 2 and 3.

Step 5: Set up on TS and sight on PI. Locate points on spiral using deflection angles.

Step 6: Locate points on circular curve:

(a) Set up on SC.

(b) Sight on intersection of $L.T.$ and $S.T.$ with telescope inverted and 0°00′ on A vernier. Or, sight on TS with telescope inverted and 2°30′ on A vernier.

(c) Set points on circular curve in normal manner.

Step 7: Set up on ST and locate spiral as in step 5.

25 STREET CURVES

Because street curves are usually short radius, field procedures in staking them may differ from those used for highway curves. All formulas used for highway curves are valid, but the choice of formulas may vary. Street curves are computed by using the arc definition just as highway curves, but stakes are usually set at 25 or 50 feet stations. The difference in the arc length and the chord length is much more pronounced on curves of short radius. Highway curves are usually designated by their degree of curve. Street curves are usually designated by a round-number radius and the degree of curve concept is not used.

From equation 16.4, the long chord for a highway curve is equal to $2R\sin\Delta/2$, Δ being the central angle. Accordingly, $C = 2R\sin D/2$ for an arc of 100 feet. For street curves, the formula is expressed as $C = 2R\sin\delta/2$, where δ is the central angle for any arc.

Some street curves are divided into 3, 4, or more equal arcs, but usually they are staked on the half- or quarter-stations. Chord lengths from quarter-station to quarter-station (or half-station to half-station) are usually measured, as is done on highway curves. For short curves, chords from the PC to each station can also be measured.

26 CURVE COMPUTATIONS

The degree of curve concept is not used. Therefore, the deflection angle will be expressed in terms of the central angle, δ. This is illustrated in example 16.9.

Example 16.9

Computations are to be made for a street curve with a deflection angle Δ of 50°00′ and a centerline radius of 120 feet. The PI is at station $8 + 72.43$.

Solution

$$T = R\tan\Delta/2 = 120\tan 25 = 55.96\,\text{ft}$$
$$L = 50/360 \times 2\pi R = 50/360 \times 2\pi 120 = 104.72\,\text{ft}$$

$$
\begin{array}{rl}
PI = & 8 + 72.43 \\
T = & -55.96 \\
\hline
PC = & 8 + 16.47 \\
L = & +1 + 04.72 \\
\hline
PT = & 9 + 21.19
\end{array}
$$

Deflection Angles

$PC\ 8 + 16.47 = 0°00′$

$$8 + 25 = \frac{25.00 - 16.47}{104.72} \times \frac{50°}{2} = \frac{8.53}{104.72} \times 25°$$
$$= 2.0364° = 2°02′$$

$$8 + 50 = \frac{8.53 + 25.00}{104.72} \times \frac{50°}{2} = \frac{33.53}{104.72} \times 25°$$
$$= 8.0047° = 8°00′$$

$$8 + 75 = \frac{8.53 + 50.00}{104.72} \times \frac{50°}{2} = \frac{58.53}{104.72} \times 25°$$
$$= 13.9730° = 13°58′$$

$$9 + 00 = \frac{8.53 + 75.00}{104.72} \times \frac{50°}{2} = \frac{83.53}{104.72} \times 25°$$
$$= 19.9413° = 19°56′$$

$$PT\ 9 + 21.19 = \frac{8.53 + 96.19}{104.72} \times \frac{50°}{2} = \frac{104.72}{104.72} \times 25°$$
$$= 25.0000° = 25°00′$$

Chord Lengths ($C = 2R\sin\delta/2$)

$8 + 16.47$ to $8 + 25.00 : C = 240\sin 2.0364° = 8.53\,\text{ft}$

$8 + 25.00$ to $8 + 50.00 : C = 240\sin(8.0047° - 2.0364°)$
$$= 24.95\,\text{ft}$$

$9 + 00.00$ to $9 + 21.19 : C = 240\sin(25.0000°$
$$- 19.9413°) = 21.16\,\text{ft}$$

Field Notes

point	station	$\delta/2$	deflection angle	C	curve data
PT	$9 + 21.19$		$25°00'$	21.16	
	$9 + 00$	$19.9412°$	$19°56'$	24.95	$\Delta = 50°00'$
	$8 + 75$	$13.9730°$	$13°58'$	24.95	$R = 120'$
	$8 + 50$	$8.0047°$	$8°00'$	24.95	$T = 55.96$
	$8 + 25$	$2.0364°$	$2°02'$	8.53	$L = 104.72$
PC	$8 + 16.7$	$0.0000°$	$0°00'$		

The length of the curve is only 104.72 feet. Therefore, the curve could be staked by measuring all chords from the *PC*. These chord lengths are computed as follows:

PC to $8 + 25.00$: $C = 240\sin 2.0364° = 8.53\,\text{ft}$
PC to $8 + 50.00$: $C = 240\sin 8.0047° = 33.42\,\text{ft}$
PC to $8 + 75.00$: $C = 240\sin 13.9730° = 57.95\,\text{ft}$
PC to $9 + 00.00$: $C = 240\sin 19.9412° = 81.85\,\text{ft}$
PC to $9 + 21.19$: $C = 240\sin 25.0000° = 104.43\,\text{ft}$

27 PARALLEL CIRCULAR ARCS

The design radius for a curve is usually to the centerline. However, it is often necessary to locate a parallel curve such as a right-of-way line, the edge of pavement, or a curb line. Because the central angle is the same for parallel arcs, deflection angles are the same.

The chord lengths are a function of the radius of the arc, and they will be different for arcs of different radius. Length of the curve is also a function of the radius. The *PC* stations for all parallel arcs will be the same. Arc lengths for inside and outside curves will be different, but the *PT* stations will be the same.

Example 16.10

Construction stakes are to be set for the curve in the preceding example. Street width is 34 feet. Stakes are to be set on 3 foot offsets, inside and outside of the edge of pavement.

Solution

Deflection angles are given in example 16.19. Chord lengths are computed from the formula $C = 2R\sin\delta/2$

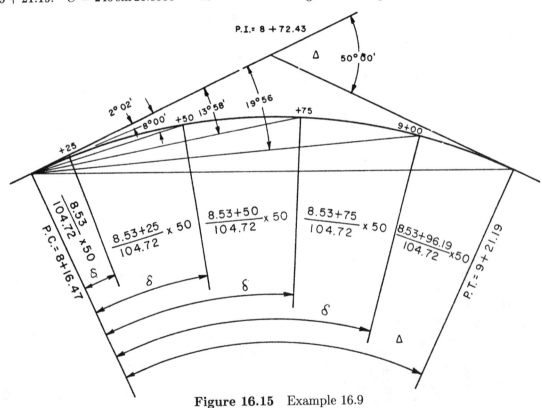

Figure 16.15   Example 16.9

and are shown in the field notes here.

point station	deflection angle	$R=100'$ C inside	$R=140'$ C outside	curve data
PT 9 + 21.19	25°00'			
		17.64	24.69	$\Delta = 50°00'$
9 + 00	19°56'			
		20.80	29.11	$R = 120$
8 + 75	13°58'			
		20.80	29.11	$T = 55.96$
8 + 50	8°00'			
		20.80	29.11	$L = 104.72$
8 + 25	2°02'			
		7.11	9.95	
PC 8 + 16.47	0°00'			

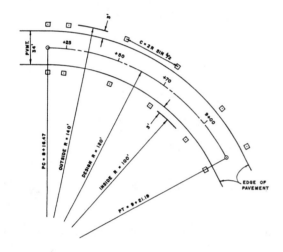

Figure 16.16   Example 16.10

28 CURB RETURNS AT STREET INTERSECTIONS

Curb returns are the arcs made by the curbs at street intersections. The radius of the arc is selected by the designer with consideration given to the speed and volume of traffic. A radius of 30 feet to the back of curb is common. Streets which intersect at right angles have curb returns of one-quarter circle.

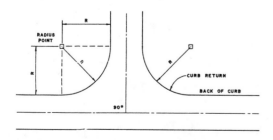

Figure 16.17   Curb returns

The arcs can be swung from a *radius point*, the center of the circle. The radius point can be located by finding the intersection of two lines, each of which is parallel to one of the curb lines and at a distance equal to the radius. One stake at the radius point is sufficient for the curb return.

Curb returns for streets which do not intersect at right angles are shown in figure 16.18. Computations for the location of PC's, PT's, and radius points will be shown in example 16.11.

Example 16.11

Elm Street and 23rd Street intersect as shown in figure 16.18. Pavement width for each street is 36 feet. Radius to edge of pavement is 30 feet. Computations for the location of radius points are to be made.

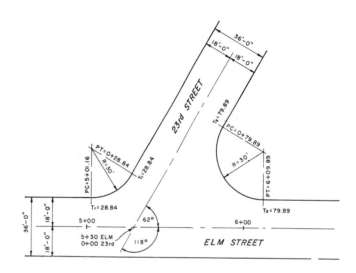

Figure 16.18   Example 16.11

Solution

PC **Stations along Elm Street**

$$\Delta_1 = 62°00'$$
$$T_1 = R \tan \Delta/2$$
$$= 48 \tan 31° = 28.84 \, \text{ft}$$

$$PI = 5 + 30.00$$
$$T_1 = \quad -28.84$$
$$PC = \overline{5 + 01.16}$$

$$\Delta_2 = 118°00'$$
$$T_2 = R \tan \Delta/2$$
$$= 48 \tan 59° = 79.89 \, \text{ft}$$

$$PI = 5 + 30.00$$
$$T_2 = \quad +79.89$$
$$PC = \overline{6 + 09.89}$$

PT Stations along 23rd Street

$$PI = 0 + 00.00$$
$$T_1 = +28.84$$
$$PT = \overline{0 + 28.84}$$

$$PI = 0 + 00.00$$
$$T_2 = +79.89$$
$$PT = \overline{0 + 79.89}$$

29 COMPOUND CURVES

A *compound curve* consists of two or more simple curves with different radii joined together at a common tangent point. Their centers are on the same side of the curve.

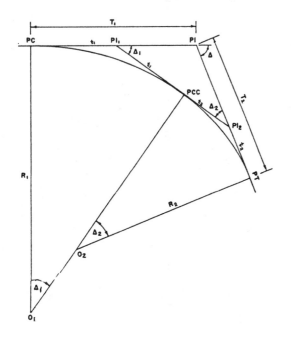

Figure 16.19   Compound curve

Compound curves are not generally used for highways except in mountainous country, because an abrupt change in degree of curve causes a serious hazard even at moderate speeds. They are sometimes used for curvilinear streets in residential subdivisions, however.

In figure 16.19, the subscript ($_1$) is used for the curve of longer radius, and the subscript ($_2$) is used for the curve of shorter radius. The point of common tangency is called the *point of common curvature, PCC*. The short tangents for the two curves are designated as t_1 and t_2.

There are seven major parameters of a compound curve: Δ, Δ_1, Δ_2, R_1, R_2, T_1, and T_2. Four of these must be known before computations can be made. Usually, Δ is measured. R_1, R_2, and either Δ_1 or Δ_2 are given.

If Δ, Δ_1, R_1, and R_2 are given, equations 16.13–16.16 can be used.

$$\Delta = \Delta_1 + \Delta_2 \qquad 16.13$$

$$\Delta_2 = \Delta - \Delta_1 \qquad 16.14$$

$$t_1 = R_1 \tan(\Delta_1/2) \qquad 16.15$$

$$t_2 = R_2 \tan(\Delta_2/2) \qquad 16.16$$

The triangle PI-PI_1-PI_2 can be solved by the law of sines. The sine of the angle at the PI equals the sine of Δ because they are related angles.

$$PI{-}PI_1 = \frac{\sin \Delta_2 (t_1 + t_2)}{\sin \Delta} \qquad 16.17$$

$$PI{-}PI_2 = \frac{\sin \Delta_1 (t_1 + t_2)}{\sin \Delta} \qquad 16.18$$

$$T_1 = t_1 + PI{-}PI_1 \qquad 16.19$$

$$T_2 = t_2 + PI{-}PI_2 \qquad 16.20$$

To stake out the curve, the PC and PT are located as is done for a simple curve. The PCC is located by establishing the common tangent from either point PI_1 or PI_2. Deflection angles for the two curves are computed separately. The first curve is staked from the PC, and the second curve is staked from the PCC using the common tangent for orientation with the vernier set on $0°00'$.

Example 16.12

PC, PCC, and PT stations, deflection angles, and chord lengths are to be computed from the following information:

$$PI = \text{Sta } 15 + 56.32$$

$$\Delta = 68°00'$$

$$\Delta_1 = 35°00'$$

$$R_1 = 600\,\text{ft}$$

$$R_2 = 400\,\text{ft}$$

Solution

$$\Delta_2 = \Delta - \Delta_1 = 68°00' - 35°00' = 33°00'$$

$$t_1 = R_1 \tan \Delta_1/2 = 600 \tan 17°30' = 189.18 \text{ ft}$$

$$t_2 = R_2 \tan \Delta_2/2 = 400 \tan 16°30' = 118.49 \text{ ft}$$

$$PI-PI_1 = \frac{(t_1+t_2)\sin\Delta_2}{\sin\Delta} = \frac{307.67\sin 33°}{\sin 68°}$$
$$= 180.73 \text{ ft}$$

$$PI-PI_2 = \frac{(t_1+t_2)\sin\Delta_1}{\sin\Delta} = \frac{307.67\sin 35°}{\sin 68°}$$
$$= 190.33 \text{ ft}$$

$$T_1 = t_1 + PI-PI_1 = 189.18 + 180.73$$
$$= 369.91 \text{ ft}$$

$$T_2 = t_2 + PI-PI_2 = 118.49 + 190.33$$
$$= 308.82 \text{ ft}$$

$$L_1 = \frac{\Delta_1}{360} \times 2\pi R = \frac{35}{360} \times 2\pi 600 = 366.52 \text{ ft}$$

$$L_2 = \frac{\Delta_2}{360} \times 2\pi R = \frac{33}{360} \times 2\pi 400 = 230.38 \text{ ft}$$

$$
\begin{array}{rl}
PI = & 15 + 56.32 \\
T_1 = & -3 + 69.91 \\
PC = & \overline{11 + 86.41} \\
L_1 = & +3 + 66.52 \\
PCC = & \overline{15 + 52.93} \\
L_2 = & +2 + 30.38 \\
PT = & \overline{17 + 83.31}
\end{array}
$$

Deflection Angles

point	station		deflection angles
PC	11 + 86.41		
	12 + 00	$\frac{13.59}{366.52} \times \frac{35}{2} =$	$0.6489° = 0°39'$
	13 + 00	$\frac{113.59}{366.59} \times \frac{35}{2} =$	$5.4235° = 5°25'$
	14 + 00	$\frac{213.59}{366.52} \times \frac{35}{2} =$	$10.1981° = 10°12'$
	15 + 00	$\frac{313.59}{366.52} \times \frac{35}{2} =$	$14.9728° = 14°59'$
PCC	15 + 52.93	$\frac{366.52}{366.52} \times \frac{35}{2} =$	$17.5000° = 17°30'$
	16 + 00	$\frac{47.07}{230.38} \times \frac{33}{2} =$	$3.3712° = 3°22'$
	17 + 00	$\frac{147.07}{230.38} \times \frac{33}{2} =$	$10.5332° = 10°32'$
PT	17 + 83.31	$\frac{230.38}{230.38} \times \frac{33}{2} =$	$16.5000° = 16°30'$

Chord Lengths

$$C = 1200 \sin 0.6489° = 13.59 \text{ ft}$$
$$C = 1200 \sin 4.7746° = 99.88 \text{ ft}$$
$$C = 1200 \sin 2.5272° = 52.91 \text{ ft}$$
$$C = 800 \sin 3.3712° = 47.04 \text{ ft}$$
$$C = 800 \sin 7.1620° = 99.74 \text{ ft}$$
$$C = 800 \sin 5.9668° = 83.16 \text{ ft}$$

Field notes showing results of computations are shown in figure 16.20.

point	station	deflection angle	chord	calculated bearing	curve data
PT	17 + 83.31	16°30'			
			83.16'		
	17 + 00.00	10°32'			
			99.74'		
	16 + 00.00	3°22.3'			$\Delta = 68°00'$
PI	15 + 56.32		47.04'		$R_1 = 600'$
PCC	15 + 52.93	17°30'			$\Delta_1 = 35°00'$
			52.91'		$R_2 = 400'$
	15 + 00.00	14°58.7'			$\Delta_2 = 33°00'$
			99.88'		$T_1 = 369.91'$
	14 + 00.00	10°12.1'			$T_2 = 308.81'$
			99.88'		$L_1 = 366.52'$
	13 + 00.00	5°25.5'			$L_2 = 230.38'$
			99.88'		
	12 + 00.00	0°38.9'			
			13.59'		
PC	11 + 86.41	0°00'			

Figure 16.20 Example 16.12—Field notes for Elm Street

PRACTICE PROBLEMS FOR CHAPTER 16

1. Write the missing word or words in each sentence.

(a) Highway curves are most often ... arcs known as simple curves.

(b) An inscribed angle is an angle which has its vertex on a ... and which has ... for its sides.

(c) An inscribed angle is measured by ... its intercepted arc.

(d) An angle formed by a tangent and a chord is measured by ... its intercepted arc.

(e) The radius of a circle is ... to a tangent at the point of tangency.

(f) A perpendicular bisector of a chord passes through the ... of the circle.

(g) By the arc definition, degree of curve D is the central angle which subtends a 100 ft ....

(h) By the chord definition, degree of curve D is the central angle which subtends a 100 ft ....

(i) By the arc definition, the radius R of a 1° curve is ... ft.

(j) The deflection angle for a full station for a 1° curve is ....

2. Place all symbols pertinent to a circular curve on the figure.

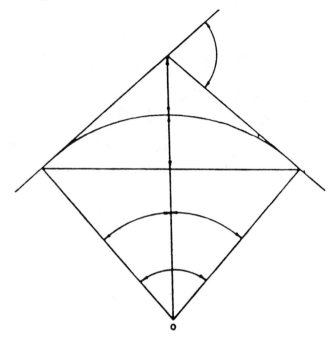

3. Using the information given, compute the PC and PT stations and deflection angles for each full station of the simple highway curve. Round-off T to the nearest foot.

$$PI = \text{Sta} \, 25 + 01$$
$$\Delta = 10°$$
$$D = 1°$$

4. Using the information given, compute the PC and PT stations and deflection angles for each full station of the simple highway curve. Express length to two decimal places.

$$PI = \text{Sta} \, 45 + 11.75$$
$$\Delta = 30°$$
$$D = 3°$$

5. Prepare field notes to be used in staking the centerline of a simple horizontal curve for a highway, using the information given.

$$PI = 29 + 62.78$$
$$\Delta = 40°21'\,\text{Lt}$$
$$D = 5°15'$$
back tangent bearing $= \text{N}\,56°12'\,\text{W}$

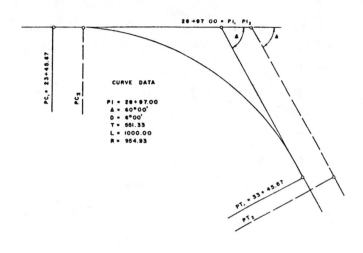

CURVE DATA

PI = 28+97.00
Δ = 60°00'
D = 6°00'
T = 551.33
L = 1000.00
R = 954.93

6. A preliminary highway location has been made by locating tangents. Deflection angles have been measured at each PI, and the station number of each PI has been established by measuring along the tangents. Circular curves have not been located, but the degree of curve has been established for each curve. The beginning point is at station $0 + 00$. Using these data, make necessary computations to establish stations for PC's and PT's of the curves and for the end of the line.

PI no.	original station	Δ	D
1	$10 + 35.27$	$13°34'\,\text{R}$	$1°30'$
2	$36 + 15.44$	$15°18'\,\text{L}$	$2°30'$
3	$52 + 98.40$	$18°05'\,\text{R}$	$3°00'$
End	$61 + 32.77$	End of line	

7. In locating a highway, the PI of two tangents fell in a lake and is inaccessible. Point A on the back tangent has been established at station $26 + 52.61$. Point B has been established on the forward tangent and is visible from point A. Angle A has been measured and found to be $23°13'$; angle B has been measured and found to be $19°55'$. Length of AB has been found to be 434.87 feet. Find the deflection angle Δ and the PC and PT stations for a $3°$ curve.

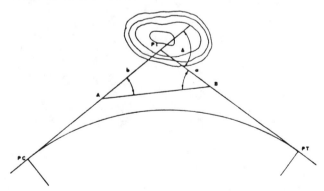

8. The forward tangent of the highway curve shown is to be shifted outward so that it will be parallel to and 50 feet from the original tangent. Data for the original curve is shown in the figure. Degree of curve is to be unchanged. Find the PC and PT stations for the new curve.

9. Find L for Δ and D indicated.

(a) $\Delta = 32°18'$

   $D = 2°30'$

(b) $\Delta = 41°27'$

   $D = 3°15'$

10. Find T for Δ and D indicated.

(a) $\Delta = 41°51'$

   $D = 1°45'$

(b) $\Delta = 39°14'$

   $D = 2°15'$

11. Find E for Δ and D indicated.

(a) $\Delta = 31°30'$

   $D = 1°30'$

(b) $\Delta = 42°21'$

   $D = 2°45'$

12. Find D for nearest full degree for Δ indicated and T to be approximately as indicated.

(a) $\Delta = 32°56'$

   $T = 600\,\text{ft}$

(b) $\Delta = 40°10'$

   $T = 1000\,\text{ft}$

13. Use trigonometric equations to solve the following problems. Show equation used for the solution.

(a) Find R for $D = 2°$.

(b) Find D for $R = 1909.86$.

(c) Find T for $\Delta = 34°44'$ and $R = 800$ ft.

(d) Find E for $\Delta = 37°20'$ and $R = 650$ ft.

(e) Find M for $\Delta = 42°51'$ and $R = 800$ ft.

(f) Find LC for $\Delta = 32°55'$ and $R = 850$ ft.

(g) Find chord length for $D = 8°$, $R = 716.20$ ft, and arc $= 50$ ft.

14. Compute deflection angles and chord lengths for quarter stations (25 ft) for a street curve. Chords are to be measured from quarter-station to quarter-station.

$$PI = \text{Sta} \, 8 + 78.22$$
$$\Delta = 28°$$
$$R = 250 \, \text{ft}$$

15. Prepare field notes to be used in staking the centerline of a horizontal street curve on the quarter-stations.

$$PI = 8 + 47.52$$
$$\Delta = 36°00' \, \text{Right}$$
$$R = 400 \, \text{ft}$$

16. Compute PC, PCC, and PT stations and deflection angles for full stations for the compound curve from the information given.

$$PI = 14 + 78.32$$
$$\Delta = 68°00'$$
$$\Delta_1 = 36°00'$$
$$R_1 = 400 \, \text{ft}$$
$$R_2 = 300 \, \text{ft}$$

17. Prepare field notes to be used in staking the centerline of the compound curve on full stations.

$$PI = 12 + 65.35$$
$$\Delta = 70°00'$$
$$\Delta_1 = 36°00'$$
$$R_1 = 900 \, \text{ft}$$
$$R_2 = 600 \, \text{ft}$$

PROFESSIONAL PUBLICATIONS, INC. • P.O. Box 199, San Carlos, CA   94070

17

TOPOGRAPHIC SURVEYING AND MAPPING

1 CARTOGRAPHY

Cartography is the profession of making maps. Topographic maps provide a plan view of a portion of the earth's surface, showing natural and man-made features such as rivers, lakes, roads, buildings and canals. The shape, or relief, of the area is shown by contour lines, hachures, or shading.

2 USES OF TOPOGRAPHIC MAPS

The planning of most construction begins with the *topographic map*, sometimes referred to as the *contour map*. A study of a topographic map should precede the planning of highways, canals, subdivisions, shopping centers, airports, golf courses, and other improvements.

3 TOPOGRAPHIC SURVEYS

Topographic surveys are made to determine the relative positions of points and objects so that the map maker can accurately represent their positions on the map.

4 TYPES OF MAPS

There are two basic types of maps: the strip map and the area map. The *strip map* is used in the development of highways, railroads, pipelines, powerlines, canals, and other projects which are narrow in width and long in length. The *area map* is used in the development of subdivisions, shopping centers, airports, and other localized projects.

5 CONTROL FOR TOPOGRAPHIC SURVEYS

Of great importance in topographic surveys is horizontal and vertical control. *Control* is the means of transferring the relative positions of points and objects on the surface of the earth to the surface of the map.

6 HORIZONTAL CONTROL

Relative position in the horizontal plane is maintained by horizontal control. Horizontal control consists of a series of points accurately fixed in position by distance and direction in the horizontal plane. For most topographic surveying, traverses furnish satisfactory control. For strip maps, the open traverse is used. For area maps, the closed traverse is used. The open traverse can be tied to fixed points at each end. The closed traverse can be closed to form a net which is accurate to the degree required.

For large areas, such as states, triangulation or trilateration furnish the most economical control.

7 VERTICAL CONTROL

Relative position in the vertical plane can be maintained by a series of bench marks in the map area. These bench marks are referred to a known *datum*, usually mean sea level.

8 HORIZONTAL TIES

After the traverse is closed to the required specifications, objects which are to be included on the map are tied to the traverse. This sometimes is called *detailing*. For large surveys, ties are made by photogrammetry. For smaller surveys, ties are made on the ground.

At least two measurements are required to tie one point to the traverse.

9 METHODS OF LOCATING POINTS IN THE FIELD

The two measurements required to tie one point to the traverse may consist of two horizontal distances, an angle and a horizontal distance, or two angles. There are several methods used to locate a point in the field. Only the four most common will be discussed.

10 RIGHT ANGLE OFFSET METHOD OF TIES

This is the most common method used in route survey-ing for preparing strip maps. The ties are made after the centerline (or traverse line) has been established. Usually, stakes are driven at each station on the center-line, a 100 foot steel tape is stretched between successive stations with the 100 foot mark on the tape forward, and points on either side of the tape are tied to the tra-verse before the tape is moved forward to the next two stakes. To tie in the corners of the house shown in Fig-ure 17.1, the surveyor moves along the tape to a point on the line where he estimates a perpendicular from the traverse line would strike a corner of the building. He observes the plus at this point by glancing at the tape on the ground and then measures the distance from the traverse line to the corner with another (usually cloth) tape, and records both measurements in the field book. With the 100 foot mark of the steel tape forward, pluses are read directly. He repeats this procedure for the next corner. All sides of the house, including the side between the tied corners are then measured with the cloth tape. A sketch of the house showing the di-mensions of all sides of the house is placed in the field book.

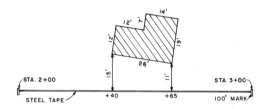

Figure 17.1   Right angle offsets

Unless the scale of the map is very large, measurements for ties are to the nearest foot. It is usually impossible to scale a distance on the map for a tenth of a foot.

A right angle mirror prism is convenient in establishing right angles. As a less-accurate method, the surveyor may stand on the transit line facing the point to be tied with arms outstretched on each side pointing along the traverse line, and then bring both arms to the front of the body. If they do not point to the object to be tied, the surveyor should move along the traverse line until they do.

11 ANGLE AND DISTANCE METHOD OF TIES

This method is the most common of those used in prepa-ration of area maps. The azimuth-stadia method allows direction and distance measurements to be made almost

simultaneously. If the object is a house or building, two corners must be tied to the traverse, and all sides must be measured and recorded in the field book. The transit does not have to be confined to the traverse stations. Intermediate stations can be set from the traverse sta-tions and ties made to the intermediate station. This method, using stadia for horizontal distance, is usually the most efficient.

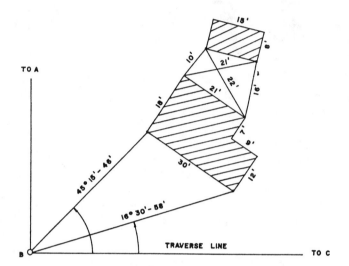

Figure 17.2   Angle/distance ties and two distance ties

12 TWO DISTANCES METHOD OF TIES

This method can be used in conjunction with the *angle and distance method*. Where barns or out-buildings lie behind a house and are obscured from view, the house can be tied to the traverse by the *angle and distance method*. The out-building can be tied to the house by the *two distances method*. Two horizontal distances are required to locate one corner of the out-building. Two corners must be tied to the house as shown in figure 17.2. All sides of the out-building are measured and recorded in the field book.

13 TWO ANGLE METHOD OF TIES

This method is used in special cases where the object to be tied is inaccessible because it lies across a river, lake or busy highway. The object is tied to the traverse by turning an angle to the object from two different points on the traverse.

14 STRENGTH OF TIES

When making horizontal ties, certain practices can be followed to reduce the error in locating points on the map.

- When the *two distances method* is used, the two distances should be as nearly at right angles as possible.

- When the *two angles method* is used, the two lines of sight to the object should be as nearly at right angles as possible.

15 VERTICAL TIES

Vertical ties can be made simultaneously with horizontal ties by stadia or by leveling. Leveling is used for strip maps with *cross-sectioning* as the usual method. For area maps, the *grid system* is common. This consists of laying out the area into a grid with 50 feet or 100 feet intervals and determining elevations at the grid intersections.

16 SUMMARY OF HORIZONTAL AND VERTICAL TIES

No one method of making ties excludes the possible use of others. The azimuth-stadia method is very efficient, does not require a large party, and it is accurate enough for most work. Combinations of methods may be used. Where ditches or streams run through the map area, a combination of azimuth-stadia and cross-sectioning may be economical. The size of the area, slope of the terrain, and the amount and size of vegetation also influence the selection of a method.

17 NOTEKEEPING

Examples of field notes for a right angle offset survey are shown figure 17.3. Most measurements are shown on the right. Transit stations and full stations are shown on the left. It is not necessary that the right half be drawn to scale, but it is often very convenient to do so. Each line space on the left represents 20 feet and the smallest line space on the right represents 10 feet. This scale makes for rapid plotting, but different topographic details require different scales. It is accepted practice to vary the scale from page to page if the amount of necessary detailing varies.

18 STADIA METHOD

The Greek word *stadia* denotes a unit of measure for horizontal distance. In surveying, the term is used to denote a system for measuring horizontal distances based on the optics of the transit telescope, theodolite, or level. This system eliminates the need for horizontal taping, and while not as accurate, it is satisfactory for

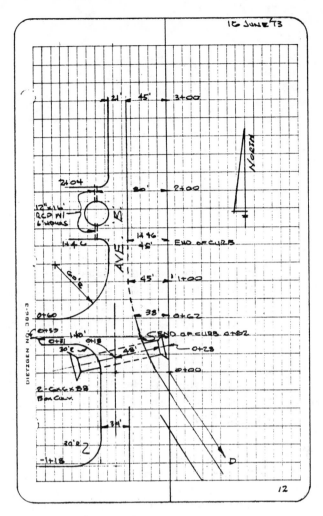

Figure 17.3   Typical field notes
for right angle offset survey

making the horizontal measurements for topographic maps. The system is also known as *tacheometry*.

When employing the stadia method to obtain horizontal distances, the horizontal circle of the transit or theodolite indicates direction, and the vertical circle and a level rod determine elevation. Horizontal and vertical ties to the traverse can therefore be made simultaneously.

19 STADIA PRINCIPLE

Stadia sighting depends on two horizontal cross-hairs, known as *stadia hairs*, within the telescope. These hairs are parallel to the horizontal cross-hair and are equally spaced above and below it. The stadia hairs are shortened so that they will not be confused with the middle horizontal cross-hair, although this may not be true in older transits. (figure 17.4)

The instrument operator sees the stadia hairs imposed on the stadia rod, as shown in figure 17.4. The distance

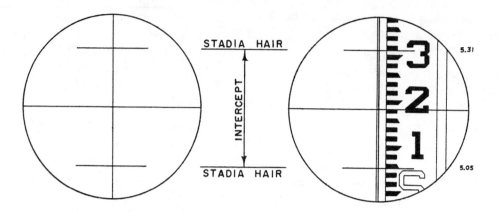

Figure 17.4   Stadia hairs and use

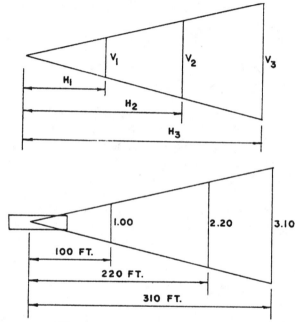

Figure 17.5   Stadia principles

on the rod between the two stadia hairs is known as the *intercept*. If the rod is vertical and the telescope is horizontal, the distance from the center of the instrument to the rod is 100 times the intercept. Actually, the diverging lines of sight to the stadia hairs are from the vertex, which may be one or two inches from the center of the instrument. This discrepancy is ignored in topographic surveying. In older transits, a constant of one foot is added to the stadia distance because the vertex is about one foot forward of the center of the instrument.

The stadia principle is illustrated in figure 17.5. V_1, V_2, and V_3 represent rod intercepts at varying distances from the vertex of the transit. H_1, H_2, and H_3 represent corresponding horizontal distances. As the lines of sight from the vertex and the intercepts form similar triangles, it can be seen that $H_1/V_1 = H_2/V_2 = H_3/V_3$. The stadia hairs are so constructed that when the stadia rod is 100 feet from the vertex, the intercept is 1.00 foot. Thus, $H_1 = 100V_1$ and $H_2 = 100V_2$. This means that the horizontal distance from the vertex to the rod is 100 times the intercept.

20 READING THE INTERCEPT

The intercept in figure 17.6a is $5.31 - 5.05 = 0.26$ ft. The intercept in figure 17.6b is also 0.26 ft. In figure 17.6b, the lower stadia hair has been placed on the nearest full foot, 5.00, by lowering the line of sight with the vertical tangent screw. The intercept can be determined easily by subtracting the full foot value from the reading at the upper stadia hair. This slight adjustment does not cause an appreciable error in horrizontal distance.

21 HORIZONTAL DISTANCE FROM INCLINED SIGHTS

In terrain steep enough to prevent all readings being made with the telescope horizontal, the horizontal dis-

tance is computed by trigonometry, using the rod intercept and the vertical angle of the line of sight.

Figure 17.7 indicates that the rod intercept is greater on inclined sights than it would be if the line of sight were perpendicular to the rod. As it would be impractical to hold the rod perpendicular to the line of sight, the rod is held plumb and the normal intercept is computed by trigonometry.

In figure 17.7, the horizontal cross-hair is placed on a rod reading equivalent to the *height of the instrument*, *h.i.* (*h.i.* is the distance from the hub to the center of the instrument. It is not to be confused with *H.I.* which is elevation above a datum.) This makes the vertical angle α at the instrument equal to the vertical angle α at the hub. In reading the intercept, the bottom stadia hair is set on a full foot mark such that the middle hair is approximately on the *h.i.* After the intercept is read,

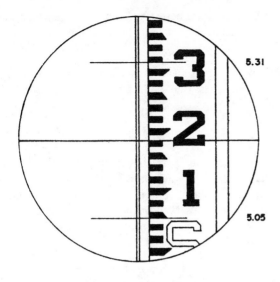

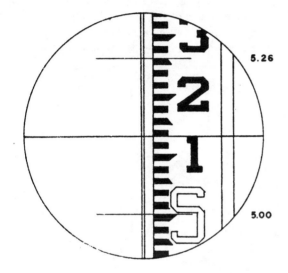

Figure 17.6 Reading intercepts

the middle hair is set on the *h.i.* with the tangent screw, and the angle α is read on the vertical circle.

If S' is the intercept actually read, S is the intercept normal to the line of sight, D is slope distance, and H is horizontal distance, then

$$S = S' \cos \alpha \qquad\qquad 17.1$$
$$D = 100S = 100S' \cos \alpha \qquad 17.2$$
$$H = D \cos \alpha = 100S' \cos^2 \alpha \qquad 17.3$$

22 VERTICAL DISTANCE TO DETERMINE ELEVATION

The difference in elevation of the hub and any point can also be determined by trigonometry, using the angle α and the rod intercept. It can be seen in figure 17.8 that if the middle cross-hair is on the *h.i.*, the vertical distance V from the horizontal through the center of the instrument to this *h.i.* reading on the rod is the same as the vertical distance from the hub to the point on the ground where the rod rests.

$$V = D \sin \alpha = 100S'(\cos \alpha)(\sin \alpha) \qquad 17.4$$

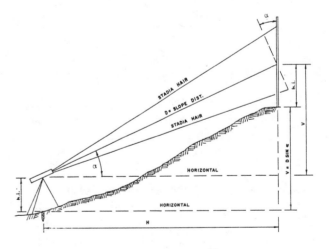

Figure 17.8 Using stadia rod to measure vertical distance

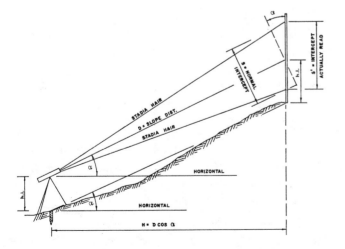

Figure 17.7 Using stadia rod to measure horizontal distance

23 USE OF STADIA REDUCTION TABLES

To avoid calculating $\cos^2 \alpha$ and $\cos \alpha \sin \alpha$, the horizontal distance H and the vertical distance V can be found in tables similar to table 17.1.

Table 17.1
Typical stadia reductions

	4°		5°	
	horizontal	vertical	horizontal	vertical
minutes	distance	distance	distance	distance
0	99.51	6.96	99.24	8.68
2	99.51	7.02	99.23	8.74
4	99.50	7.07	99.22	8.80
6	99.49	7.13	99.21	8.85
8	99.48	7.19	99.20	8.91
10	99.47	7.25	99.19	8.97
12	99.46	7.30	99.18	9.03
14	99.46	7.36	99.17	9.08

Table 17.1 lists horizontal and vertical distances for a rod intercept of 1.00 foot for 4° and 5° vertical angles. Vertical angles to the nearest full degree are shown at the top of the columns and minutes are shown in the left column. Interpolation is required for angles of an odd number of minutes. To find horizontal and vertical distance for a particular vertical angle and intercept not equal to 1.00 foot, the intercept is multiplied by the distance found in the table.

Example 17.1

Find the horizontal distance H and vertical distance V for a rod intercept of 3.68 feet and vertical angle α of 4°12′.

Solution

$$H = 3.68 \times 99.46 = 366\,\text{ft}$$
$$V = 3.68 \times 7.30 = 26.9\,\text{ft}$$

It can be seen from the table that for vertical angles up to 4°, the slope distance is very nearly the horizontal distance. For topographic detailing, no correction is needed. Of course, the vertical distance must still be computed .

Also used in reducing stadia notes are stadia slide rules and special attachments to the transit. The Beaman arc, which attaches to the vertical circle, is widely used.

24 AZIMUTH

Azimuth is the most efficient method of determining direction where a number of shots are taken from one station. By measuring all angles from the same reference line, plotting points on the map is simplified. A full circle protractor is oriented to north on each transit station. Azimuth readings from each transit station are plotted with one setting of the full circle protractor.

After the control traverse has been closed, the direction of all legs of the traverse should be recorded in azimuth

in the field book. These directions are used in orienting the horizontal circle of the transit at each transit station.

To orient the horizontal circle, first set the vernier on the azimuth of the line along which the transit is to be sighted.

Set the vertical cross-hair on that line by using the lower clamp, and then release the upper clamp. The circle will be oriented, but as a check, turn the alidade until the vernier reads 0° and observe the direction of the telescope to see that it is pointing to the north. In figure 17.9, the transit is set up on point B. The azimuth of AB is 150°00′. The horizontal circle is to be oriented for backsighting on point A.

To orient the circle, the vernier must be set on the azimuth of the line from the transit station B to the backsight A. The azimuth of BA is the back-azimuth of AB, so the vernier must be set on 330° before the vertical cross-hair is set on point A with the lower clamp.

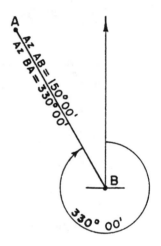

Figure 17.9 Azimuth measurements

In locating points to be tied to the control net by azimuth, it is not necessary to read the vernier to determine the azimuth. In plotting with a protractor, it is impossible to plot to one-minute accuracy. Plotting to the nearest quarter of a degree is all that is practical. The horizontal circle can be read to ten minutes without the use of the vernier. In establishing intermediate transit points not previously located, more care should be exercised in reading the circle.

After moving the transit to a new station, the circle will be oriented by setting on the back-azimuth of the line just established. The intercept from the new station to the previous station should be read as a check against the previous intercept. It is also good practice to take a shot on a known point as a check.

Where a number of shots are taken from one station, it is good practice to observe the backsight again before moving to a new station to see that the transit has not been disturbed.

25 ALGEBRAIC SIGN OF VERTICAL ANGLE

An *angle of elevation* is given a plus sign, and an *angle of depression* is given a minus sign. For small vertical angles, care should be taken that the wrong sign is not recorded. As a check, if the telescope bubble is forward, the sign is plus.

26 ELEVATION

As an example of determining vertical distance from the transit station to a point, assume in figure 17.8 that the elevation of the hub is 465.8, the *h.i.* is 5.2, the rod intercept is 4.22, and the vertical angle is +4°12′ with the middle cross-hair on the *h.i.* Using the tables,

$$V = 7.30 \times 4.22 = 30.8 \,\text{ft}$$
$$\text{elevation} = 465.8 + 5.2 + 30.8 - 5.2 = 496.6$$
$$= 465.8 + 30.8 = 496.6$$

It can be seen that when the middle horizontal cross-hair is on the *h.i.* when the vertical angle is read, the two *h.i.*'s will cancel. *V* can be added (or subtracted when α is minus) to the elevation of the transit station to obtain the elevation of a point. However, the reading of the *h.i.* on the rod may be obscured. In this case, the middle hair can be placed on the next higher (or lower) full foot mark and the vertical angle read. The *h.i.* and *V* will be added to the hub elevation and the rod reading subtracted.

Where possible, rod readings should be taken with the telescope horizontal to eliminate work in reducing field notes. In this case, the *h.i.* is added to the elevation of the station and the rod reading is subtracted from this elevation just as in leveling. This eliminates the need to read the vertical angle.

27 FINDING THE h.i.

The *h.i.* can be found by placing the rod near the transit and moving a target along the rod until the horizontal line of the target lines up with the horizontal axis of the telescope.

28 SELECTING POINTS TO BE USED IN LOCATING CONTOURS

Contours are located on the map by assuming that there is a uniform slope between any two points which have been recorded in the notes. Elevations of points are written on the map and contour lines are interpolated between any two points. In order to insure that the slope between any two points is uniform, shots must be taken in the field at certain key points.

29 KEY POINTS FOR CONTOURS

Key points are any points which will show breaks in the slope of the ground, just as in cross-sectioning. The most important of these are

- summits or peaks
- stream beds or valleys
- saddles (between two summits)
- depressions
- ridge lines
- ditch bottoms and tops of cuts
- tops of embankments and toes of slope

30 SPECIAL SHOTS

It is sometimes impossible for the rodperson to place the rod exactly on the point to be located, such as a corner of a house with a wide roof overhang. In this case, the rod is held near the house in such position that the rod is the same distance from the transit as the house corner, the intercept is read at this point, and the house corner is used to determine horizontal direction.

For long shots where the intercept does not fall entirely on the rod or portion of the rod observed, the intercept between the upper cross-hair and the middle cross-hair can be observed and doubled for the intercept. On long shots, a stadia rod with bold markings and different colors is more suitable than a level rod.

31 EFFICIENCY OF THE SURVEY PARTY

The efficiency of the field party depends on the number of points located during a period of time. Time can be saved by following the order of taking readings on a point as follows:

Step 1: Set the vertical cross-hair on the rod.

Step 2: Set the lower stadia cross-hair on a full foot mark such that the middle cross-hair is approximately on the *h.i.*

Step 3: Read the upper cross-hair, subtract the reading of the lower cross-hair and record the intercept.

Step 4: With the tangent screw, place the middle cross-hair on the *h.i.*

Step 5: Wave the rodperson to the next point.

Step 6: Read and record the horizontal angle.

Step 7: Read and record the vertical angle.

It is important that the rodperson be waved on (step 5) before the two angle readings are made and recorded. Many shots in a day's work will be lost by forgetting to do this.

32 COMPUTATIONS FROM FIELD NOTES

Field notes for an azimuth-stadia survey are shown in figure 17.10. In the notes, the transit is at station *B* at the beginning of the day. A backsight is made on station *A* (which was set at the close of the preceding day.) Referring to the field notes of the preceding day, the azimuth of *AB* was 316°22′. Therefore, when the sight on station *A* is made, the vernier is set on 136°22′, which is the azimuth of *BA* and the back azimuth of *AB*.

The intercept at station *A* is read as a check against what was read at *B* from *A* on the previous day. The vertical angle, −1°18′, is also read as a check.

The rod reading for Station 1 was made with the telescope level. Elevation of Station 1 is

$$467.2 + 5.0 - 8.0 = 464.2$$

The rod reading on Station 4 was made with a vertical angle of +4°34′. This is large enough to require a correction for horizontal distance. The correction factor for 4°34′, found in stadia reduction tables, is 99.37. The horizontal distance *B* to 4 is

$$H = 99.37 \times 3.40 = 338\,\text{ft}$$

sta.	rod int	az	∠ or rod	h.-dist.	v.-dist.	elev.
		Λ @ **B** elev. 467.2 *h.i.* 5.0				
A	6.82	136°22′	−1°18′			
1	1.15	88°10′	8.0			
2	1.84	96°30′	7.8			
3	2.28	124°45′	+3°21′			
4	3.40	206°20′	+4°34′			
5	3.25	318°00′	+4°20′			
6	4.36	345°30′	+2°47′			
C	6.25	318°52′	+1°12′			
		Λ @ **C** elev. 480.3 *h.i.* 5.2				
B	6.24	138°52′	−1°13′			
1	1.78	96°10′	+2°41′			
2	2.49	128°45′	9.2			
3	3.12	186°00′	−4°56′			
4	3.40	232°40′	9.5			
D	7.18	285°30′	−1°16′			
		Λ @ **D** elev. 464.4 *h.i.* 5.1				
C	7.18	105°30′	+1°16′			
1	2.42	145°15′	7.7			
2	3.18	180°20′	10.6			

Figure 17.10   Field notes
for an azimuth-stadia survey

The vertical angle was read when the horizontal cross-hair was set on the *h.i.* If it had not been, the notes would indicate this fact. The factor for the vertical distance for an angle of 4°34′ is found to be 7.94. Vertical distance is

$$V = 7.94 \times 3.40 = 27.0\,\text{ft}$$

Elevation of Station 4 is

$$467.2 + 27.0 = 494.2$$

If horizontal cross-hair could not be placed on 5.0 (the *h.i.*) at Station 4 because a limb of a tree obstructed the view. The horizontal cross-hair was placed on 6.0, and the vertical angle was +4°45′. Then,

$$V = 8.25 \times 3.40 = 28.0\,\text{ft}$$
$$\text{elevation} = 467.2 + 5.0 + 28.0 - 6.0 = 494.2$$

PROFESSIONAL PUBLICATIONS, INC. • P.O. Box 199, San Carlos, CA   94070

33 CONTOURS AND CONTOUR LINES

A contour is an imaginary line on the surface of the earth which connects points of equal elevation. A contour line is a line on a map which represents a contour on the ground.

34 CONTOUR INTERVAL

The contour interval of a map is the vertical distance between contour lines. The contour interval is selected by the map-maker. In flat country it may be one foot and in mountainous country it may be 100 feet, depending on the scale of the map and the character of the terrain. The contour interval can be too small, making the map a maze of lines which are not legible; the contour interval can be too large, not showing the true relief. The more accurate the contours, the more costly the map. The intended use of the map is a basic consideration in the selection of the contour interval.

Figure 17.11 shows that the vertical distance between contour lines is constant, but the horizontal distance varies with the steepness of the ground.

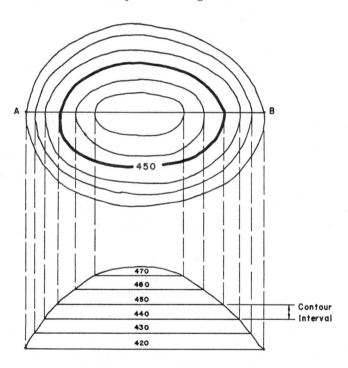

Figure 17.11   Contour line intervals

35 INDEX CONTOURS

To facilitate reading a topographic map, every fifth contour line may be darker. The elevation of that contour line is written in a break in the line as shown in figure 17.12. There are five spaces between any two heavy lines, called *index contours*, so that the contour interval can be computed by dividing the difference in elevation between two index contours by five. In figure 17.12, there are five spaces between the 700 foot contour and the 750 foot contour. The contour interval is $(750 - 700) \div 5 = 10$ feet.

36 CLOSED CONTOUR LINES

Contour lines which close represent either a hill or depression. Whether the closed contour lines represent a hill or depression can be determined by reading the elevations of the index contours. For a hill, the elevations increase as the contour lines become shorter. Depressions are often indicated by short *hachures* on the down slope side of the contour line.

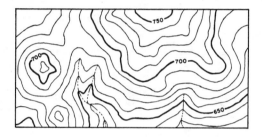

Figure 17.12   Typical contour map

37 SADDLE

The name *saddle* is given to the shape of contours which define two summits in the same vicinity. A profile view of the two summits looks somewhat like the profile view of a horse saddle, as shown in figure 17.13.

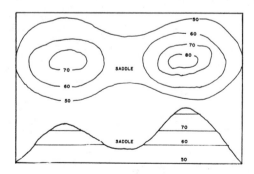

Figure 17.13   Saddle

38 CHARACTERISTICS OF CONTOURS

Certain fundamental characteristics of contours should be kept in mind when plotting contour lines and reading a contour map. The characteristics are illustrated in figure 17.14.

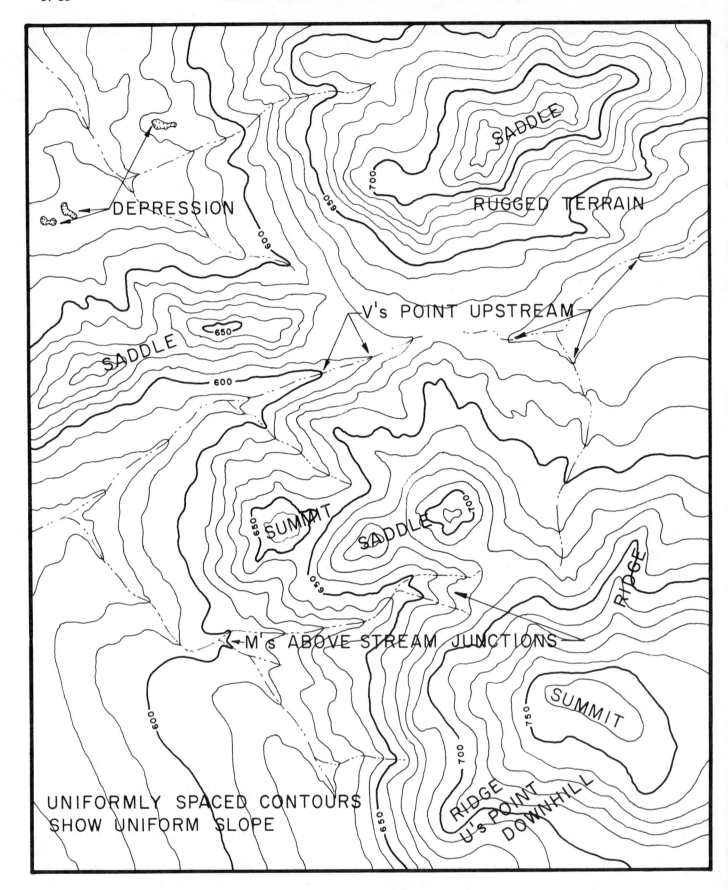

Figure 17.14   Typical contour map features

- Each contour line must close upon itself either within or outside the borders of the map. Since all land areas on the surface of the earth rise above the sea, it can be seen that each contour will, no matter how long, finally close. This means that a contour line cannot end abruptly on a map.

- Contour lines cannot cross or meet, except in unusual cases of waterfalls or cliff overhangs. (If it were possible for two contour lines to cross, the intersection would represent two elevations for the same point.)

- A series of closed contour lines represents either a hill or depression. The elevations of the index contour lines will indicate which series represents a hill and which represents a depression.

- Contour lines crossing a stream form V's which point upstream.

- Contour lines crossing a ridge form U's which point down the ridge.

- Contour lines tend to parallel streams. Rivers usually have a flatter gradient than do intermittent streams. Therefore, contour lines along rivers will be more nearly parallel than contours along intermittent streams, and they will run parallel for a longer distance.

- Contour lines form M's just above stream junctions.

- Contour lines are uniformly spaced on uniformly sloping ground.

- Irregularly spaced contour lines represent rough, rugged ground.

- The horizontal distance between contour lines indicates the slope of the ground. Closely spaced contour lines represent steeper ground than widely spaced contour lines.

- Contours are perpendicular to the direction of maximum slope. The direction of rainfall run-off in a map area can be determined from this characteristic.

39 METHODS OF LOCATING CONTOURS

Several methods of locating contours are used in topographic surveying and mapping. They are the grid method, controlling points method, cross-section method, and trace contours method. All methods depend on the assumption that there is a uniform slope between any two ground points located in the field.

40 GRID METHOD

The grid method is very effective in locating contours in a relatively small area of fairly uniform slope. The area is divided into squares or rectangles of 25 to 100 feet (depending on the scale of the map and the contour interval desired). Stakes are set at each intersection and at any other points of slope change, such as at ridge lines or valleys.

The location of the point where the contour line crosses each side of each square is determined by interpolation (either by estimation or mathematical proportion).

In figure 17.15a, contours are to be plotted on a two foot interval. Starting at A-1, figure 17.15b, it can be seen that the 440 foot contour will cross between A-1 and B-1. The vertical distance between A-1 and B-1 is 4 feet, so the 440 foot contour will cross halfway between A-1 and B-1. The 442 foot contour will cross at B-1, so a mark is placed at each of these points.

The 444 foot contour will cross between B-1 and C-1. The vertical distance between B-1 and C-1 is 444.3 − 442.1 = 2.3 feet. The vertical distance between B-1 and the 444 foot contour is 2.0 feet. Therefore, the horizontal distance will be 2.0/2.3 of the way, or about 0.9 of the way. A mark is made at this point.

The 440, the 442, and the 444 foot contours will cross between A-2 and B-2. The crossing points are found in a similar manner and marked.

After the crossing points are located by interpolation, the crossing points for each contour are connected as shown in figure 17.15c.

After all crossing points are connected, the contour lines are smoothed. Small irregularities are taken out so that the contour lines are more like the contours on the ground.

In following a particular contour line using the grid method, an inspection must be made of each grid line between each intersection to see if the contour line can cross. If it cannot, another line must be inspected. For example, in figure 17.15c, the 444 feet contour line can cross between B-1 and C-1, between B-1 and B-2, between A-2 and B-2, or between B-2 and B-3, but not between B-2 and C-2. Each contour line must close or reach the border of the map at two points.

After contour lines are smoothed, index contour lines must be made heavier than the other lines, and the elevation of the index contour must be written in a break in the line.

41 CONTROLLING POINTS METHOD

The controlling points method is suitable for maps of large area and small scale. The selection of ground points is very important. The accuracy of the contours depends on the knowledge and experience of the survey party. Shots should be taken at stream junctions, at intermediate points in stream beds between junctions, and along ridge lines. Field notes should indicate these points so that ridges and streams can be plotted before interpolations are made. Interpolations are made in much the same way as they are made using the grid method.

Figures 17.16a, 17.16b, 17.16c, and 17.16d show the progressive steps in plotting contour lines by the controlling points method.

42 CROSS-SECTION METHOD

The cross-section method is satisfactory for the preparation of strip maps. It can be accomplished by using level and tape or azimuth-stadia. Cross-sections are taken at right angles to a centerline or base line. Elevations of each cross-section shot are written on the strip map and interpolation of contour lines is performed as in the grid and controlling points methods.

43 TRACING CONTOURS METHOD

This method is used when the exact location of a particular contour line is needed. It is effectively performed by use of the plane table, but can be done by the azimuth-stadia survey method.

44 MAPPING

The first step in preparing a map is the selection of a scale and a contour interval. This selection is influenced by the size of the sheet to be used, the purpose of the map, and the required accuracy.

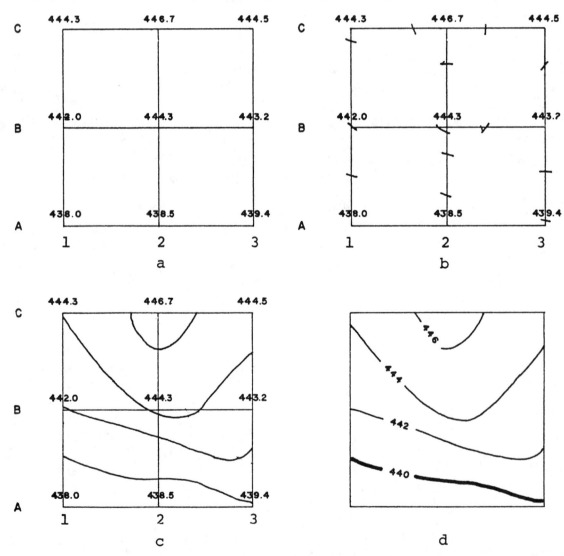

Figure 17.15   Grid method of contour location

PROFESSIONAL PUBLICATIONS, INC. ● P.O. Box 199, San Carlos, CA 94070

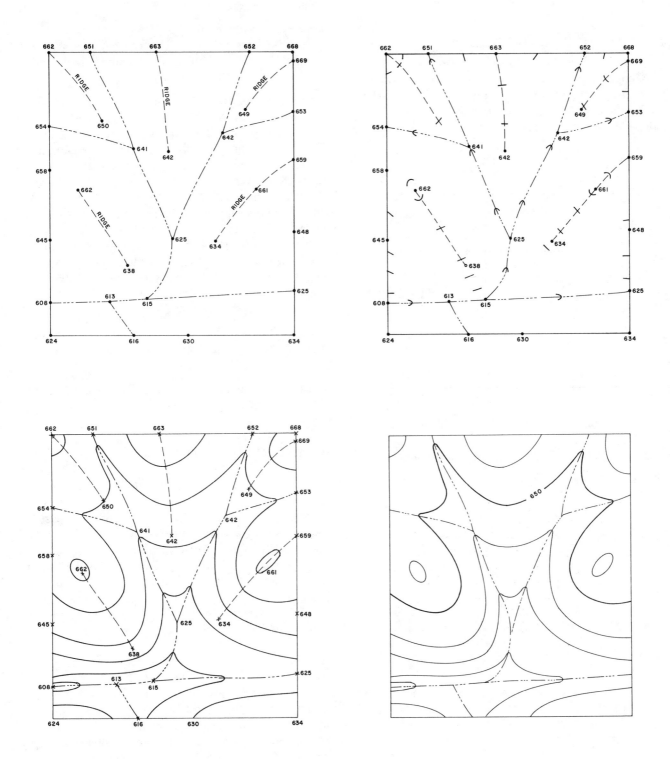

Figure 17.16 Controlling points for contour location

The traverse can be plotted by the coordinate method, the tangent method, or the protractor method.

45 COORDINATE METHOD

This method is the most accurate for plotting a traverse. Any error in plotting one point does not affect the location of the other points. Each point is plotted independently of the others.

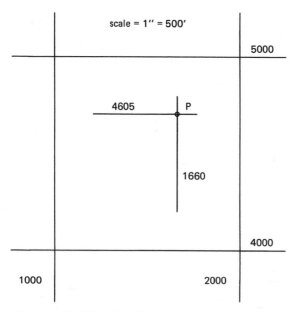

Figure 17.17 Coordinates of a traverse point

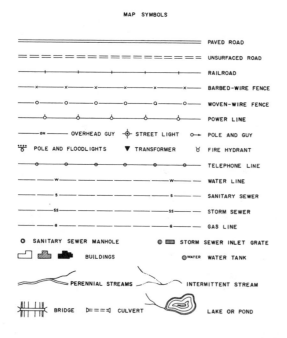

Figure 17.18 Typical map symbols

Coordinates of each traverse station are computed prior to plotting. Each point is plotted on the grid system using the coordinates.

Before a traverse is plotted, the sheet is laid out with perpendicular grid lines. Distance between grid lines can be 50, 100, 500, 1000 feet, or any other multiple which suits the scale of the map.

As an example, figure 17.17 shows a grid system laid out to a scale of 1 inch = 500 feet with grid lines 1000 feet apart. The point P has the coordinates $x = 1660$, $y = 4605$. In plotting the point, the 50 scale is laid on the paper horizontally so that the 10 mark on the scale lines up with the vertical grid line marked 1000. A pencil dot is then made at 1660 on the scale. With a straight edge, a temporary vertical line is drawn through the pencil dot. The scale is then laid vertically along this vertical line with the 40 mark on the scale lined up with the horizontal line on the paper marked 4000. A pencil dot is carefully made on the vertical line at 4605 on the scale (as close as can be read).

After all points of the traverse are plotted, they are connected with lines. Distances between points are scaled and checked against distances which were recorded in the field.

Detail points can be plotted with a protractor using grid lines to orient the protractor.

46 TANGENT METHOD

This method is very convenient and accurate where deflection angles have been turned for a route survey. At any traverse station, the back line is produced past the points. A convenient distance (such as 10 inches) is laid off from the traverse point along this prolongation. A perpendicular is erected at this point and the tangent distance for the deflection angle is marked on this line. A line from the traverse point to this point defines the next leg of the traverse. Any error made in this plotting will be carried on to the next plotting.

47 PROTRACTOR METHOD

Ths method is the fastest but least accurate. An error in plotting an angle or a distance will be carried on throughout the traverse. The protractor is commonly used for detailing and is sufficiently accurate.

PRACTICE PROBLEMS FOR CHAPTER 17

1. Complete the topography field notes using the right angle offset method for ties to the road, stream, and buildings. Consider the enclosed area to be the right half of a page in the field notes. The vertical line is the base line of the survey. Scale: $1/2'' = 100'$.

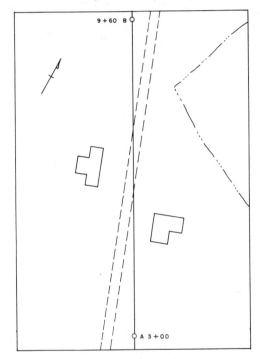

2. Complete the topography field notes using the angle and distance method and the two distances method. Use two distances method to tie the small building to the adjacent building; use angle and distance method for other ties. Consider transit to be set up at station A on the base line with foresight on station B on the base line. Scale: $1/2'' = 100'$.

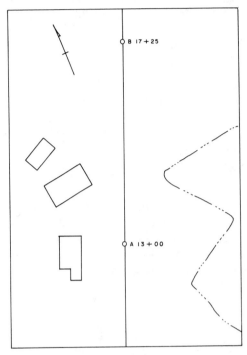

3. Tie the streets to the base line by the right angle offset method. Scale: $1/2'' = 100'$.

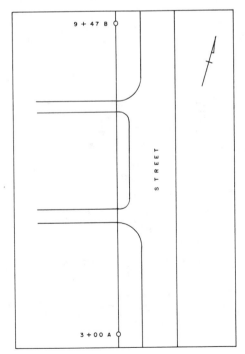

4. Find the horizontal distance H and the vertical distance V between two points when the stadia rod intercept and vertical angle α are as shown.

	rod int	α
(a)	3.48	4°00′
(b)	4.68	5°09′
(c)	5.75	5°12′
(d)	3.91	4°11′

5. Indicate the angle to be set on the vernier to orient the transit for an azimuth-stadia survey for each station of the traverse $ABCDEA$. AB : S 53°18′ E; BC : N 42°00′ E; CD : N 72°47′ W; DE; N 36°27′ W; EA : S 30°39′ W.

6. Considering the transit to be set up at station A, elevation 463.2, with $h.i. = 5.1$, compute the horizontal distance from A to points 1,2,3, and 4 and the elevation of points 1, 2, 3, and 4. (Consider middle cross-hair on $h.i.$ when vertical angle is read.)

pt	int	v ang
1	3.85	+4°05′
2	2.98	+5°08′
3	2.56	−5°09′
4	5.44	−4°03′

7. Complete the Azimuth-Stadia Survey field notes.

azimuth-stadia survey

sta	int	az	v ang or rod	h dist	elev
TRAN @ B ELEV 467.2 $h.i.$ 5.0					
o A	6.74	148°04′	−4°14′		
1	0.91	90°45′	8.1		
2	1.66	120°20′	−5°12′		
3	2.10	135°15′	−5°07′		
4	3.25	143°00′	+4°11′		
o C	4.00	60°10′	+5°08′		
TRAN @ C ELEV 502.8 $h.i.$ 5.2					
o B	4.01	240°10′	−5°08′		
5	3.15	286°00′	−4°00′		
6	2.21	36°20′	9.2		

8. Plot one foot contours. Indicate index contours.

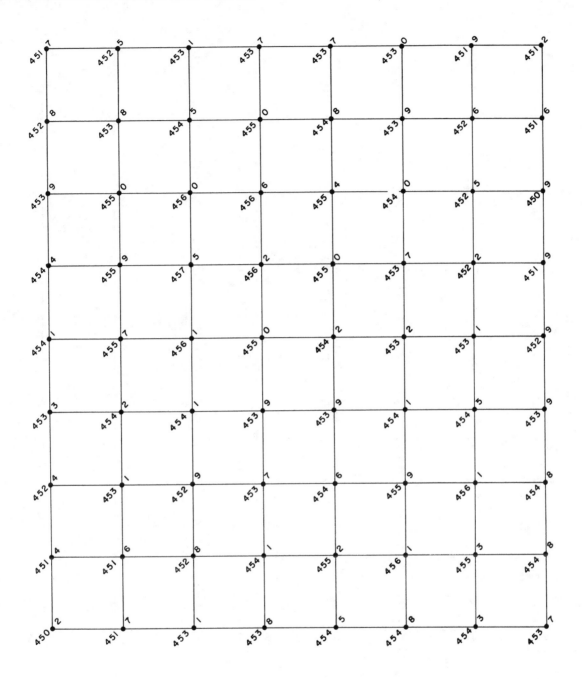

9. Plot one foot contours. Indicate index contours.

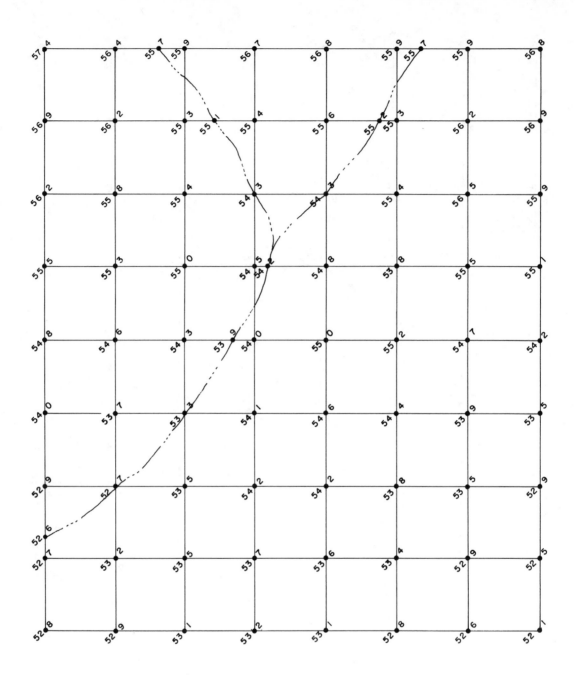

10. Plot five feet contours.

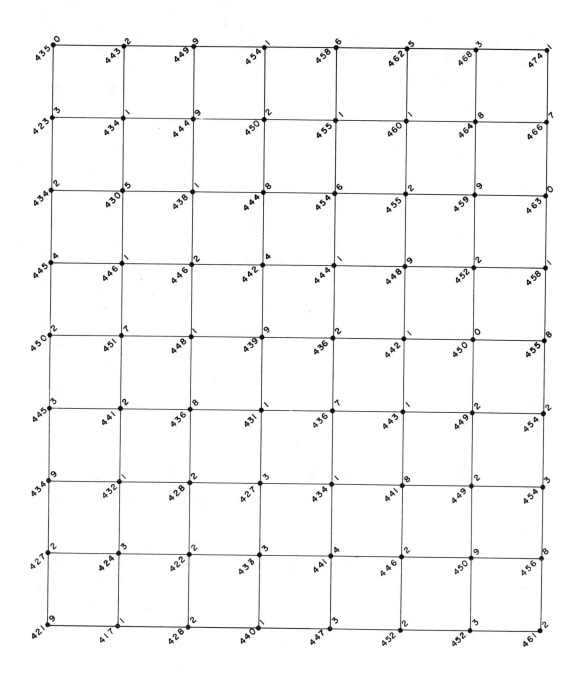

11. Plot five feet contours.

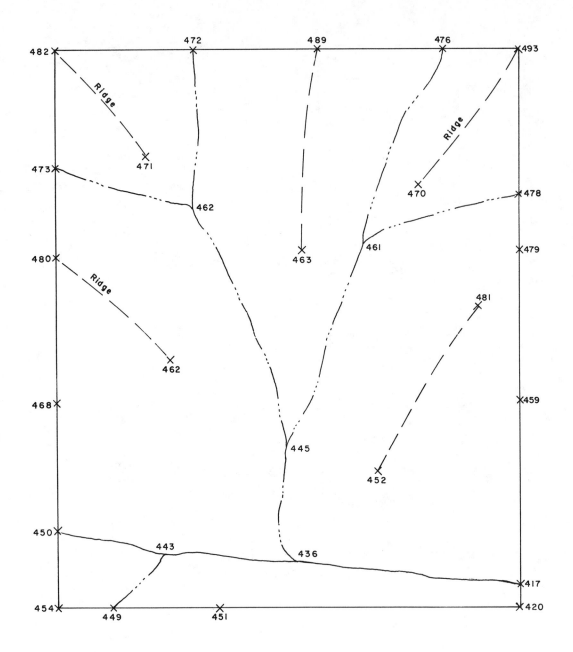

18

GRADE LINES AND VERTICAL CURVES

1 GRADE OR STEEPNESS

The grade, or steepness, of a road is the ratio of elevation to horizontal distance. If a highway rises 6 feet for every 100 feet of horizontal distance, the grade of the highway is $6/100 = 0.06\,\text{ft/ft}$.

Figure 18.1 Calculation of grade

2 SLOPE OF A LINE

The grade of a line representing the profile of a highway is also known as the slope of the line.

$$\text{slope} = \frac{\text{rise}}{\text{run}} \qquad 18.1$$

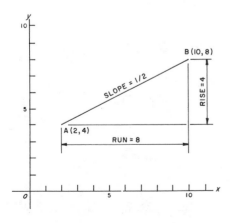

Figure 18.2 Slope of a line

In figure 18.2, points $\mathbf{A}(2,4)$ and $\mathbf{B}(10,8)$ are connected by the straight line AB.

$$\text{slope of } AB = \frac{\text{ordinate of } \mathbf{B} - \text{ordinate of } \mathbf{A}}{\text{abscissa of } \mathbf{B} - \text{abscissa of } \mathbf{A}}$$
$$= \frac{8-4}{10-2} = \frac{1}{2} \qquad 18.2$$

The symbol m is used to denote the slope of the line between points 1 and 2.

$$m = \frac{y_2 - y_1}{x_2 - x_1} \qquad 18.3$$

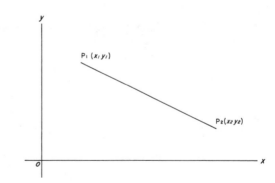

Figure 18.3 Line between two points

In figure 18.4, the slope of line AB is

$$m = \frac{y_2 - y_1}{x_2 - x_1} = \frac{8-2}{6-3} = +2$$

Line BC has a slope of

$$m = \frac{y_2 - y_1}{x_2 - x_1} = \frac{2-8}{9-6} = -2$$

A line rising from left to right has a positive slope, and a line falling from left to right has a negative slope.

A horizontal line has a slope of zero, as shown in figure 18.5.

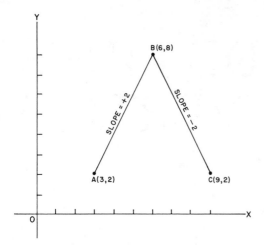

Figure 18.4 Positive and negative slopes

Figure 18.5 Slope of horizontal line

3 GRADE OR GRADIENT

In highway construction, the slope of a line which is the profile of the centerline is known as the *grade* or *gradient.*

The grade of a highway is computed in the same way as the slope of a line is computed. Horizontal distances are usually expressed in stations; vertical distances are expressed in feet.

Example 18.1

Determine the gradient of a highway which has a centerline elevation of 444.50 at station 20 + 75.00 and a centerline elevation of 472.20 at station 32 + 25.00.

Figure 18.6 Example 18.1

Solution

$$\text{gradient} = \frac{472.20 - 444.50}{3225.00 - 2075.00} = \frac{27.70}{1150.00}$$
$$= +0.02409 \, \text{ft/ft}$$

If the numerator is expressed in feet and the denominator is expressed in stations, the decimal point in the gradient will move two places to the right. The gradient can then be expressed as a percent. In this form, gradient expresses change in elevation per station. For example 18.1,

$$\text{gradient} = \frac{27.70 \, \text{ft}}{11.50 \, \text{sta}} = +2.409\%$$

4 POINTS OF INTERSECTION

Vertical alignment for a highway is located similarly to horizontal alignment. Straight lines are located from point to point, and vertical curves are inserted. The points of intersecting gradients are known as points of intersection. These lines, after vertical curves have been inserted, are the centerline profile of the highway. Usually the profile is the finish elevation (pavement) profile, but may be the subgrade (earth-work) profile.

5 TANGENT ELEVATIONS

After points of intersection have been located and connected by tangents (straight lines), elevations of each station on the tangent need to be determined before finding elevations on the vertical curve.

Since the gradient is the change in elevation per station, if the station number and elevation of each *PI* is known, the elevation at each station can be calculated. The gradient should be computed to three decimal places in percent. Finish elevations should be computed to two decimal places in feet (hundredths of a foot).

Example 18.2

From the information shown in figure 18.7, compute the gradient of each tangent and the elevation at each full station on the tangents.

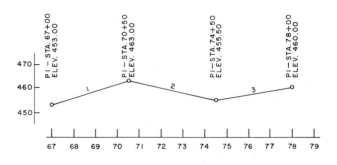

Figure 18.7 Example 18.2

Solution

$$\text{gradient 1} = \frac{463.00 - 453.00}{70.50 - 67.00} = +2.857\%$$

$$\text{gradient 2} = \frac{463.00 - 455.50}{74.50 - 70.50} = -1.875\%$$

$$\text{gradient 3} = \frac{460.00 - 455.50}{78.00 - 74.50} = +1.286\%$$

Table 18.1

Example 18.2   Elevation computations

point	station	computations	elevation
PI	67 + 00		= 453.00
	68 + 00	453.00 + (1)(2.857)	= 455.86
	69 + 00	453.00 + (2)(2.857)	= 458.71
	70 + 00	453.00 + (3)(2.857)	= 461.57
PI	70 + 50	453.00 + (3.5)(2.857)	= 463.00
	71 + 00	463.00 − (0.5)(1.875)	= 462.06
	72 + 00	463.00 − (1.5)(1.875)	= 460.19
	73 + 00	463.00 − (2.5)(1.875)	= 458.31
	74 + 00	463.00 − (3.5)(1.875)	= 456.44
PI	74 + 50	463.00 − (4.0)(1.875)	= 455.50
	75 + 00	455.50 + (0.5)(1.286)	= 456.14
	76 + 00	455.50 + (1.5)(1.286)	= 457.43
	77 + 00	455.50 + (2.5)(1.286)	= 458.72
PI	78 + 00	455.50 + (3.5)(1.286)	= 460.00

6 VERTICAL CURVES

Just as horizontal curves connect two tangents in horizontal alignment, vertical curves connect two tangents in vertical alignment. However, although the horizontal curve is usually an arc of a circle, the vertical curve is usually a parabola.

PC–beginning of curve

PT–end of curve

PI–point of intersection of two tangents

g_1–gradient of first connecting tangent

g_2–gradient of second connecting tangent

y_1–ordinate of any station less than PI station, in feet

x_1–horizontal distance in stations from PC

y_2–ordinate of any station greater than PI station, in feet

x_2–horizontal distance in stations from PT

L–length of curve in stations

e–ordinate at PI in feet

The simplest equation of a parabola is $y = ax^2$, where y is vertical distance and x is horizontal distance. This means that the vertical distances y from tangent to

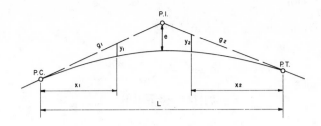

Figure 18.8   A vertical curve

curve vary as the square of the horizontal distances x, measured from either the PC or PT.

Equations for computing y distances can be derived.

$$e = \frac{(g_1 - g_2)L}{8} \qquad 18.4$$

$$a = \frac{(g_1 - g_2)}{2L} \qquad 18.5$$

$$y = ax^2 \qquad 18.6$$

7 COMPUTATIONS FOR FINISH ELEVATIONS

Elevations along the tangent are first computed. y distances are then computed for stations on the vertical curve and added to or subtracted from the tangent elevations to determine the finish elevation. The y distance is added on sag curves (figure 18.9) and subtracted on crest curves (figure 18.8).

Example 18.3

A −1.500% grade meets a +2.250% grade at station 36 + 50, elevation 452.00. A vertical curve of length 600 feet (6 stations) will be used. Compute the finish elevations for each full station and the PC and PT from station 32 + 00 to station 41 + 00. Refer to figure 18.3.

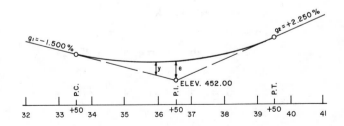

Figure 18.9   Example 18.3

---

[1] The symbol C is often used in place of a in equation 18.6. E may also be used instead of e.

Solution

From equations 18.5 and 18.4,

$$a = \frac{g_1 - g_2}{2L} = \frac{-1.500 - 2.250}{(2)(6)} = 0.312$$

$$e = \frac{(g_1 - g_2)L}{8} = \frac{(3.75)(6)}{8} = 2.81\,\text{ft}$$

Table 18.2

Example 18.3   Tangent
elevation computations

point	station	computations	elevation
	32 + 00	$452.00 + (4.5)(1.500) =$	458.75
	33 + 00	$452.00 + (3.5)(1.500) =$	457.25
PC	33 + 50	$452.00 + (3.0)(1.500) =$	456.50
	34 + 00	$452.00 + (2.5)(1.500) =$	455.75
	35 + 00	$452.00 + (1.5)(1.500) =$	454.25
	36 + 00	$452.00 + (0.5)(1.500) =$	452.75
PI	36 + 50	$=$	452.00
	37 + 00	$452.00 + (0.5)(2.250) =$	453.12
	38 + 00	$452.00 + (1.5)(2.250) =$	455.37
	39 + 00	$452.00 + (2.5)(2.250) =$	457.62
PT	39 + 50	$452.00 + (3.0)(2.250) =$	458.75
	40 + 00	$452.00 + (3.5)(2.250) =$	459.88
	41 + 00	$452.00 + (4.5)(2.250) =$	462.13

Table 18.3

Computations for $y = 0.312x^2$

station	computations	y
33 + 50		$= 0.00$
34 + 00	$= (0.312)(0.5)^2$	$= 0.08$
35 + 00	$= (0.312)(1.5)^2$	$= 0.70$
36 + 00	$= (0.312)(2.5)^2$	$= 1.95$
36 + 50		$= 2.81$
37 + 00	$=$ same as $36 + 00$	$= 1.95$
38 + 00	$=$ same as $35 + 00$	$= 0.70$
39 + 00	$=$ same as $34 + 00$	$= 0.08$
39 + 50		$= 0.00$

The curve is symmetrical about the *PI*, so the distance x to station $37 + 00$ (from the *PT*) is the same as the distance x to station $36 + 00$ (from the *PC*). Therefore, the y distances are the same. Likewise, the y distances for stations $38 + 00$ and $35 + 00$ are the same, as well as for stations $39 + 00$ and $34 + 00$.

Table 18.4

Finish elevations
for example 18.3

point	station	tangent elevation	y	finish elevation
	32 + 00	458.75		458.75
	33 + 00	457.25		457.25
PC	33 + 50	456.50	0.00	456.50
	34 + 00	455.75	+0.08	455.83
	35 + 00	454.25	+0.70	454.95
	36 + 00	452.75	+1.95	454.70
PI	36 + 50	452.00	+2.81	454.81
	37 + 00	453.12	+1.95	455.07
	38 + 00	455.37	+0.70	456.07
	39 + 00	457.62	+0.08	457.70
PT	39 + 50	458.75	0.00	458.75
	40 + 00	459.88		459.88
	41 + 00	462.13		462.13

8 PLAN-PROFILE SHEETS

On construction plans, a profile of the natural ground along the centerline of a highway project and the profile of the finish grade along the centerline are shown on plan-profile sheets. A plan view of the centerline with surrounding topography is shown on the top half of the sheet. The profiles are shown on the bottom half, together with *PI*'s, *PC*'s, *PT*'s, gradients and finish elevations. A plan-profile sheet is shown in figure 18.10.

Plan-profile sheets are also used in construction plans for streets, sanitary sewers, and storm sewers.

Example 18.4

Using the following information, compute finish elevations for each full station. Plot the finish elevation profile and show pertinent information needed for construction.

- Gradient, sta $32 + 00$ to $34 + 00 = -2.000\%$

- Gradient, $39 + 50$ to $43 + 00 = -2.125\%$

- *PI*, sta $34 + 00$, elev. 470.00, 300 ft VC

- *PI*, sta $39 + 50$, elev. 482.00, 500 ft VC

Solution

Refer to footnote number 1 for an explanation of the change in variables.

$$\text{gradient} = \frac{482.00 - 470.00}{39.5 - 34.00} = +2.182\%$$

$$C_1 = \frac{-2.000 - 2.182}{(2)(3.00)} = 0.697$$

$$C_2 = \frac{2.182 - (-2.125)}{(2)(5.00)} = 0.431$$

$$E_1 = \frac{(-2.000 - 2.182)(3.00)}{8} = 1.57 \text{ ft}$$

$$E_2 = \frac{[2.182 - (-2.125)][5.00]}{8} = 2.69 \text{ ft}$$

Prior to performing finish elevation computations, the approximate finish elevation profile should be plotted on the plan-profile sheet. Vertical curves are symmetrical. To locate the *PC* and *PT*, measure one-half the length of the vertical curve in each direction from the *PI*. The midpoint of the vertical curve can be found by drawing a straight line from the *PC* to the *PT* and measuring one-half the distance from the *PI* to this line.

In determining whether y distances should be added to or subtracted from tangent elevations, look at the plotted profile to see whether the curve is higher or lower than the tangent at a particular station.

Figure 18.10 would normally show the profile of the natural ground along the centerline, and the elevation at each station would be shown just under the finish elevation at the bottom of the sheet.

Table 18.5

Example 18.4 Computations for finish elevations

point	station	x	x^2	tangent elevation	y	finish elevation
	32 + 00			474.00		474.00
PC	32 + 50					473.00
	33 + 00	0.5	0.25	472.00	+0.17	472.17
PI	34 + 00	1.5	2.25	470.00	+1.57	471.57
	35 + 00	0.5	0.25	472.18	+0.17	472.35
PT	35 + 50					
	36 + 00			474.36		474.36
PC	37 + 00			476.55		476.55
	38 + 00	1.0	1.00	478.73	-0.43	478.30
	39 + 00	2.0	4.00	480.91	-1.72	479.19
PI	39 + 50	2.5	6.25	482.00	-2.69	479.31
	40 + 00	2.0	4.00	480.94	-1.72	479.22
	41 + 00	1.0	1.00	478.81	-0.43	478.38
PC	42 + 00			476.69		476.69
	43 + 00			474.56		474.56

9 TURNING POINT ON SYMMETRICAL VERTICAL CURVE

The highest point on a crest curve (or the lowest point on a sag curve) is not usually vertically below (or above) the *PI*. This point is called the *turning point*. The distance x from the *PC* to the turning point can be

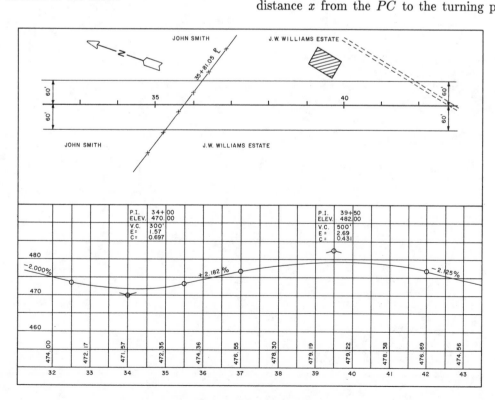

Figure 18.10 Example of 18.4

found from equation 18.7.

$$x = \frac{g_1 L}{g_1 - g_2} \qquad\qquad 18.7$$

Example 18.5

A +1.500% grade meets a −2.500% grade at station 12 + 50. Determine the distance from the *PC* to the turning point if a 600 feet vertical curve is used.

Solution

$$x = \frac{g_1 L}{g_1 - g_2} = \frac{(1.500)(6)}{+1.500 - (-2.500)}$$
$$= 2.25\,\text{sta} = 225\,\text{ft}$$

PRACTICE PROBLEMS FOR CHAPTER 18

1. Determine the gradients between the points on the highway profiles in percent to three decimal places.

Example:

$$PI = 5 + 50$$
$$EL = 452.00$$

$$PI = 8 + 50$$
$$EL = 455.00$$

$$PI = 11 + 00$$
$$EL = 453.00$$

Solution:

$$g_1 = \frac{455.00 - 452.00}{8.50 - 5.50} = +1.000\%$$
$$g_2 = \frac{455.00 - 453.00}{11.00 - 8.50} = -0.800\%$$

(a) $PI = 20 + 70$
$EL = 504.00$

$PI = 23 + 50$
$EL = 498.00$

$PI = 26 + 60$
$EL = 503.00$

(b) $PI = 40 + 00$
$EL = 461.00$

$PI = 46 + 00$
$EL = 459.00$

$PI = 52 + 00$
$EL = 465.00$

(c) $PI = 55 + 00$
$EL = 474.00$

$PI = 59 + 00$
$EL = 469.00$

$PI = 64 + 00$
$EL = 477.50$

$PI = 67 + 00$
$EL = 477.50$

(d) $PI = 67 + 00$
$EL = 453.00$

$PI = 70 + 50$
$EL = 463.00$

$PI = 74 + 50$
$EL = 455.50$

$PI = 79 + 00$
$EL = 461.70$

(e) $PI = 29 + 25$
$EL = 445.00$

$PI = 32 + 50$
$EL = 432.00$

$PI = 37 + 75$
$EL = 432.00$

$PI = 41 + 00$
$EL = 437.70$

2. Compute the gradient for each tangent of the highway profile and elevation of each full station on the tangents.

point	station	tangent elevation
PI	25 + 00	466.00
PI	31 + 00	458.00
PI	35 + 50	472.00
PI	39 + 00	465.00
PI	43 + 00	472.00

3. Compute a, e, and y for each station on the vertical curve.

(a) A +2.234% grade meets a −1.875% grade at station 28 + 50, elevation 436.00. Vertical curve = 600 ft.

(b) A −3.467% grade meets a +2.250% grade at station 45 + 00, elevation 515.00. Vertical curve = 800 ft.

4. From the information given, compute finish elevation
for each full station.

$$PI = 20 + 00$$

$$EL = 455.00$$

No VC

$$PI = 23 + 50$$

$$EL = 448.50$$

$400'$ VC

$$PI = 33 + 00$$

$$EL = 469.00$$

$1000'$ VC

$$PI = 40 + 00$$

$$EL = 455.50$$

No VC

station	finish elevation
32 + 00	476.00
33 + 00	474.00
34 + 00	472.18
35 + 00	471.65
36 + 00	472.58
37 + 00	474.80
38 + 00	476.82
39 + 00	478.09
40 + 00	478.61
41 + 00	478.37
42 + 00	477.37
43 + 00	475.63

6. Determine the distance x from the PC of the symmetrical curve to the high point of the crest curve or to the low point of the sag curve.

(a) A $+4.000\%$ grade meets a -3.000% grade at sta $35 + 00$. Vertical curve is 600 feet in length.

(b) A $+3.125\%$ grade meets a -2.250% grade at sta $12 + 00$. Vertical curve is 800 feet in length.

(c) A -2.750% grade meets a $+3.500\%$ grade at sta $22 + 00$. Vertical curve is 500 feet in length.

(d) A -1.275% grade meets a $+3.250\%$ grade at sta $15 + 00$. Vertical curve is 600 feet in length.

5. From the information given, complete the profile half of the plan-profile sheet.

gradient sta $32 + 00 - 35 + 00 = -2.000\%$ $E_1 = 1.65$ $C_1 = 0.733$

gradient sta $35 + 00 - 40 + 00 = +2.400\%$ $E_2 = 3.39$ $C_2 = 0.377$

gradient sta $40 + 00 - 43 + 00 = -2.125\%$ $VC_1 = 300'$ $VC_2 = 600'$

$PI = 35 + 00$, elev. 470.00

$PI = 40 + 00$, elev. 482.00

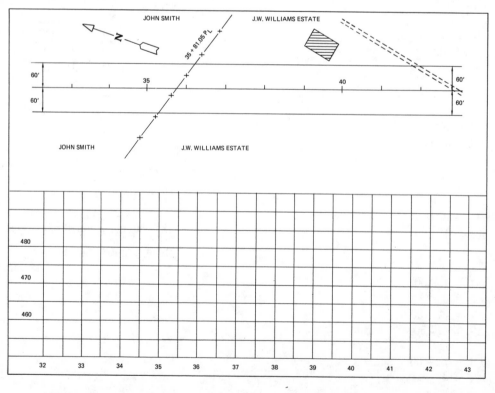

19 CONSTRUCTION SURVEYING

1 DEFINITION

Construction surveying involves locating and marking locations of structures which are to be built. It is often referred to as *giving line and grade*. A transit or theodolite is used in establishing line (horizontal alignment), and a level is used in establishing grade (elevation).

2 CONVERSION BETWEEN INCHES AND DECIMALS OF A FOOT

Engineering plans usually show dimensions of structures in feet and inches, while elevations are established in feet and decimals of a foot. It is necessary to make the conversions to establish finish elevations. Construction stakes are usually set to the nearest hundredth of a foot for concrete, asphalt, pipelines, etc. For earthwork, stakes are set to the nearest tenth of a foot.

In converting measurements in feet, inches, and fractions to feet and decimals of a foot, it may be easier to convert the inches and fractions of an inch separately. Then, add the parts.

Example 19.1

Convert the measurements to feet and decimals of a foot.

(a) $1'4''$

(b) $11'9\frac{1}{3}''$

(c) $7'5\frac{3}{4}''$

(d) $2'8\frac{7}{8}''$

(d) $5'11\frac{1}{2}''$

---

[1] The word *grade* is not consistently used, sometimes meaning slope, and sometimes meaning elevation above a datum. In this text, *gradient* will be used for rate of slope, and *finish elevation* will be used for the elevation above a datum to which a part of the structure is to be built.

Solution

(a) 1.33 ft

(b) 11.76 ft

(c) 7.48 ft

(d) 2.74 ft

(e) 5.96 ft

Example 19.2

Convert the following measurements to feet and inches.

(a) 3.79 ft

(b) 6.34 ft

(c) 5.65 ft

(d) 3.72 ft

Solution

(a) $3'-9\frac{1}{2}''$

(b) $6'-4\frac{1}{8}''$

(c) $5'-7\frac{3}{4}''$

(d) $3'-8\frac{5}{8}''$

3 STAKING OFFSET LINES FOR CIRCULAR CURVES

Stakes set for the construction of pavement or curbs must be set on an offset line so that they will not be destroyed by construction equipment. However, they must be close enough for short measurements to the actual line. The offset line may be 3 or 5 feet, or any convenient distance from the edge of pavement or back of curb. Stakes are set at 25 or 50 feet intervals, and tacks are set in the stakes to designate the offset line.

In setting stakes on a parallel circular arc, the central angle is the same for parallel arcs. The radius to the centerline of the street or road is usually the design radius. The *PC* of the design curve and the *PC* of a parallel offset curve, whether right or left, will fall on the same radial line. Likewise, the *PT*'s of the parallel arcs will fall on the same radial line. In computing the stations for *PC*'s and *PT*'s, the design curve data (design radius) should be used. Then, the *PC* and *PT* stations for an offset line will be the same as for the design curve (centerline of road or street), even though the lengths of offset curves will not be the same as the length of the centerline curve.

The design curve data will be used to compute deflection angles. These angles will be the same for offset lines, since the central angle between any two radii is the same for the parallel arcs.

Because chord lengths are a function of the radius of an arc $(C = 2R\sin(\Delta/2))$, the chord length between two stations on the design curve and two corresponding stations on the offset line will not be the same.

By using design curve data in computing *PC* and *PT* stations, deflection angles for curves can be recorded in the field book. Such angles will be the same whether a right offset line, a left offset line, or both right and left offset lines are used.

In performing field work, centerline *PI*'s, *PC*'s, and *PT*'s are located on the ground before construction. Offset *PC*'s and *PT*'s are located at right angles to the centerline *PC*'s and *PT*'s. Offset *PC*'s and *PT*'s should be carefully referenced.

Example 19.3

Stakes are to be set on 4 foot offsets for each edge of pavement, which is 36 feet wide. The curve has a deflection angle Δ of 60° to the right, a centerline radius of 300 feet, *PI* is at station $12 + 44.32$, and stakes are to be set for each full station, each half station, and at the *PC* and *PT*.

Compute *PC* and *PT* stations, deflection angles, and chord lengths. Set up field notes for the curve.

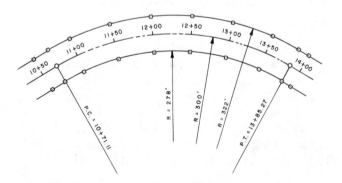

Figure 19.1   Example 19.3

Solution

$$\Delta = 60°00'$$

$$T = R\tan\Delta/2 = 300\tan 30° = 173.21\,\text{ft}$$

$$L = \frac{60}{360} \times 2\pi R = \frac{60}{360} \times 2\pi 300 = 314.16\,\text{ft}$$

$$
\begin{aligned}
PI &= 12 + 44.32 \\
T &= -1 + 73.21 \\
PC &= \overline{10 + 71.11} \\
L &= 3 + 14.16 \\
PT &= \overline{13 + 85.27}
\end{aligned}
$$

Deflection Angle Computations

Sta $10 + 71.11 : 0°00'$

$$\text{Sta } 11 + 00 : \frac{(11 + 00 - 10 + 71.11)}{314.16} \times \frac{60°}{2}$$

$$= \frac{28.89}{314.16} \times 30° = 2.7588° = 2°45'$$

$$\text{Sta } 11 + 50 : \frac{(28.89 + 50)}{314.16} \times \frac{60°}{2}$$

$$= \frac{78.89}{314.16} \times 30° = 7.5334° = 7°32'$$

$$\text{Sta } 12 + 00 : \frac{128.89}{314.16} \times 30° = 12.3081° = 12°18'$$

$$\text{Sta } 12 + 50 : \frac{178.89}{314.16} \times 30° = 17.0827° = 17°05'$$

$$\text{Sta } 13 + 00 : \frac{228.89}{314.16} \times 30° = 21.8573° = 21°51'$$

$$\text{Sta } 13 + 50 : \frac{278.89}{314.16} \times 30° = 26.6320° = 26°38'$$

$$\text{Sta } 13 + 85.27 : \frac{314.16}{314.16} \times 30° = 30.0000° = 30°00'$$

Outside Chord Lengths

station to station	computation	length
$10 + 71.11$ to $11 + 00$:	$(2)(322)\sin 2.7588°$	$= 31.00\,\text{ft}$
$11 + 00$ to $11 + 50$:	$(2)(322)\sin 4.7746°$	$= 53.60\,\text{ft}$
$13 + 50$ to $13 + 85.27$:	$(2)(322)\sin 3.3680°$	$= 37.83\,\text{ft}$

Inside Chord Lengths

station to station	computation	length
$10 + 71.11$ to $11 + 00$:	$(2)(278)\sin 2.7588°$	$= 27.72\,\text{ft}$
$11 + 00$ to $11 + 50$:	$(2)(278)\sin 4.7746°$	$= 47.94\,\text{ft}$
$13 + 50$ to $13 + 85.27$:	$(2)(278)\sin 3.3680°$	$= 33.84\,\text{ft}$

Field Notes

point	station	angle	C-out	C-in	curve data
PT	13 + 85.27	30°00′			
			37.83	33.84	
	+50.00	26°38′			
			53.60	47.94	
	13 + 00.00	21°51′			$\Delta = 60°00′$ Rt
			53.60	47.94	
	+50.00	17°05′			$R = 300′$
			53.60	47.94	
	12 + 00.00	12°18′			$T = 173.21$
			53.60	47.94	
	+50.00	7°32′			$L = 314.16$
			53.60	47.94	
	11 + 00.00	2°45′			
			31.00	27.72	
PC	10 + 71.11	0°00′			

4 CURB RETURNS AT STREET INTERSECTIONS

Curb returns are the arcs made by the curbs at street
intersections. The radius of the arc is selected by the designer with consideration given to the speed and volume
of traffic. A radius of 30 feet to the back of curb is common. Streets which intersect at a right angle have curb
returns of one-quarter circle. The arcs can be swung
from a radius point (center of circle). The radius point
can be located by finding the intersection of two lines,
each of which is parallel to one of the centerlines of the
streets and at a distance from the centerline equal to
half the street width plus the radius, as shown in figure
19.2. One stake at the radius point is sufficient for a
curb return.

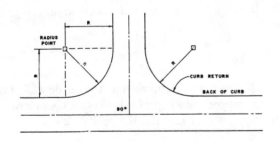

Figure 19.2   Curb returns

5 STAKING OFFSET LINES AT STREET INTERSECTIONS

In setting stakes for curb and gutter for street construction on an offset line, stations for the PC and PT of a
curb return at a street intersection are computed along
the centerline of the street. In example 19.4, stakes
are to be set on a 5 foot offset line from the back of
the left curb along Elm Street. In setting stakes at the
intersection of 24th Street, the PC's and PT's of the
curb return are computed from the centerline stations.
However, it must be remembered that in computing the
long chord on the offset line, the radius R is not the design radius, but is the design radius minus the offset
distance. Stakes are set at the PC, the PT, and the
radius point of each curb return arc.

At street intersections not at 90°, there will be two deflection angles, one being the supplement of the other.

Example 19.4

Elm Street and 24th Street intersect as shown in figure
19.3. Both streets are 26 feet wide, back to back of
curb.

Compute PC and PT stations, deflection angles from
PC to PT, and long chord measured from the offset
line for each return.

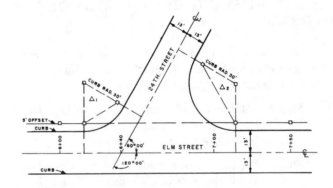

Figure 19.3   Example 19.4

Solution

PC stations along Elm Street

$$\Delta_1 = 60°00′$$
$$T_1 = R_1 \tan \Delta_1/2$$
$$= 43 \tan 30°$$
$$= 24.83 \, \text{ft}$$
$$PI = 6 + 40.00$$
$$T_1 = \quad -24.83$$
$$PC = \overline{6 + 15.17}$$
$$\Delta_2 = 120°00′$$
$$T_2 = R_2 \tan \Delta_2/2$$
$$= 43 \tan 60°$$
$$= 74.48 \, \text{ft}$$
$$PI = 6 + 40.00$$
$$T_2 = \quad +74.48$$
$$PC = \overline{7 + 14.48}$$

PT stations along 24th Street

$$PI = 0 + 00.00$$
$$\underline{T_1 = \quad +24.83}$$
$$PT = \overline{0 + 24.83}$$

$$PI = 0 + 00.00$$
$$\underline{T_2 = \quad +74.48}$$
$$PT = \overline{0 + 74.48}$$

Deflection angles and long chords

$$LC = 2R \sin \Delta_1/2$$
$$= (2)(25) \sin 30°$$
$$= 25.00 \, \text{ft}$$
$$\text{Defl. ang.} = 60°/2 = 30°$$
$$LC = 2R \sin \Delta_2/2$$
$$= (2)(25) \sin 60°$$
$$= 43.30 \, \text{ft}$$
$$\text{Defl. ang.} = 120°/2 = 60°$$

6 ESTABLISHING FINISH ELEVATIONS OR "GRADE"

Establishing the elevation above a datum to which a structure, or a part of a structure, is to be built is usually accomplished by the following steps:

Step 1: Set the top of a grade stake to that exact elevation (nearest one hundredth). Mark the top of the stake with blue keel.

Step 2: Set the top of a grade stake at an exact distance above or below finish elevation. Mark the top of it with blue keel, and mark this exact distance above or below (called *cut* or *fill*) on another stake known as a *guard stake*, usually driven at an angle, beside the grade stake.

Step 3: Use the line stake as a grade stake driven to a random elevation. Compute the difference in elevation between that elevation and finish elevation, and mark this difference as cut or fill on a guard stake.

Marks on a wall, such as the wall of forms for a concrete structure, may be used instead of the tops of stakes. This is illustrated in figure 19.5.

7 GRADE ROD

A grade rod is the rod reading determined by finding the difference in elevation between the height of instrument, (height of the level) and the finish elevation. In figure 19.4, the finish elevation is 441.23, and the H.I. of the

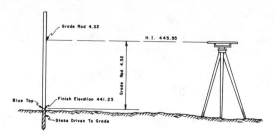

Figure 19.4   Use of grade rod

level is 445.55. The grade rod is the difference in these two numbers, 4.32. A stake is driven so that when the level rod is placed on the top of it, the rod reading is 4.32.

The procedure to set the stake is to place the rod on the ground where the stake is to be driven and determine the distance (in tenths) which the top of the stake should be above the ground. Then drive the stake until the top of the stake is at finish elevation, stopping to check the rod reading so that the top of the stake will not be too low. When the grade rod reading is reached, the top of the stake is marked with blue keel, and thus the name *blue top* is given to this type of stake. A guard stake is driven beside the blue top in a slanting position. The guard stake is marked "G" to indicate the stake is driven to grade (finish elevation).

If the finish elevation is just below ground level, the blue top can be left above ground. The cut from the top of the stake to finish elevation should be marked on a guard stake.

Where line stakes are also used as grade stakes driven to random elevations, a grade rod is not used. The elevation of the top of the stake is determined by leveling. The difference in elevation between top of stake and finish elevation is determined and marked on the guard stake.

Example 19.5

Finish elevation is to be marked on the inside wall of the form for the concrete cap of a bridge. Finish elevation of the cap is 466.97 and the H.I. is 468.72.

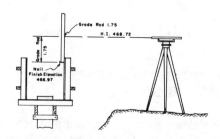

Figure 19.5   Example 19.5

PROFESSIONAL PUBLICATIONS, INC. • P.O. Box 199, San Carlos, CA   94070

Solution

$$H.I. = 468.72$$
$$\text{finish elevation} = 466.97$$
$$\text{grade rod} = \overline{1.75}$$

The rod is held against the side of the form and raised or lowered until the rod reading is 1.75. A nail is driven at the bottom of the rod, and the rod is placed on the nail so that the rod reading can be checked to see that the nail is correctly placed. Another finish elevation nail is driven at the other end of the form, a string line is drawn between the two nails, is chalked, and snapped to mark the grade line on the form. A chamfer strip is nailed on the form along this line, and the strip is used to finish the concrete to grade (finish elevation).

8 SETTING STAKES FOR CURB AND GUTTER

Separate stakes are often set for line and grade. In figure 19.6, a hub stake is driven so that a tack is exactly 3 feet from the back of a curb. These line stakes are set on any convenient offset to avoid disturbance by construction equipment. A separate grade stake is driven so that the top of the stake is either at finish elevation or at an elevation which makes it an exact distance above or below finish elevation.

A guard stake is driven near the grade stake and marked to show this exact distance as cut or fill, and the top of the grade stake is marked with blue keel. Grade stakes can be driven so that the cut or fill is in multiples of a half foot. If the grade stake is driven to finish elevation, the guard stake is marked "G" for grade. The cut or fill can be determined by considering the finish elevation and a ground rod reading at each station.

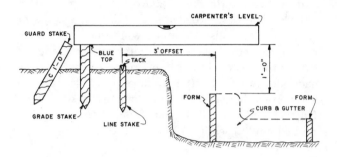

Figure 19.6 Line and grade stakes for curb and gutter

The top of the grade stake should be above ground. Then, the builder can lay a carpenter's level on top of

station	+	H.I.	−	ROD	elevation	fin. elev.	grade rod	ground	mark stake
BM#1	3.42	455.78			452.36	r.r.spike in 12″ oak-100′ lt. sta 0 + 00			
0 + 00				0.18	455.60	454.10	1.68	0.4	C 1′-6″
0 + 50				1.20	454.58	454.58	1.20	1.6	grade
1 + 00				2.72	453.06	455.06	0.72	3.2	F 2′-0″
1 + 50				2.75	453.03	455.53	0.25	3.3	F 2′-6″
2 + 00				2.27	453.51	456.01	−0.23	2.5	F 2′-6″
2 + 50				1.79	453.99	456.49	−0.71	1.8	F 2′-6″
3 + 00				0.31	455.47	455.97	−1.19	0.6	F 1′-6″
T.P.	8.21	460.24	3.75		452.03				
3 + 50				3.30	456.94	457.44	2.80	3.6	F 0′-6″
4 + 00				1.41	458.83	457.83	2.41	1.7	C 1′-0″
4 + 50				0.70	459.54	458.04	2.20	1.0	C 1′-6″
5 + 00				1.16	459.08	458.08	2.16	1.5	C 1′-0″
5 + 50				0.31	459.93	457.93	2.31	0.7	C 2′-0″
6 + 00				1.04	459.20	457.70	2.54	1.4	C 1′-6″
BM#2			2.06		458.18				
		$\overline{11.63}$	$\overline{5.81}$						

$$\begin{array}{cc} 11.63 & 458.18 \\ 5.81 & 452.36 \\ \overline{5.82} & \overline{5.82} \end{array}$$

Figure 19.7 Example 19.6 Field notes for curb and gutter grades

PROFESSIONAL PUBLICATIONS, INC. • P.O. Box 199, San Carlos, CA 94070

the stake and measure from the established level line to establish the top of curb forms. In figure 19.6, the guard stake is marked "C 1'-0''" so that the builder will measure 1'-0'' down from the level line to the top of the forms. Horizontal alignment will be maintained by measuring 3 feet from each tack point to the back of curb line.

Example 19.6

Grade stakes for curb and gutter have been driven to grade or to a multiple of 6'' above or below grade. Part of the level notes recorded in setting the stakes is shown. Also shown are finish elevations which have been taken from construction plans. Computations for grade rod and for the cut or fill marks on guard stakes are to be made and recorded. Rod readings on grade stakes as driven are to be recorded in the column marked "Rod." (Note: A more detailed explanation of this procedure can be found in section 19.14.)

19.1

Example 19.6   Curb & gutter grades

station	+	−	elevation	finish elevation	ground rod
BM-1	3.42		452.36		
0 + 00				454.10	0.4
+ 50				454.58	1.6
1 + 00				455.06	3.2
+ 50				455.53	3.3
2 + 00				456.01	2.5
+ 50				456.49	1.8
3 + 00				456.97	0.6
TP	8.21	3.75			
+ 50				457.44	3.6
4 + 00				457.83	1.7
+ 50				458.04	1.0
5 + 00				458.08	1.5
+ 50				457.93	0.7
6 + 00				457.70	1.4
BM-2		2.06			

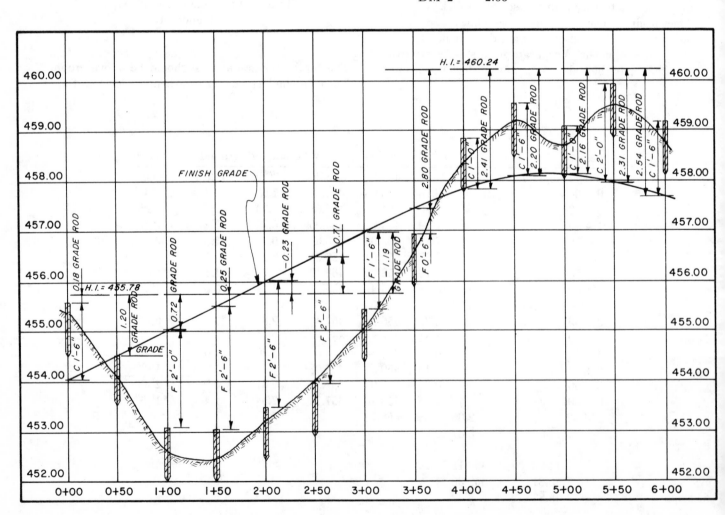

Figure 19.8   Example 19.6 Finished grade line

PROFESSIONAL PUBLICATIONS, INC. • P.O. Box 199, San Carlos, CA   94070

Solution

Finished field notes are shown in figure 19.7. Graphical solution is shown in figure 19.8. Solutions for stations 0 + 00, 0 + 50, 1 + 00, and 2 + 50 are shown in figure 19.9.

9 STAKING CONCRETE BOX CULVERTS ON HIGHWAYS

Tack points for concrete box culverts can be set on off-sets from the outside corners of the culvert headwalls. For normal culverts (centerline of culvert at right angle to centerline of roadway), the distance from the center-line of roadway to the outside of the headwall is equal to one-half the clear roadway width plus the width of the headwall. Tacks should also be set on the centerline of roadway, offset from the outside of the culvert wall. The offset distance from the outside walls depends on the depth of cut. Stakes for wingwalls and aprons are not necessary, although stakes to establish the center-line of culvert can be set if desired. Cuts to the flowline of the culvert can be marked on guard stakes at the tack points.

In staking skewed culverts, the distance from the cen-terline of roadway to the outside of headwall (along the

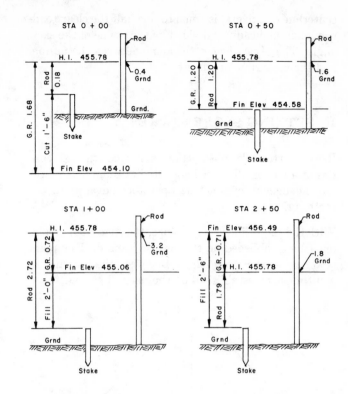

Figure 19.9   Example 19.6 Sample staking

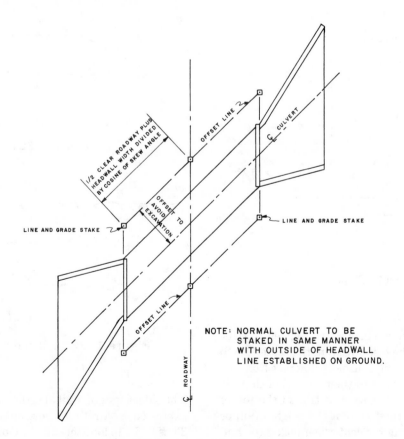

Figure 19.10   Staking box culverts

PROFESSIONAL PUBLICATIONS, INC. ● P.O. Box 199, San Carlos, CA   94070

centerline of culvert) is equal to one-half the clear roadway plus the headwall divided by the cosine of the skew angle. The *skew angle* is the angle between the normal and the centerline of culvert.

10 SETTING SLOPE STAKES

Before earthwork construction is started, the extremities of a cut or fill must be located at numerous places for the benefit of machine operators engaged in the earthwork.

With the centerline as a reference, the edge (*toe*) of a fill must be established on the natural ground. This point is known as the *toe of slope*. Likewise, the top edge of a cut must be established on the natural ground.

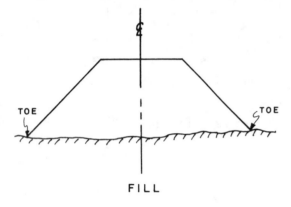

Figure 19.11   Slope staking

Because the natural ground may slope from left to right or from right to left, the distance from the centerline to the left toe of slope of a fill is usually different from the distance from the centerline to the right toe of slope at any particular station. The same is true of the top of a cut. This fact, plus the fact that the height of fill or depth of cut varies along the centerline, makes toe and top lines irregular when seen in plan view, as is shown in figure 19.12.

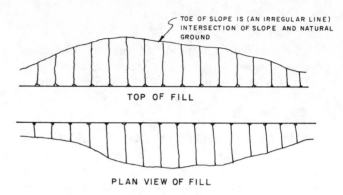

Figure 19.12   Fill views

The toe of a fill or the top of a cut is found by a measure and try method. The horizontal distance from centerline to toe or top is determined by horizontal tape measurements combined with vertical distance measurements derived by use of level and rod.

Dimensions of the top of a fill or bottom of a cut, and the slope of the sides of the fill or cut must be known. These are used in the measure and try method. They are shown on the 'Typical Sections Sheet' of construction plans, as in figure 19.13.

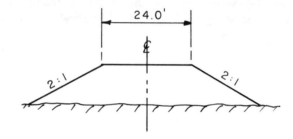

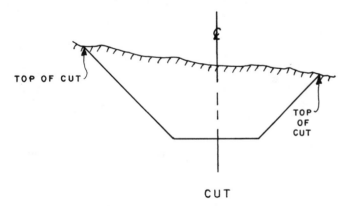

Figure 19.13   Fill and cut dimensioning

The side slopes of a fill and the back slopes of a cut are expressed as a ratio of horizontal to vertical distance. Thus, a 4 : 1 slope means a rise or fall of 1 foot for each 4 feet of horizontal distance. Slopes of 1 : 1, 2 : 1, and 3 : 1 are illustrated in figure 19.14.

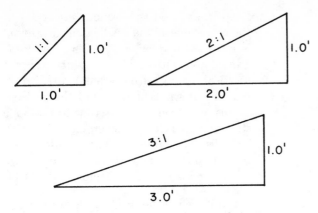

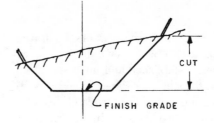

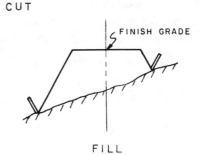

Figure 19.14   Calculation of slopes

With the centerline finish elevation, width of top of fill or bottom of cut, and side slopes all known, the intersection of the side slopes and the natural ground is located at each station or intermediate point.

When the intersection is found, it is marked by a *slope stake*. The stake is driven so that it slopes away from the fill or cut, and is marked with its horizontal distance (left or right) from the centerline, and the vertical distance from the ground at the stake to the finish elevation. A stake marked "C 3.2-48.2" means that the stake is 48.2 feet from the centerline, and the ground at the stake is 3.2 feet above the finish elevation. The station number is shown on the side of the stake facing the ground.

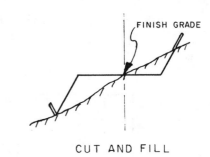

Figure 19.15   Stake orientations

11 GRADE ROD

In setting slope stakes, as in setting finish elevation for pavement, sewer lines, etc., the grade rod is used to determine the difference in elevation between the H.I. and the finish elevation. To determine the cut at a particular point, the rod is read on the ground, and the ground rod is subtracted from the grade rod at that point. To determine the fill at a particular point, the grade rod is subtracted from the ground rod if the H.I. is above the finish elevation. The grade rod is added to the ground rod if the H.I. is below the finish elevation. (see figure 19.16, 19.19, and 19.20.)

12 SETTING SLOPE STAKES AT CUT SECTIONS

An explanation of setting slope stakes without the benefit of a demonstration in the field is difficult. In example 19.7, a scale drawing is used at a cut section at which the H.I. and finish elevation are known and plotted on the drawing. The width of the ditch bottom and the side slopes (also referred to as *back slopes*) are also known. In this example, the level and rod are replaced

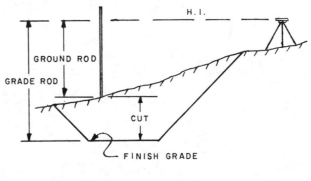

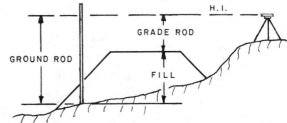

Figure 19.16   Use of grade rod to determine
cut and fill

by the plotted H.I. and an engineer's scale. The scale is used to measure vertical distance from H.I. to ground, just as the level and rod are used.

PROFESSIONAL PUBLICATIONS, INC. • P.O. Box 199, San Carlos, CA   94070

Example 19.7

Figure 19.17 shows the ground cross section at a station at which slope stakes are to be set for a ditch to be excavated. H.I. has been established and finish elevation, width of ditch bottom, and side slopes have been obtained from construction plans. Centerline of ditch has also been established at this station. In this example, two unsuccessful trys to locate the stakes are shown in the space below the cross-section for each side of the centerline. Known information is tabulated.

> finish elevation of ditch bottom $= 470.45$
>
> bottom width $= 12\,\text{ft}$
>
> side slopes $= 2:1$
>
> H.I. $= 479.24$

Solution

Step 1: Compute grade rod (G.R.):

$$\text{G.R.} = \text{H.I.} - \text{finish elevation}$$
$$= 479.24 - 470.46 = 8.78$$

Step 2: Read the rod on the ground (use scale) at the centerline. A rod reading on the ground is known as a *ground rod*. This centerline ground rod will enable us to find the cut (vertical distance from ground to finish elevation) at the centerline and the horizontal distance from the centerline to the slope stake (on each side) if the ground were level. This distance will be used as a guide to find the actual distance to the slope stake where the ground is not level. Ground rod = 4.1

$$\text{Cut at centerline} = \text{G.R.} - \text{ground rod}$$
$$= 8.78 - 4.1 = 4.7$$

Step 3: Find cut and horizontal distance from centerline to slope stake on the left side.

(a) The horizontal distance from centerline to left stake is equal to one-half the width of the ditch bottom plus the horizontal distance from the left edge of ditch bottom to stake.

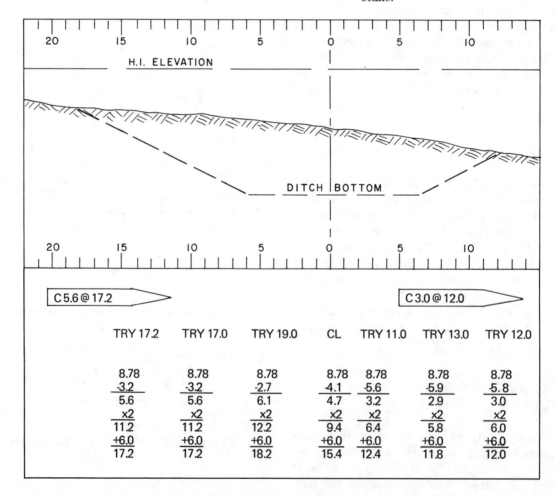

Figure 19.17   Example 19.7 Cut cross section

PROFESSIONAL PUBLICATIONS, INC. • P.O. Box 199, San Carlos, CA   94070

(b) The slope is 2 : 1. Therefore, the side slope will rise (from ditch bottom) one foot vertically for each two feet horizontally. For level ground, the vertical rise is the cut at the centerline which has been found to be 4.7. Therefore, the horizontal distance for level ground is

$$2 \times 4.7 = 9.4$$

The distance from the centerline is

$$6 + 2 \times 4.7 = 15.4$$

(c) The ground is not level. The slope is down from left to right, and the left slope stake will be at a greater distance from the centerline than the right slope stake.

(d) Use the horizontal distance computed for level ground (15.4) as a guide. Make a first try beyond it because of the slope of the ground.

(e) Try a distance of 19.0 feet (chosen arbitrarily) from the centerline and read the rod on the ground at this point. (Use scale for rod.) The Ground Rod is 2.7. Then,

$$\text{cut} = 8.78 - 2.7$$
$$= 6.1$$
$$\text{distance from centerline} = 6 + 2 \times 6.1$$
$$= 18.2$$

This is not the correct location because the measured distance (19.0) does not agree with the computed distance (18.2).

(f) For the next try, move toward the centerline because 19.0 was too far. Try 17.0 where the Ground Rod is 3.2. Then

$$\text{cut} = 8.78 - 3.2$$
$$= 5.6$$
$$\text{distance from centerline} = 6 + 2 \times 5.6$$
$$= 17.2$$

(g) Try 17.2, where the Ground Rod is 3.2 again.

$$\text{distance from centerline} = 6 + 2 \times 5.6 = 17.2$$

We have found the correct location for the slope stake, so we mark it **C 5.6 @ 17.2** on one side and the station number on the

other. We drive the stake with the station number down and sloping away from the cut.

Step 4: Find cut and horizontal distance on the right side.

The slope stake on the right is set in the same manner. In arbitrarily selecting a horizontal distance for the first try, we select a distance less than the 15.4 computed for level ground because the slope is down from left to right.

The correct cut and distance for the right slope stake is shown on the stake marking in figure 19.17.

Example 19.8

Slope stakes are to be set at station 3+00. The bottom of the cut is to be at elevation 462.00 and is 10 feet wide. The side slopes are 2 : 1. (See figure 19.18).

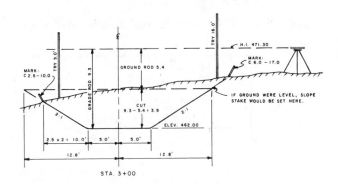

Figure 19.18   Example 19.8

Solution

Step 1: Establish the level near station 3 + 00 and determine H.I. (471.30 in this example.)

Step 2: Compute grade rod by subtracting elevation at bottom of cut from H.I.: 471.30−462.00 = 9.30.

Step 3: Determine ground rod by placing rod on ground at centerline: Read 5.4.

Step 4: Compute cut at centerline by subtracting ground rod from grade rod: 9.3 − 5.4 = 3.9.

Step 5: Compute distance to left slope stake from centerline as if the ground were level at this station: 5 + 2 × 3.9 = 12.8.

Step 6: Note that the ground on the left slopes down and the side of the cut slopes up, indicating that the distance to the stake will be less than that for level ground.

Step 7: Try a distance less than 12.8, say 9.0, and read rod at this distance. Rod reading is 6.6. Grade Rod − Ground Rod = 9.3 − 6.6 = 2.7. Distance computed from this rod reading is 5 + 2 × 2.7 = 10.4. Move toward 10.4; try 10.0. (Move less because slopes are opposite.)

Step 8: Ground Rod at 10.0 is 6.8 : 9.3 − 6.8 = 2.5. Computed distance is 5 + 2 × 2.5 = 10.0. Computed distance agrees with measured distance.

Step 9: Set stake at 10.0 feet left of centerline and mark: **C 2.5 @ 10.0′** on top face of stake and **3 + 00** on bottom.

Step 10: Move to the right side. Try a distance greater than that for level ground because ground and sides both slope up.

Step 11: Try 16.0; Ground Rod is 3.4 : 9.3 − 3.4 = 5.9. 5 + 2 × 5.9 = 16.8.

Step 12: Try 17.0. Move beyond 16.8 because slopes are in the same direction. Ground Rod is 3.3 : 9.3 − 3.3 = 6.0. 5 + 2 × 6.0 = 17.0.

Step 13: Set stake at 17.0 feet. Mark: C 6.0′ @ 17.0′.

13 SETTING SLOPE STAKES AT FILL SECTION

In setting slope stakes for fills, two situations may arise: (a) The H.I. may be below the finish elevation as shown in figure 19.19 and 19.20. (b) The H.I. may be above the finish elevation as shown in figure 19.16. If the H.I. is below the finish elevation, the fill is the sum of the Grade Rod and the Ground Rod. If the H.I. is above the finish elevation, the fill is the difference between the Ground Rod and the Grade Rod.

Example 19.9

Figure 19.19 shows the ground cross-section at a station at which slope stakes are to be set for a fill. H.I. has been established and finish elevation, width of top of

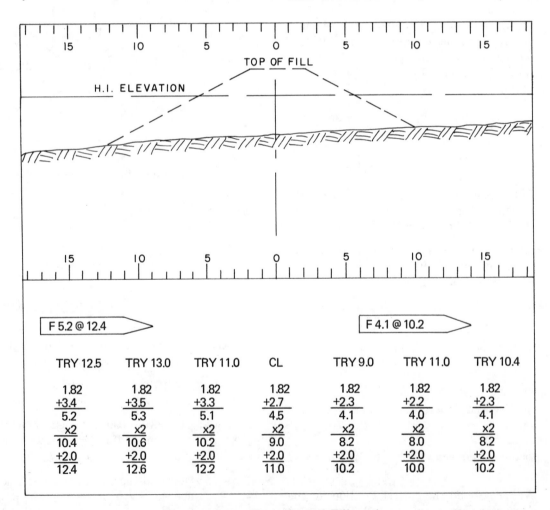

Figure 19.19   Example 19.9 Fill section

PROFESSIONAL PUBLICATIONS, INC. • P.O. Box 199, San Carlos, CA 94070

fill, and side slopes have been obtained from construction plans. Centerline of fill has also been established. Known information is tabulated:

$$\text{finish elevation of top of fill} = 452.36$$
$$\text{top of fill width} = 4\,\text{ft}$$
$$\text{side slopes} = 2:1$$
$$\text{H.I.} = 450.54$$

Solution

The solution is shown in figure 19.19. The correct cut and distance can be found on the marked stake.

Example 19.10

In figure 19.20, slope stakes are to be set at station $10+00$. The top of fill is to be at elevation 468.00 and is 10 feet wide. Side slopes are $1\frac{1}{2}:1$.

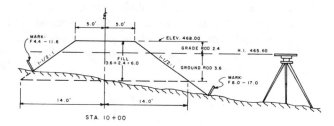

Figure 19.20   Example 19.10

Solution

Step 1: Establish level near station $10+00$ and determine H.I. (465.60 in this example.)

Step 2: Compute Grade Rod by subtracting H.I. from elevation of top of fill: $468.00 - 465.60 = 2.4$.

Step 3: Determine Ground Rod by placing rod on ground at centerline. Read 3.6.

Step 4: Compute fill at centerline by adding Grade Rod and Ground Rod: $3.6 + 2.4 = 6.0$.

Step 5: Compute distance to left slope stake from centerline as if the ground were level at this station: $5 + 1.5 \times 6.0 = 14.0\,\text{ft}$.

Step 6: Note that slopes are opposite, indicating distance will be less than that for level ground.

Step 7: Try 11.0. Rod reads 2.2 : $2.2 + 2.4 = 4.6$. $5 + 1.5 \times 4.6 = 11.9$. Move toward 11.9, but less because slopes are opposite.

Step 8: Try 11.5. Ground Rod is 2.0 : $2.0 + 2.4 = 4.4$. $5 + 1.5 \times 4.4 = 11.6$. Close enough.

Step 9: Set stake at 11.6 feet left of centerline and mark **F 4.4 @ 11.6**. Mark **10 + 00** on bottom face of stake.

Step 10: Move to right side. Try a distance greater than that for level ground because ground and slope are in same direction.

Step 11: Try 15.0. Ground Rod is 5.3. $5.3 + 2.4 = 7.7$. $5 + 1.5 \times 7.7 = 16.6$. Move toward 16.6 and beyond because slopes are both down.

Step 12: Try 17.0. Ground Rod is 5.6 : $5.6 + 2.4 = 8.0$. $5 + 1.5 \times 8.0 = 17.0$.

Step 13: Set stake at 17.0 feet and mark: **F 8.0′ @ 17.0′**.

14 SETTING STAKES FOR UNDERGROUND PIPE

Stakes for line and grade for underground pipe, like stakes for roads and streets, are set on an offset line. One hub stake with tack can be used at each station for both line and grade, or separate stakes can be set for line and grade. If only one stake is to be used, the elevation of the top of that stake is determined. Cut from the top of the stake to the flowline. *Invert* of the pipe is computed and marked on a guard stake.[2] This method is faster.

It is often desirable to set a grade stake close to the tacked line stake. This may be set so that the cut from the top of stake to the flowline is at some multiple of a half-foot. (Constructors use foot and inch rules. So, in deference to them, engineering technicians or surveyors set the stakes for their convenience.)

In setting a cut stake for underground pipe, the surveyor first sets up the level and determines its H.I. Using the flowline of the pipe at a particular station, the Grade Rod at that station is computed and recorded in the field book. A rod reading on the ground is taken at the point where the stake is to be driven. This is the *Ground Rod*. Using the Grade Rod and the Ground Rod, the rod reading on top of the stake which will give a half-foot cut from the top of the stake to the flowline is computed. A stake is driven to the rod reading which gives this cut. The stake is blued, and the cut is marked on the guard stake.

Example 19.11

A grade stake is to be set to show the cut to the flowline of a pipe. H.I. is 472.36, flowline is 462.91, and Ground Rod at the point of stake is 5.1. Determine the Grade Rod and rod reading which will give a half-foot cut to the flowline.

---

[2] *Flowlines* are the lines used as finish elevation for pipes.

Solution

$$H.I. = 472.36$$
$$flowline = 462.91$$
$$Grade\ Rod = \overline{9.45}$$

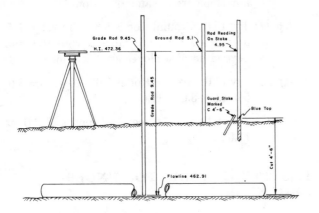

Figure 19.21   Example 19.11

The rod reading on the grade stake to give a half-foot cut from the top of the stake to the flowline could be 8.95, 8.45, ..., 5.45, 4.95, etc., and the corresponding cuts would be 0'-6", 1'-0", 4'-0", 4'-6", etc. The Ground Rod is 5.1; therefore, the cut is approximately $9.5 - 5.1 = 4.4$. Therefore, the rod reading for this cut will be either 5.45, which will give a cut of 4'-0", or 4.95, which will give a cut of 4'-6".

A rod reading of 5.45 cannot be used because the top of the stake would be 0.3' below the surface of the ground. A rod reading of 4.95 would place the top of the stake about 0.1' above ground, which is satisfactory. The stake is driven so that the rod reading is 4.95. The top is marked with blue keel, and the guard stake is marked **C 4'-6"**. $(9.45 - 4.95 = 4.50 = 4'\text{-}6".)$ It can be seen that the rod reading on the stake must be less than the Ground Rod in order that the top of the stake be above ground.

15 FLOWLINE AND INVERT

The bottom inside of a drainage pipe is known as the *flowline*. It is also referred to as the *invert*. Invert is more commonly used to describe the bottom of the flow channel within a manhole.

Vertical control is of prime importance in laying pipe for gravity flow, especially sanitary sewer pipe. In order to facilitate vertical alignment, excavation of the trench often extends a few inches below the bottom of the pipe so that a bedding material, such as sand, is placed in the trench for the pipe to lay on. Because of various methods of using bedding material in laying pipe, stakes are always set for the flowline, or invert, of the pipes. Excavation depth allows for the amount of bedding specified.

16 MANHOLES

Sanitary sewers are not laid along horizontal or vertical curves. Horizontal and vertical alignments are straight lines. Where a change in horizontal alignment or a change in slope is necessary, a manhole is required at the point of change. Therefore, a vertical drop within the manhole is needed. In staking, two cuts are often recorded on guard stakes; one for the incoming sewer and one for the outgoing sewer.

Gravity lines, such as sanitary sewers, flow only partially full. The slope of the sewer determines the flow velocity, and the velocity and size of the pipe determine the quantity of flow. Manholes are used to provide a point of change in conditions. Sewers must be deep enough below the surface of the ground to prevent freezing of their contents and damage to the pipe by construction equipment.

PRACTICE PROBLEMS FOR CHAPTER 19

1. Stakes are to be set on 4 foot offsets for each edge of pavement (which is 28 feet wide), for a curve which has a deflection angle Δ of $55°00'$ and a centerline radius of 250 feet. *PI* is at station $8 + 56.45$. Stakes are to be set on full stations, half stations, and at the *PC* and *PT*. Calculate T and L. Determine the deflection angles used to stake the curve. Calculate the outside and inside chord lengths.

2. Prepare a set of field notes to be used in staking a street curve on the quater-stations from 3 foot offset lines on both sides of the street.

$$PI = 4 + 55.00$$
$$\Delta = 60°00' \text{ left}$$
$$R = 100\,\text{ft (centerline)}$$
$$\text{pavement width} = 28\,\text{feet}$$

3. The intersection of Ash Lane and 32nd Street is to be staked for paving from an offset line 4 feet left of the left edge of pavement. Pavement width is 28 feet and radius-to-edge-of-pavement is 30 feet. From this information and information shown on sketch, compute *PC* and *PT* stations and deflection angles and chord lengths from *PC* to *PT*. Scale: $1/2'' = 30'$.

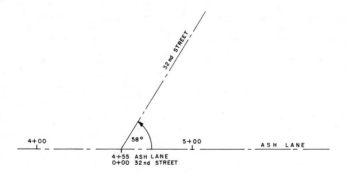

20 EARTHWORK

1 DEFINITION

Earthwork is the excavation, hauling, and placing of soil, rock, gravel, or other material found below the surface of the earth. The definition also includes the measurement of such material in the field, the computation in the office of the volume of such material, and the determination of the most economical method of performing such work.

2 UNIT OF MEASURE

The *cubic yard* is the unit of measure for earthwork. However, the volume and density of earth changes under natural conditions and during the operations of excavation, hauling, and placing.

3 SWELL AND SHRINKAGE

A cubic yard of earth measured in its natural position will be more than a cubic yard after it is excavated. If the earth is compacted after it is placed, the volume may be less than a cubic yard.

The volume of the earth in its natural state is known as *bank-measure*. The volume in the vehicle is known as *loose-measure*. The volume after compaction is known as *compacted-measure*.

The change in volume from its natural to loose state is known as *swell*. Swell is expressed as a percent of the natural volume.

The change in volume from its natural state to its compacted state is known as *shrinkage*. It also is expressed as a percent of the natural state.

As an example, one cubic yard in the ground may become 1.2 cubic yards loose-measure and 0.85 cubic yards after compaction. The swell would be 20%, and the shrinkage would be 15%. Swell and shrinkage vary with soil types.

4 CLASSIFICATION OF MATERIALS

Excavated material is usually classified as *common excavation* or *rock excavation*. Common excavation is soil.

In highway construction, common road excavation is soil found in the roadway. *Common borrow* is soil found outside the roadway and brought in to the roadway. Borrow is necessary where there is not enough material in the roadway excavation to provide for the embankment.

5 CUT AND FILL

Earthwork which is excavated, or is to be excavated, is known as *cut*. Excavation which is placed in embankment, or is to be placed in embankment, is known as *fill*.

Payment for earthwork is normally either for cut and not for fill, or for fill and not for cut. In highway work, payment is usually for cut; in dam work, payment is usually for fill. To pay for both would be paying for the same earth twice.

6 FIELD MEASUREMENT

Cut and fill volumes can be computed from slope-stake notes, from plan cross-sections, or by photogrammetric methods.

7 CROSS-SECTIONS

Cross-sections are profiles of the earth taken at right angles to the centerline of an engineering project (such as a highway, canal, dam, or railroad.) A cross-section for a highway is shown in figure 20.1.

PROFESSIONAL PUBLICATIONS, INC. • P.O. Box 199, San Carlos, CA 94070

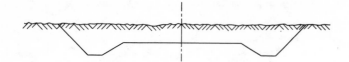

Figure 20.1 Typical cross section

8 ORIGINAL AND FINAL CROSS-SECTIONS

In order to obtain volume measurement, cross-sections are taken before construction begins and after it is completed. By plotting the cross-section at a particular station both before and after construction, a sectional view of the change in the profile of the earth along a certain line is obtained. The change along this line appears on the plan as an area. By using these areas at various intervals along the centerline, and by using distance between the areas, volume can be computed.

9 ESTIMATING EARTHWORK

Earthwork quantities for a highway, canal, or other project can be estimated by superimposing a template on the original plotted cross-section which is drawn to represent the final cross-section. The template is obtained from the *typical section sheet* of the construction plans.

10 TYPICAL SECTIONS

Typical sections show the cross-section view of the project as it will look on completion, including all dimensions. Highway projects usually show several typical sections including cut sections, fill sections, and sections showing both cut and fill. Interstate highway plans also show access-road sections and sections at ramps.

11 DISTANCE BETWEEN CROSS-SECTIONS

Cross-sections are usually taken at each full station and at breaks in the ground along the centerline. In taking cross-sections, it must be assumed that the change in the earth's suface from one cross-section to the next is uniform, and that a section halfway between the cross-sections is an average of the two. If the ground breaks appreciably between any two full stations, one or more cross-sections between full stations must be taken. This is referred to as *taking sections at pluses*. Figure 20.3 shows the stations at which cross-sections should be taken.

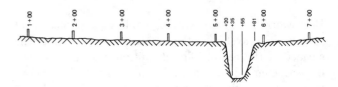

Figure 20.3 Cross section distances

In rock excavation, or any other expensive operation, cross-sections should be taken at intervals of 50 feet or less. Cross-sections should always be taken at the *PC* and *PT* of a curve. Plans should also show a section on each end of a project (where no construction is to take place) so that changes caused by construction will not be abrupt.

Where a cut section of a highway is to change to a fill section, several additional cross-sections are needed. Such stations are shown in figure 20.4.

12 GRADE POINT

The point where a fill section meets the natural ground (where a cut section begins) is known as a *grade point*.

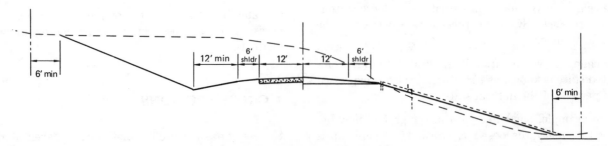

Figure 20.2 Typical completed section

PROFESSIONAL PUBLICATIONS, INC. • P.O. Box 199, San Carlos, CA 94070

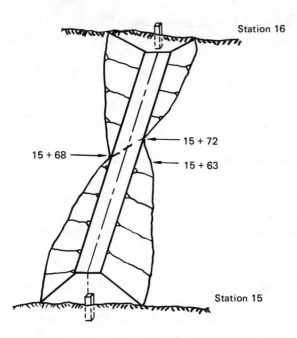

Figure 20.4 Cut changing to fill

13 METHODS FOR COMPUTING VOLUME

The most common method for computing volume of earthwork is by the *average end area method*. (The *Prismoidal formula* furnishes more accurate results but is much more cumbersome. It can be found in most surveying textbooks but will not be discussed herein.) The average end area method is accurate enough for most work.

14 AVERAGE END AREA METHOD

The average end area method is based on the assumption that the volume of earthwork between two vertical cross-sections A_1 and A_2 is equal to the average of the two end areas multiplied by the horizontal distance L between them. Area is expressed in square feet, and distance is expressed in feet. So, the volume in cubic yards is

$$V = \frac{L(A_1 + A_2)}{2 \times 27} = \frac{L}{54}(A_1 + A_2) \qquad 20.1$$

15 FIELD NOTES

Figure 20.5 shows a sample sheet from cross-section field notes. The left half of the page is the same as for any set of level notes. The right half shows rod readings over horizontal distance measured from the centerline for each point on the ground which requires a reading. These readings should always include shots on the centerline, at each break in the ground, and at the right-of-way on each side.

16 PLOTTING CROSS-SECTIONS

Cross-sections are plotted on specially printed cross-section paper. A scale of $1'' = 5'$ is usually used for both the horizontal and vertical. For wide sections, a scale of $1'' = 10'$ or $1'' = 20'$ can be used. The vertical scale can also be exaggerated if necessary.

X-section fm hwy 123

sta	+	H.I.	−	rod	elev
BM#12	3.30	468.21			464.91
11 + 00					
TP$_4$	5.27	462.97		10.51	457.70
12 + 00					
13 + 00					
TP$_5$	1.76	458.22		6.51	456.46
13 + 50					
14 + 00					
15 + 00					
BM#13				5.15	453.07
	10.33			22.17	464.91
				10.33	453.07
				11.84	11.84

r.r. spik in 12″ elm 125 rt

sta   16 + 75

4.4 7.1	4.9 7.3	7.9	9.1 12.0	9.7 11.2
5.0 15	12 5		20 25	30 50
2.0 4.0	1.3 4.5	5.0	6.0 10.1	5.7 8.0
50 20	15 10		15 20	27 50
4.8 6.0	4.2 7.7	9.9	9.8 12.6	11.0 13.0
50 25	20 15		8 15	24 50
2.3 4.1	1.2 5.0	6.7	7.1 12.2	8.3 11.1
50 30	25 20		3 10	19 50
5.2 6.0 10.2	7.9 10.1 11.0	8.1		
50 42 35	20	20 50		
5.0 5.8	9.6 7.3	9.2 10.1	7.5	
50 48	40 20	22	50	

Figure 20.5 Typical field notes for cross section work

A vertical line in the center of the sheet is drawn to represent the centerline of the project. Shots taken in the field are plotted to the proper elevation and distance from the centerline.

Each cross-section is plotted as a separate section. Sufficient space is allowed between cross-sections so that they do not overlap. The station number for each cross-section is recorded just under the centerline shot, and the elevation at the centerline is recorded in a vertical direction just above the centerline shot.

The heavy lines on the paper are used to represent an elevation ending in 0 or 5 feet. With the elevation of the centerline recorded, these heavy lines can be identified as the elevation they represent.

The notes can be plotted by first reducing the level shots to elevation and plotting by elevation. Alternatively, the rod shots can be plotted directly from a line on the paper representing the H.I. As an example, if the H.I. is 447.6 (rounded off) and the rod shot is 5.4, subtracting 5.4 from 7.6 gives an elevation of 442.2 (which can be found quickly because of the printing of the paper). It is also convenient to lay a straight-edge along the H.I. and plot down from it, mentally calculating to get the decimal plotting.

17 DETERMINING END AREAS

End areas are commonly determined by planimetry or by dividing the area into triangles and trapezoids.

Cut areas and fill areas must be kept separate. After the areas have been determined, the sum of each two adjacent areas is placed in a column. The distance between two sections is recorded, and the volume for each sum is computed from equation 20.1.

After the volume has been computed, shrinkage must be added to fill quantities to balance with cut quantities. Shrinkage will vary from 30% for light cuts and fills to 10% for heavy cuts and fills.

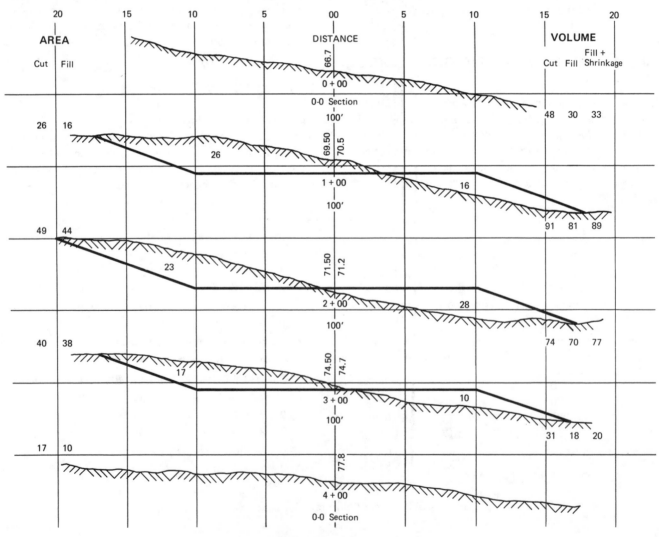

Figure 20.6 Plotting cross sections

Excavated rock will occupy larger volume when placed in a fill, and this swell will be subtracted from the fill quantity.

18 VOLUMES FROM PROFILES

For preliminary estimates of earthwork, volumes can be computed from the centerline profiles. After the ground profile and finish grade profile are plotted, the area of cut can be planimetered and the average determined by dividing by the length of cut. Using the average cut, a template can be drawn, and the end area can also be planimetered. This area times the length of the cut will give the volume.

19 BORROW PIT

As mentioned previously, it is often necessary to borrow earth from an adjacent area to construct embankments.

Normally, the borrow pit area is laid out in a rectangular grid with 10, 50, or even 100 foot squares. Elevations are determined at the corners of each square by leveling before and after excavation so that the cut at each corner can be computed.

Points outside the cut area are established on the grid lines so that the lines can be re-established after excavation is completed.

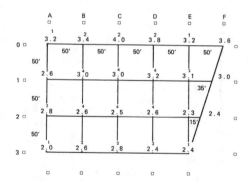

Figure 20.7 Borrow pit areas

As an example, volumes for two of the prisms shown in figure 20.7 are computed by multiplying the average cut by the area of the figure. The volume of the prism A0-B0-B1-A1 is

$$V = \frac{50 \times 50}{27} \times \frac{3.2 + 3.4 + 3.0 + 2.6}{4} = 282 \text{ cu. yds.}$$

The volume of the triangle E2-F2-E3 is

$$V = \frac{50 \times 15}{2 \times 27} \times \frac{2.3 + 2.4 + 2.4}{3} = 33 \text{ cu. yds.}$$

Instead of computing volumes of prisms represented by squares separately, all square based prisms can be computed collectively by multiplying the area of one square by the sum of the cut at each corner times the number of times that cut appears in any square, divided by 4. For instance, on the second line from the top in figure 20.7, which is line 1, 2.6 appears in two squares, 3.0 appears in 4 squares, 3.0 appears in 4 squares, 3.2 appears in 4 squares and 3.1 appears in 2 squares. In the figure, the small number above the cut indicates the number of times the cut is used in averaging the cuts for the prisms.

Example 20.1

Calculate the volume of earth excavated from the borrow pit shown in figure 20.7.

Solution

Volume of squares:

$$\begin{aligned} V = \frac{50 \times 50}{(27)(4)} &(3.2 + 2 \times 3.4 + 2 \times 4.0 + 2 \times 3.8 + 3.2 \\ &+ 2 \times 2.6 + 4 \times 3.0 + 4 \times 3.0 + 4 \times 3.2 + 2 \times 3.1 \\ &+ 2 \times 2.8 + 4 \times 2.6 + 4 \times 2.5 + 4 \times 2.6 + 2 \times 2.3 \\ &+ 2.0 + 2 \times 2.6 + 2 \times 2.8 + 2 \times 2.4 + 2.4) = 3{,}194 \end{aligned}$$

Volume of trapezoids:

$$V = \frac{50 + 35}{2 \times 27} \times 50 \times \frac{3.2 + 3.6 + 3.1 + 3.0}{4} = 254$$

$$V = \frac{35 + 15}{2 \times 27} \times 50 \times \frac{3.1 + 3.0 + 2.3 + 2.4}{4} = 125$$

Volume of triangle:

$$V = \frac{15 \times 50}{2 \times 27} \times \frac{2.3 + 2.4 + 2.4}{3} = \quad 33$$

total cu. yd.   3,606

20 HAUL

In some contracts for highways and railroads, the contractor is paid per cubic yard for excavation (which includes the cost of excavation, hauling, placing in embankment, and compaction of embankment.) However, the cost of hauling one cubic yard of earth over a long distance can easily become greater than the cost of excavation, so that it is often practical to pay a contractor for excavating and hauling earth.

21 FREE HAUL

It is common not to pay for hauling if the material is hauled less than a certain distance, usually 500 to 1000 feet. An additional price is paid for hauling the earth beyond the prescribed limit. The haul distance for which no pay is received is known as *free haul*.

22 OVERHAUL

The hauling of material beyond the free haul limit is known as *overhaul*. The unit of overhaul measure is yard-stations or yard-quarters. A *yard-quarter* is the hauling of one cubic yard of earth one-quarter of a mile. For example, if six yards of earth were hauled one mile, the overhaul would be 24 yard-quarters.

Thus, the word *haul* may have two meanings. It may mean linear distance or volume times distance.

It should be mentioned that the distance is measurement along the centerline. Distance from the extremity of the right-of-way to the centerline is not considered.

23 BALANCE POINTS

It is important in planning and construction to know the points along the centerline a particular section of cut which will balance a particular section of fill. For example, assume that a cut section extends from station $12+25$ to station $18+65$, and a fill section extends from station $18+65$ to station $26+80$. Also, assume that the excavated material will exactly provide the material needed to make the embankment. We can say, then, that cut balances fill, and that stations $12+25$ and $26+80$ are balance points.

24 MASS DIAGRAMS

A method of determining economical handling of material, quantities of overhaul, and location of balance points is the mass-diagram method.

The mass diagram is a graph which has distance in *stations* as the abscissa and the algebraic sums of cut and fill as ordinate. The x-axis parallels the centerline, and the cut and fill (plus shrinkage) quantities are taken from the cross-section sheets. Often, the mass diagram is plotted below the centerline profile so that like stations are vertically in line.

In order to add cut and fill algebraicly, cut is given a plus sign, and fill is given a minus sign.

25 PLOTTING THE MASS DIAGRAM

After volumes of cut and fill between stations have been computed, they are tabulated as shown in table 20.1. The cuts and fills are then added, and the cumulative yardage at each station is recorded in the table. It is this cumulative yardage which is plotted as an ordinate. In figure 20.8, the baseline serves as the x-axis and cumulative yardage which has a plus sign is plotted above the baseline. Cumulative yardage which has a minus sign is plotted below the baseline.

The scale is not important. In figure 20.8, the horizontal scale is $1'' = 5$ stations, and the vertical scale is $1'' = 5000$ cubic yards. A larger scale would be more practical in actual computations.

In figure 20.8, the mass diagram is plotted on the lower half of the sheet, and the centerline profile of the project is plotted on the upper half.

Table 20.1
Typical cut and fill calculations

(All volumes are in cubic yards)

sta	cut +	fill & shr −	cum sum	sta	cut +	fill & shr −	cum sum
0			0	23			−4710
1	184		+184	24	1377		−3034
2	622		+806	25	1676		−1358
3	1035		+1841	26	1860		+502
4	1268		+3109	27	1917		+2419
5	1231		+4340	28	1839		+4258
6	919		+5259	29	1611		+5869
7	503		+5762	30	1338		+7207
8	164	21	+5905	31	1029		+8236
9	12	190	+5727	32	652		+8888
10		616	+5111	33	357		+9245
11		942	+4169	34	150	39	+9356
12		1150	+3019	35	52	236	+9172
13		1500	+1519	36		465	+8707
14		1773	−254	37		712	+7995
15		1755	−2009	38		904	+7091
16		1540	−3549	39		904	+6187
17		1262	−4811	40		757	+5430
18		932	−5743	41		516	+4914
19		546	−6289	42		280	+4724
20		203	−6461	43		127	+4913
21		101	−6283	44		98	+5455
22		18	−5715	45		20	+6206

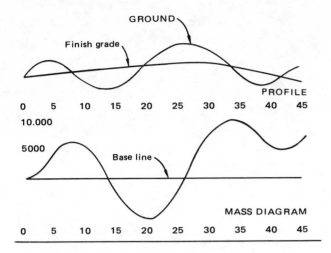

Figure 20.8 Baseline and centerline profile mass diagram

26 BALANCE LINE

Any horizontal line cutting off a loop of the mass curve intersects the curve at two points, between which the cut is equal to the fill. Figure 20.9 shows a portion of the mass diagram of figure 20.8 at enlarged scale.

27 SUB-BASES

Sub-bases are horizontal balance lines which divide an area of the mass diagram between two balance points

into trapezoids for the purpose of more accurately computing overhaul. In figure 20.9, the top sub-base is less than 600 feet in length. Therefore, the volume of earth represented by the area above this line will be hauled a distance less than free haul distance, and no payment will be made for overhaul. All the volume represented by the area below this top sub-base will receive payment for overhaul.

The area between two sub-bases is very nearly trapezoidal. The average length of the bases of a trapezoid can be measured in feet, and the altitude of the trapezoid can be measured in cubic yards of earth. The product of these two quantities can be expressed in yard-quarters. If free haul is subtracted from the length, the quantity can be expressed as overhaul.

Sub-bases are drawn at distinct breaks in the mass curve. Distinct breaks in figure 20.9 can be seen at stations $1+00(184)$, $2+00(806)$, and $5+00(4340)$. After sub-bases are drawn, a horizontal line is drawn mid-way between the sub-bases. This line represents the average haul for the volume of earth between the two sub-bases. If a horizontal scale of $1'' = 100$ feet is used, the length of the average haul can be determined by scaling. This line, shown as a dashed line in figure 20.9, scales 875 feet for the area between the top sub-bases. The free haul is subtracted from this in table 20.1.

The volume of earth between the two sub-bases is found by subtracting the ordinate of the lower sub-base from the ordinate of the upper sub-base. These ordinates are found in table 20.1.

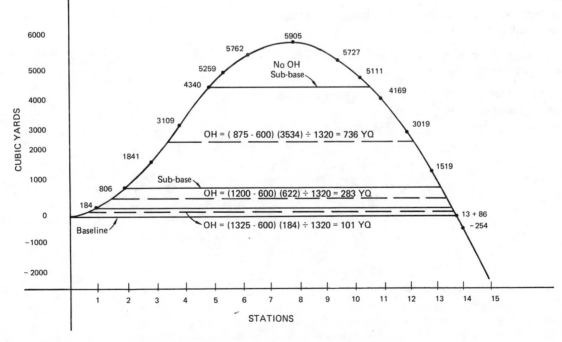

Figure 20.9 Mass diagram showing sub-bases

Multiplying average haul minus free haul in feet by volume of earth in cubic yards gives overhaul in yard-feet. Dividing by 1320 feet gives yard-quarters.

28 LOCATING BALANCE POINTS

A balance point occurs where the mass curve crosses the base line. It can be seen that a balance point falls between stations 13 and 14. The ordinate of 13 is +1519; the ordinate of 14 is −254. Therefore, the curve fell $1519 + 254 = 1773$ cubic yards in one hundred feet or 17.73 cubic yards per foot. The curve crosses the base line at a distance of $1519 \div 17.73 = 86$ feet from station 13 (station 13 + 86).

29 CHARACTERISTICS OF THE MASS DIAGRAM

Important characteristics which should be considered in using the mass diagram in planning the economical hauling of earth are listed here:

- A horizontal line connecting two points on the mass curve cuts off a loop in which the cut equals the fill.

- A loop which rises and then falls from left to right indicates that the haul from cut to fill will be from left to right.

- A loop which falls and then rises from left to right indicates the haul will be from right to left.

- A high point on a mass curve indicates a change from cut to fill.

- A low point on a mass curve indicates a change from fill to cut.

- High and low points on the mass curve occur at or near grade points on the profile.

RESERVED FOR FUTURE USE

PROFESSIONAL PUBLICATIONS, INC. ● P.O. Box 199, San Carlos, CA 94070

RESERVED FOR FUTURE USE

21 FIELD PRACTICE

PART 1: Taping

1 STEEL TAPES

Steel tapes are made in lengths of 50, 100, 200, 300, and 500 feet, but the 100 foot tape is the most common. Tapes are also made in 30, 50, and 100 meter lengths.

Some 100 foot tapes measure 100 feet from the outer edges of the end loops, but most tapes are in excess of 100 feet from end loop to end loop, and they have a line graduation for 0 and 100 foot marks on the tape itself. Graduations for every foot are marked from 0 to 100 feet.

An *add tape* has an extra graduated foot beyond the 0 mark. The extra foot is usually graduated in tenths, but is sometimes graduated in tenths and hundredths of a foot.

A *cut tape* does not have the extra graduated foot, but it has the last foot at each end graduated in tenths or tenths and hundredths of a foot.

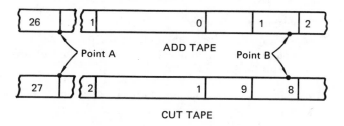

Figure 21.1 Add and cut tapes

Figure 21.1 shows the distance between points A and B being measured by both add and cut (subtract) tapes. The distance, using the add tape, is $26 + 0.18 = 26.18$ feet. The distance, using the cut tape, is $27 - 0.82 = 26.18$ feet.

The Gunter's chain was used in the United States prior to the introduction of the steel tape. It was 66 feet long and had 100 links. The 66 foot chain was 1/80 mile. An area of 10 square chains equaled an area of 1 acre ($10 \times 66^2 = 43{,}560$ sq ft). The chain is seldom used today, but the term *chaining* remains.

2 HORIZONTAL TAPING

In surveying, the distance between two points is the horizontal distance, regardless of the slope.

3 TAPING WITH TAPE SUPPORTED THROUGHOUT ITS LENGTH

The rear chainperson wraps the leather thong at the end of the tape tightly around his right hand near the knuckles.[1] He then faces at right angles to the line of measurement. He kneels with his left knee near the pin (or other mark) and braces the right arm against the right leg near the knee with the heel of the right hand firmly against the ground. In order to bring the end mark of the tape exactly on the pin, he shifts his weight to the left knee or right foot as desired, keeping the heel of the right hand firmly braced on the ground. In this position, he is off the line of sight and his eyes are directly over the end mark of the tape and the pin.

The forward chainperson wraps the leather thong around his left hand, faces at right angles to the line of measurement, and kneels on his right knee. He increases or decreases the pull on the tape by shifting his body weight. With his right hand, he sticks the pin at the zero mark on a call from the rear chainperson indicating the 100 foot mark is on the pin. In this position, he is also off the line of sight.

---

[1] Left-handed chainpersons should use opposite positioning.

4 TAPING ON SLOPE WITH TAPE SUPPORTED AT ENDS ONLY

When taping downhill, the rear chainperson proceeds as in taping on level ground.

The forward chainperson wraps the leather thong around his left hand and takes a position facing at right angles to the line of sight (as he did in taping on level ground) but remains standing. With his right hand, he makes one loop of the plumb bob string around the tape. With the forefinger under the tape and the thumb on top of the tape, he can roll the string to the proper mark with his thumb. The last two fingers of the right hand grasp the loose end of the string. To get the proper length of string, the plumb bob is rested on the ground and the string is fed with the right hand.

The tape is held as nearly horizontal as possible. The chainperson's feet should be placed well apart and the left elbow should be braced against the body. To apply tension, the left knee is bent so that the weight of his body pushes against the arm holding the tape. The plumb bob is steadied by lowering it to the ground. When the plumb bob is just slightly above ground and is steady, the tape is horizontal, and the chainperson feels the proper tension, he lets the plumb bob drop and then marks the point with a pin.

Taping uphill, the rear chainperson holds the plumb bob over the pin or point on the ground, and the forward chainperson proceeds as he does when taping on even ground. The rear chainperson holds the tape and the plumb bob as the forward chainperson did in taping downhill.

5 STATIONING WITH PINS AND RANGE POLE ON LEVEL GROUND

A set of pins consists of eleven pins. A station is 100 feet (one tape) in length. In route surveying, stationing is carried along continuously from a starting point designated as station $0 + 00$.

The rear chainperson stations himself at the point of beginning with one pin in hand (or in the ground if the beginning point was not previously marked).

The forward chainperson takes ten pins. With the zero end of the tape and the range pole, he advances in the direction of the stationing. He counts his paces from the point of beginning so that, if he does not hear the rear chainperson's call, he knows when he has advanced approximately one station.

The rear chainperson watches the tape pass his beginning station and when the end is about 6 feet from his station, calls to the forward chainperson, "chain." He

grabs the leather thong on the end of the tape as it nears him and proceeds as explained in section 21.2.

The forward chainperson, on hearing "chain", immediately turns and faces the rear. He observes the rear chainperson grab the leather thong and, with the tape in his left hand and the range pole in his right hand, he puts tension on the tape, and flips it to straighten it. Holding the range pole vertically, he places it near and slightly to the rear of the zero mark. He immediately drops the tape and, with legs spread fairly wide apart, takes the range pole between the forefinger and thumb of each hand and observes the transitperson (or rear chainperson) for alignment. He keeps the range pole vertical and his legs apart so that, on long shots, the transitperson will have a clear view of the range pole between his legs. When he receives an "ok" from the transitperson, he presses the point of the range pole in the ground and then removes it, placing a chaining pin in the hole. He then wraps the thong around his left hand, flips the tape for alignment, and pulls the edge of the tape over to the pin.

On observing the forward chainperson reach this point, the rear chainperson checks the 100 foot mark to see that it is on the pin and then calls out his station number, such as "eight." Besides keeping up with the station, this call is also saying to the forward chainperson, "I am on my mark."

On hearing the rear chainperson call "eight," the forward chainperson quickly and carefully sticks his pin at the zero mark and calls his station "nine." The call "nine," besides keeping up with the station number, says to the rear chainperson, "I have marked my point, so drop the tape and start walking forward."

The rear chainperson should never hang onto the tape as he moves forward, but he should keep the end of the tape in view.

The system whereby both chainpersons call out the stations is a double-check on counting the pins. And it is a simple way to communicate as to when the forward pin should be set.

In chaining long distances, both chainpersons should pick out distant objects on their line to walk toward.

Chaining pins should be stuck at an angle of 45° with the ground and at right angles to the line of measurement.

6 STATIONING WHEN DISTANCE IS MORE THAN 10 TAPE LENGTHS

When the forward chainperson has set his last pin in the ground, he should have just heard the rear chainperson call "nine." He should have replied "ten." His last pin in the ground indicates he has taped 10 stations or 1000 feet. He waits at this last pin until his rear chainperson comes forward and hands him his pins. Both chainpersons count the pins to be sure there are 10 in hand and one in the ground. As taping is resumed, the situation is the same as at the the point of beginning: one pin is in the ground in front of the rear chainperson, and 10 pins are in the hand of the forward chainperson.

7 STATIONING AT END OF LINE OR WHEN PLUS IS DESIRED AT POINT ON LINE

Using an add tape. The rear chainperson moves to the forward station and holds a foot mark on the pin.

If the forward chainperson needs more tape, he calls "Give a foot."

The rear chainperson slides the next larger foot mark to the pin.

If the forward chainperson has too much tape, he calls "Take a foot."

The rear chainperson slides the next smaller foot mark to the pin.

The forward chainperson then calls "What are you holding?"

The rear chainperson calls "Holding 46."

The forward chainperson then calls "Reading 46.32."

The forward chainperson then calls "Station?"

The rear chainperson counts the pins in his possession, but does not count the pin in the ground at the last full station. The station number is the same as the number of pins in his hand if the station is less than 10. If it is more than 10, the station number is the same as the number of pins in his hand plus 10 for each exchange of 10 pins. He calls out the station number.

The forward chainperson checks the rear chainperson's count. The difference between 10 and the number of pins in his hand is the station number plus 10 for each exchange of 10 pins.

The forward chainperson calls out the full station number and plus. Both chainpersons record it.

Using a cut tape. The procedure is the same for the rear chainperson in placing a foot mark on the pin.

The forward chainperson calls "What are you holding?"

The rear chainperson calls "Holding 47."

The forward chainperson then calls "Cut 68."

The rear chainperson then calls "46.32."

The forward chainperson repeats "46.32." "Station?"

Both chainpersons then check the number of pins in hand and call the station number.

8 BREAKING TAPE

Where the slope is so great that a 100 foot tape cannot be held horizontally without plumbing above the shoulders, a procedure known as *breaking tape* can be used.

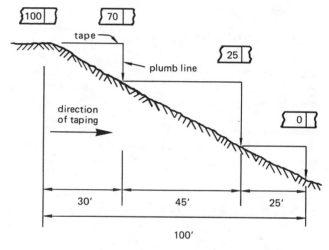

Figure 21.2 Breaking tape

The forward chainperson pulls the tape forward a full length as usual.

He puts the tape approximately on line and walks back along the tape to a point where the tape can be held horizontal below the shoulder level.

He then picks up a foot mark ending in 0 or 5 (70, for instance) and measures a partial tape length (30 feet, for instance), using the plumb bob as described in section 4, and marking the point with a pin.

After the forward chainperson has placed the pin, he waits for the rear chainperson to come forward and tells him what foot mark he was holding, as "Holding 70."

The rear chainperson repeats: "Holding 70." He hands the forward chainperson a pin to replace the one which was used to mark the intermediate point.

The chainpersons continue the procedure at as many intermediate points as necessary. The rear chainperson always picks up the intermediate pin and when he moves forward to an intermediate point, he always hands the forward chainperson a pin. (He does not hand a pin to the forward chainperson when he moves forward to the zero mark.)

9 TAPING AT AN OCCUPIED STATION

When taping at a station which is occupied by an instrument, chainpersons must be extremely careful not to hit the leg of the instrument. If a plumb bob is needed at the point, the plumb bob string hanging from the instrument can be used. In some cases, it may be necessary to use the point on top of the instrument on the vertical axis as a measuring point.

10 CARE OF THE TAPE

The tape will not be broken by pulling on it unless there is a kink (loop) in it. Chainpersons should always be alert to kinking.

When the tape has been used in wet grass or mud, it should be cleaned and and oiled lightly by pulling it through an oily rag.

11 SLOPE MEASUREMENTS

On fairly even ground where the slope is uniform, it is sometimes easier to determine the slope and make corrections for changing the slope measurement to horizontal measurement rather than break tape every few feet. To determine horizontal distance, the correction will be subtracted from the slope distance.

In figure 21.3, H is the horizontal distance from A to B. S is the slope distance from A to B. V is the difference in the elevation from A to B, and C is the required correction.

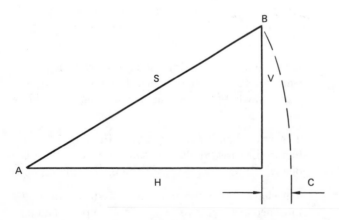

Figure 21.3 Measurements on slope

From the Pythagorean theorem,
$$V^2 = S^2 - H^2 = (S - H)(S + H) \qquad 21.1$$
Where the slope difference is small, $S + H$ is approximately $2S$.
$$V^2 \approx 2S(S - H) \qquad 21.2$$
Since $S - H = $ correction C,
$$C \approx \frac{V^2}{2S} \qquad 21.3$$
Where $S = 100$ feet (one chain length),
$$C \approx \frac{V^2}{200} \qquad 21.4$$
The approximate value will be within 0.007 feet of the actual value when the difference in elevation per 100 feet slope distance is not more than 15 feet. For steeper slopes, more exact formulas may be used.

Example 21.1

The difference in elevation between two points is 4.0 feet and the slope distance is 100.00 feet. What is the horizontal distance?

Solution

The correction is given by equation 21.4.
$$C \approx \frac{(4)^2}{200} = 0.08 \, \text{ft}$$
$$H = 100.00 - 0.08 = 99.92 \, \text{ft}$$

12 TENSION

Tapes are not guaranteed by their manufacturers to be exact length. The National Bureau of Standards will, for a fee, compare any tape with a standard tape or distance, and it will certify the exact length of the tape under certain conditions.

In the United States, steel tapes are standardized for 68°F. The standard pull for a 100 foot tape with the tape supported throughout its length is 10 pounds. Usually, when tapes are standardized, they are standardized for use in two conditions: (1) supported throughout and (2) supported only at the ends. When supported only at the ends, the pull is usually 20 pounds, but this can be varied by request.

When the tape is pulled with more or less than the standard amount of tension, the actual distance is more or less than 100.00 feet. However, variation in pull when the tape is supported does not affect the distance greatly. (If a pull of 20 pounds is exerted on a tape which was standardized for a pull of 10 pounds, the increased length is 0.006 feet.)

Chainpersons, by use of spring-balance handles, can get the feel of a 10 pound pull. In ordinary taping, with tape supported, the error due to variation in tension is negligible.

13 CORRECTION FOR SAG

When the tape is supported only at the ends, it sags and takes the form of a catenary. The correction for sag can be determined by formula or can be offset by increased tension. For a medium weight tape standardized at 10 pounds, a pull of 30 pounds will offset the difference in length due to sag. If the tape is supported at 25 foot intervals, the pull can be reduced to 14 pounds.

Chainpersons should use the spring-balance handle to familiarize themselves with various "pulls."

14 EFFECT OF TEMPERATURE ON TAPING

Steel tapes are standardized for use at 68°F. For a change in temperature of 15°F, a steel tape will undergo a change in length of about 0.01 foot, introducing an error of about 0.5 foot per mile.

The coefficient of thermal expansion for steel is approximately 0.0000065 per unit length per degree Fahrenheit.

For a 100 foot tape where T is the temperature (°F) at time of measurement, the correction (C) is

$$C = 0.0000065(T - 68°) \times \text{measured length} \quad 21.5$$

Example 21.2

A line was measured as 675.48 feet at 30° with a 100 foot tape (standardized at 68°F.) What is the true distance?

The change in the recorded length is given by equation 21.5.

$$C = 0.0000065(30 - 68) \times 675.48 = -0.17\,\text{ft}$$

The corrected length is

$$L = 675.48 - 0.17 = 674.31\,\text{ft}$$

15 EFFECT OF IMPROPER ALIGNMENT

Improper alignment is probably the least important error in taping. The linear error when one end of the tape is off line can be computed in the same manner as slope correction. For example, for a 100 foot tape out of alignment one foot at an end, the correction is

$$C = \frac{V^2}{200} = \frac{1}{200} = 0.005\,\text{ft}$$

When the error in alignment is 0.5 foot, the linear error is 0.001 foot per tape length or about 0.05 foot per mile.

16 INCORRECT LENGTH OF TAPE

A standardized tape can be used to check other tapes. If a 100 foot tape is known to be of incorrect length, the correction factor to be used for measurements made with the incorrect tape per 100 feet is

$$C = \text{actual length} - 100.00$$

For example, the correction for a tape found to be 100.02 feet long after comparison with a standardized tape is

$$C = 100.02 - 100.00 = +0.02\,\text{ft per }100\,\text{ft}$$

If a line is measured to be 662.35 feet with this tape, the corrected length would be

$$662.35 + 6.6235(+0.02) = 662.48$$

For a tape found to be 99.98 feet long after comparison, the correction would be

$$C = 99.98 - 100.00 = -0.02\,\text{ft per }100\,\text{ft}$$

If a line is measured to be 662.35 with this tape, the corrected length would be

$$662.35 + 6.6235(-0.02) = 662.22$$

A line measured with a tape which is longer than 100 feet is actually longer than the measurement shown. A line measured with a tape which is shorter than 100 feet is actually shorter than the measurement shown.

Here is a rule that can be remembered: For a tape too long, add; for a tape too short, subtract. This rule can also be applied to temperature corrections.

17 COMBINED CORRECTIONS

Corrections for incorrect length of tape, temperature, and slope can be combined.

Example 21.3

A tape which is 100.03 feet long was used to measure a line which was recorded as 1238.22 feet when the temperature was 18°F. The difference in elevation from beginning to end was 12.1 feet. What is the corrected length?

Solution

Tape correction = 12.3822(+0.03) = +0.37
Temp. correction = 0.0000065(18 − 68)(1238.22) = −0.40
Slope correction = $\frac{(12.1)^2}{(2)(1238.22)}$ = −0.06

Total correction = $\overline{-0.09}$

Corrected length = 1238.22 − 0.09 = 1238.13

PART 2: Leveling

1 DEFINITIONS

Leveling requires a vocabulary of words and terms used in the study of the earth's surface.

Vertical Line. A line from any point on the earth to the center of the earth.

Plumb Line. A vertical line, usually established by a pointed metal bob hanging on a string or cord.

Level Surface. Because the earth is round, a level surface is a curved surface. Although a lake appears to have a flat surface, it follows the curvature of the earth. A level surface is a curved surface which, at any point, is perpendicular to a plumb line.

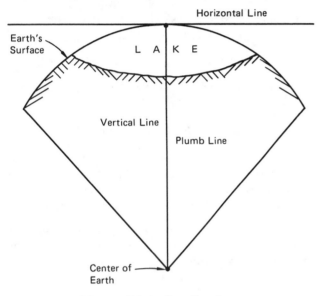

Figure 21.4 Leveling terms

Horizontal Line. A line perpendicular to the vertical.

Datum. Any level surface to which elevations are referred. *Mean sea level* is usually used for a datum.

Elevation. The vertical distance from a datum to a point on the earth.

Leveling. The process of finding the difference in elevation of points on the earth.

Spirit Level. A device for establishing a horizontal line by the centering of a bubble in a slightly curved glass tube (*vial*) filled with alcohol or other liquid.

Bench Mark. A marked point of known elevation from which other elevations may be established.

Turning Point. A temporary point on which an elevation has been established and which is held while an engineer's level is moved to a new location.

Height of Instrument. Vertical distance from the datum to the line of sight of the level.

2 DIFFERENTIAL LEVELING

Differential leveling is the process of finding the difference in elevation between two points. An engineer's level and a level rod are used. The rod is a piece of wood, fiberglass, or aluminum which is marked in feet, tenths of a foot, and hundredths of a foot (or in meters). It is held in a vertical position. The level is a telescope with cross-hairs attached to a spirit level. By keeping the level bubble centered in its vial, the horizontal cross-hair in the telescope can be kept on the same elevation as the telescope turns to any direction in the horizontal plane. It establishes a horizontal plane in space from which measurements can be made with the rod. As the levelperson focuses on the level rod, a measurement can be made from the horizontal plane to the point on which the rod rests merely by reading the marking on the rod where the horizontal cross-hair is imposed.

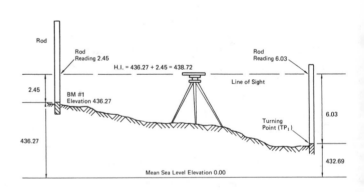

Figure 21.5 Differential leveling

Figure 21.5 illustrates how the level and rod can be used to find the difference in elevation between two points.

Bench mark 1 (BM #1) is a semi-permanent object, the elevation of which is 436.27 feet above mean sea level.

Turning point 1 (TP_1) is a temporary object (in this case, the top of a stake), the elevation of which is to be determined.

The level is set up so that both objects can be seen through the telescope.

The rod is first placed on BM #1, and a reading of the rod is made and recorded with the bubble centered. This reading is known as a *backsight* (BS), and is added to the elevation of BM #1 to find the height of instrument (H.I.). Backsight is commonly called a *plus sight* because it is always added to the known elevation.

The rod is then placed on TP_1, and a reading of the rod is made and recorded with the bubble centered. This reading is known as a *foresight* (FS) and is subtracted from the H.I. to find the elevation of TP_1. It is commonly called a *minus sight* because it is always subtracted from a known H.I.

The elevation of a continuing line of objects can be determined by moving the level along the line. While the level is being moved forward, the rod person must hold the turning point so that the levelperson may make a backsight reading on it from the new location of the level.

Figure 21.6 shows a profile view of differential levels between BM #3 and BM #4 and field notes which were recorded at the time the levels were run. It can be seen that differential leveling is a series of vertical measurements which alternate in sequence from a plus sight to a minus sight.

The field notes show columns for plus readings, minus readings, H.I.'s, and elevations of bench marks and turning points.

Notice that a plus reading is shown on the same horizontal line with BM #3, but a minus reading is not shown on that line. Only one reading was taken on BM #3. Also, notice that a plus reading and a minus reading are shown on the same horizontal line with each turning point. The minus reading is subtracted from H.I. on the line above it to determine the elevation of the turning point, which is also shown on the same horizontal line.

After the elevation of the turning point has been determined, the level is moved forward. Then, a backsight is read on the same turning point. This plus reading is added to the elevation just determined to give a new H.I. Notice that a minus reading is shown on the same horizontal line as BM #4, but a plus reading is not shown on that line. BM #4 is the end of the level line; only one reading was made on it.

At the bottom of the field notes, the plus column and the minus column are totalled. The smaller total is subtracted from the larger total. The difference should be the same as the difference in the beginning and ending elevations. If it is not, a mistake has been made in arithmetic.

The "rod" column is not used for differential leveling.

3 PROFILE LEVELING

In planning highways, canals, and pipelines, a vertical section of the earth is needed to determine the vertical location of the centerline of the project. This vertical section is known as a *profile*. It is plotted on paper from field notes. *Profile leveling* is similar to differential

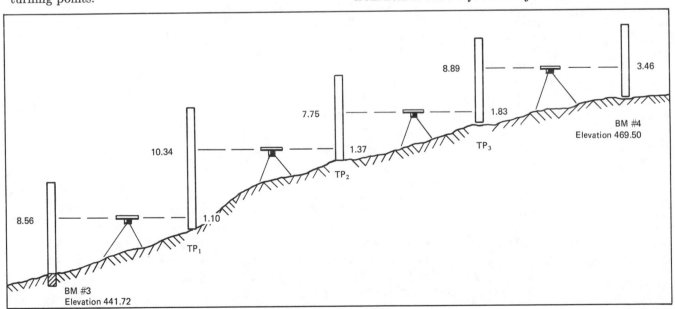

Figure 21.6 Continuous differential leveling

PROFESSIONAL PUBLICATIONS, INC. ● P.O. Box 199, San Carlos, CA 94070

leveling except that many minus readings are taken in addition to the usual plus and minus readings taken on bench marks and turning points.

At each set-up of the level, readings are taken on the ground along the centerline at each full station and at each *break* on the ground. (A break on the ground is a point on the ground where the slope changes.) These readings are all minus readings. They are measurements made to determine the elevation at each point on the profile, and are subtracted from the H.I. at each level set up. For clarity, these ground readings are recorded in the "rod" column. Bench mark and turning point readings are recorded in the normal manner as in differential leveling. To determine elevations of ground points, all readings taken at one level set-up are subtracted from the H.I. at that level set-up.

After elevations at each ground point are determined, they are plotted on special profile paper or profile sheets.

Figure 21.8 shows the profile which was plotted from the field notes of figure 21.7.

PROFILE LEVELS   FM ROAD 123

sta	+	H.I.	−	rod	elev	
BM#4	4.87	483.13			478.26	r.r. spike in 12″ Oak 75′L sta 33 + 50
32 + 00				11.5	471.6	
33 + 00				9.4	473.7	
+75				10.1	473.0	
34 + 00				8.2	474.9	
35 + 00				3.0	480.1	
+15				1.9	481.2	
+70				2.3	480.8	
36 + 00				5.2	477.9	
+50				6.8	476.3	
37 + 00				5.9	477.2	
38 + 00				13.3	469.8	
TP_1	4.54	476.95	10.72		472.41	*T*/stake − 38 + 00 Lt
38 + 60				13.2	463.8	
39 + 00				12.0	465.0	
40 + 00				3.9	478.1	
41 + 00				1.2	475.8	
42 + 00				0.8	476.2	
+70				0.7	476.3	
+80				1.5	475.5	
43 + 00				0.4	476.6	
BM#5			0.17		476.78	r.r. spike in 16″ Elm, 100′ Rt sta 42 + 50
	9.41		10.89			

10.89 − 9.41 = 1.48   478.26 − 476.78 = 1.48

Figure 21.7 Typical profile leveling field notes

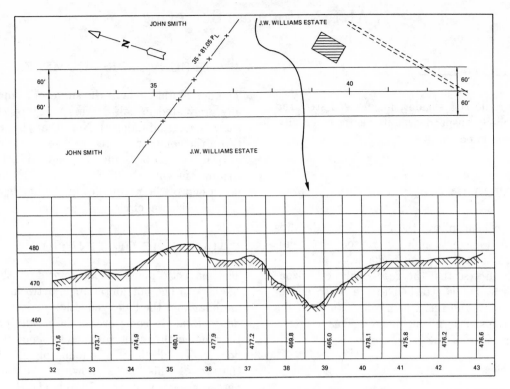

Figure 21.8   A typical profile

PART 3: Compass Surveying

1 MAGNETIC NEEDLE

A magnetic needle is a slender, magnetized steel rod which, when freely suspended at its center of gravity, points to magnetic north.

2 MAGNETIC DIP

In the northern hemisphere, the magnetic needle dips toward the north magnetic pole. In the southern hemisphere, it dips toward the south magnetic pole. To counteract the dip (so that the needle will be horizontal), a counterweight is attached to the south end of needles used in the northern hemisphere (and to the north end in the southern hemisphere). This weight is usually a short piece of fine brass wire.

3 THE MAGNETIC COMPASS

The magnetic compass consists of a magnetic needle mounted on a pivot at the center of a graduated circle in a metal box covered with a glass plate. It is constructed so that the angle between a line of sight and the magnetic meridian can be measured. The compass housing can be turned in the horizontal plane while the needle continues to point to magnetic north. The needle indicates the angle made by the magnetic meridian and the line of sight.

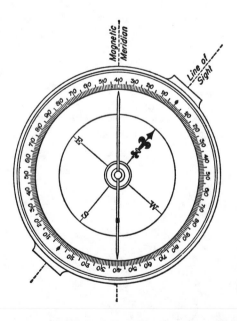

Figure 21.9 A typical compass

4 THE SURVEYOR'S COMPASS

The horizontal circle in the surveyor's compass is usually graduated in half degrees. (Figure 21.9 does not show these graduations.) Note that the letters E and W on the compass are reversed so that direct readings of bearings can be made. In the figure, the bearing of the line of sight is N 40° E. By reversing the letters, the north end of the needle lies in the northeast quadrant of the horizontal circle.

5 MAGNETIC DECLINATION

The magnetic poles do not coincide with the axis of the earth. The horizontal angle between the magnetic meridian and the true (geodetic) meridian is known as *declination*. In some areas, the needle points east of the true north, and in some areas it points west of true north. Zero declination is found along a line between the areas. This line is known as the *agonic line*. It passes, generally, through Florida and the Great Lakes, but is constantly changing its location. East of this line, declination is west (−); west of this line, declination is east (+). In the United States, declination varies from 0° to 23°.

6 VARIATIONS IN DECLINATION

Declination at any one point varies daily. Declination for a particular location for a particular year can be obtained from the United States Geological Survey. Figure 21.10 illustrates a typical declination map.

7 IMPORTANCE OF COMPASS SURVEYING

Compass surveying is as obsolete as the Gunter chain, but it is important for the modern surveyor to understand it when retracing old lines. The modern surveyor needs to be able to convert magnetic bearings to true bearings. In order to do so, he must know whether the declination is east or west and what the declination is, or was, on a certain day.

Example 21.4

Convert the following magnetic bearings to true bearings.

(a) N 68°20′ E, decl. 8°00′ W.

(b) S 12°30′ W, decl. 3°45′ E.

(c) S 20°30′ E, decl. 6°30′ W.

(d) N 3°15′ W, decl. 4°20′ E.

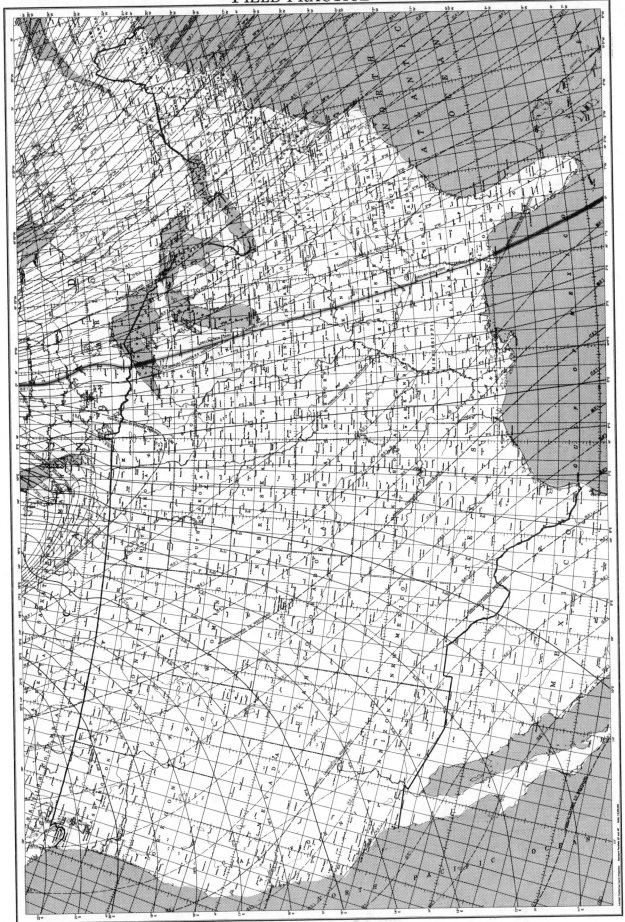

Figure 21.10 Typical declination map

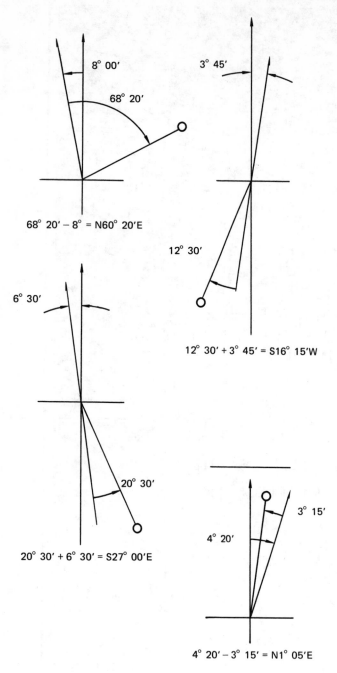

Solution

$$\begin{aligned}
\text{magnetic bearing} &= \text{S}\,85°15'\,\text{W}\\
\text{magnetic azimuth} &= 265°15'\\
\text{declination} &= +8°30'\\
\text{true azimuth} &= 273°45'\\
\text{true bearing} &= \text{N}\,86°15'\,\text{W}
\end{aligned}$$

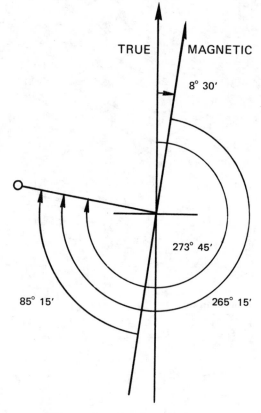

Figure 21.12   Example 21.5

Figure 21.11   Example 21.4

Solution

Magnetic bearings can also be converted to true bearings by use of azimuths.

Example 21.5

The magnetic bearing of a line was S 85°15′ W at a location where the declination was 8°30′ E. Find the true bearing.

RESERVED FOR FUTURE USE

PROFESSIONAL PUBLICATIONS, INC. • P.O. Box 199, San Carlos, CA   94070

RESERVED FOR FUTURE USE

22 ASTRONOMICAL OBSERVATIONS[1]

1 TRIANGULATION

Triangulation is a method of surveying in which the positions of survey points are determined by measuring the angles of triangles. In triangulation the survey lines form a network of triangles. Each survey point (monument) is at a corner of one or more triangles. The three angles of each triangle are measured. Lengths of triangle sides are trigonometrically calculated through the chain between measured base lines. The positions of the points are established from the measured angles and the computed sides.

Triangulation is used primarily for geodetic surveys, such as those performed by the National Geodetic Survey. Most first- and second-order control points in the national control network have been established by triangulation procedures.[2] The use of triangulation for transportation surveys is minimal. Generally, its use is limited to strengthening traverses for control surveys. Schemes involved are relatively simple. Rarely will triangulation be used for non-control surveys. A possible exception might be that of establishing the position of a single point by a single triangle. However, even this use has been supplanted, for the most part, by electronic distance measurement (EDM).

2 TRILATERATION

Trilateration is similar to triangulation in that the survey lines form triangles. But, in trilateration the lengths of the triangles' sides are measured, and the angles are computed. Orientation of the survey is established by selected sides whose directions are known or measured. The positions of trilaterated points are determined from the measured distances and the computed angles.

Today, the availability of EDM instruments has made trilateration economically feasible. With proper survey design and field procedures, the positions of points can be established as accurately and as economically as by triangulation.

Generally, trilateration is limited to reinforcing triangulation schemes. Such reinforced triangulation schemes, which usually involve only a few triangles, are sometimes used to strengthen control traverses.

3 SOLAR OBSERVATIONS

The direction of a survey line can be determined by making horizontal and vertical angular measurements to the sun at a known time and from a point of known latitude. Such measurements are called *solar observations*.

Generally, solar observations are required only in areas where horizontal control points are sparse or control data is unavailable. In such areas, solar observations provide azimuth control (starting, closing, and check azimuths). In some cases, the entire orientation of a survey might be based on solar observations. In other surveys, they might simply provide orientation checks. For example, long traverses can require intermediate azimuth checks where record points do not exist. Another common use is orientation for retracement surveys when searching for corners in rugged terrain or in heavy vegetation.

Azimuths determined by solar observations using the procedures herein are adequate for third-order surveys. Greater accuracy is impractical because exact pointings can not be made on the relatively fast-moving sun. If more accurate azimuths must be determined by astronomical observations, Polaris sightings should be used.

---

[1] Edited by Michael R. Lindeburg. This chapter is largely based on Chapter 5, section 5-05 of *Surveys Manual*, issued by the Office of Geometronics Surveys Branch of State of California Department of Transportation.
[2] There are three orders of accuracy. First order is the highest accuracy.

Azimuths should be reliable within 12 seconds. However, many factors can affect the accuracy. Two of the more important factors are the skill of the observer and the care used in making the observations. (Care in leveling the instrument and pointing on the sun are especially important.) Other factors are the hour of the day and the time of the year the observations are made, the precision of the theodolite, the accuracy of the time measurements, and the latitude of the point of observation and the accuracy to which it is determined.

4 CELESTIAL BASIS

An azimuth determined by solar observations is based on the solution of a spherical triangle, the *astronomical triangle*. Figure 22.1 shows the points on the celestial sphere which form this triangle. These three points are

- **P**-north pole of the celestial sphere
- **Z**-observer's zenith point. This is the point on the celestial sphere directly above the point of observation.
- **S**-sun

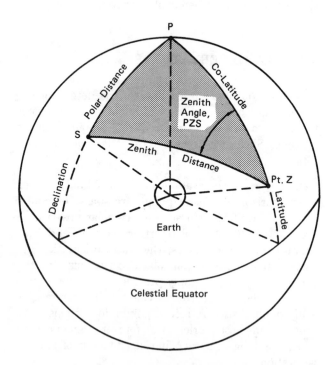

Figure 22.1 Astronomical triangle for solar observation

In figure 22.1, triangle PZS is the *astronomical triangle*. Angle PZS is known as the *zenith angle*.

5 SIDES (ANGLES) OF THE ASTRONOMICAL TRIANGLE

Side S-P is known as the *polar distance*. It can be found as 90° minus the sun's declination. *Declination* is the angular distance the sun is above or below the celestial equator. The value is typically determined from an *ephemeris*.

Side P-Z is known as the *co-latitude*. It is found as 90° minus the latitude of the point from which the observation is made.

Side Z-S is known as the *zenith distance*. It is 90° minus the true altitude of the sun. The *true altitude* is the measured altitude corrected for refraction and parallax. Tables of corrections are also available in ephemerides. Figure 22.2 illustrates how the refraction correction (r) is added to the field-measured zenith angle, and the parallax correction (p) is subtracted from it.

$$\text{true altitude} = \text{measured altitude} - \text{refraction} + \text{parallax} \qquad 22.1$$

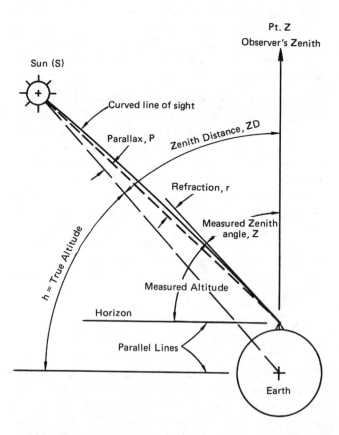

Figure 22.2 Zenith distance and true altitude

6 ZENITH ANGLE

With three sides of the spherical triangle known, the angle between the north pole and the sun at the observer's zenith can be calculated. This angle is known as the *zenith angle*, *PZS*.[3] It establishes the azimuth of the sun (line *Z-S*), and therefore is also known as the *azimuth angle*.

Two equations can be used to calculate the zenith angle *PZS* from its trigonometric functions. In equations 22.2 and 22.3, the following nomenclature is used:

PZS = zenith angle.

d = sun's declination.

h = sun's true altitude.

l = latitude of the point of observation.

p = polar distance $= 90° - d$.

$s = 1/2(l + h + p)$

$$\cos PZS = \frac{(\sin d) - [(\sin h)(\sin l)]}{(\cos h)(\cos l)} \qquad 22.2$$

If cos *PZS* is negative, *PZS* is greater than 90 degrees.

$$\tan\left(\frac{PZS}{2}\right) = \sqrt{\frac{[\sin(s-l)][\sin(s-h)]}{[\cos s][\cos(s-p)]}} \qquad 22.3$$

The cosine formula is easier to use. However, the cosine function is less precise than the tangent when the angle is near zero or 180 degrees.

If the observations are made in the morning, the sun's azimuth is equal to the zenith angle. For afternoon observations, the sun's azimuth is obtained by subtracting the zenith angle from 360 degrees.

7 LINE AZIMUTH

The azimuth of the survey line is referred to the sun's azimuth by simultaneously measuring the horizontal angle from the line to the sun and the vertical angle to the sun. When the horizontal angle is measured clockwise, the azimuth of the survey line is determined as follows:

- If the horizontal angle is equal to or less than the sun's azimuth, subtract the horizontal angle from the sun's azimuth.

- If the horizontal angle is greater than the sun's azimuth, subtract the horizontal angle from 360 degrees and add the remainder to the sun's azimuth.

Example 22.1

In figure 22.3, the zenith angle *PZS* was 100°. (Sun azimuth is the same as zenith angle.) The horizontal angle measured was 336°. Therefore, the azimuth of line *A-B* was

---

[3] Note that the term "zenith angle" also refers to the angle *Z* in a vertical plane which is measured with a theodolite which has a vertical circle oriented at zero degrees when the telescope is pointed on the zenith. See figure 22.2.

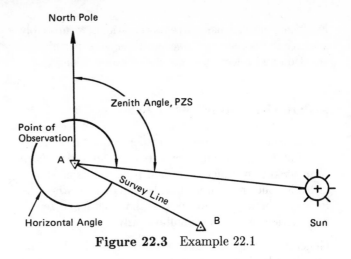

Figure 22.3   Example 22.1

Solution

azimuth $A\text{-}B = 100 + 360 - 336 = 124°$

8 PRACTICAL CONSIDERATIONS IN SOLAR SIGHTING

A one-minute theodolite is sufficiently precise for solar observations (third-order or less). A one-second instrument will produce slightly better results. However, pointing errors inherent with solar observations will somewhat nullify the increased instrument precision.

Make solar observations between the hours of 8 and 10 a.m. or 2 and 4 p.m., standard time. Results are not as reliable when the time is near noon because the solution of the astronomical triangle is weak during this period. In fact, when the sun is at the meridian of the observation point (this occurs at local apparent noon), the astronomical triangle becomes a straight line. Thus, it is indeterminate at this instant. Furthermore, do not observe during the early morning or late afternoon periods because the correction for refraction becomes large and very uncertain at low altitudes.

Best results are possible if the observations are made when the sun's altitude is between 25 and 40 degrees.

For third-order azimuths with a one-minute theodolite, average the calculated azimuths from three sets of one DIRECT and one REVERSE observation.

Observe the sets in a continuous sequence. In DIRECT, point the backsight; point the sun three successive times; and repoint the backsight. Immediately, follow the same sequence in REVERSE to complete the set.

Reject and reobserve a set if its calculated azimuth differs by more than 18 seconds from the average azimuth of the three sets.

For most situations, three observations is probably somewhat excessive. However, three sets permit the isolation and rejection of erroneous observations.

9 POLARIS OBSERVATIONS

Azimuths of survey lines can be determined by making horizontal angular measurements to *Polaris*, the *North Star*, at known, exact times and from points of known latitude. Such measurements also can be used when the latitude is unknown. In this case, horizontal and zenith observations are made simultaneously.

Generally, Polaris observations are required only in areas where horizontal control points are sparse or control data is unavailable. In such areas, Polaris observations can be used to provide azimuth control. In some cases, the entire orientation of a survey might be based on Polaris observations. In other surveys, they might simply provide azimuth checks. For example, Polaris observations might be used to provide azimuth checks on a long traverse which has numerous courses between existing control points.

10 ACCURACY OF POLARIS AZIMUTHS

Azimuth control can be determined by observations on the sun. Generally, solar observations are more convenient than Polaris observations. Therefore, solar observations are usually made when the accuracy requirements are third-order or less. Polaris observations must be used for second-order accuracy.

Polaris azimuths should be reliable within five seconds. However, many factors can affect the accuracy. The more important factors (in approximate order of importance) are:

- The skill of the observer and the care used in making the observations. Care in leveling the instrument and noting when the star is bisected by the cross hair are especially important.

- The accuracy of the time measurement.

- The hour of the observations.

- The latitude, and its accuracy, of the occupied point. Or, if the star's altitude is measured, the accuracy of the zenith angle measurements.

11 CELESTIAL BASIS

Polaris is a relatively bright star in the northern sky. Its apparent position is always very close to the vertical projection of the north pole. The actual angular distance from the pole (the *polar distance*) is approximately 51 minutes and is almost a constant. It varies less than one minute during the year.

As viewed from the earth, Polaris appears to revolve counter-clockwise around the pole. One "revolution" is completed each sidereal day (time measured relative to the stars). Actually Polaris does not orbit the earth's pole. The apparent motion is caused by the rotation of the earth. Unlike the sun and some other stars, Polaris does not "rise" and "set." It is always visible. Because of its location and relatively slow apparent motion, Polaris is the star commonly used for determining azimuths and latitudes by stellar observations.

An azimuth determined by Polaris observations is based on the solution of a spherical triangle which is called the *astronomical triangle*. Figure 22.4 shows the points on the celestial sphere which form this triangle. These three points are:

- **P**-north pole of the celestial sphere

- **Z**-observer's zenith point. This is the point on the celestial sphere directly above the point of observation.

- **S**-pole star Polaris

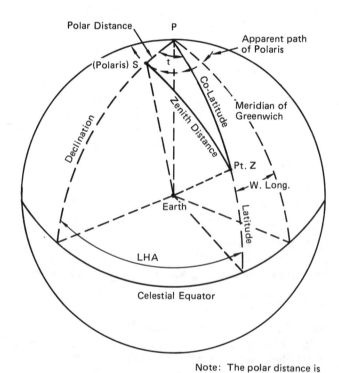

Note: The polar distance is greatly exaggerated.

Figure 22.4 Astronomical triangle for Polaris observations

12 SIDES (ANGLES) OF THE ASTRONOMICAL TRIANGLE

In figure 22.4, side S-P is known as the *polar distance*. It is calculated as 90° minus Polaris' declination. This *declination* is the angular distance that Polaris is above or below the celestial equator. It is found from an ephemeris.

Side P-Z is the *co-latitude*. It is calculated as 90° minus the latitude of the point of observation.

Side S-Z is the *zenith distance*. It is 90° minus the true altitude of Polaris. The *true altitude* is the measured altitude corrected for refraction. The *measured altitude* is 90° minus the zenith angle of polaris. Tables for refraction are contained in ephemerides.

In addition to the three sides, the angle at the pole between the zenith point and Polaris can be determined. This angle (SPZ) is called the *meridian angle*. The meridian angle is commonly designated by the letter "t" as shown in figure 22.4. It is computed by comparing the actual time of the observation with the time at which upper culmination of Polaris occurs at the observer's meridian.[4] (*Upper culmination* is the upper crossing of a meridian by a celestial body.) The meridian angle is computed as $t = LHA$ or $t = LHA - 360$, whichever is smaller in magnitude.

LHA is the *local hour angle*, equal to the time of observation minus the time of upper culmination (LCT) plus the correction for sidereal time. (*Sidereal time* is based on the rotation of the earth relative to the stars.)

LCT is the *local civil time*, which is the mean time based on the observer's meridian. Mean time is based on the average rotation of the earth relative to the sun.

13 ZENITH ANGLE FOR POLARIS OBSERVATIONS

With three elements of the spherical triangle known, the angle between the north pole and Polaris at the observer's zenith can be calculated. This angle is known as the *zenith angle*, PZS.[5] It establishes the azimuth of Polaris (line Z-S), and is, therefore, sometimes called the *azimuth angle*.

The zenith angle PZS can be calculated from equations 22.4 and 22.5. In both equations, the algebraic sign of t is ignored. (It should be remembered that the sign

---

[4] The times at which upper culmination of Polaris occur are tabulated in ephemerides. Some ephemerides use a different procedure to compute the meridian angle. Thus, they will tabulate some value other than the time of upper culmination.
[5] The term "zenith angle" can also be used for the angle in a vertical plane which is measured with a theodolite which has a vertical circle oriented at zero degrees when the telescope is pointed on the zenith.

of the cosine function is negative when the angle is between 90 and 270 degrees.) The following nomenclature is used in these two equations.

PZS = zenith angle
d = Polaris' declination
l = latitude of the point of observation
t = meridian angle
h = Polaris' true altitude

$$\tan(PZS) = \frac{\sin t}{(\cos l)(\tan d) - (\sin l)(\cos t)} \qquad 22.4$$

$$\sin(PZS) = \frac{(\sin t)(\cos d)}{\cos h} \qquad 22.5$$

Because of the size and limited variance in the angles involved, the sine equation can be very closely approximated by equation 22.6. In equation 22.6, P is the polar distance, $90° - d$.

$$\angle PZS = \frac{\sin t}{\cos h}(90 - d) = \frac{\sin t}{\cos h}p \qquad 22.6$$

When t is negative (0 to 12 hours prior to upper culmination), the azimuth of polaris is equal to the zenith angle. When t is positive (0 to 12 hours after upper culmination), the azimuth is obtained by subtracting the zenith angle from 360 degrees.

14 LINE AZIMUTH FROM POLARIS OBSERVATIONS

The azimuth of the survey line is referred to the azimuth of Polaris by the horizontal angle measurements. When the horizontal angle is measured clockwise, the azimuth of the survey line is determined as follows:

1. If the horizontal angle (H) is equal to or less than the azimuth of Polaris (Az), subtract the horizontal angle from the azimuth of Polaris.

2. If the horizontal angle is greater than the azimuth of Polaris, then subtract the horizontal angle from 3 degrees and add the result to the azimuth of P

Example 22.2

Referring to figure 22.5, suppose the 0°44'0" and the horizontal angle would be the azimuth of the su

Solution

$360° - H = 55°30'0$

azimuth of line

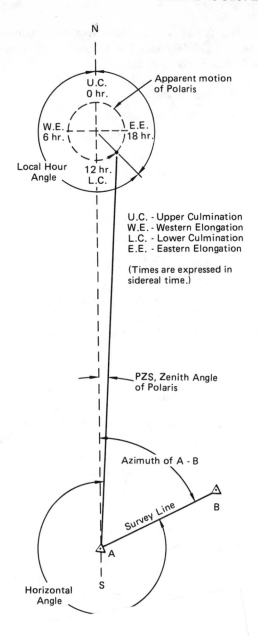

N

U.C.
0 hr.

Apparent motion
of Polaris

W.E.
6 hr.

E.E.
18 hr.

Local Hour
Angle

12 hr.
L.C.

U.C. - Upper Culmination
W.E. - Western Elongation
L.C. - Lower Culmination
E.E. - Eastern Elongation

(Times are expressed in
sidereal time.)

PZS, Zenith Angle
of Polaris

Azimuth of A - B

B

Survey Line

A

Horizontal
Angle

S

slightly different computing procedure. Values obtained from the ephemeris are as follows:

- The declination (d) of Polaris

- The time Polaris is at upper culmination for the meridian of Greenwich.

- If the altitude of Polaris is measured, the correction for refraction which is added to the measured zenith angle and this sum is subtracted from 90 degrees to obtain the true altitude.

Normally, the latitude can be determined with sufficient accuracy by scaling on a U.S. Geological Survey 7 1/2-minute quadrangle map. The accuracy requirements of the latitude measurement are not high. For uniformity, estimate latitude to the nearest second.

If it is impossible to scale the latitude on a map, the altitude of Polaris must be measured. This procedure is inferior to using the latitude because simultaneous horizontal and vertical measurements are required. Generally this is more difficult and less precise than simply measuring the horizontal angle alone. The latitude value is also useful as an approximate zenith angle for rough sighting on Polaris.

Normally, longitude is also scaled on a quad map. The longitude is used to calculate the time (LCT) of the observation. Thus, its required accuracy will depend upon the accuracy requirements of the time measurement. An error of 15 seconds in longitude will cause a one-second error in the time of observation. For uniformity, estimate longitude to the nearest second.

If a quad map is not available, or if the point of observation can not be identified with sufficient accuracy, longitude can be determined by astronomic observations.

The accuracy of the time measurements can be critical for Polaris observations. Thus, determine the watch correction to the nearest second by comparing with an accurate time source.

If observations are made when Polaris is at elongation (approximately six hours before or after upper culmination as shown in figure 22.5), the accuracy of the time measurement is not as great.

At upper or lower culmination, the azimuth of Polaris is changing rapidly and time is very critical. At these times, a three-second error in time will cause a one-second error in the azimuth of Polaris.

zenith angle was
304°30'0''. What
urvey line?

A-B = 55°30'0'' + 0°44'0'' = 56°14'0''

ox 199, San Carlos, CA   94070

• P.O. Box 199, San Carlos, CA   94070

RESERVED FOR FUTURE USE

PROFESSIONAL PUBLICATIONS, INC. • P.O. Box 199, San Carlos, CA   94070

RESERVED FOR FUTURE USE

PROFESSIONAL PUBLICATIONS, INC. • P.O. Box 199, San Carlos, CA 94070

23

MAP PROJECTIONS AND STATE COORDINATE SYSTEMS

1 GEODESY

Map makers have always had to face the problem of projecting the curved surface of the earth onto a plane surface. Representing the true shape of the lands and waters and the relative positions of points on the earth on a plane surface without distortion is impossible. *Geodesy* is the science of measuring the size and shape of the earth.

The earth is not a true sphere but a *spheroid*. Spinning on its axis, tilted 23-1/2 degrees to the plane of its orbit around the sun, it bulges at the equator. The equatorial diameter of 7,927 miles is some 27 miles greater than the polar diameter of 7,900 miles.

2 THE CLARKE SPHEROID OF 1866

The Clarke Spheroid of 1866 is a theoretical spheroid representing the earth. It is the basis for the National Ocean Survey (U.S. Department of Commerce) measurements and tables referenced in many state plane coordinate systems.[1]

3 GERARDUS MERCATOR

As discoveries of new lands were made, commerce increased accordingly. Ships needed charts which would

---

[1] Formerly U.S. Coast and Geodetic Survey.

guide them to their destination, but captains became aware that a straight line on a chart is not the same as a straight line on the globe of the earth. They needed charts on which a straight line drawn from port to port could be used to take them to their destination.

Gerardus Mercator, born Gerhard Kremer in Flanders in 1512, gave them what they wanted. Mercator wrote of his charts, "If you wish to sail from one port to another, here is a chart, and a straight line on it, and if you follow this line carefully you will certainly arrive at your destination. But the length of the line may not be correct. You may get there sooner or may not get there as soon as you expected, but you will certainly get there."

4 JOHANN HEINRICH LAMBERT

Probably the greatest contributor to modern cartography is Johann Heinrich Lambert, born in Alsasce in 1728. A philosopher and mathematician, Lambert demonstrated that maps could be made with truer shapes by using mathematics. His conic conformal projection is the basis of many state plane coordinate systems.

5 LATITUDE AND LONGITUDE

Dividing the earth into a huge grid are lines of latitude and longitude. Unlike the rectangular coordinate system in which distances from x- and y-axes are measured in linear units, the unit of measure for this grid system is the degree.

6 LATITUDE

Lines of *latitude* measure the distance from the equator to the north and south poles in degrees. As is shown in figure 23.1, the 30th parallel of north latitude, known as the 30th parallel, is measured by an angle formed by a line from the center of the earth to a point on the equator and a line from the center of the earth to a point on the 30th parallel. All lines of latitude are parallel to the equator. Parallels of north latitude measure the 90° from the equator to the north pole. Parallels of south latitude measure from the equator to the south pole. Each degree of latitude represents about 69 miles on the surface of the earth.

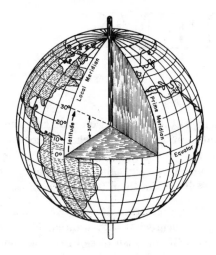

Figure 23.1   Lines of latitude

7 LONGITUDE

Measuring distance east and west around the earth at right angles to the equator are lines of *longitude*. These lines are also known as *meridians*, which are defined as great circles of the earth passing through both poles. They differ from lines of latitude in that they are not parallel but converge at the poles. Distance is measured east and west from a line of reference known as the *prime meridian* which passes through the observatory in Greenwich, England. This reference line (longitude zero) is used by agreement by all nations. The measurement of longitude is either east or west from the prime meridian at Greenwich.

Along the equator, each degree of longitude represents about 69 miles. Due to the convergence of the meridians, this distance becomes smaller toward the poles. The National Geodetic Survey furnishes tables which give distances on the surface of the earth for one degree along any parallel of latitude.

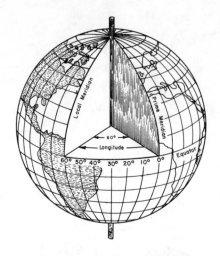

Figure 23.2   Lines of longitude

8 PROPERTIES OF MAPS

Mercator was able to make a map which allowed sailors to plot a line on a chart from port to port and to follow that line to their destination. However, he admitted that his map would not scale to correct distances. In order to achieve one property for his map he had to sacrifice another. The main properties of maps are shape, area, distance, and direction. Map makers may attain one of the properties and combinations of some of the others. Accurate representation of all properties is not possible if the map represents large areas such as continents or the whole world.

9 CONFORMAL MAPS

To show the true shape of the earth on a flat surface is highly desirable but impossible if we are considering large areas. If we show small areas such as cities, counties, or even states, it may be possible to maintain their true shapes. The shape of a small area on the map will *conform* to the same area on the earth. Thus, the word *conformal*. Conformal projections have the property that the scale at any point is the same in all directions.

10 MAP PROJECTIONS

A map projection is a representation of the surface of the earth on a flat sheet. Just as movie projectors "project" an image on a screen, map makers project points from a spherical surface onto a plane surface. Various methods are used in map projection depending on whether the map maker wants his map to be conformal (showing true shape) to be equal-area, (having the area shown on the map in proportion to the area on the earth), or to show true distance or true direction. Map projections can be developed by using a cylinder, cone, or plane.

11 CYLINDERS AS DEVELOPABLE SURFACES

A flat sheet of paper can be rolled into a cylinder. If the sheet of paper is rolled around a globe, it will be in contact with the globe along the equator. If the equator is inked when the sheet is wrapped around the globe, a straight line will be printed on the sheet when it is unwrapped. The scale on this line will be exactly correct, but any point on the sphere not on the equator will not touch the sheet and will have to be "projected" onto the cylinder.

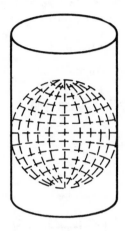

Figure 23.3   Cylindrical projections

12 CONES AS DEVELOPABLE SURFACES

Suppose we take a sheet of paper, form a cone with it, and then place this cone over a sphere. The cone comes in contact with the sphere only along the parallel at 45° N latitude.[2] In other words, it is tangent to the sphere along the parallel. If we ink the parallel before placing the cone, a line will be printed on the cone, but when the cone is unwrapped, the line will not be a straight line but will be an arc of a circle. The scale

----
[2] Cones can be made to contact the sphere at different parallels.

on this line will be exact, but points not on the 45° N parallel will have to be projected onto the cone. The 45° N parallel in this case will be known as *the standard parallel.*

13 PLANES AS DEVELOPABLE SURFACES

If we pass a plane parallel to the equator and tangent at the north pole, we can project points on the globe onto the plane. On this map the parallels of latitude appear as concentric circles and the meridians are straight lines radiating from the pole. Points can be projected from the center of the earth or from the opposite pole as shown in figure 23.5.

14 THE MERCATOR PROJECTION

The *Mercator projection* is known to most people. Mercator wanted navigators to be able to plot a line on a map and sail in that direction to their destination. He was successful, but in order to be so, he had to stretch the meridians and parallels on his projections, as can be seen in figure 23.6. On his map, meridians are straight parallel lines, as are parallels. Parallels and meridians are at right angles. Meridians are equally spaced but parallels are proportionally farther as the latitude increases. By comparing figures 23.5 and 23.6, it can be seen that the area around the North Pole is greatly distorted in the Mercator projection, but the area along the equator is not.

15 TRANSVERSE MERCATOR PROJECTION

The *transverse mercator projection* is a conformal cylinder projection. Instead of a cylinder wrapped around the earth and tangent to it, a cylinder slightly less in diameter than the globe is passed through the globe, at right angles to the poles, so as to cut a band or strip which produces a strip map. It is the basis for plane

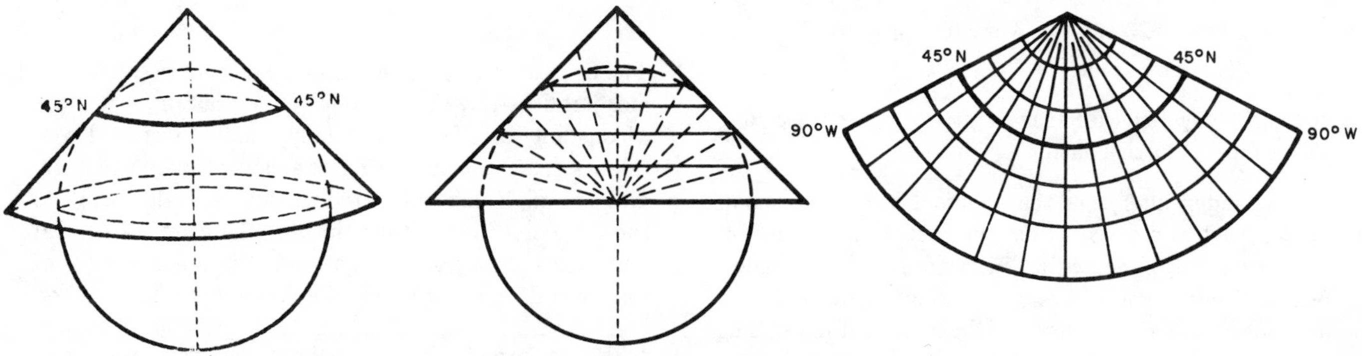

Figure 23.4   Conical projections

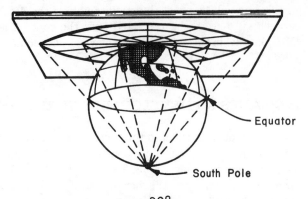

Figure 23.7　The transverse mercator projection

Given the geodetic latitude ϕ_p and geodetic longitude λ_p, the coordinates of a point **P** can be calculated.

$$X' = \lambda'' \pm ab \qquad\qquad 23.1$$

$$X = X_o + X' \qquad\qquad 23.2$$

$$X = Y_o + V\left(\frac{\Delta\lambda''}{100}\right)^2 \pm C \qquad\qquad 23.3$$

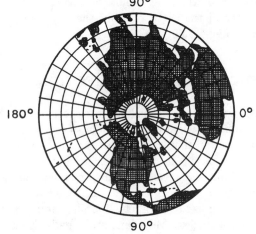

Figure 23.5　Plane projections

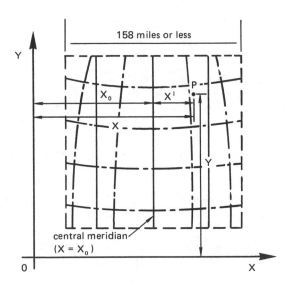

Figure 23.8　A transverse mercator projection of a small area

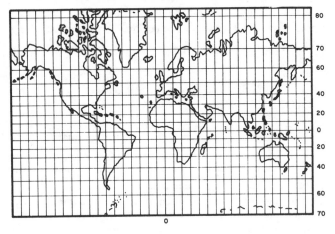

Figure 23.6　The Mercator projection

The constants in these equations are obtained from tables published by the National Geodetic Survey.[3] X_o is a constant for the zone, usually 500,000 feet and represents the x-coordinate for the central meridian.[4] H, a, Y_o, and V are tabulated versus latitude. $\Delta\lambda''$ is the difference in longitude (in seconds) from the central meridian to the point (negative if the point is west of the central meridian). b and c depend on $\Delta\lambda''$. When

coordinate systems for many states which have a long axis running north–south.

The cylinder intersects the spheroid along two small circles, each equidistant from the central meridian. On the developed plane surface in figure 23.8, all parallels of latitude and meridians except the central meridian are curves. They are shown as light broken lines. The central meridian establishes the north direction. x- and y-coordinates of points are measured perpendicularly and parallel to the central meridian.

---

[3] Refer to *Special Publication 8*, National Geodetic Survey. Other special NGS publications are available for individual states.
[4] For Rhode Island, $X_o = 600,000$ feet. For Georgia's eastern and western zones, $X_o = 2,000,000$ feet.

Xp is less than 158 miles, the error ratio between grid and geodetic lengths is less than $1/10,000$. Multiple zones within a state can be used to minimize the error further.

Except where a line of reference azimuth exceeds about five miles in length, its grid azimuth can be calculated from geodetic azimuth from equation 23.4.

$$\text{grid azimuth} = \text{geodetic azimuth} - \Delta\alpha'' \quad 23.4$$

$$\text{In equation 23.4, } \Delta\alpha'' = \Delta\lambda'' \sin\phi + g. \quad 23.5$$

The value of g is tabulated versus $\Delta\lambda''$. (g may be zero in some states.)

To convert from coordinates X and Y to geodetic latitude ϕ and longitude λ, g equations 23.6 through 23.10 can be used.

$$X' = X - X_o \quad 23.6$$

$$Y_o = Y - P\left(\frac{X'}{10,000}\right)^2 - d \quad 23.7$$

$$\Delta\lambda'' \approx \frac{X'}{H} \quad 23.8$$

$$\Delta\lambda'' = \frac{X' \pm ab}{H} \quad 23.9$$

$$\lambda_P = \lambda_C - \Delta\lambda'' \quad 23.10$$

The values of P and d in equation 23.7 are obtained by interpolation from tables where they are tabulated versus Y and X' respectively. Once equation 23.7 is solved, the value for ϕ is obtained by interpolation in the tables. By interpolating from Y_o, values for H and a are obtained. Equation 23.8 then gives an approximate value of $\Delta\lambda''$. Using $\Delta\lambda''$, a value for b can be interpolated from the tables. Equations 23.9 and 23.10 then give the geodetic longitude of the point. (In equation 23.9, the sign of the product ab is reversed from its tabulated value.)

16 THE LAMBERT CONIC PROJECTION

In many states which have their long axis oriented east-to-west, the basis for the state plane coordinate system is the *Lambert conformal projection*. Lambert's conformal projection is produced by mathematics. In theory, a cone, too small to cover one-half of the earth, is passed through the earth so that it intersects the earth at two parallels of latitude as shown in figure 23.9. The two parallels are known as the *standard parallels.*

A Lambert conformal conic projection passing through the earth at the parallels of latitude N 33° and N 45° would produce the map projection of the United States shown in figure 23.10. Notice that the parallels of

Table 23.1
Projections used by states

Lambert System	Transverse Mercator System	Both Systems
Arkansas	Alabama	Alaska
California	Arizona	Florida
Colorado	Delaware	New York
Connecticut	Georgia	
Iowa	Idaho	
Kansas	Illinois	
Kentucky	Indiana	
Louisiana	Maine	
Maryland	Michigan	
Massachusetts	Mississippi	
Minnesota	Missouri	
Montana	Nevada	
Nebraska	New Hampshire	
North Carolina	New Jersey	
North Dakota	New Mexico	
Ohio	Rhode Island	
Oklahoma	Vermont	
Oregon	Wyoming	
Pennsylvania		
South Carolina		
South Dakota		
Tennessee		
Texas		
Utah		
Virginia		
Washington		
West Virginia		
Wisconsin		
Long Island		
Nantucket and Martha's Vineyard		

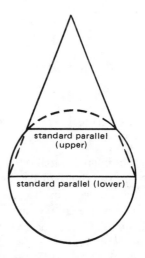

Figure 23.9 Conic two-parallel projection

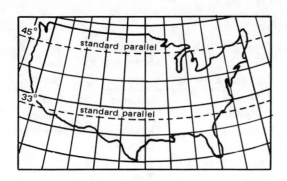

Figure 23.10 A Lambert projection of the United States

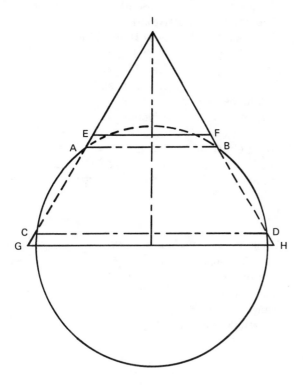

Figure 23.11

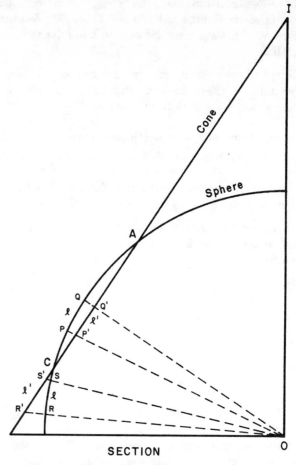

SECTION

Figure 23.12 Distortion from projection

latitude appear as curved lines and the meridians as straight, converging lines.

In figure 23.11, the sphere is intersected along lines AB and CD, which are the standard parallels for this projection. Along the standard parallels the scale is exact, but the projection of any point not on the standard parallels will be distorted. It can be seen that the wider apart are the standard parallels, the greater will be the distortion both north and south of the parallels. It can also be seen that if a projection is made south of line CD or north of line AB, the farther away a point is

from these lines of intersection, the greater will be the distortion.

Distortion simply means that the distance between two points on the sphere is not the same as the distance between the same points on the projection. Variation in the amount of distortion at different parallels within the projection can be seen in figure 23.12, which is a section through IAC of figure 23.11 showing the intersection of the sphere and the cone.

Points A and C lie on the standard parallels and distortion at these points is zero. But when any other two points, such as P and Q on the sphere, are projected onto the cone as P' and Q', the distance PQ on the sphere is greater than $P'Q'$ on the projection. If we designate any distance PQ as l and its projection $P'Q'$ as l', the ratio l'/l is known as the *scale factor*. The scale factor at points A and C is unity, and at any point between them it is less than unity. It is a minimum halfway between the standard parallels.

Likewise, if we consider two points outside the standard parallels, such as R and S, the projection $R'S'$ is greater than RS on the sphere and the scale factor is greater than unity.

Greatest accuracy (minimum distortion) is obtained when the distance between the standard parallels is two-thirds of the distance between the north and south limits of the map. In other words, 1/6 of the map lies outside the standard parallel on the north and 1/6 lies south of the south parallel.

The scale factor varies in a north-south direction only—not in an east-west direction. This means that the scale factor is the same along any parallel. Figure 23.13 shows the lambert conformal projection cone laid out flat.

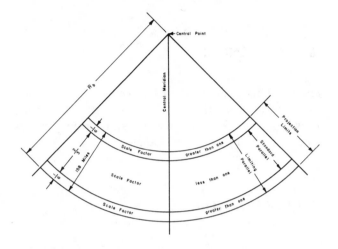

Figure 23.13   Flattened Lambert cone

It can now be seen that the accuracy of the Lambert Projection can be controlled by changing the distance

between the standard parallels and that the closer the standard parallels, the greater the accuracy. By setting the permissible distortion at not more than one part in ten thousand, the maximum distance between the limiting parallels is 158 miles.

In establishing a Lambert projection for a given zone in a state plane coordinate system, a meridian is chosen near the center of the zone and called the *central meridian*. Next, two limiting parallels are chosen, if possible, within the 158 mile limit. Standard parallels are spaced so that the distance between them is about two-thirds the distance between the limiting parallels. (There is no limit to the projection in an east-west direction.)

Next, a rectangular coordinate system is laid over the projection with the y-axis lying along the central meridian and x-axis lying so that it just touches the lower limit of the projection, as in figure 23.14. Note that the vertical grid lines are not parallel to meridians projected on the system except for the central meridian. If we consider the y-axis to be a distance C from the origin rather than at the origin, we can avoid negative numbers because all the system will be in the first quadrant (north-east). The value of C is usually considered to be 2,000,000 feet. The coordinates of the intersection of the y-axis and the lower limit of the projection are then $X = 2,000,000$ and $Y = 0$.

17   GENERAL USE

For most surveying work state plane coordinates may be used in plane surveying, but it must be remembered

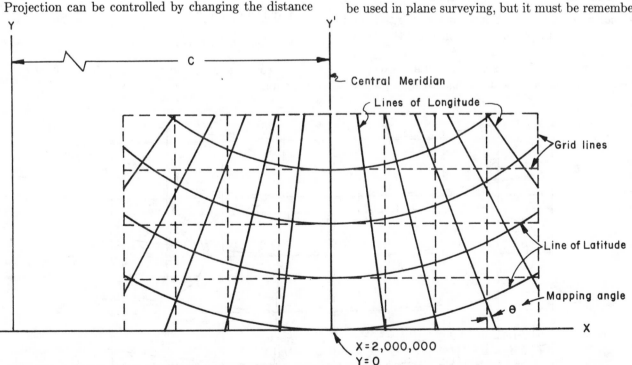

Figure 23.14   Conical projection with superimposed grid

that azimuths are referred to grid north and not to geodetic north. Only at the central meridian does grid north coincide with geodetic north. Horizontal control data distributed by the NGS show both the state coordinates and the latitude and longitude for most triangulation stations. If two points of known coordinates are available, no computations are required except those used in traverse computations in plane surveying. The difference in x-coordinates is treated as departure, and the difference in y-coordinates is treated as latitude. The direction of a line is found by remembering "the tangent of bearing equals departure divided by latitude". Bearing is grid bearing, but it can be converted to geodetic bearing. Areas can be computed by the coordinate method.

18 GEODETIC AZIMUTH TO GRID AZIMUTH

Although state plane coordinates can be used without making special computations for most work, it may be necessary to convert the geodetic azimuth from a triangulation station to an azimuth point to grid azimuth. For short lines (less than about 5 miles), azimuth can be found by using equation 23.11

$$\text{grid azimuth} = \text{geodetic azimuth} - \theta + \text{second term} \qquad 23.11$$

The second term is very small and can be ignored except for precise work over longer distances.

The *mapping angle*, θ, is a function of longitude and can be found in projection tables in the NGS special publications. (As has been stated previously, geodetic meridians and grid lines are not parallel except at the central meridian. The angle formed by a geodetic meridian and a vertical grid line at any point is the mapping angle.)

It can be seen in figure 23.15 that θ increases moving either east or west from the central meridian. In the projection tables, values of θ are shown for each minute of longitude. Values of θ are positive east of the central meridian and negative west of it. As the grid azimuth is the algebraic difference of the geodetic azimuth and θ, east of the central meridian θ is subtracted numerically. West of the central meridian θ is added numerically.

19 FIELD ANGLES

The mapping angle θ is used to convert geodetic azimuth to grid azimuth, but when working within a zone using points of known coordinates, no conversion is needed. Field angles are grid angles.

20 PRECISE WORK

For precise surveys and surveys over long distances it may be necessary to make corrections of three different types:

- reduction to sea level
- reduction for scale
- reduction for curvature

These corrections should be made prior to other computations.

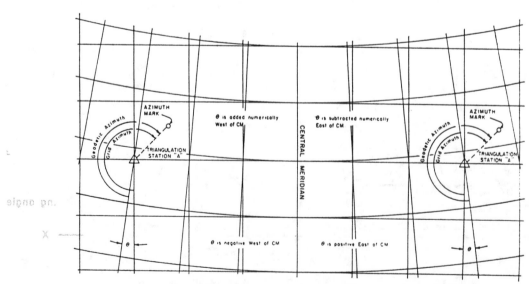

Figure 23.15 Grid versus geodetic azimuth

21 REDUCTION TO SEA LEVEL

Most map projection systems used for state plane coordinates project points from the Clarke spheroid at sea level. The distance between two points of known coordinates on the surface of the earth is not the same as the distance between the two points when reduced to sea level. The difference in the two distances is a function of the elevation of the points.

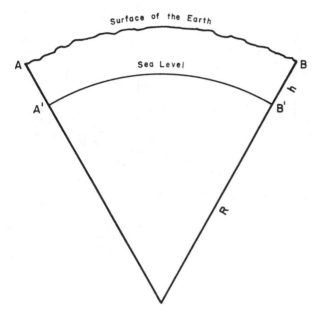

Figure 23.16 Reduction to sea level

In figure 23.16, points A and B are two points on the surface. A is projected along a radius of the earth to point A' at sea level, and B is projected to point B'. If h is the mean elevation of AB, and R is the mean radius of the earth,

$$\frac{A'B'}{AB} = \frac{R}{R+h} \qquad 23.12$$

$$A'B' = \left(\frac{R}{R+h}\right) AB \qquad 23.13$$

For most computations, a value of $R = 20,906,000$ feet is sufficiently accurate.

Consider two points at mean elevation of 500 feet which measure 5,000.00 feet apart on the surface. Then, the sea level separation is

$$A'B' = \frac{20,906,000}{20,906,000 + 500} \times 5,000.00 = 4999.88 \text{ ft.}$$

It can be seen that the correction at low elevations is negligible for short distances.

Table 23.2
Multiplicative factors for reduction to sea level

elevation (h)	sea level factor
500 ft	0.999976
1,000 ft	0.999952
1,500 ft	0.999928
2,000 ft	0.999904
2,500 ft	0.999880
3,000 ft	0.999857
3,500 ft	0.999833
4,000 ft	0.999809
4,500 ft	0.999785
5,000 ft	0.999761
5,500 ft	0.999737
6,000 ft	0.999713
6,500 ft	0.999689
7,000 ft	0.999665
7,500 ft	0.999641
8,000 ft	0.999617
8,500 ft	0.999593
9,000 ft	0.999570
9,500 ft	0.999546
10,000 ft	0.999522

22 REDUCTION FOR SCALE

To convert surface distance to grid distance, the surface distance must be multiplied by a scale factor which is a function of the latitude. Scale factors can be found in Table 1 of the NGS Projection tables. Latitude can be found in USGS topographic maps.

23 REDUCTION FOR CURVATURE OF EARTH (SECOND TERM)

A straight line on a sphere when projected on the Lambert cone will be a curved line. The angle formed by the straight line and the tangent to the curved line is very small and need not be considered for most traverses. For surveys over 5 miles in length, it may become a factor but it will not be discussed here. A formula for this correction will be found in the NGS Projection tables. The correction is known as the *second term*.

24 CONVERTING BETWEEN GEOGRAPHIC AND GRID POSITIONS

Horizontal control data furnished by National Ocean Survey show latitude and longitude as well as coordinates for triangulation stations. This would seem to indicate that there is no necessity to convert from geographic position to plane coordinates and vice versa.

But, for surveys which cross from one zone to another, it is convenient to compute from grid position in one zone to geographic position and then to compute from geographic position to grid position in the other zone.

The following symbols are used in making computations:

ϕ = latitude

λ = longitude

θ = mapping angle

l = a constant for each zone

h = elevation

L = slope distance

D = horizontal distance

S = geodetic distance (sea level distance)

CM = central meridian or Y axis

C = a constant for each zone, usually 2,000,000 ft.

X' = distance east or west of CM

R_b = distance from vertex of cone to origin

R = distance from vertex of cone to a point

25 LATITUDE AND LONGITUDE TO X AND Y

The radius of the limiting parallel on the south which passes through the origin can be found in the projection tables. This radius is known as R_b and is constant in any zone. It is shown in figure 23.17.

If the longitude of any point on the earth within the zone is known, the angle θ at this point can be determined from the projection tables.

If the latitude and longitude of a point are known, the rectangular coordinates on the state grid can be determined.

$$X = R \sin \theta + C = X' + C \qquad 23.14$$

$$Y = R_b - R \cos \theta \qquad 23.15$$

26 X AND Y TO LATITUDE AND LONGITUDE

$$\tan \theta = \frac{X - C}{R_b - Y} \qquad 23.16$$

$$R = \frac{R_b - Y}{\cos \theta} \qquad 23.17$$

When determining latitude and longitude, the *convergence constant* for the zone l must be used. l represents

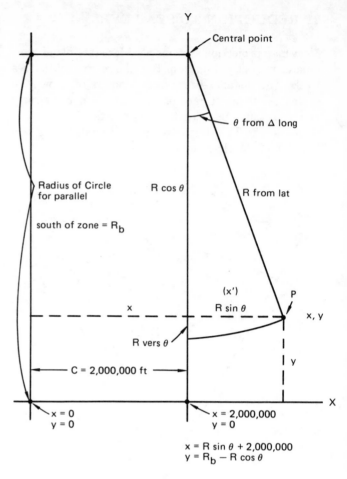

$$x = R \sin \theta + 2,000,000$$
$$y = R_b - R \cos \theta$$

Figure 23.17

the ratio of a change in angle on the plane to a change in angle on the sphere.

$$l = \theta / \Delta\lambda \qquad 23.18$$

$$\Delta\lambda = \theta / l \qquad 23.19$$

l is a constant for each zone which can be found in the projection tables. $\Delta\lambda$ represents the difference in longitude between any point P and the central meridian.

$$\text{longitude of } P = \text{longitude of } CM - \Delta\lambda \qquad 23.20$$

27 USE OF TRIANGULATION STATION AND AZIMUTH MARK

State plane coordinates and area of a traverse can be computed by tying the traverse to an NGS triangulation station of known coordinates using an azimuth mark associated with the station.

All azimuths for NGS stations are measured from the south. The azimuth to the mark is converted to north by adding or subtracting 180°. Surface distances are converted to grid (Clarke spheroid) distances by applying the combined scale factor (sea level and scale). Sea level distance is computed by using the average elevation of the area taken from a USGS quadrangle map.

A calculator permits computations for latitudes and departures to be made from azimuths without converting to bearing. Azimuths are determined for each course by turning angles to the right.

Area produced by use of the coordinate formula will be the area on the Clarke spheroid (grid area). It can be converted to surface area by dividing by the combined scale factor squared.

28 CONVERSION OF COORDINATES FROM ZONE TO ZONE

Surveys which involve two zones often require conversion of coordinates in one zone to coordinates in the other. Computations involve conversion of plane coordinates to latitude and longitude in one zone and conversion of latitude and longitude to plane coordinates in the other zone. All procedures have been outlined previously.

PROFESSIONAL PUBLICATIONS, INC. • P.O. Box 199, San Carlos, CA 94070

RESERVED FOR FUTURE USE

PROFESSIONAL PUBLICATIONS, INC. • P.O. Box 199, San Carlos, CA   94070

RESERVED FOR FUTURE USE

PROFESSIONAL PUBLICATIONS, INC. ● P.O. Box 199, San Carlos, CA 94070

RESERVED FOR FUTURE USE

24 RESTORING LOST CORNERS[1]

1 JURISDICTION

The Bureau of Land Management, under the supervision of the Secretary of the Interior, has complete jurisdiction over the survey and resurvey of the public lands of the United States.

After title to a piece of land is granted by the United States, jurisdiction over the property passes to the State. The Federal Government retains its authority only with respect to the public lands in Federal ownership. Where the lands are in private ownership, it is a function of the county or local surveyor to restore lost corners and to subdivide the sections. Disputes concerning these questions must come before the local courts unless settled by joint survey or agreement. It should be understood, however, that no adjoining owner can make a valid encroachment upon the public lands.

2 RESURVEYS

Public and privately owned lands may both be resurveyed by the Bureau of Land Management in certain cases, under the authority of an act of Congress approved March 3, 1909, and amended June 25, 1910:

That the Secretary of the Interior may, in his discretion, cause to be made, as he may deem wise under the rectangular system now provided by law, such resurveys or retracements of the surveys of public lands as, after full investigation, he may deem essential to properly mark the boundaries of the public lands remaining undisposed of.

The 1909 act is generally invoked where the lands are largely in Federal ownership and where there may be

---

[1] This chapter was edited by Michael Lindeburg and is based largely on the publication *Restoration of Lost or Obliterated Corners and Subdivision Sections*, a supplement to the Manual of Surveying Instructions. In all cases where public lands are not involved, surveying procedures must be in harmony with state laws and court opinions. In such cases, the methods and explanations of the Bureau of Land Management must be regarded as advisory only, as that Bureau is without jurisdiction.

extensive obliteration or other equally unsatisfactory conditions.

Another act of Congress approved September 21, 1918, provides authority for the resurvey by the Government of townships previously ineligible for resurvey by reason of the disposals being in excess of 50 percent of the total area.

The 1918 act may be invoked where the major portion of the township is in private ownership, where it is shown that the need for retracement and remonumentation is extensive, and especially if the work proposed is beyond the scope of ordinary local practice. The act requires that the proportionate costs be carried by the landowners.

3 PROTECTION OF BONA FIDE RIGHTS

Under the above laws, and in principle as well, it is required that no resurvey or retracement shall be so executed as to impair the bona fide rights or claims of any claimant, entryman, or owner of land so affected.

Likewise in general practice, the local surveyor should be careful not to exercise unwarranted jurisdiction, nor to apply an arbitrary rule. He should note the distinction between the rules for original surveys and those that relate to retracements. The disregard of these principles, or for acquired property rights, may lead to unfortunate results.

In unusual cases where the evidence of the survey cannot be identified with ample certainty to enable the application of the regular practices, the surveyor may submit his questions to the proper State office of the Bureau of Land Management, or to the Director.

4 ORIGINAL SURVEY RECORDS

The township plat furnishes the basic data relating to the survey and the description of all areas in the particular township. All title records within the area of the former public domain are based upon a Government grant or patent, with description referred to an official plat. The lands are identified on the ground through the retracement, restoration, and maintenance of the official lines and corners.

The plats are developed from the field notes. Both are permanently filed for reference purposes and are accessible to the public for examination or making of copies.

Many supplemental plats have been prepared by protraction to show new or revised lottings within one or more sections. These supersede the lottings shown on the original township plat. There are also many plats of the survey of islands or other fragmentary areas of public land which were surveyed after the original survey of the township.

These plats should be referred to as governing the position and description of the subdivisions shown on them.

5 RESURVEY RECORDS

The plats and field notes of resurveys which become a part of the official record fall into two principal classes according to the type of resurvey.

A *dependent resurvey* is a restoration of the original survey according to the record of that survey, based upon the identified corners of the original survey and other acceptable points of control, and the restoration of lost corners in accordance with proportional measurement as described herein. Normally, the subdivisions shown on the plat of the original survey are retained on the plat of the dependent resurvey, although new designations and areas for subdivisions still in public ownership at time of the resurvey may be shown to reflect true areas.

An *independent resurvey* is designed to supersede the original survey and creates new subdivisions and lottings of the vacant public lands. Provision is made for the segregation of individual tracts of privately-owned lands, entries, or claims that may be based upon the original plat, when necessary for their protection, or for their conformation, if feasible, to the regular subdivisions of the resurvey.

6 RECORDS TRANSFERRED TO STATES

In those states where the public land surveys are considered as having been completed, the field notes, plats, maps, and other papers relating to those surveys have been transferred to an appropriate state office for safekeeping as public records. No provision has been made for the transfer of the survey records to the State of Oklahoma, but in the other States the records are filed in offices where they may be examined and copies made or requested.

7 GENERAL PRACTICES

The rules for the restoration of lost corners have remained substantially the same since 1883, when they were first published. These rules are in harmony with the leading judicial opinions and the most approved surveying practice. They are applicable to the public land rectangular surveys and to the retracement of those surveys, (as distinguished from the running of property lines that may have legal authority only under State law, court decree, or agreement.)

In the New England and Atlantic Coast States, except Florida, and in Pennsylvania, West Virginia, Kentucky, Tennessee, and Texas, jurisdiction over the vacant lands remained in the states. The public land surveys were not extended in these states, and it follows that the practices outlined herein are not applicable there, except as they reflect sound surveying methods.

The practices outlined herein are in accord with the related provisions of the Manual. They have been segregated for convenience in order to separate them from the instructions pertaining only to the making of original surveys.

For clarity, the practices, as such, are set in bold face type. The remainder of the text is explanatory and advisory only, the purpose being to exemplify the best general practice.

In some states, the substance of the practices for restoration of lost or obliterated corners and subdivision of sections as outlined herein has been enacted into law. It is incumbent on the surveyor engaged in practice of land surveying to become familiar with the provisions of the laws of the State, both legislative and judicial, as affecting his work.

8 GENERAL RULES

The general rules followed by the Bureau of Land Management, which affect all public lands, are summarized in the following paragraphs:

- **The boundaries of the public lands, when approved and accepted, are unchangeable.**

- **The original township, section, and quarter-section corners must stand as the true corners which they were intended to represent, whether in the place shown by the field notes or not.**

- **Quarter-quarter-section corners not established in the original survey shall be placed on the line connecting the section and quarter-section corners, and midway between them, except on the last half mile of section** lines closing on the north and west boundaries of the township, or on the lines between fractional or irregular sections.

- The center lines of a section are to be straight, running from the quarter-section corner on one boundary to the corresponding corner on the opposite boundary.

- In a fractional section where no opposite corresponding quarter-section corner has been or can be established, the center line must be run from the proper quarter-section corner as nearly in a cardinal direction to the meander line, reservation, or other boundary of such fractional section, as due parallelism with the section boundaries will permit.

Corners established in the public land surveys remain fixed in position and are unchangeable, and lost or obliterated corners of those surveys must be restored to their original locations from the bast available evidence of the official survey in which such corners were established.

9 RESTORATION OF LOST OR OBLITERATED CORNERS

The restoration of lost corners should not be undertaken until after all control has been developed. Such control includes both original and acceptable collateral evidence. However, the methods of proportionate measurement will be of material aid in the recovery of evidence.

An existent corner is one whose position can be identified by verifying the evidence of the monument, or its accessories, by reference to the description that is contained in the field notes, or where the point can be located by an acceptable supplemental survey record, some physical evidence, or testimony.

Even though its physical evidence may have entirely disappeared, a corner will not be regarded as lost if its position can be recovered through the testimony of one or more witnesses who have a dependable knowledge of the original location.

An obliterated corner is one at whose point there are no remaining traces of the monument, or its accessories, but whose location has been perpetuated, or the point for which may be recovered beyond reasonable doubt, by the acts and testimony of the interested landowners, competent surveyors, or other qualified local authorities, or witnesses, or by some acceptable record evidence.

A position based upon collateral evidence should be duly supported, generally through proper relation to known corners, and agreement with the field notes regarding distances to natural objects, stream crossings, line trees, and off-line tree blazes, etc., or unquestionable testimony.

A lost corner is a point of a survey whose position cannot be determined, beyond reasonable doubt, either from traces of the original marks or from acceptable evidence or testimony that bears upon the original position, and whose location can be restored only by reference to one or more interdependent corners.

If there is some acceptable evidence of the original location of the corner, that position will be employed.

Decision that a corner is lost should not be made until every means has been exercised that might aid in identifying its true original position. The retracements, which are usually begun at known corners and run according to the record of the original survey, will indicate the probable position for the corner, and show what discrepancies may be expected. Any supplemental survey record or testimony should then be considered in the light of the facts thus developed. A line will not be regarded as doubtful if the retracement affords recovery of acceptable evidence.

In cases where the probable position for a corner cannot be made to harmonize with some of the calls of the field notes, due to errors in description or to discrepancies in measurement developed in the retracement, it must be ascertained which of the calls for distances along the line are entitled to the greater weight. Aside from the technique of recovering traces of the original marks, the main problem is one that treats with the discrepancies in alinement and measurement.

Existing original corners cannot be disturbed; consequently, discrepancies between the new and the record measurements will not in any manner affect the measurements beyond the identified corners, but the differences will be distributed proportionately within the several intervals along the line between the corners.

10 PROPORTIONATE MEASUREMENT

The ordinary field problem consists of distributing the excess or deficiency in measurement between existent corners in such a manner that the amount given to each interval shall bear the same proportion to the whole difference as the record length of the interval bears to the whole record distance. After having applied the proportionate difference to the record length of each interval the sum of the several parts will equal the new measurements of the whole distance.

A proportionate measurement is one that gives concordant relation between all parts of the line, i.e.—the new values given to the several parts, as determined by the remeasurement, shall bear the same relation to the record lengths as the new measurement of the whole line bears to that record. Lengths of proportioned lines are comparable only when reduced to their cardinal equivalents.

Discrepancies in measurement between those recorded in the original survey and those developed in the retracements should be carefully verified with the object to placing each such difference properly where it belongs. This is quite important at times, because, if disregarded, the result may be the fixing of a corner position where it is obviously improper. Accordingly, wherever possible, the manifest errors in the original measurements should be segregated from the general average difference, and placed where the blunder was made. The accumulated surplus or deficiency that then remains is the quantity that is to be uniformly distributed by the methods of proportionate measurement.

11 SINGLE PROPORTION

The term "single proportionate measurement" is applied to a new measurement made on a line to determine one or more positions on that line.

In single proportionate measurement, the position of two identified corners controls the direction of the line between those corners, and intermediate positions on that line are determined by proportionate measurement between those controlling corners. The method is sometimes referred to as a *two-way proportion.* Examples are: a quarter-section corner on the line between two section corners; all corners on standard parallels; and all intermediate positions on any township boundary line.

12 DOUBLE PROPORTION

The term "double proportionate measurement" is applied to a new measurement made between four known corners, two each on intersecting meridional and latitudinal lines, for the purpose of relating the intersection to both.

By double proportionate measurement, the lost corner is reestablished on the basis of *measurement* only, disregarding the record directions. An exception will be found in those cases where there is some acceptable survey record, some physical evidence, or testimony, that

may be brought into the control. The method may be referred to as a *four-way proportion.* Examples are: a corner common to four townships, or one common to four sections within a township.

The double proportionate measurement is the best example of the principle that existent or known corners to the north and to the south should control any intermediate latitudinal position, and that corners east and west should control the position in longitude.

As between single or double proportionate measurement, the principle of precedence of one line over another of less original importance is recognized in order to harmonize the restoring process with the method followed in the original survey, thus limiting the control.

13 STANDARD PARALLELS AND TOWNSHIP BOUNDARIES

Standard parallel will be given precedence over other township exteriors, and ordinarily the latter will be given precedence over subdivisional lines; section corners will be relocated before the position of lost quarter-section corners can be determined.

In order to restore a lost corner of four townships, a retracement will first be made between the nearest known corners on the meridional line, north and south of the missing corners, and upon that line a temporary stake will be placed at the proper proportionate distance; this will determine the latitude of the lost corner.

Next, the nearest corners on the latitudinal line will be connected, and a second point will be marked for the proportionate measurement east and west; this point will determine the position of the lost corner in departure (or longitude).

Then, through the first temporary stake run a line east or west, through the second temporary stake a line north or south, as relative situations may determine; the intersection of these two lines will fix the position for the restored corner.

In figure 24.2, the points A, B, C, and D represent four original corners. Point E represents the proportional measurement between A and B; and, similarly, F represents the proportional measurement between C and D. The point X satisfies the first control for latitude and the second control for departure.

A lost corner of four townships should not be restored, nor the township boundaries reestablished, without first considering the full field note record of the four intersecting lines and the plats of the township involved. In

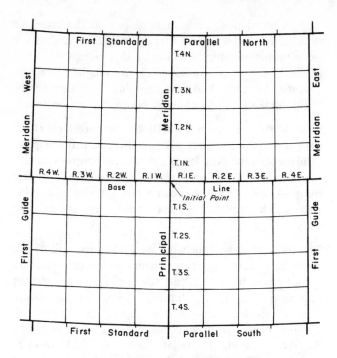

Figure 24.1 Standard parallel
and township boundaries

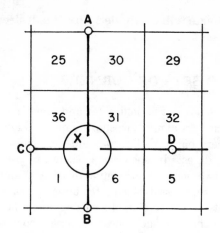

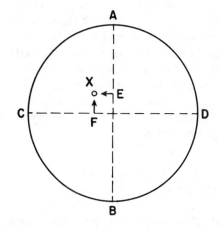

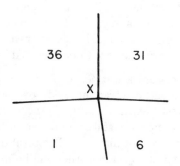

Figure 24.2 Lost township

most cases, there is a fractional distance in the half-mile to the east of the township, corner, and frequently in the half-mile to the south. The lines to the north and to the west are usually regular, i.e.—quarter-section and section corners at normal intervals of 40.00 and 80.00 chains, but there may be closing-section corners on any or all of the boundaries. So, it is important to verify all of the distances by reference to the field notes.

14 INTERIOR CORNERS

A lost interior corner of four sections will be restored by double proportionate measurement.

When a number of interior corners of four sections, and the intermediate quarter-section corners, are missing on all sides of the one sought to be reestablished, the entire distance between the nearest identified corners both north and south, and east and west, must be measured. Lost section corners on the township exteriors, if required for control, should be relocated.

15 RECORD MEASUREMENT

Where the line has not been established in one direction from the missing township or section corner, the record distance will be used to the nearest identified corner in the opposite direction.

Thus, in figure 24.2, if the latitudinal line in the direction of the point D has not been established, the position of the point F in departure would have been determined by reference to the record distance from the point C. The point X would then be fixed by cardinal offsets from the points E and F as already explained.

Where the intersecting lines have been established in only two of the directions, the record distances to the nearest identified corners on these two lines will control the position of the temporary points; then from the latter the car-

dinal offsets will be made to fix the desired point of intersection.

16 TWO SETS OF CORNERS

In many surveys the field notes and plats indicate two sets of corners along township boundaries and section lines where parts of the township were subdivided on different dates. In such cases, there are usually corners of two sections at regular intervals, and closing section corners established later upon the same line at the points of intersection of a closing line. The quarter-section corners on such lines usually are controlling for one side only.

In the more recent surveys, where the record calls for two sets of corners, those that are the corners of the two sections first established and the quarter-section corners relating to the same sections will be employed for the retracement, and they will govern both the alinement and the proportional measurements along that line. The closing section corners set at the intersections will be employed in the usual way, (i.e.—to govern the direction of the closing lines).

17 RESTORATION BY SINGLE PROPORTION

The method of single proportionate measurement is generally applicable to the restoration of lost corners on standard parallels and other lines established with reference to definite alinement in one direction only. Intermediate corners on township exteriors and other controlling boundary lines are to be included in this class.

In order to restore a lost corner by single proportionate measurement, a retracement will be made connecting the nearest identified regular corners on the line in question. A temporary stake will be set on the trial line at the original record distance. The total distance and the falling at the objective corner will be measured.

On meridional township lines, an adjustment will be made at each temporary stake for the proportional distance along the line. The temporary stake then will be set over to the east or to the west for falling, counting its proportional part from the point of beginning.

On east-and-west township lines and on standard parallels, the proper adjustment should be made at each temporary stake for the proportional distance along the line for the falling. (The temporary stake will either be advanced or set back for the proportional part of the difference between the record distance and the new measurement. It will then be set over for the curvature of the line, and eventually corrected for the proportional part of the true falling.)

The adjusted position is thus placed on the true line that connects the nearest identified corners, and at the same proportional interval from either as existed in the original survey. Any number of intermediate lost corners may be located on the same plan by setting a temporary stake for each when making the retracement.

Lost standard corners will be restored to their original positions on a base line, standard parallel or correction line, by single proportionate measurement on the true line connecting the nearest identified standard corners on opposite sides of the missing corner or corners, as the case may be.

The term "standard corners," as used above, will be understood to mean all corners which were established on the standard parallel during the original survey of that line, including, but not limited to, standard township, section, quarter-section, meander, and closing corners. Closing corners, or other corners purported to be established on a standard parallel after the original survey of that line will not control the initial restoration of lost standard corners.

Corners on base lines are to be regarded the same as those on standard parallel. In the older practice, the term "correction line" was used for what later has been called the "standard parallel." The corners first set in the running of a correction line will be treated as original standard corners. Those that were set afterwards at the intersection of a meridional line will be regarded as closing corners.

All lost section and quarter-section corners on the township boundary lines will be restored by single proportionate measurement between the nearest identified corners on opposite sides of the missing corner, north and south on a meridional line, or east and west on a latitudinal line, after the township corners have been identified or relocated.

An exception to this rule will be found in the case of any exterior the record of which shows deflections in alinement between the township corners

A second exception to this rule is found in those occasional cases were there may be persuasive proof of a deflection in the alinement of the township boundary, even though the record shows the line to be straight. For example, measurements east and west across a range line, or nouth and south across a latitudinal township line, counting from a straight-line exterior adjustment, may show distances to the nearest identified subdivisional corners to be materially long in one direction and correspondingly short in the opposite direction as compared to the record measurements. This condition, when supported by corroborative collateral evidence as might generally be expected, would warrant an exception to the straight-line or two-way adjustment under the rules for the acceptance of evidence, i.e.—the evidence outweighs the record. The rules for a four-way or double proportionate measurement would then apply, provided there is conclusive proof.

All lost quarter-section corners on the section boundaries within the township will be restored by single proportionate measurement between the adjoining section corners, after the section corners have been identified or relocated.

This practice is applicable in the majority of the cases. However, in those instances where other corners such as meander corners, sixteenth-section corners, etc., were originally established between the quarter-section and the section corners, such minor corners, when identified, will exercise control in the restoration of lost quarter-section corners.

Lost meander corners, originally established on a line projected across the meanderable body of water and marked upon both sides will be relocated by single proportionate measurement, after the section or quarter-section corners upon the opposite sides of the missing meander corner have been duly identified or relocated.

Under ordinary conditions, the actual shore line of a body of water is considered the boundary of lands included in an entry and patent, rather than the meander line returned in the field notes. It follows that the restoration of a lost meander corner would be required only infrequently. Under favorable conditions, a lost meander corner may be restored by treating the shoreline as an identified natural feature controlling the measurement to the point for the corner. This is particularly applicable where it is evident that there has been no change in the shore line.

A lost closing corner will be reestablished on the true line that was closed upon, and at the proper proportional interval between the nearest regular corners to the right and left.

In order to reestablish a lost closing corner on a standard parallel or other controlling boundary, the line that was closed upon will be retraced, beginning at the corner from which the connecting measurement was originally made. A temporary stake will be set at the record connecting distance, and the total distance and falling will be noted at the next regular corner on that line on the opposite side of the missing closing corner. The temporary stake will then be adjusted as in single proportionate measure.

A closing corner not actually located on the line that was closed upon will determine the direction of the closing line, but not its legal terminus. The correct position is at the true point of intersection of the two lines.

18 IRREGULAR EXTERIORS

Some township boundaries, not established as straight lines, are termed "irregular" exteriors. Parts were surveyed from opposite directions, and the intermediate portion was completed later by random and true line, leaving a fractional distance. Such irregularity follows some material departure from the basic rules for the establishment of original surveys. A modified form of single proportionate measurement is used in restoring lost corners on such boundaries. This is also applicable to a section line or a township line which has been shown to be irregular by a *previous* retracement.

In order to restore one or more lost corners or angle points on such irregular exteriors, a retracement between the nearest known corners is made on the record courses and distances to ascertain the direction and length of the closing distance. A temporary stake is set for each missing corner or angle point. The closing distance is then reduced to its equivalent latitude and departure.

On a meridional line, the *latitude* of the closing distance is distributed among the courses in proportion to the *latitude* of each course. The *departure* of the closing distance is distributed among the courses in proportion to the *length* of each course. That is, after the excess or deficiency of latitude is distributed, each temporary stake is moved east or west an amount proportional to the total distance from the starting point.

On a latitudinal line, the temporary stakes should be placed to suit the usual adjustments for the curvature. The *departure* of the closing distance is distributed among the courses in proportion to the *departure* of each course. Then, each temporary stake is moved north or south an amount proportional to the *total distance* from the starting point.

Angle points and intermediate corners are treated alike.

19 ONE-POINT CONTROL

Where a line has been terminated with measurement in one direction only, a lost corner will be restored by record bearing and distance, counting from the nearest regular corner, the latter having been duly identified or restored.

Examples will be found where lines have been discontinued at the intersection with large meanderable bodies of water, or at the border of what was classified as impassable ground.

20 INDEX ERRORS FOR ALINEMENT AND MEASUREMENT

Where the original surveys were faithfully executed, it is to be anticipated that retracement of many miles of the lines in a given township will develop a definite and consistent difference in measurement and in a bearing between original corners. Under such conditions, it is proper that allowance be made for the average differences in the restoration of a lost corner where control is lacking on one direction. The adjustment will be taken care of automatically where there is a suitable basis for proportional measurement.

21 SUBDIVISION OF SECTIONS

The ordinary unit of administration of the public lands under the rectangular system of surveys is the quarter-quarter section of 40 acres. Usually the sections are not subdivided on the ground in the original survey. The boundaries of the legal subdivisions generally are shown by protraction on the plats.

On the plat of the original survey of a normal township, it is to be expected that the subdivision of sections will be indicated (by protraction) according to standard procedures.

The sections bordering the north and west boundaries of the township, except section 6, are subdivided into two regular quarter-sections, two regular half-quarter sections, and four fractional quarter-quarter units which are usually designated as lots. In section 6, the subdivision will show one regular quarter-section, two regular half-quarter sections, one regular quarter-quarter section, and seven fractional quarter-quarter units. This is the result of the plan of subdivision, whereby the excess or deficiency in measurement is placed against the north and west boundaries of the township.

The plan of subdivisions and controlling measurements employed is illustrated in figure 24.4.

In a normal section which is subdivided by protraction into quarter sections, it is not considered necessary to indicate the boundaries of the quarter-quarter sections on the plat. Such subdivisions are aliquot parts of the quarter sections, based on mid-point protraction.

Sections which are invaded by meanderable bodies of water or by private claims which do not conform to the regular legal subdivisions are subdivided by protraction into regular and fractional parts as nearly as practicable in conformity with the uniform plan already outlined. The meander lines, and the boundary lines of the private claims, are platted according to the field note record. The subdivision-of-section lines are terminated at the meander line or claim boundary, as the case

may be, but their positions are controlled precisely as though the section had been completed regularly. For the purpose of protracting the subdivisional lines in a section whose boundary lines are partly within the limits of a meanderable body of water or private claim, the fractional section boundaries are completed in theory.

The protracted position of the subdivision-of-section lines is controlled by the theoretical points so determined. (See figure 24.3.)

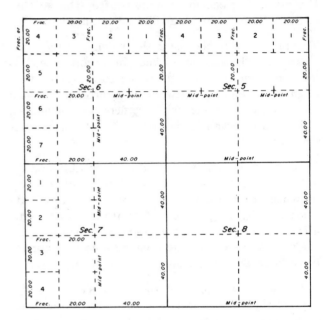

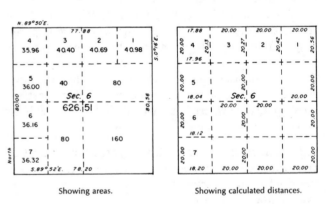

Showing areas.   Showing calculated distances.

Figure 24.3   Examples of subdivision by protraction

22 ORDER OF PROCEDURE IN SURVEY

The order of procedure is first to identify or reestablish the corners on the section boundaries, including determination of the points for the necessary one-sixteenth section corners. Next, fix the boundaries of the quarter section. Finally, form the quarter sections or small tracts by equitable and proportionate division.

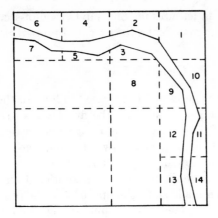

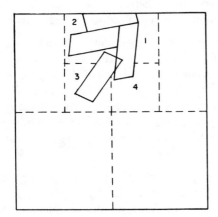

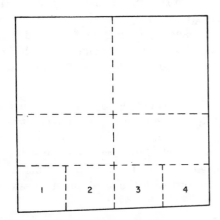

Figure 24.4 Examples of subdivision
of fractional sections

23 SUBDIVISION OF SECTIONS INTO QUARTER SECTIONS

To subdivide a section into quarter sections, run straight lines from the established quarter-section corners to the opposite quarter-section corners. The point of intersection of the lines thus run will be the corner common to the several quarter sections, or the legal center of the section.

Upon the lines closing on the north and west boundaries of a regular township, the quarter-section corners were established originally at 40 chains to the north or west of the last interior section corners. The excess or deficiency in measurement was thrown into the half-mile next to the township or range line, as the case may be. If such quarter-section corners are lost, they should be reestablished by proportionate measurement based upon the original record.

Where there are double sets of section corners on township and range lines, the quarter-section corners for the sections south of the township line and east of the range line usually were not established in the original surveys. In subdividing such sections, new quarter-section corners are required. They should be placed to suit the calculations of the areas that adjoin the township boundary, as indicated upon the official plat, adopting proportional measurements where the new measurements of the north or west boundaries of the section differ from the record distances.

24 SUBDIVISION OF FRACTIONAL SECTIONS

The law provides that where opposite corresponding quarter-section corners have not been or cannot be fixed, the subdivision-of-section lines shall be ascertained by running from the established corners north, south, east, or west, as the case may be, to the water course, reservation line, or other boundary of such fractional section, as represented upon the official plat.

In this, the law presumes that the section lines are due north and south, or east and west lines, but usually this is not the case. Hence, in order to carry out the spirit of the law, it will be necessary in running the center lines through fractional sections to adopt mean courses, where the section lines are not on due cardinal, or to run parallel to the east, south, west, or north boundary of the section, as conditions may require, where there is no opposite section line.

25 SUBDIVISION OF QUARTER SECTIONS

Preliminary to the subdivision of quarter sections, the quarter-quarter, or sixteenth-section corners will be established at points midway between the section and quarter-section corners, and the center of the section, except on the last half mile of the lines closing on township boundaries, where they should be placed at 20 chains, proportionate measurement, counting from the regular quarter-section corner.

The quarter-quarter, or sixteenth-section corners having been established as directed above,
the center lines of the quarter section will be run
straight between opposite corresponding quarter-quarter, or intersection of the lines thus run
will determine the legal center of a quarter section. (See figure 24.5.)

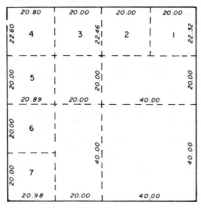

Official measurements.

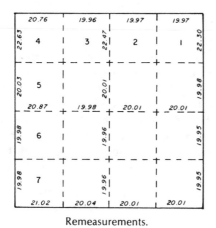

Remeasurements.

Figure 24.5 Example of subdivision by survey

26 SUBDIVISION OF FRACTIONAL
QUARTER SECTIONS

The subdivisional lines of fractional quarter sections will be run from properly established quarter-quarter, or sixteenth-section corners, with
courses governed by the conditions represented
upon the official plat, to the lake, water-course,
reservation, or other irregular boundary which
renders such sections fractional.

What has been written on the subject of subdivision of
sections relates to the procedure contemplated by law,
and refers to the methods to be followed in the initial subdivision of the areas, prior to development and
improvement. Care should be exercised to avoid dis-
turbing satisfactory improvements such as roads, fences,

or other in features making subdivision-of-section lines
and which may define the extent of property rights.

27 RETRACEMENTS

When it is necessary to retrace the lines of the rectangular public-land surveys, the first step is to assemble
copies of the field notes and plats, and determine the
names of the owners who will be concerned in the retracement and survey. A thorough search and inquiry
with regard to the record of any additional surveys that
have been made since the approval of the original survey should be made. The county surveyor, county clerk,
register of deeds, practicing engineers and surveyors,
landowners, and others who may furnish useful information should be consulted as to such features.

The matter of boundary disputes should be carefully reviewed, particularly as to whether claimants have based
their locations upon evidence of the original survey and
a proper application of surveying rules. If there has
been a boundary suit, the record testimony and the
court's opinion and decree should be carefully examined
insofar as these may have a bearing upon the problem
at hand.

The law requires that the position of original corners
must not be changed. There is a penalty for defacing
corner marks, or for changing or removing a corner.
The corner monuments afford the principal means for
identification of the survey, and accordingly, the courts
attach the greatest weight to the evidence of their location. Discrepancies that may be developed in the directions and length of lines, as compared with the original
record, do not warrant any alteration of a corner position.

Obviously, on account of roadways or other improvements, it is frequently necessary to reconstruct a monument in some manner in order to preserve its position.
Alterations of this type are not regarded as changes
in willful violation of the law, but rather as being in
complete accord with the legal intent to safeguard the
evidence.

Therefore, whatever the purpose of the retracement
may be—if it calls for the recovery for the true lines
of the original survey, or for the running of the subdivisional lines of a section, the practices outlined require
some or all of certain definite steps.

> *Step 1*: Secure a copy of the original plat and field
> notes;
>
> *Step 2*: Secure all available data regarding subse-
> quent surveys;
>
> *Step 3*: Secure the names and contact the owners of
> the property adjacent to the lines that are
> involved in the retracement;

Step 4: Find the corners that may be required:

- First: By the remaining physical evidence;

- Second: By collateral evidence, supplemental survey records, or testimony, if the original monument is regarded as obliterated, but not lost, or;

- Third: By application of the rules for proportionate measurement, if lost;

Step 5: Reconstruct the monuments as required, including the placing of reference markers where improvements of any kind might interfere, or if the site is such as to suggest the need for supplemental monumentation;

Step 6: Note the procedure for the subdivision of sections where these lines are to be run; and

Step 7: Prepare and file a suitable record of what was found, the supplemental data that was employed; a description of the methods used, the direction and length of lines, the new markers, and any other facts regarded as important.

A knowledge of the practices and instructions in effect at the time of the original survey will be helpful. These should indicate what was required and how it was intended that the original survey should be made.

The data used in connection with the retracements should not be limited to the section or sections under immediate consideration. It should also embrace the areas adjacent to those sections. The plats should be studied carefully. Fractional parts of sections should be located on the ground as indicated on the plats.

28 DOUBLE SETS OF CORNERS

The technique of making field astronomical observations for determinations of the true meridian, and methods for establishment of the true latitudinal curve, were developed many years after the inception of the rectangular survey system. Without these refinements, accumulated discrepancies were bound to develop.

As a result, in order to maintain rectangularity in some of the older surveys, two sets of corners were established on the township boundaries. The section and quarter-section corners established in the survey of the boundary itself are the corners to be adopted in retracement and for control of proportionate measurements. These corners control the subdivisions on one side only of the township boundary. The second set of corners on these

boundaries are the closing section corners for the subdivisional surveys on the opposite side of the boundary. The descriptions of these closing corners, and the connecting distances to the regular township boundary corners, will be found in the field notes of the subdivisional survey in which they were established. These closing section corners should be considered and evaluated as evidence in the solution of the whole problem.

Where the section corners on the township boundaries are of minimum control, the quarter-section corners have the same status for the same side of the boundary. In the older surveys, quarter-section corners usually were not established for the opposite side of the boundary. Subsequent to 1919, it was the practice to establish the second set of quarter-section corners. These are at midpoint for distances between the closing section corners, except where the plan of subdivision dictates otherwise, in which event the quarter-section corner is placed at 40.00 chains from the controlling closing corner.

These conditions merit careful study of the plats to the end that the subdivisions shown on the plats be given proper protection. The plats will indicate whether these quarter-section corners should be at mid-point between the closing corners, or if they should be located with regard to a fractional distance. The surveyor should make sure that the position is determined for all corners necessary for control in his work.

There is nothing especially different or complicated in the matter of one or two sets of corners on the township boundary lines. It is merely a matter of assembling complete data and of making a proper interpretation of the status of each monument.

The same principles should be applied in the consideration of the data of the subdivisional surveys, where for any of several causes, there may be two sets of interior corners.

29 THE NEEDLE COMPASS AND SOLAR COMPASS

Very simple needle-compass equipment in the hands of men skilled as surveyors, coupled with natural woodcraft and faithfulness in doing their work, satisfied the requirements of the early public-land surveys.

A large proportion of surveys made prior to 1890 are of the needle compass type. It should be noted that retracements may be made, i.e.—the evidence of the marks can be developed by needle-compass methods if properly employed. Some surveyors maintain that you can "follow the steps" of the original surveyor more closely by use of the needle compass than by more precise methods.

In addition to the uncertainties of local attraction and temporary magnetic disturbances, the use of the needle compass is exceedingly unreliable in the vicinity of power lines, pipelines, steel rails, steel-framed structures of all kinds, wire fences, etc. Its use is now much more restricted because of these improvements. The needle compass is rapidly becoming obsolete since it fails to satisfy the present need for more exact retracements.

30 EXCESSIVE DISTORTION

The needle-compass surveys, before being discontinued, had extended into the region of magnetic ore deposits of the Lake Superior watershed in northern Michigan, Wisconsin, and Minnesota. Here many townships were surveyed, and the lands patented, in which the section boundaries are now found to be grossly distorted. There is no way in which to correct these lines, nor to make an estimate (except by retracement), of the extent of the irregularities, which involve excessive discrepancies both in the directions and lengths of lines.

Considerable experience is required in retracing and successfully developing the evidence of the lines and corners in these areas of excessive distortion. However, the procedure for restoration of lost corners and for the subdivision of sections is the same as in areas of more regularity.

Another feature to be considered in connection with the retracements is that the record may show that one surveyor ran the south boundary, a second the east boundary, and others the remaining exteriors and subdivisional lines. All of these lines may be reported on cardinal, but may not be exactly comparable, i.e.—the east boundary may not be truly normal to the south boundary, etc. It was customary to retrace one or more miles of the east boundary of the township to determine the "variation" of the needle. This value was then adopted in the subdivision of the township. It follows that the meridional section lines should be found to be reasonably parallel with the east boundary. Under that plan of operation, it should be anticipated that the latitudinal section lines will be reasonably parallel with the south boundary. However, discrepancies in measurement on the meridional lines frequently affect such parallelism. For these reasons, the index corrections for bearings may not be the same for the east and west as for the north and south lines. The two classes should be considered separately in this respect.

Before 1900, most lines were measured with the Gunter's link chain. The present surveyor must realize the difficulties of keeping that chain at standard length and the inaccuracy of measuring steep slopes by this method. It is to be expected that the retracements will

show various degrees of accuracy in the recorded measurements which were intended to reflect true horizontal distances.

31 INDEX ERRORS

Where the original surveys were faithfully made, there will generally be considerable uniformity in the directions and lengths of the lines. Frequently, this uniformity is so definite as to indicate "index errors" which, if applied to the record bearings and distances, will place the trial lines in close proximity to the true positions and aid materially in the search for evidence. With experience, the present surveyor will become familiar with the work of the original surveyor and know about what to expect in the way of such differences.

32 COLLATERAL EVIDENCE

The identified corners of the original survey constitute the main control for the surveys to follow. After those corners have been located, and before resorting to proportionate measurement for restoration of lost corners, the other calls of the field notes should be considered. The recorded distances to stream crossings or to other natural objects which can be identified often lead to the position for a missing corner. At this stage, the question of acceptance of later survey marks and records, the location of roads and property fences, and the reliability of testimony, are to be considered.

A line tree, connection to some natural object, or an improvement recorded in the original field notes which can be identified may fix a point of the original survey. The calls of the field notes for the various items of topography may assist materially in the recovery of the lines. The mean position of a blazed line, when identified as the original line, will identify a meridional line for departure or a latitudinal line for latitude. These are matters which require the exercise of considerable judgment.

33 ORIGINAL MARKS

Original line-tree marks, off-line tree blazes, and scribe marks on bearing trees and tree corner-monuments whose age exceeds 100 years, are found occasionally. Such marks of later surveys are recovered in much greater number. Different surveyors used distinctive marks. Some surveyors used hacks instead of blazes, and some used hacks over and under the blazes; some employed distinctive forms of letters and figures. All these will be recognized while retracing the lines of the same survey and will serve to verify the identification of the work of a particular surveyor.

The field notes give the species and the diameter of the bearing trees and line trees. Some of the smooth-barked trees were marked on the surface, but most of the marks were made on a flat smoothed surface of the live wood tissue. The marks remain as long as the tree is sound. The blaze and marks will be covered by a gradual overgrowth, showing a scar for many years. The overgrowth will have a lamination similar to the annual rings of the tree, which may be counted in order to verify the date of marking, and to distinguish the original marks from later marks and blazes. On the more recent surveys, it is to be expected that the complete quota of marks should be found, clear cut and plainly legible. This cannot be expected in the older surveys, however.

It is advisable not to cut into a marked tree except as necessary to secure proof. The evidence is frequently so abundant, especially in the later surveys, that the proof is conclusive without inflicting an additional injury that would hasten the destruction of the tree.

Finding original scribe marks, line-tree hacks, and off-line tree blazes, furnishes the most convincing identification that can be desired.

It is not intended to disturb satisfactory local conditions with respect to roads and fences. The surveyor has no authority to change a property right that has been acquired legally. On the other hand, he should not accept the location of roads and fences as evidence of the original survey without something to support these locations. This supporting evidence may be found in some intervening survey record, or the testimony of individuals who may be acquainted with the facts.

34 RULES ESTABLISHED BY STATE LAW OR DECISIONS

Other factors that require careful consideration are the rules of the state law and the state court decisions, as distinguished from the methods followed by the Bureau of Land Management. Under State law, property boundaries may be fixed by agreement between owners, acquiescence, or adverse possession. Such boundaries may be defined by roads, fences, or survey marks, disregarding exact conformation with the original section lines. The rights of adjoining owners may be limited to such boundaries.

In many cases, due care has been exercised to place the property fences on the lines of legal subdivision. It has been the general practice in the prairie states to locate the public roads on the section lines. These are matters of particular interest to the adjoining owners. It is reasonable to presume that care and good faith were exercised in placing such improvements with regard to the evidence of the original survey in existence at the time. Obviously, the burden of proof to the contrary must be borne by the party claiming differently. In many cases, at the time of construction of a road, the positions for the corners were preserved by subsurface deposits of marked stones or other durable material. These are to be considered as exceptionally important evidence of the position of the corner, when duly recovered and verified.

The replacement of those corners that are regarded as obliterated, but not lost, should be based on such collateral evidence as has been found acceptable. All lost corners can be restored only by reference to one or more interdependent corners.

35 ADEQUATE MONUMENTATION ESSENTIAL

The surveyor will appreciate the great extent to which his successful retracements has depended upon an available record of the previous surveys, and upon the markers that were established by those who preceded him. The same appreciation will apply in subsequent retracements. It is essential to the protection of the integrity and accuracy of the work, the reputation of the surveyor, and the security of the interested property owners, that durable new corner markers be constructed in all places where required, and that a record be filed of the survey as executed.

The preferred markers are of stone, concrete block, glazed sewer-tile filled with concrete, cast-iron or galvanized-iron pipe, and similar durable materials. Many engineers and surveyors, counties, and landowners employ specially designed markers with distinctive lettering, including various cast-iron plates or bronze tablets.

The Bureau of Land Management has adopted a standard monument for use on the public-land surveys. This is made of wrought iron-pipe, zinc coated, 2 1/2 inches diameter and 28 inches long, with one end split and spread to form flanges or foot plates. A brass cap is securely attached to the top on which appropriate markings for the particular corner are inscribed by use of steel dies.

Frequently, on account of roadway or other improvements, it is advisable to set a subsurface marker and, in addition, to place a reference monument where it may be found readily, selecting a site that is not likely to be disturbed.

36 MEANDER LINES AND RIPARIAN RIGHTS

The traverse run by a survey along the bank of a stream or lake is termed a *meander line*. The meander line is not generally a boundary in the usual sense, as ordinarily the bank itself marks the limits of the survey. All navigable bodies of water are meandered in the public-land surveying practice, as well as many other important streams and lakes that have not been regarded as navigable in the broader sense.

All navigable rivers within the territory occupied by the public lands remain, and are deemed to be, public highways. Unless otherwise reserved for federal purposes, the beds of these waters were vested in the states at time of statehood. Under Federal law, in all cases where the opposite banks of any stream not navigable belong to different persons, the stream and the bed thereof become common to both.

Grants by the United States of its public lands, including lands bounded by streams or other waters, are construed as to their effect according to Federal law. This includes lands added to the grants by accretion.

The Government conveyance of title to a fractional subdivision fronting upon a nonnavigable stream, unless specific reservations are indicated in the patent from the Federal Government, carries ownership to the middle of the stream.

Where surveys purport to meander a body of water where no such body exists, or where the meanders may be considered grossly erroneous, the United States may have a continuing public land interest in the lands within the segregated areas.

Where partition lines are to be run *across accretions*, the ordinary Federal rule is to apportion the new frontage along the water boundary in the same ratio as the frontage along the line of the record meander courses. There are many variations to this rule where local conditions prevail and the added lands are not of great width or extent. The application of any rule, when surveying private lands, should be brought into harmony with the state law.

Where there is occasion to define the partition lines *within the beds* of nonnavigable streams, the usual rule is to begin at the property line at its intersection with the bank. From that point, run a line normal to the medial line of that stream that is located midway between the banks. Where the normals to the medial lines are deflecting rapidly, owing to abrupt changes in the course of the stream, suitable locations are selected above and below the doubtful positions, where acceptable normals may be placed. The several intervals along the medial line are then apportioned in the same ratio as the frontage along the bank.

The partition of the bed of nonnavigable lakes, whether water-covered or relicted, presents a more difficult problem because of the wide range of shapes of lake beds. In the simplest case of a circular bed, the partition lines can be run to the centroid, thus creating pie-shaped tracts fronting the individual holdings at the edge of the lake bed. Where odd shaped beds are concerned, ingenuity will be required to divide the lake bed in such a manner that each shore proprietor will receive an equitable share of land in front of his holding. Any consideration of riparian rights insuring to private lands should be brought into line with appropriate state laws or decisions.

RESERVED FOR FUTURE USE

PROFESSIONAL PUBLICATIONS, INC. • P.O. Box 199, San Carlos, CA   94070

RESERVED FOR FUTURE USE

PROFESSIONAL PUBLICATIONS, INC. • P.O. Box 199, San Carlos, CA 94070

25

PROPERTY LAW

PART 1: Colonization History

1 ENGLISH COMMON LAW

English common law consists of those ideas of right and wrong determined by court decisions over many centuries. Such ideas have been accepted by generations as they sought to establish rules to meet social and economic needs. Blackstone, eighteenth century author of *Commentaries on the Law of England* called it "unwritten law" in the sense that it was not enacted by a legislative body. It was legal custom expressed by the decisions of judges.

English common law was evolutionary. It changed slowly. Judges made decisions based on former decisions, but they also modified their decisions to reflect changing times.

Some of the important parts of our present law which came from English common law are the grand jury, trial by jury, freedom of press, habeas corpus, and oral testimony.

2 STATUTE LAW

Laws enacted by legislative bodies are known as *statute laws.* In contrast with common law, it is written law. Laws of France, Germany, Spain, and other countries on the continent of Europe are largely statute laws. After the French Revolution, France adopted the Code Napoleon to clarify its laws.

3 COLONIAL LAW

Colonial law was not evolutionary because there was nothing on which to base precedence. Therefore, settlers of the original thirteen colonies adopted English common law. But they did not adopt it in its entirety.

It did not fit their new social and economic environment entirely. Many of their laws were statutory (written).

Furthermore, the same parts of English common law were not adopted in each of the colonies. The parts of the law which seemed to fit the needs of the particular situation in each colony were the parts which were adopted.

4 SPAIN AND FRANCE IN THE NEW WORLD

Spain acquired title to land in the New World by grant from Pope Alexander VI in 1493. The grant conveyed all lands not held by a Christian prince on Christmas day of 1492 from a meridian known as the "Line of Demarcation," 100 leagues west of the Azores and Cape Verde Islands. Later, a treaty with Portugal moved the Line of Demarcation 270 leagues west, with Portugal to have rights to the east and Spain to the west. Possession came from conquest, and by 1600, the territory extended from New Mexico and Florida on the north to Chile and Argentina on the south. The first seat of government was at Santa Domingo.

In 1511, Don Diego Velasquez led an expedition for the conquest of Cuba which was accomplished without serious opposition. Velasquez was appointed Governor of the island. During his rule, he promoted settlement of the land by Spaniards.

Hernandez de Cordova set out in 1517 from Cuba on an expedition to the Bahama Islands to obtain Indian slaves, but a storm drove him off his course. Three weeks later, he landed on the coast of Yucatan. Cordova returned to Cuba with tales of a more advanced civilization than previously found in the New World. And of great interest to Velasquez, he brought tales of gold and fine cotton garments. Valasquez, in 1518, sent his nephew, Juan de Grijalva, on an expedition to explore the coast of Mexico. Grijalva also landed on Yucatan and was impressed by the advanced civilization. He returned to Cuba with many gold ornaments he had received in trade.

After approval from Spain, Velasquez decided on the conquest and colonization of the new land. He chose Hernando Cortes as the commander of his expedition.

After many skirmishes and battles, Cortes reached the capital of the Aztecs, Tenochtitlan, now Mexico City, on November 8, 1519. He quickly conquered the forces of Montezuma, the Aztec emperor, and placed him under house arrest where he was treated very cordially by Cortes and allowed to retain his many luxuries.

After the conquest of the new land, Spanish statute law was introduced. Legislation for this new land was codified in Recopilacion de las Leyes de Indias in 1680. The Crown of Spain had complete authority which was administered through the Minister of the Indies, and all the land belonged to the King of Spain.

Spain ruled Mexico for three hundred years. However, in 1821, the people of Mexico revolted and declared their independence from Spain. Augustin de Iturbide, leader of the revolt, was crowned as Augustin I, Emperor of Mexico in 1822. In 1824, he was deposed by Lopez de Santa Anna who established a constitutional government.

France established Quebec in 1608. From there, settlement moved south along the Mississippi to its mouth. Rene Robert Cavalier, Sieur de la Salle, claimed all the Mississippi Valley for France and named it Louisiana in honor of Louis XIV.

Napoleon, in 1803, sold all of Louisiana to the United States "with the same extent that it now has in the hands of Spain, and that it had when France possessed it." President Thomas Jefferson claimed Texas as part of the purchase but Spain protested vigorously. The boundary between Texas and the United States was established by treaty between the King of Spain and the United States in 1819.

An act barring emigration from the United States and other restrictive land laws caused unrest in Texas, which resulted in Texas winning its independence from Mexico. In 1836, the Republic of Texas was established.

PART 2: A Brief History of Property Law

5 EARLY HISTORY OF PROPERTY LAW

The earliest record of property ownership goes back to the Babylonians in 2500 B.C. The Bible also furnishes us with many references to property ownership. Numerous references are found in the book of Genesis, including a passage relating to the purchase of land by Abraham on which to bury his wife, Sarah.

In the book of Jeremiah is "Thou shall not remove thy neighbor's landmark which they of old times have set in thine inheritance which thou shall inherit in the land that the Lord thy God giveth thee to possess it." Also in the book of Jeremiah is "Cursed be to he that removeth his neighbor's landmark and all the people shall say Amen."

Historical records show that in about 1400 B.C. the King of Egypt divided the land into squares of equal size and gave each Egyptian one square. The King in turn levied taxes on each person. This land was in the fertile Nile valley where the river overflowed and destroyed parts of these plots. The owner of the destroyed land was required to report his loss to the King and request a reduction of taxes. The need to determine the actual losses of property resulted in the beginning of surveying in this part of the world. These early surveyors are referred to as "rope stretchers" in the Bible. Drawings on the walls of tombs show these rope stretchers accompanied by officials who recorded the measurements.

The Greeks and Romans also recognized individual property ownership and the Romans used taxes on land to support the cost of government.

At the beginning of the Christian era in Europe, the ownership of land was usually determined by conquest. After conquest, the ruler took over all the land and earlier land titles were extinguished. Often, sovereign rights were vested in the ruler by the Pope, but in non-Christian lands, the practice of the ruler having rights to all the land was much the same. The land belonged to the Sovereign, often referred to as the Crown.

6 FEUDAL SYSTEM

Great Britain in the 11th century was conquered by William the Conqueror (William I). He claimed all the land and ruled over all the people. Later, he introduced the feudal system which was the social and political system in both Great Britain and Europe during the 11th, 12th, and 13th centuries. The feudal system was the basis of real property law in medieval times. Many of its principles found their way into American law.

In England, the system was an arrangement between the King, noblemen, and vassals. The arrangement included an intricate set of rules for the tenure and transfer of real property. The noblemen ruled over the tenants (*vassals*) of the land. These landlords protected their vassals, but expected them to pay rent on the land they used and to pay allegiance to their Lord. Often, they were required to fight for their Lord. The land was in possession of the Lord and could not be sold. It passed to the eldest son by inheritance. Vassals had no chance to own property.

7 COMMON LANDS

Almost all the peasant class at the time of William the Conqueror was engaged in farming. Tenant farmers acquired strips of land from the Lord for row crops and for producing hay for the cattle. While the crops were growing, these farmers needed pasture for their cattle, and they acquired from their Lord *right in common* on land used for permanent pasture. These lands eventually became known as *commons*, and the word was brought to this country by early settlers. In the United States, *common* means a park.

8 DOMESDAY BOOK

In 1086, William the Conqueror made a survey for tax purposes which included every farm, farm owner, and all common land. This was known as the "Domesday Book." Besides being useful for collecting taxes, it allowed him to demand allegiance from everyone in his kingdom. The Domesday Book was a complete record for doomsday, the modern spelling.

A modern edition of the Domesday Book, which lists these common lands, has recently been published. Over the centuries, they have become private property subject to certain rights by claimants to rights in common. Parliament, with the new book, is trying to pin down just what may be done on this land (which comprises four percent of the land in England).

9 TREND TO PRIVATE OWNERSHIP

As people turned from agriculture to crafts and trade in villages and cities, they curbed the power of Kings and began to think of ownership of land without fealty to the Lord, without obligations of service, and with the right to dispose of the land as they saw fit. This is known as *fee ownership*.

Early grants of the Kings were not recorded. The grantee received a packet of papers as evidence of his ownership. These documents became so voluminous that loss of them became common. Parliament began to take steps to correct the confusion caused by lost or stolen documents. Laws were established which abolished the practice of passing title from father to eldest son, and steps were taken to better describe the property transferred.

10 MAGNA CARTA

In 1215, powerful English noblemen forced the King of England to sign the Magna Carta, a document which forced the King to share authority with the nobles. By 1700, Parliament had gained supremacy over the King.

11 STATUTE OF FRAUDS

In 1677, the English Parliament passed the Statute of Frauds which, among other things, prohibited any transfer of land or any transfer of interest in land by oral agreement. All conveyances were required to be in writing, a requirement which is very much a part of our present day law.

12 PROPERTY LAW IN THE UNITED STATES

Ideas of property ownership in existence in England at the time were brought to this country by English settlers. Lands in America were granted to settlers by Kings and Queens of England. The idea of ownership by the sovereign still prevailed and some tribute to the crown was required.

The settlers themselves still had the idea of ownership by the conqueror as they displaced the Indians from their homes and lands with little compensation to them.

As land was cheap in the early days of our country, exact descriptions and exact locations were not necessary. Some grants to the original colonists extended from the Atlantic to the Pacific.

13 TYPES OF PROPERTY

Property is divided into two classes: real property and personal property. *Real property* is immovable and can be recovered. It has been defined as "the interest that a man has in lands, tenements, or hereditaments." These terms include land, buildings, trees, and the right to use them. Anything which grows on the land or any structure which is fixed to the land is real property.

Real property law is, for the most part, state law rather than federal law. It, therefore, varies among the states.

Personal property is movable and often cannot be recovered. Action to recover such things as money and valuable goods is often taken against the person who removed them illegally.

PART 3: Title

14 DEFINITION

Title is the right to own real property and the evidence of that right. Right to ownership is not enough, however. There must also be possession of property. Title, then, is the outward evidence of the right to ownership.

15 CLEAR TITLE, GOOD TITLE, MERCHANTABLE TITLE

These terms are essentially synonymous. *Clear title* means the property is free from encumbrances. *Good title* is one free from litigation.

16 RECORD TITLE

A title entered on the public records is referred to as a *record title.*

17 COLOR OF TITLE

Any written instrument, such as a forgery, which appears to convey title but in fact does not, establishes *color of title.* A consecutive chain of transfers of title down to the person in possession in which one or more of the written instruments is not registered may also establish color of title.

18 CLOUD ON TITLE

A claim on land which, if valid, would impair the title to the land creating a *cloud on title.* The claim may be any encumbrance such as a lien, judgment, tax-levy, mortgage, or conveyance.

19 CHAIN OF TITLE

The change in ownership of a piece of property in sequence is known as the *chain of title.* Any defective conveyance of title in the chain adversely affects the title from that point on.

20 ABSTRACT OF TITLE

Before buying real property, a buyer should institute a search of title—a review of all documents affecting the ownership of the property to determine if the person selling the property has a good and clear title. A compilation of abstracts of deeds, deeds of trust, or any other estate or interest, together with all liens or liabilities which affect the title to the property, may be obtained from an abstract or title company. This condensed history of the title to the land in chronological order is known as an *abstract of title* or simply an *abstract.*

21 ATTORNEY'S OPINION

After an attorney secures an abstract of title, he examines the various transfers of title and writes an opinion for his client as to whether he thinks the grantor has a good and clear title to the property. The attorney does not guarantee the title; he merely states his opinion from the facts shown in the abstract. He cannot guarantee that there has not been fraud or forgery.

An attorney may point out errors which were made in the execution of conveyances but which, in his opinion, will not affect the title. For instance, a deed was dated October 5, 1938, and the acknowledgment was dated October 4, 1938. The attorney's opinion was that the instrument had been recorded for more than ten years and the acknowledgment was cured by limitation. He further stated that, if necessary, an affidavit could be secured from the notary public involved to the effect that a stenographic error had been made in dating the acknowledgment.

22 AFFIDAVIT

An affidavit is a statement made under oath in the presence of a notary public or other authorized person. In the case of the misdated acknowledgment mentioned above, the notary public made a sworn statemet, in the presence of another notary public, to the effect that the acknowledgment was actually made after the grantor had signed the instrument.

23 TITLE INSURANCE POLICY

In recent years, the practice of preparing an abstract of title and the practice of submitting an opinion on the title has been replaced by the issuance of a title insurance policy, often referred to as a *title policy*. Title insurance policies assure the purchaser of real property that he has good title to the land he has purchased. These policies are issued by title abstract companies operating under the insurance laws of the state.

The amount the assured is guaranteed is usually limited to the amount he is paying for the property. If the property is enhanced in later years, he would receive no extra compensation for the increased value of the property if his title were defeated.

The policy is usually issued subject to certain exceptions: taxes, easements, encumbrances, oil royalties, etc.

24 HOMESTEAD RIGHTS

Many of the colonists in Stephen F. Austin's colony had left the United States because of a financial crisis of the time. Austin knew they needed time to establish themselves in Texas, and he appealed to the legislature of Coahuila-Texas for legislation to protect them from property seizure for old debts in the United States. A bill was passed in 1829 which exempted, without limitation, lands acquired by virtue of a colonization law from seizure for debts incurred before the acquisition of the land. The law had some precedence in the laws of Spain in the 15th century.

In 1839, the Third Congress of the Republic of Texas passed a law which read as follows: "Be it enacted ... that from and after the passage of this act, there shall be reserved to every citizen or head of a family in this Republic, free and independent of the power of a writ of fire facies, or other execution issuing from any court of competent jurisdiction whatever, fifty acres of land or one town lot, including his or her homestead, and improvements not exceeding five hundred dollars in value, all household and kitchen furniture (provided it does not exceed in value two hundred dollars), all implements of the husbandry (provided they shall not exceed fifty dollars in value), all tools, apparatus and books belonging to the trade or profession of any citizen, five milch cows, one yoke of work oxen or one horse, twenty hogs, and one year's provisions: ...."

This was the first law of its kind. It has since been adopted by most of the states, with realistic revisions in the protected quantities.

PART 4: Transfer of Ownership of Real Property

25 CONVEYANCE

A *conveyance* is a written instrument which tranfers ownership of property. It includes any instrument which affects the ownership of property. The term not only refers to a written document, but also means a method of transfer of property.

Changes in title law in early U.S. history were accompanied by changes in methods of conveying land. Deeds had to be in writing, but they were shortened and simplified. No longer was it necessary that deeds be written by lawyers learned in English law.

26 ESTATE

An *estate* in real property is an interest in real property. It can be complete and inclusive without limit or duration; it can be partial and of limited duration; it can be for the life of one person or for the life of several; it can include surface and all minerals below, or surface and no minerals below, or minerals but not the surface. It can be acquired in many ways: by purchase, by inheritance, by power of the state, by gift.

Several people may hold an interest in the same property. Consider a person who buys a house with a mortgage, and then leases the house to someone else. The lessee, the owner, the mortgagee, and various taxing agencies have an interest in the house.

27 FEE

The word *fee* comes from the feudal era and refers to an estate in land. The true meaning of the word is the same as that of "feud" or "fief." Under the feudal system, a freehold estate in lands came from a superior lord as a reward for services and on the condition that services would be rendered in the future. A fee and a freehold estate are the same.

28 FEE TAIL

Under the feudal system, a fee or freehold estate was passed on to the eldest son on the death of the fee holder. An estate in which there is a fixed line of heirs to inherit the estate is known as a *fee tail*.

29 ESTATE IN FEE SIMPLE ABSOLUTE

In U.S. law, an *estate in fee simple absolute* (also called an *estate in fee simple*) is the highest type of interest. It is an estate limited absolutely to a man, his heirs, and assigns forever without limitation. In other words a person who owns a parcel of land "in fee" can hold it, sell it, or divide it without limitations.

30 DEED

The most important document in the transfer of ownership of real property is the *deed*, which is evidence in writing of the transfer of an estate. A deed is a formal document. It needs not only to be in writing but also to be written by a person versed in the law. Deeds are of two principle types: Warranty deeds and quitclaim deeds.

31 WARRANTY DEEDS

In a *warranty deed*, the grantor proclaims that he is the lawful owner of the real estate and binds himself, his heirs, and assigns to warrant and forever defend the property unto the grantee, his heirs, and assigns against every person who lawfully claims it or any part of it. The warranty deed is the instrument used to convey an estate in fee simple absolute (in fee).

32 QUITCLAIM DEEDS

The *quitclaim deed* passes on to the grantee whatever interest the grantor has. If the grantor has a complete title, he passes on a complete title. If his title is incomplete, he passes on whatever interest he has.

33 ESSENTIALS OF A DEED

Because a deed is evidence of the transfer of an estate, the evidence must be clear and concise. The wording of the deed must clearly state the intent of the parties involved in the transfer. It is not sufficient that the grantor and grantee understand the terms of the transfer. In order to protect the rights of the real property owner and to establish an orderly method of transfer of real property, state legislatures and courts have adoted requirements for conveyance of such interest.

- A deed must be in writing. As previously mentioned, this requirment originated in the *statute of frauds* and is now found in the statutes of all states.

- A deed must be in legal terminology.

- Parties to a deed must be competent. A person of unsound mind or a minor cannot execute a deed.

- There must be a grantor and a grantee, and they must be clearly identified.

- There must be a valid consideration, although the total amount of the consideration need not be shown. Deeds containing the phrase "ten dollars and other consideration" provide evidence that the grantor received remuneration for his property.

- A deed must contain a description of the property being conveyed and clearly show the interest conveyed.

- A deed must be signed. In the case of joint ownership by husband and wife, both must sign.

- A deed must be acknowledged. The signer or signers of the deed must sign in the presence of a registered notary public who must know the identity of the signer or signers. The notary must sign the acknowledgment and affix a seal to it.

- A deed must be delivered. Centuries ago, land was conveyed by a ceremony known as *livery of seisin*. Parties to the transfer of ownership met on the property to be conveyed and performed such acts as handing over twigs and soil, driving stakes in the ground and shouting. The ceremony was practiced in England as late as 1845. Today, delivery of the deed is considered to be the delivery of the property.

34 RECORDING DEEDS

It is important that deeds recorded in order to constitute notice to the public. Unrecorded deeds may be valid, but to avoid future controversy, deeds should be recorded as soon after execution as possible. It is not necessary for the grantor to actually carry the deed to the grantee.

35 PATENT

A *patent* is a conveyance or deed from the sovereign for the sovereign's interest in a tract of land. Most, but not all, land in the United States was patented by the United States. The original thirteen colonies received grants from the King of England. Owners of land in Texas have received patents (grants) from the King of Spain, the Republic of Mexico, the Republic of Texas, and the State of Texas. The lands of Texas have never come under the ownership of the United States, and no patents have been conveyed from that source.

36 WILL

A will is a declaration of a person's wishes for the distribution of his property after his death. These wishes are carried out by a probate court. A *devise* transfers real property, whereas a *bequest* transfers personal property. The devisee is the person receiving the real property. The *probate court* will distribute the property according to the wishes of the *testator* (the deceased) if a will exists. If no will exists, the court will distribute the property in accordance with the law of descent and distribution. Widows and children come first in this succession.

A will may devise certain property to certain individuals, or it may devise an entire estate to several heirs. In the latter case, the heirs will own the undivided property jointly.

Before property can be transferred under the terms of the will, the heirs must submit the will to a *probate court* or a county court which has probate jurisdiction. If the will designates an *executor*, the court will recognize him. If no will exists, the court will appoint an administrator. Heirs of an estate must then file an inventory of the property of the estate. Public notice must be given to creditors of the estate and these claims must be paid, if valid. State and federal taxes must also be paid before final settlement of the estate.

In Louisiana, the *forced heirship law*, based on the Napoleonic Civil Code, decrees that children are entitled to a testator's property regardless of the terms of the will. Under this law, children are entitled to one-fourth of the estate. If there are no children, parents are entitled to this one-fourth.

37 HOLOGRAPHIC WILL

A *holographic will* is a will in the handwriting of the deceased.

38 EASEMENT

An *easement* is the right which the public or an individual has in the lands of another. An easement does not give the grantee a right to the land—only a right to use the land for a specified purpose. The owner of the land may also use it for any purpose that does not interfere with the specified use by the grantee.

Utilities wishing to install power lines, underground pipe, canals, drainage ditches, etc., sometimes do not require fee title to land but need only the use of the land to install and maintain the facility. The owner of the land retains title to it, subject to the terms of the easement.

39 LEASE

A lease must be for a certain term, and there must be a consideration. It is a contract for exclusive possession of lands or tenements though use may be restricted by reservations. The person who conveys is known as the *lessor* and the person to whom the property is conveyed is known as the *lessee*. Both parties must be named in the lease.

In many cases a tenant holds real estate without a lease, paying rent every week, month, or year. This is known as *tenancy without lease*.

In general, whatever buldings or improvements stand upon the land and whatever grows upon the land belongs to the landlord. Under a lease, the tenant is entitled to the crops of annual planting.

40 SHARECROPPER'S LEASE

A lease of farmland wherein the landlord and tenant each receive a predetermined share of the total income from crops on the land is known as a *sharecropper's lease*. The share to each is usually determined by custom in certain areas, but it can be set at any figure by agreement between the two.

41 OIL LEASE

An oil company or private individual may enter into an agreement with a land owner to remove oil, gas, or other minerals from the land and to share the profits from the sale of the minerals with the landowner. This is known as a *mineral lease* or *oil lease*. The shares are usually set by the company which removes the minerals, and this share has been accepted by custom.

An oil lease is for a definite number of years (often five), stipulating rental on a per acre, per year basis. In addition to the yearly rental, the agreement usually includes a bonus paid by the oil company at the beginning of the lease period. This also is usually on a per acre basis. If drilling has not commenced by the end of a specified period, the lease expires.

42 MORTGAGE

A *mortgage* is a conditional conveyance of an estate as a pledge for the security of a debt. A person borrowing money to purchase property guarantees that he will repay the lender by making a conditional conveyance to the lender. If he repays the loan as specified, the mortgage becomes null and void. If he does not pay the loan as specified (a *default*), he must deliver the property to the lender.

43 DEED OF TRUST

A *deed of trust* is a mortgage which gives the creditor the right to sell property, in case of default, through a third person known as the trustee. Early American law regarding mortgages included a complicated system of equitable foreclosure to give the debtor protection. It included the debtor's "equity" which gave him the right to redeem his land after it had been foreclosed on. This "equity" created difficulties for the lendor, and in time, laws in many states were modified so that if the debtor agrees in advance, the creditor can sell the property through a third person, known as the trustee, without going through court, in case of default.

44 CONTRACT OF SALE

Often the sale of a large estate involves many complexities which are time consuming for the parties involved and their attorneys. A thorough examination of the complexities of the transaction in advance can save time which otherwise might be spent in court settling a dispute.

In order that the buyer may express his intent to buy and the seller may express his intent to sell, the two parties may enter into a *contract of sale* which describes the property involved and the terms of the sale and specifies a date, not later than which the transfer of property must be completed. This contract usually stipulates that the seller will furnish a good and merchantable title to the buyer by a warranty deed, and that if the seller cannot furnish such deed, the contract is null and void. The contract also often provides for an *escrow fund*, which the buyer will forfeit if he does not carry out the terms of the contract.

In some instances, real property is sold by contract of sale and all payments are made before the deed is executed. Throughout the period of the payments, the title remains in the name of the seller and his name appears on the tax roll as owner of the property. Contracts of sale are frequently not recorded.

45 UNWRITTEN TRANSFERS OF LAND OWNERSHIP

The *statute of frauds* requires that all conveyances of real property be in writing. But, if that statute would deprive the rightful owner of his property, then the law may be set aside and an unwritten transfer of real property may take place.

This transfer may take place by expressed or implied agreement such as by the principle of recognition and

PROFESSIONAL PUBLICATIONS, INC. • P.O. Box 199, San Carlos, CA 94070

acquiescence over a long period of time, by dedication, by adverse (hostile) relationships, or by acts of nature.

A legal unwritten transfer of title supercedes written title and will extinguish written title. Evidence to prove the location of a written title will not overturn a legal unwritten title.

46 RECOGNITION AND ACQUIESCENCE

Acquiescence in a boundary line is evidence from which it may be inferred that the parties by agreement established that line as the true line. From such acquiescence, a jury or court may find that the line used is the true line. Acquiescence in a line other than the true line will not support a finding of an agreement establishing the line as the boundary when there is no evidence of agreement other than acquiescence and where it is shown that the use of the line resulted, not from agreement, but only from a mistaken belief of the parties that it was the true line.

47 DEDICATION

Dedication is the giving of land or rights in land to the public. It must be given voluntarily, either expressed or implied. It may be written or unwritten, but there must be acceptance of the dedication. A consideration is not necessary.

Common law dedication may be expressed, as when the intention to dedicate is expressed by a written document or by an act which makes the intent obvious. It may be implied, as when some act or acts of the donor make it reasonable to infer that he intended to dedicate.

Dedication made in accordance with the provisions of a statute is called *statutory dedication* and usually requires that the donor sign and acknowledge the dedication.

The developer of a subdivision may subdivide a tract of land, lay out streets, lay sewer lines and water lines, and pave streets. He may then turn over the use of these facilities to the public. The facilities must be accepted for use by the public by the state, city, town, or other governing body. Other examples of land dedicated to the public include parks, cemeteries, and schools.

48 ADVERSE POSSESSION AND TITLE BY LIMITATION

Transfer of property may occur without the agreement of the owner by the method known as *adverse possession*. Adverse possession is the acquisition of title to property belonging to another by performing certain acts. The rights to acquire property in this manner are often referred to as *squatter's rights*.

Requirements for transfer of title by adverse possession vary among the states but are essentially the same.

- Possession of the land by the person claiming it from another must be actual. Generally, possession must be such that the owner will be aware of the possession if he visits the property.

- Possession must by open and notorious. Possession so open, visible, and notorious that it will raise the presumption of an adverse claim is the equivalent of actual knowledge. The land must be occupied in a straightforward, not clandestine, manner.

- Possession must be continuous. Statutes vary among the states as to the period necessary to establish title from adverse possession, but the land in question must be held continuously for the period required by statute.

- Possession is required to be exclusive. This means that the person making the claim cannot share the possession with the owner or others. He must have complete control of the property.

- The possession must be hostile. The claimant must possess the land as if he were the owner in defiance of the owner.

49 ADVERSE POSSESSION USED TO CLEAR TITLE

In modern times, people seldom squat on land with the intention to acquire title by adverse possession (although there have been many instances in the past when this has occurred.) The importance of adverse possession today is in its use to clear up defects in title or to settle boundary disputes between adjacent land owners.

Honest differences may occur between adjacent owners as to where the boundary between them actually is. Monuments and landmarks may be obliterated; changes in the location of fences, ditches, and roadways may have occurred. Adverse possession allows a landowner to establish his boundary without a costly survey.

50 RIGHT OF THE STATE AGAINST ADVERSE POSSESSION

Title to state or public land generally cannot be acquired by adverse possession.

51 TRESPASS TO TRY TITLE ACTION

The action usually taken by the record owner of land against a person in adverse possession of land is known as *trespass to try title*. The record owner brings suit against the person in possession for recovery of the land and for damages for any trespass committed. If the court rules in favor of the plaintiff, the person in possession is evicted, but if the court rules in favor of the defendant, the defendant acquires a good title to the land.

52 PRESCRIPTION

The method of obtaining easement rights from long usage is known as *prescription*. A person may travel across a tract or parcel of land for a period of time required by the statute of limitations and acquire a right to continue the act of using the land. The act of using the land must have been open, continuous, and exclusive for the period of time required.

A highway right of way can be acquired by the state if it has been used by the public for a long period of time. As with individual acquisition, the use must be open and continuous for the required period of time.

53 RIGHT OF EMINENT DOMAIN

When the owner of land refuses to sell, and the improvement is of public character, the law allows that land shall be taken under what is called the right of *eminent domain*. Eminent domain gives the state, or others delegated, the right and power to condemn private property for public use.

The constitution of the United States and state laws limit eminent domain. The owner is guaranteed adequate compensation for his property, and he may not be deprived of his property without *due process of law*.

The power to exercise eminent domain must be authorized by the state legislature by statute, and the legislature may delegate this power to such agencies as it deems proper. Counties, incorporated cities and towns, water districts, and school districts have been delegated the power of eminent domain. The power is also given to private corporations which are engaged in public service.

The owner of condemned property must be fully compensated for the property. When only a part of his property is taken, he is entitled to compensation for *consequential damage*. (A highway which cuts off access to a watering tank for cattle might create consequential damage.)

The owner is entitled to know the precise boundaries of the land to be condemned. It is the obligation of the agency executing the acquisition to furnish an adequate description of the boundaries.

Before the right of eminent domain can be exercised, it is essential that no purchase agreement be reached between the parties. It is necessary that the state or city or other governing body make the owner an offer which he refuses, and that the owner shall name his price, which the state or city refuses, or else that the owner refuses to name the price. There must be a definite failure to agree. After disagreement, the state or other body must initiate *condemnation proceedings*.

54 ENCROACHMENT

An *encroachment* is a gradual, stealthy, illegal, acquisition of property. By moving a fence a small amount over a period of years, an adjoining owner may acquire from the lawful owner a strip of land.

55 ACTION TO QUIET TITLE

Where the boundary between adjacent landowners is not clear, or where there is a dispute over the location of the boundary line, one of the parties can sue the other to determine the location of the line. Either party may employ a surveyor as an expert witness. The judgment in the lawsuit becomes a public record and will be reflected in abstracts of title.

56 COVENANT

An agreement on the part of the grantee to perform certain acts or to abstain from performing certain acts regarding the use of property which has been conveyed to him is known as a *covenant*. Developers of residential property, in order to assure buyers that their neighborhood will be pleasing to the eye and pleasant to live in, require the buyer to accept certain restrictions as to the use of property he buys. These restrictions include such things as type of building construction, minimum distance between house and property line, minimum number of square feet in floor plan, use of the property, and kinds of animals allowed on premises. These covenants are sometimes called *deed restrictions*.

57 LIEN

A *lien* is a claim or charge on property for payment of a debt or obligation. It is not the right of possession and enjoyment of property, but it is the right to have the property sold to satisfy a debt. Mortgages and deeds of trust constitute liens.

58 TAX LIEN

States, counties, and other governing bodies impose taxes on real property which gives them a first lien on the property. Failure to pay taxes gives the governing body the right to have the property sold to satisfy the tax debt. Failure to pay income taxes gives the Federal government the same right. Before purchasing real estate, the buyer may obtain a *tax certificate* in which the tax collector certifies that there are no unpaid taxes on the property up to a certain date.

59 PROMISSORY NOTE

A *promissory note* is the written promise of the borrower to pay the lender a sum of money with interest. The principal sum, the interest rate, and a schedule of dates of payment are included on the face of the note. Also included in some notes is a listing of the *security* for the note—the property which is to be mortgaged to guarantee payment of the note. The same information as to principal, interest, and schedule of payments shown on the note is, where applicable, shown in the deed of trust, and reference is made in the deed of trust to the promissory note between the two parties involved.

A deed of trust does not necessarily accompany all promissory notes, and a promissory note need not list any security. A lender may, if he wishes, lend simply on a person's *personal note*, which is his promise to pay. But this does not prevent the lender from legal action to collect the amount of the note or to obtain property of equal value in the event the borrower does not pay the note.

PART 5: Ownership of Beds of Rivers and Streams

60 COMMON LAW

Both English common law and Spanish civil law pertaining to ownership of the beds of rivers and streams are based on Roman civil law and follow the same rules in certain particulars. American common law has not followed English common law exactly because there are wide variations in the interpretation of English common law.

It appears that early English law was concerned only with the water in streams and the public right to use it, such as for fishing. Ownership of the beds of streams is not clear except for the beds of streams which are affected by the tide. Under English common law, title to the beds of streams is retained by the sovereign in so far as the waters of the stream are affected by the tide. Ownership beyond the point where the waters of the stream are affected is not clearly defined, and *navigability of streams* seems to be of no consequence.

One theory as to the reason for the lack of classification of streams as navigable or non-navigable is that in England there were no navigable streams except where the tide affected them.

It has been stated previously that, for the most part, property law is state law. This is also true in the matter of ownership of the beds of streams and the water in streams. Because of the vagueness of the English common law, different states have made different interpretations. Even on the same river, adjoining states may have different laws as to the ownership of the bed.

Differences between states pertain mostly to ownership of the beds of navigable streams. Under both English common law and Spanish law, grants bordering on non-navigable streams extend to the center of the stream. In other words, beds of non-navigable streams are privately owned. But in most states, the beds of navigable streams are owned by the state.

Some states consider streams to be navigable in law if navigable in fact. Other states have defined a navigable stream by a legislative act.

61 SPANISH AND MEXICAN LAW

In Spanish law, the sovereign owns both the water and beds of *perennial streams* whether they are navigable or not. The beds of non-perennial streams, called *torrential streams*, belong to the adjacent owners, each owning to the center of the stream. Perennial streams are considered to have continuous flow, except in periods of drought. Non-perennial streams flow after heavy rains or melting snow.

62 FEDERAL LAW

While the waters and beds of navigable streams are the property of the states, the Federal government has absolute supremacy over navigation on navigable streams within a state.

The Federal government can take action against a state or individual for any acts that might interfere with or diminish the navigability of streams. To insure commerce between the states, state legislatures are barred from enacting legislation which would interfere.

63 ISLANDS

The ownership of an island in a stream is with the owner of the bed of the stream. If the stream is navigable, the island belongs entirely to the state. If an island is on one side of a non-navigable stream—between one bank and the middle of the stream—the island belongs to the riparian owner on that side. If an island is formed in the middle of a non-navigable stream, the island is owned by both riparian owners to the middle of the stream.

64 THREAD OF A STREAM

The *thread of a stream* is a line equidistant from its shores when the stream is at ordinary stage. It is not necessarily in the middle of the main channel. The U. S. Supreme Court found the thread of the river to be midway between the cut banks when the river is at a level such that it just covers the stream bed.

65 MEANDER LINES

Surveyors run meander lines in order to plat a stream. Where a grant or patent calls for the shore of a stream or for the middle of the stream, it is this line which is the boundary and not the meander line. Meander lines, run for patents to United States lands, were used in computing the acreage to be paid for by the grantee. Often no charge was made for land between the meander line and the stream itself.

PART 6: Ownership of Tidelands and Lake Beds

66 TIDES

Waters of the oceans are attracted by the sun, moon, and planets. This attraction causes the rise and fall of the surface of the sea (the tide). When the sun, moon, and the earth are in line and pulling in the same direction, the tides are highest and called *spring tides*. When the attractions of the sun and the moon are in opposite directions, the tides are lowest and are called *neap tides*.

Along the Pacific and Atlantic coasts, there are two high and two low tides during a tidal day. Along the Gulf of Mexico, there is usually one high and one low tide each tidal day.

67 TIDAL WATERS

Waters in which the tide ebbs and flows are known as *tidal waters* or *coastal waters*. Salt water flats which are alternately covered and uncovered as the tide ebbs and flows are also considered to be covered by tidal waters. A salt water marsh which is not an integral part of a bay but which is affected by the ebb and flow of the tide only because of a ditch or ditches which have been excavated for the purpose of drainage are not considered to be tidal waters.

Rivers and streams are considered to be tidal waters to the extent that the waters in these streams are affected by the ebb and flow of the tide.

68 OWNERSHIP OF TIDELANDS

Lands under tidal waters, known as *tidelands*, are generally owned by the state.

69 DIVIDING LINE BETWEEN TIDELANDS AND PRIVATE LANDS

Under common law, the line between privately-owned uplands and state-owned tidelands, known as the *shoreline*, is the line of mean high tide. Under Spanish law, the shoreline is the line of the high tide during the year.

70 OWNERSHIP OF OFFSHORE SUBMERGED LANDS

Until 1947, coastal states claimed ownership of offshore submerged lands. In a suit brought against California, Louisiana, and Texas by the United States, the U. S. Supreme Court held that these lands belonged to the Federal Government. In 1953, by an act of Congress, these lands were returned to the states.

71 OWNERSHIP OF LAKE BEDS

The common law rule is that ownership of the bed of a non-navigable lake is with the littoral owner, extending to the center of the lake unless the conveyance specifies otherwise.

Beds of navigable lakes belong to the state. To be navigable, the lake must be navigable in fact to the extent that it is used as a highway for commerce (not just for pleasure boating).

PART 7: Riparian and Littoral Rights

72 RIPARIAN AND LITTORAL OWNERS

A person who owns land abutting on a body of moving water is known as a *riparian owner*, and he has certain rights to, or in, the water. His rights rest solely in the fact that his land abuts on the water.

A person whose land does not abut on a body of water, even though there is a very small tract of land between his land and the water, is not considered to be a riparian owner.

A person who owns land on a body of water not in motion, such as a pond, lake, gulf, or ocean, is known as a *littoral owner*. He also has rights in the body of water.

73 RIPARIAN RIGHTS

The riparian owner may use the water in the stream for any reasonable use to which the stream is adapted. These include domestic water, water for animals, fishing, boating, and swimming. Many of these rights are rights in common, and one owner may not interfere with another's use of the water. Riparian owners have the right to use water for irrigation of crops within the limits of the amount of water available. Riparian rights also include rights to stream beds and to the alluvium below the water.

74 LITTORAL RIGHTS

Littoral rights are very much the same as riparian rights but differ among the states. Littoral owners on the ocean or gulf have the right to use the water and beach for bathing and boating, but they do not have exclusive rights.

75 NATURAL CHANGES IN STREAMS, LAKES AND TIDELANDS

Rivers are constantly changing their boundaries with the rise and fall of the water. Lakes also change their shoreline with the rise and fall of the surface of the water, and the shoreline of tidelands changes with the tide and storm waters. These changes have made it necessary for the law to provide rules pertaining to these changes as they apply to land ownership.

Both English common law and Spanish law follow Roman law in regard to changing boundaries in streams, lakes, and tidal waters caused by nature. So, courts have not found it necessary to distinguish between English common law and Spanish law in cases of this kind.

76 ALLUVIUM

Rivers carry silt, pebbles, and rocks as they travel to the sea. As the flow of the river rises and falls, the amount of material it carries rises and falls. The deposits made by water on a shore are known as *alluvium*.

77 ACCRETION

When land is formed slowly and imperceptibly by alluvium, the process is known as *accretion*. This build-up of alluvium may be on the banks of rivers and streams, or on the shores of lakes and tidal waters.

78 RELICTION

The gradual withdrawal of water which leaves land uncovered, such as the shore of a lake gradually receding, is known as *reliction*.

79 AVULSION

The sudden and perceptible change of the course of a river or stream such as the stream forming a new channel across a horseshoe bend is known as *avulsion*.

80 BOUNDARY CHANGES CAUSED BY ACCRETION, EROSION, RELICTION, AND AVULSION

Where accretion, erosion, or reliction occurs along a river or stream, the boundary of the riparian owner does not remain fixed but changes with the change in the gradient boundary. This means that where accretion occurs, the riparian owner, along whose land the accretion is joined, gains this land.

Where erosion occurs, the riparian owner loses the land which may have formed accretion down stream.

Where reliction occurs, the boundary of the riparian or littoral owner changes with the boundary of the stream or with the shoreline of a lake or of tidal waters. Land left uncovered by reliction is owned by the riparian or littoral owner.

When avulsion occurs along a navigable stream, the owner of the land across which the new channel was formed loses title to the bed of the new channel to the state, but ownership of the land between the new and old channels does not change. Title to the bed of the old channel passes to the riparian owners from the state, unless the stream uses both channels, in which case the state has title to the bed of both channels.

It is probably true that land adjoining tidal waters that has been formed by accretion or reliction is owned by the party owning the adjoining land and not by the state, although this has not been established by the courts. Likewise, it is probably true that land belonging to the private owner which has been eroded or submerged by tidal waters belongs to the state.

PROFESSIONAL PUBLICATIONS, INC. • P.O. Box 199, San Carlos, CA 94070

PART 8: Surveys of the Public Lands

81 GENERAL

Thirty states of the United States were subdivided into rectangular tracts by a system known as the United States System of Rectangular Surveys. In 1785, Congress enacted a law which provided for the subdivision of the public lands into townships six miles square, with townships subdivided into thirty-six (36) sections, most of which are one mile on a side. Sections were subdivided into half-sections, quarter sections, and quarter-quarter sections (the quarter-quarter section being 40 acres in area).

The other states did not pass title to vacant lands to the United States. These states are Texas, West Virginia, Kentucky, Tennessee, the Colonial States, and the other New England and Atlantic Coast states except Florida.

82 QUADRANGLES

A quadrangle is approximately twenty-four (24) miles square and consists of sixteen (16) townships. Quadrangles were laid out from an initial point through which was established a *principal meridian* and a base line extending east and west which is a true parallel of latitude. All north-south township lines are true meridians, and all east-west township lines are circular curves which are parallels of latitude. Because of the convergence of meridians, quadrangle corners do not coincide except along the principal meridian.

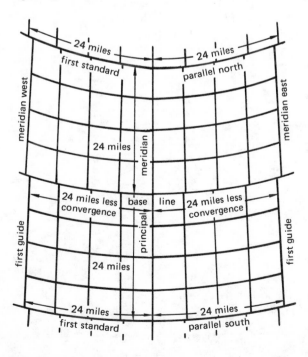

Figure 25.1   Quadrangle divided into townships

83 SUBDIVISION OF TOWNSHIPS

Townships were divided into sections numbered from 1 to 36 beginning in the northeast corner and ending in the southeast corner as shown in figure 25.2. As many sections as possible with one mile on a side (640 acres) were laid out in the township. But, due to the convergence of the east and west boundaries of a township, it was impossible for all thirty-six sections to be one mile on each side. East and west section boundaries were laid out parallel and not as true meridians. They were laid out parallel to the east boundary of the township. This made it impossible for all sections to be one mile on a side and at the same time coincide with the township lines.

In order to produce as many sections as possible one mile on a side, the sections along the north township line and the west township line were of varying dimensions to compensate for errors and the convergence of the west township line. The errors were actually thrown into the north one-half of the sections along the north township line and into the west one-half of the sections along the west township line. Thus, sections 1–6, 7, 18, 19, 30, and 31 were not regular sections. When a section was limited by a lake, river, or old survey, part of it was eliminated, but the existing section was numbered as if the whole section were laid out.

	changing error compensated				
6	5	4	3	2	1
7	8	9	10	11	12
18	17	16	15	14	13
19	20	21	22	23	24
30	29	28	27	26	25
31	32	33	34	35	36

(left side: chaining and convergence compensated)

Figure 25.2   Township subdivided into sections

84 SUBDIVISION OF SECTIONS

Sections may be divided into half sections, quarter sections, half-quarter sections, or quarter-quarter sections

as shown in figure 25.3. Locating small fractions on a
sketch from the description is made easier by following
the description in reverse order.

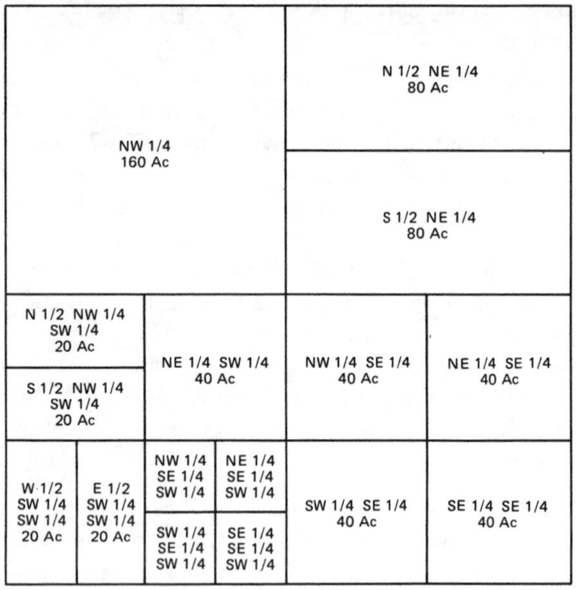

Figure 25.3 Regular section divided into fractions

A section subdivided into quarter-quarter sections is
shown in figure 25.4. Each quarter-quarter section is
40 acres. They are numbered from 1 to 16 starting
in the northeast corner of the section and ending in
the southeast corner, just as townships are numbered.
Sections subdivided in this manner are sometimes called
blocks, and the quarter-quarter sections are called *lots*.
Thus, the northeast quarter-quarter of Section 8 would
be called Lot 1, Block 8. The southeast quarter-quarter
would be called Lot 16, Block 8.

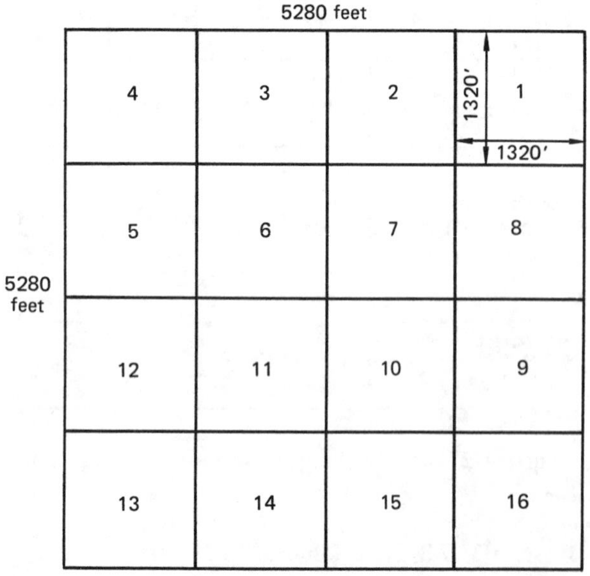

Figure 25.4 Subdivision of section
into 40 acre lots

Table 25.1
Monument markings

Mark	Meaning
AM	Amended monument
AMC	Auxiliary meander corner
AP	Angle point
BO	Bearing object
BT	Bearing tree
C	Center
CC	Closing corner
E	East
LM	Location monument
M	Mile
MC	Meander corner
N	North
NE	Northeast
NW	Northwest
PL	Public land (unsurveyed)
R	Range
RM	Reference monument
S	Section
S	South
SC	Standard corner
SE	Southeast
SMC	Special meander corner
SW	Southwest
T	Township
TR	Tract
W	West
WC	Witness corner
WP	Witness point
1/4	Quarter-section
1/16	Sixteenth-section
1916	Date (year)

PART 9: Boundary Law

85 RESURVEYING

One of the difficult tasks for a surveyor is the resurveying of lands, the relocation of the boundary lines between privately-owned lands or the relocation of the boundary between two political entities. The original descriptions of boundary lines found in deeds and grants are often vague or in error.

English kings who made land grants to the original colonists often were vague in describing the boundaries. This often resulted in overlapping boundaries. Some of the original colonies claimed land to the Pacific coast. Old descriptions included reference to landmarks which disappeared many years before resurveys were made.

Old surveyors were not equipped with sophisticated instruments, and units of measurements used were not the same as those of today. All these things have presented the surveyor with problems which, in many cases, have solutions only in the Courts. Court decisions, in many boundary disputes over many years, have made a pattern which has established general rules for the location of boundaries.

86 RULES FOR LOCATING BOUNDARIES

When a new tract of land is established from an old tract of land by subdivision, the boundaries of the subdivision become, in part, old boundaries and new boundaries. New boundaries are dependent on old boundaries. Since property laws are state laws, the rules for locating boundaries may differ from state to state.

87 INTENT

Of first importance in boundary location is the intent of the parties to the deed. What did the grantor intend to grant to the grantee? This intent must be determined from the wording of the deed and the description of the land contained therein. In court cases involving boundary disputes, decisions of the court have repeatedly upheld the principle that the written intentions of the parties to the original transaction are of first importance.

88 METES AND BOUNDS DESCRIPTIONS

Metes and bounds refer to the measurements of the limits or boundaries of a tract. This type of description identifies a beginning point and then describes each course of the tract in sequence in either a clockwise or a counter-clockwise direction and returns to the beginning point. The description includes not only the direction and length of each course, but includes calls for monuments and adjoiners.

89 RETRACEMENT

In locating lost or disputed boundaries, the surveyor's job is to determine where the boundaries were actually located, not where they should have been located. The primary objective in locating a survey is to 'follow the footsteps of the surveyor'. In a suit involving boundary question, search must be made for the footsteps of the original surveyor, and when found, the case is solved.

After boundaries have once been fixed by monuments, they cannot be changed because of errors of distance or direction unless fraud is involved.

90 CALLS

A *call*, sometimes referred to as a *deed call*, is a phrase in the written description of the location of a parcel of land contained in the body of a conveyance.

Calls usually start with the word BEGINNING, for the first call, or THENCE for each succeeding call, written in capital letters with each call separated from another as a paragraph. Calls may refer to the direction and distance of a side of the perimeter of a tract of land, to an adjoining tract of land, to a monument at a corner, to the area contained in a tract of land, or to natural objects along the way which a surveyor may use to better identify the location of the land.

91 LOCATIVE CALLS

Calls which give the exact location of a point or line are known as *locative calls*. A call for a monument referenced to witness trees or other objects is a locative call.

92 PASSING CALL

A call which refers to a creek, highway, fence, tree, or other object which is crossed or passed in a survey is called a *passing call*. It does not serve as a locative call but better identifies the location of the land.

PROFESSIONAL PUBLICATIONS, INC. • P.O. Box 199, San Carlos, CA 94070

93 CALL FOR A MONUMENT

A *call for a monument* describes a natural or artificial object at the corner of a survey or describes a course of the survey as being along a natural or artificial monument such as a stream or highway.

94 MONUMENTS

Monuments may be natural or artificial. Natural monuments include rivers, lakes, oceans, gulfs, bays, trees, and large boulders. Artificial monuments include stakes, fences, concrete markers, etc.

95 CALL FOR ADJOINER

Calls for the lines and corners of an adjacent or adjoining survey are known as *calls for adjoiner*.

96 CALL FOR DIRECTION AND DISTANCE

A call giving the bearing and distance of a line in a survey is known as a *call for direction and distance*.

97 CALL FOR AREA

A *call for area* gives the area contained in the survey.

98 RULES OF COMPARATIVE DIGNITY

As mentioned previously, the intent of the parties to a deed or other instrument must be determined from the wording of the deed. The description of the location of the land is contained in the various calls. These calls are often in conflict. The decisions of courts over many years have established an order of precedence for the various types of calls in a description of the location of a tract of land. This order of importance is expressed as "rules of comparative dignity." These rules are the rules of logic because the courts have carefully weighed the possibility of error in various types of calls and have given a value of reliability to each type of call. The order of importance of calls, where there is conflict between them, is given here in order of decreasing importance.

- Call for natural monument
- Call for artificial monument
- Call for adjoiner
- Call for direction
- Call for distance

- Call for area

The courts found it logical to assume that the most important call is the call for a natural object because of the permanence of the object. The next most important call is the artificial object because it can be identified by the description in the call. Courts are aware that errors can be made in determining direction and distance. If a description gives a bearing and distance which should bring the surveyor to a river which is called for as the boundary, but in fact does not bring the surveyor to the river, it is logical to assume that the call for direction and distance is incorrect if the intent of the parties was to make the river the boundary.

Courts also have found that it is logical to assume that a call for direction is of higher importance than a call for distance because direction is determined by the surveyor, whereas distance is determined by his chainpersons who may not be of the same competence.

A call for adjoiner is usually given more importance than a call for direction and distance because where an adjoining line is well established, it can be considered to be a monument.

Area is of least importance, but where a call for area brings other calls into harmony, it may be considered of higher importance.

Where no calls for monuments exist or where monuments cannot be found, a call for direction and distance may control.

A call for a monument does not always control if there are two monuments in conflict. The monument which is in harmony with direction, distance, and the other calls will control.

99 PAROL AGREEMENT

When the boundary between two owners is not distinct but there is no disagreement as to its location, the boundary can be fixed by *parol agreement* between the two owners—the word "parol" meaning "by word of mouth."

100 POSSESSION AND SENIOR RIGHTS

Possession that ripens into a fee right, such as acquisition by adverse possession, may make calls meaningless. For instance, property acquired by the statute of limitations would have the boundaries actually acquired under the statute and would not necessarily conform to any previously described boundaries.

If two parties both have deeds to the same land (such as an overlap), the *senior right* holds. In other words, the person with the first deed has the best deed.

101 MAP OR PLAT AS A REFERENCE

A map or plat referred to in a deed becomes a part of
the description of the land conveyed as if it were a part
of the deed. Monuments, bearings, and distances shown
on the map become a part of the description.

PART 10: Subdivisions

102 DEFINITION

The act of subdivision is the division of any tract or parcel of land into two or more parts for the purpose of sale or building development.

103 REGULATION

Resurveys are made to determine what has taken place in the past. On the other hand, surveys made for the purpose of planning subdivisions are creative in nature, and the care and imagination used in planning affect the entire community for many years to come. There are many instances in the United States in the past where developers have created subdivisions which are a credit to their foresight and integrity without any regulatory laws. But, increasing population and decreasing availability of land for development have made it necessary for the states and the Federal government to adopt laws regulating land divisions.

104 SUBDIVISION LAW AND PLATTING LAW

Subdivision law includes regulations for land use, types of streets and their dimensions, arrangement of lots and their sizes, land drainage, sewage disposal, protection of nature, and many other details. Platting law includes regulations for recording the subdivision plat, monumenting the parcels, establishing the accuracy of the survey, and means of identifying the parcels and their dimensions.

105 PURPOSE OF SUBDIVISION LAW

Creation of a subdivision involves more than furnishing a location for a home for a new member of a community and a profit for the developer. It requires planning for traffic, transportation, the location of schools, churches, and shopping centers, and the health and happiness of the citizens of the community. Poorly planned subdivisions have caused cities to spend excessive amounts for street widening and resurfacing, reconstruction of sewer lines, establishment of additional drainage facilities and increase in size of water mains. Poor planning in the past has caused an acute awareness among state, county, and municipal officials of the need for the regulation of subdivision development.

106 THE CITY AS THE REGULATORY AUTHORITY

The authority to enforce subdivision regulations lies for the most part with cities. Counties have legal authority to regulate subdivisions, but most of the regulation is enforced by cities. Cities have the authority to enforce regulations by means of an ordinance adopted by the city council.

107 CERTAINTY OF LAND LOCATION

Monuments set by the original surveyor and called for by the conveyance have no error of position. An interior monument in a subdivision, after it is used, is correct, but boundary monuments of a subdivision marking the line between an adjacent owner, if not original monuments, can be located in error. Establishing a subdivision does not take away the rights of the adjoiner. It is the surveyor's role to eliminate future boundary difficulties. An important part in eliminating future disputes is the establishment of a precise control traverse. It should be established before any detailed planning is made.

108 MONUMENTS

When the control traverse has been established and preliminary approval of street right-of-way widths and locations has been obtained from municipal authorities, monuments should be set on property corners and street right-of-way lines based on the control survey. Control monuments which are to be used to relocate lost corners should be permanent and indestructible.

109 BOUNDARY SURVEY

The first step in subdividing is the establishment of the control traverse which is the base for determining the boundaries of the tract. Investigation as to any conflict with senior title holders should be made. After certainty of location is established, the corners should be monumented.

110 TOPOGRAPHIC MAP

The next step is to establish a system of bench marks for vertical control and to prepare a topographic map. The map is essential in planning the subdivision, especially in regard to drainage and sanitary sewer plans.

111 THE PLANNING COMMISSION

Most cities deal with developers of subdivisions through a planning commission. The next step for the developer is to contact the planning commission of the city (or any other designated approving agency) for consultation. Many planning commissions require, at the outset, a subdivider's data sheet which indicates the general features of his plan and a location map which locates the proposed subdivision in relation to zoning regulation and to existing community facilities.

112 GENERAL DEVELOPMENT PLAN

After preliminary discussion, the developer should submit a penciled sketch showing contours, street locations, and lots. If the planning commission feels that the development plans fit into the city's overall plan and into the surrounding neighborhood, it will ask for a preliminary plat.

113 PRELIMINARY PLAT

The *preliminary plat* is actually the detailed plan for the subdivision and includes names of the subdivider, engineer, or surveyor, a legal description of the tract, location and dimensions of all streets, lots, drainage structures, parks, and public areas, easements, lot and block numbers, contour lines, scale of map, north arrow, and date of preparation. After approval of the preliminary plat, the developer can proceed with stake-out and construction operations.

114 FINAL PLAT

The final plat conforms to the preliminary plat except for any changes imposed by the planning commission. It must be prepared and filed for record in accordance with platting laws of the state. It establishes a legal description of the streets, residential lots, and other sites in the subdivision.

PART 11: Planning the Residential Subdivision

115 STORM DRAINAGE

The first step in planning subdivision drainage is a careful study of the contour map. Storm sewers and sanitary sewers are designed for gravity flow. Streets act as drainage collectors for storm runoff. Therefore, a general plan for storm sewers and sanitary sewers should be formulated before the streets are finally located.

Lots should drain to the street or to some open drainage system in the area. Lots should never receive drainage from the street. The Federal Housing Administration has set up requirements for lot drainage for various types of topography where this agency guarantees loans. These requirements call for the lot to slope away from the house for some distance in all directions. The ideal situation is for the lot to slope from the house to the street in the front of the lot and from the house to some drainage collector at the rear of the lot if one exists. Concrete alleyways with inverted crowns sometimes serve this purpose. Topography sometimes makes it necessary for the lot to slope from front to rear or from rear to front, but the house is always located so that the slope is away in all directions.

Storm water is carried along the gutter which necessitates a curb. To avoid water collecting in pools, streets should have minimum grade of 0.3% and preferably 0.5%. Water should not run across streets in valleys. Inlets should be planned for both sides of the street at intersections.

The design of storm sewers depends on the amount of surface runoff to be carried in the storm sewers. The amount of rainfall which is absorbed by the surface soil depends on the perviousness of the soil, the intensity of the rainfall, the duration of the storm, and the slope of the surface. Water which is not absorbed is known as *surface runoff*. Prior to subdivision, farm or grazing land may absorb a large percent of a given rainstorm. Streets, driveways, sidewalks, and rooftops, however, will cause most of the rain water to run off rather than be absorbed by the soil. The developer must be conscious of this fact in order to prevent damage to his own development and to property downstream.

Factors included in the design of storm drains include area of drainage area in acres, shape of drainage area, slope of land in drainage area, use of land in drainage area (present and future), maximum intensity of rainfall, and frequency of maximum intensity.

116 SANITARY SEWERAGE

Septic tanks are undesirable in subdivisions and should

not be used. The most desirable solution to sewage disposal is a collecting system connected to a municipal sewage disposal system. The second most desirable solution is a private collecting system and disposal plant.

Sanitary sewers flow by gravity, and their location depends on the topography of the land. They are often located under the streets, but they may also be located at the back of lots or in alleyways. The slope of the sewer should be such that the velocity will be between 2 and 10 feet per second. Manholes must be located at change of grade, junctions, and other points for inspection. Sewers must be laid in a straight line between manholes.

117 STREETS

Of major importance in the subdivision is the street system. Streets not only furnish an avenue for vehicle passage, they also furnish access to property for pedestrians, right-of-way for utility lines, channels for drainage, and access to fire plugs, garbage cans, etc. The right street in the right place contributes to pleasant living. Insufficient street width creates traffic hazards, while excess width adds to the cost of construction and uses land which could be used for lots.

Local residential streets (also called *minor streets*) are designed to furnish access to private property. They are not designed to carry through traffic and should be designed to discourage it. Curved streets, loops, and cul-de-sacs all discourage speed. Long, curvilinear streets with block lengths up to 1800 feet have been found to be satisfactory. Where off-street parking is adequate, a roadway width of 26 feet (27 feet back-to-back of curbs), and a right-of-way of 50 feet are adequate for single-family residential neighborhoods. For multifamily neighborhoods, street widths should be 31 feet minimum (32 feet back-to-back of curbs) with 60 feet right-of-way. A street width of 26 feet will not provide parking on both sides of the street with two lanes for traffic, but many homeowners find weaving in and out between parked cars slows traffic.

Collector streets carry traffic between local streets and arterial streets. They also furnish access to private property along the street. A width of 36 feet (37 feet back-to-back of curbs) will furnish two parking lanes and two traffic lanes. Right-of-way width should be 60 feet.

Arterial streets (also called *major streets*) move heavy traffic at relatively high speeds. Intersections are usually controlled by traffic lights. Many arterial streets provide six moving lanes with no parking. A 100 foot right-of-way will provide two 33 foot (back-to-back of curbs) lanes and a 14 foot median.

Cul-de-sacs are dead-end streets with turnarounds at the end. They are popular for single-family residences because of the privacy and freedom from noise they provide. They should not be more than 1000 feet in length because of the long turnaround. The turnaround circle should be not less than 40 feet in radius.

Loop streets have the same advantages as cul-del-sacs, plus the advantage of better circulation for fire trucks, delivery trucks, and police cars. Loop streets and cul-de-sacs can be used in odd corners of subdivisions.

118 BLOCKS

Until recent years, most subdivisions were laid out on a grid iron pattern. Modern subdivisions use curved streets with blocks 1400 to 1800 feet in length. These have proven to be more economical, using less street area by eliminating many cross-streets, and safer because of fewer intersections and slower speeds on curved streets. They also provide more lots per undeveloped acre.

119 LOTS

Lot size should depend on the type of development, the topography of the land, and the expected cost per housing unit. Lot dimensions vary in different parts of the U.S. but a minimum width of 60 feet and a minimum depth of 100 feet (or about 6,000 square feet) is considered desirable. Large, ranch-style houses with multi-car garages opening on the front require more than 75 feet in width.

120 COVENANTS

In order to protect the interests of future property owners, developers often include certain restrictive covenants as a part of the deed to the property. These include such things as type of construction, minimum size and cost, set-back distance, restrictions against advertising signs, raising of animals, parking of mobile homes, conducting certain types of commercial enterprises, and any other restrictions which will insure that the neighborhood is used for the purpose for which it was designed.

121 SET-BACK LINES

To prevent locating buildings in such a way as to mar the general view of the neighborhood, restrictive covenants often designate the minimum distance from the front property line to the front of the building. To prevent monotony, staggered set-back lines are sometimes used.

122 DENSITY ZONING

Many cities which have formerly controlled crowding in new neighborhoods by controlling the dimensions of lots have now adopted *density control* which controls the maximum number of dwellings per acre. This has allowed more imagination in planning and better use of natural features.

123 CLUSTER PLANNING

A recent development in subdivision planning is the *cluster pattern* where residences are clustered together in small, private sections of the subdivision with common open space. Cul-de-sacs and loop streets are adaptable to the cluster plan.

RESERVED FOR FUTURE USE

RESERVED FOR FUTURE USE

RESERVED FOR FUTURE USE

PROFESSIONAL PUBLICATIONS, INC. • P.O. Box 199, San Carlos, CA 94070

26 STATISTICAL ANALYSIS

1 STATISTICAL ANALYSIS OF EXPERIMENTAL DATA

Experiments can take on many forms. An experiment might consist of measuring the weight of one cubic foot of concrete. Or, an experiment might consist of measuring the speed of a car on a roadway. Generally, such experiments are performed more than once to increase the precision and accuracy of the results.

Of course, the intrinsic variability of the process being measured will cause the observations to vary, and we would not expect the experiment to yield the same result each time it was performed. Eventually, a collection of experimental outcomes (observations) will be available for analysis.

One fundamental technique for organizing random observations is the frequency distribution. The frequency distribution is a systematic method for ordering the observations from small to large, according to some convenient numerical characteristic.

Example 26.1

The number of cars that travel through an intersection between 12 noon and 1 p.m. is measured for 30 consecutive working days. The results of the 30 observations are:

79, 66, 72, 70, 68, 66, 68, 76, 73, 71, 74, 70, 71, 69, 67, 74, 70, 68, 69, 64, 75, 70, 68, 69, 64, 69, 62, 63, 63, 61

What is the frequency distribution using an interval of 2 cars per hour?

Solution

cars per hour	frequency of occurrence
60–61	1
62–63	3
64–65	2
66–67	3
68–69	8
70–71	6
72–73	2
74–75	3
76–77	1
78–79	1

In example 26.1, 2 cars per hour is known as the step interval. The step interval should be chosen so that the data is presented in a meaningful manner. If there are too many intervals, many of them will have zero frequencies. If there are too few intervals, the frequency distribution will have little value. Generally, 10 to 15 intervals are used.

Once the frequency distribution is complete, it may be represented graphically as a histogram. The procedure in drawing a histogram is to mark off the interval limits on a number line and then draw bars with lengths that are proportional to the frequencies in the intervals. If it is necessary to show the continuous nature of the data, a frequency polygon can be drawn.

Example 26.2

Draw the frequency histogram and frequency polygon for the data given in example 26.1.

If it is necessary to know the number or percentage of observations that occur up to and including some value, the cumulative frequency table can be formed. This procedure is illustrated in the following example.

Example 26.3

Form the cumulative frequency distribution and graph

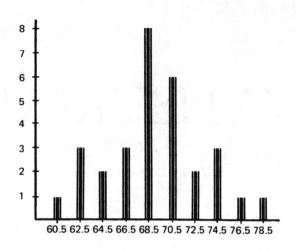

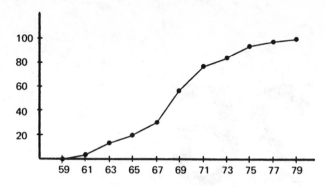

Figure 26.2 Example 26.3

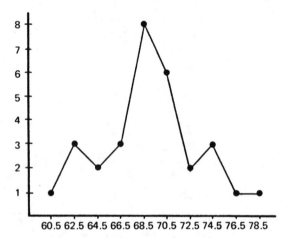

Figure 26.1 Example 26.2

for the data given in example 26.1.

cars per hour	frequency	cumulative frequency	cumulative per cent
60–61	1	1	3
62–63	3	4	13
64–65	2	6	20
66–67	3	9	30
68–69	8	17	57
70–71	6	23	77
72–73	2	25	83
74–75	3	28	93
76–77	1	29	97
78–79	1	30	100

It is often unnecessary to present the experimental data in its entirety, either in tabular or graphical form. In such cases, the data and distribution can be represented by various parameters. One type of parameter is a measure of central tendency. The mode, median, and mean are measures of central tendency. The other type of parameter is a measure of dispersion. Standard deviation and variance are measures of dispersion.

The mode is the observed value which occurs most frequently. The mode may vary greatly between series of observations. Therefore, its main use is as a quick measure of the central value since no computation is required to find it. Beyond this, the usefulness of the mode is limited.

The median is the point in the distribution which divides the total observations into two parts containing equal numbers of observations. It is not influenced by the extremity of scores on either side of the distribution. The median is found by counting up (from either end of the frequency distribution) until half of the observations have been accounted for. The procedure is more difficult if the median falls within an interval, as is illustrated in example 26.4.

The arithmetic mean is the arithmetic average of the observations. The mean may be found without ordering the data (which was necessary to find the mode and median). The mean can be found from the following formula:

$$\bar{x} = \left(\frac{1}{n}\right)(x_1 + x_2 + \cdots + x_n) = \frac{\sum x_i}{n} \qquad 26.1$$

The geometric mean is occasionally used when it is necessary to average ratios. The primary application of the geometric mean in this book involves coliform counting. The geometric mean is calculated as

$$\text{geometric mean} = \sqrt[n]{x_1 x_2 x_3 \ldots x_n} \qquad 26.2$$

Example 26.4

Find the mode, median, and arithmetic mean of the distribution represented by the data given in example 26.1.

Solution

The mode is the interval 68–69, since this interval has the highest frequency. If 68.5 is taken as the interval center, then 68.5 would be the mode.

PROFESSIONAL PUBLICATIONS, INC. • P.O. Box 199, San Carlos, CA 94070

Since there are 30 observations, the median is the value which separates the observations into 2 groups of 15. From example 26.3, the median occurs some place within the 68–69 interval. Up through interval 66–67, there are 9 observations, so 6 more are needed to make 15. Interval 68–69 has 8 observations, so the mean is found to be (6/8) or (3/4) of the way through the interval. Since the real limits of the interval are 67.5 and 69.5, the median is located at

$$67.5 + \frac{3}{4}(69.5 - 67.5) = 69$$

The mean can be found from the raw data or from the grouped data using the interval center as the assumed observation value. Using the raw data,

$$\overline{x} = \frac{\sum x}{n} = \frac{2069}{30} = 68.97$$

Similar in concept to the median are percentile ranks, quartiles, and deciles. The median could also have been called the 50th percentile observation. Similarly, the 80th percentile would be the number of cars per hour for which the cumulative frequency was 80%. The quartile and decile points on the distribution divide the observations or distribution into segments of 25% and 10% respectively.

The most simple statistical parameter which describes the variation in observed data is the range. The range is found by subtracting the smallest value from the largest. Since the range is influenced by extreme (low probability) observations, its use is limited as a measure of variability.

The standard deviation is a better estimate of variability because it considers every observation. The standard deviation can be found from:

$$\sigma = \sqrt{\frac{\sum (x_i - \overline{x})^2}{n}} = \sqrt{\frac{\sum x_i^2}{n} - (\overline{x})^2} \qquad 26.3$$

The above formula assumes that n is a large number, such as above 50. Theoretically, n is the size of the entire population. If a small sample (less than 50) is used to calculate the standard deviation of the distribution, the formulas are changed. The sample standard deviation is

$$s = \sqrt{\frac{\sum (x_i - \overline{x})^2}{n-1}} = \sqrt{\frac{\sum x_i^2 - (\sum x_i)^2/n}{n-1}} \qquad 26.4$$

The difference is small when n is large, but care must be taken in reading the problem. If the 'standard deviation of the sample' is requested, calculate σ. If an estimate of the 'population standard deviation' or 'sample standard deviation' is requested, calculate s. (Note that the standard deviation of the sample is not the same as the sample standard deviation.)

Example 26.5

Calculate the range, standard deviation of the sample, and population variance from the data given in example 26.1.

Solution

$$\sum x = 2069$$
$$\left(\sum x\right)^2 = 4280761$$
$$\sum x^2 = 143225$$
$$n = 30$$
$$\overline{x} = 68.97$$
$$\sigma = \sqrt{\frac{143225}{30} - (68.97)^2} = 4.16$$
$$s = \sqrt{\frac{143225 - (4280761)/30}{29}} = 4.29$$
$$s^2 = 18.4$$
$$R = 79 - 61 = 18$$

Referring again to example 26.1, suppose that the hourly through-put for 15 similar intersections is measured over a 30 day period. At the end of the 30 day period, there will be 15 ranges, 15 medians, 15 means, 15 standard deviations, and so on. These parameters themselves constitute distributions.

The mean of the sample means is an excellent estimator of the average hourly through-put of an intersection:

$$\overline{\overline{x}} = \left(\frac{1}{15}\right)\sum \overline{x} \qquad 26.5$$

The standard deviation of the sample means is known as the standard error of the mean to distinguish it from the standard deviation of the raw data. The standard error is written as $\sigma_{\overline{x}}$.

The standard error is not a good estimator of the population standard deviation.

In general, if k sets of n observations each are used to estimate the population mean (u) and the population standard deviation (σ'), then

$$u \approx \left(\frac{1}{k}\right)\sum \overline{x} \qquad 26.6$$
$$\sigma' \approx \sqrt{k}\,\sigma_{\overline{x}} \qquad 26.7$$

2 MEASUREMENTS OF EQUAL WEIGHT

There are many opportunities for errors in surveying, although calculators and modern equipment have reduced the magnitudes of most errors. The purpose of error analysis is not to eliminate errors, but rather to estimate their magnitudes and to assign them to the appropriate measurements.

The expected value (also known as the most likely value or the probable value) of a measurement is the value which has the highest probability of being correct. If a series of measurements is taken of a single quantity, the most probable value is the average (mean) of those measurements. That is, if $x_1, x_2, \ldots, x_n$ are values of some measurement, then the most probable value is

$$x_p = \frac{x_1 + x_2 + \cdots + x_n}{n} \qquad 26.8$$

For related measurements whose sum should equal some known quantity, the most probable values are the observed values corrected by an equal part of the total error. This is illustrated in example 26.6.

Example 26.6

The interior angles of a traverse were measured as 63°, 77°, and 41°. Each measurement was made once, and all angles were measured with the same precision. What are the most probable interior angles?

Solution

The sum total of angles should equal 180. The error in the measurements is $(63 + 77 + 41 - 180) = +1°$. Therefore, the correction required is $-1°$ which is proportioned equally among the three angles. The most probable values are 62.67°, 76.67°, and 40.67°.

Measurements of a given quantity are assumed to be normally distributed. If a quantity has a mean u and a standard deviation s, the probability is 50% that a measurement of that quantity will fall within the range of $u \pm (.6745)s$. The quantity $(.6745)s$ is known as the *probable error*. The probable *ratio of precision* is $u/(.6745)s$. The interval between the extremes is known as the *confidence interval*.

The standard deviation, s, is the small sample standard deviation.

The probable error of the mean of k observations of the same quantity is given by equation 26.9.

$$E_{\text{mean}} = \frac{.6745s}{\sqrt{k}}$$
$$= \frac{E_{\text{total, } k \text{ measurements}}}{\sqrt{k}} \qquad 26.9$$

Example 26.7

12 tapings were made of a critical distance. The mean value was 423.7 feet with a standard deviation (s) of .31 feet. What are the 50% confidence limits for the distance?

Solution

From equation 26.9, the standard error of the mean value is

$$E_{\text{mean}} = \frac{(.6745)(.31)}{\sqrt{12}} = .06$$

Therefore, the probability is 50% that the true distance is within the limits of $423.7 \pm .06$ feet.

Example 26.8

The true length of a tape is 100 feet. The most probable error of a measurement with this tape is .01 feet. What is the expected error if the tape is used to measure out a distance of one mile?

Solution

The number of tapings will be $(5280/100) = 52.8$, or 53 tapings. The most probable error will be $(.01)(\sqrt{53}) = .073$ feet.

3 MEASURMENTS OF UNEQUAL WEIGHT

Some measurements may be more reliable than others. It is not unreasonable to weight each measurement with its relative reliability. Such weights may be determined subjectively, but more frequently, they are determined from relative frequencies of occurrence or from the relative inverse squares of the probable errors.

Example 26.9

An angle was measured five times by five equally competent crews on similar days. Two of the crews obtained a value of 39.77°, and the remaining three crews obtained a value of 39.74°. What is the probable value of the angle?

Solution

$$\theta = \frac{(2)(39.77) + (3)(39.74)}{5} = 39.75°$$

Example 26.10

A distance has been measured by three different crews. The lengths and their probable error intervals are given below. What is the most probable value?

$$\begin{aligned}
\text{crew 1:} \quad & 1{,}206.40 \pm .03 \text{ feet} \\
\text{crew 2:} \quad & 1{,}206.42 \pm .05 \text{ feet} \\
\text{crew 3:} \quad & 1{,}206.37 \pm .07 \text{ feet}
\end{aligned}$$

Solution

The sum of the squared probable errors is

$$(.03)^2 + (0.5)^2 + (.07)^2 = .0083$$

The weights to be applied to the three measurements are

$$.0083/(.03)^2 = 9.22$$
$$.0083/(.05)^2 = 3.32$$
$$.0083/(0.7)^2 = 1.69$$

The most probable length is

$$\frac{(1206.40)(9.22) + (1206.42)(3.32) + (1206.37)(1.69)}{9.22 + 3.32 + 1.69}$$
$$= 1206.40$$

The probable error and 50% confidence interval for weighted observations can be found from equation 26.10. x_i represents the ith observation and w_i represents its weight. The number of observations is n.

$$E_{p,\text{weighted}} = .6745\sqrt{\frac{\sum[w(\bar{x} - x_i)^2]}{(\sum w)(n - 1)}} \qquad 26.10$$

Example 26.11

What is the 50% confidence interval for the measured distance in example 26.10?

Solution

It is easier to work with the decimal part only.

$$\bar{x} = \left(\frac{1}{3}\right)(.40 + .42 + .37) = .40 \text{ approximately}$$

i	x_i	$\bar{x} - x_i$	$(\bar{x} - x_i)^2$	w	$w(\bar{x} - x_i)^2$
1	.40	.00	.0000	9.22	.0000
2	.42	−.02	.0004	3.32	.0013
3	.37	.03	.0009	1.69	.0015
				14.32	.0028

From equation 26.10

$$E_{p,\text{weighted}} = .6745\sqrt{\frac{.0028}{(14.23)(3 - 1)}} = .0067$$

The 50% confidence interval is $1206.40 \pm .0067$.

For related weighted measurements whose sum should equal some known quantity, the most probable weighted values are corrected inversely to the relative frequency of observation.

Example 26.12

The interior angles of a triangular traverse were repeatedly measured, with the results shown below. What is the most probable value for angle #1?

angle	value	number of measurements
1	63°	2
2	77°	6
3	41°	5

Solution

The total of the angles is $(63 + 77 + 41) = 181$. So $(-1°)$ must be divided among the three angles. These corrections are inversely proportional to the number of measurements. The sum of the measurement inverses is

$$\left(\frac{1}{2}\right) + \left(\frac{1}{6}\right) + \left(\frac{1}{5}\right) = .867$$

The most probable value of angle #1 is

$$63° + \left(\frac{\frac{1}{2}}{.867}\right)(-1) = 62.42°$$

Weights may also be calculated when the probable errors are known. These weights are the relative squares of the probable errors. This is illustrated in example 26.13.

Example 26.13

The interior angles of a triangular traverse were measured, with the results shown below. What is the most probable value of angle #1?

angle	value
1	63° ± .01°
2	77° ± .03°
3	41° ± .02°

Solution

The total of the angles is $(63 + 77 + 41) = 181$. So $(-1°)$ must be divided among the three angles. The corrections are proportional to the square of the probable errors.

$$(.01)^2 + (.03)^2 + (.02)^2 = .0014$$

The most probable value of angle #1 is

$$63 + \frac{(.01)^2}{.0014}(-1) = 62.93°$$

4 ERRORS IN COMPUTED QUANTITIES

When quantities with known errors are added or subtracted, the error of the result is given by equation 26.11. The squared errors under the radical are added regardless of whether the calculation is addition or subtraction.

$$E_{\text{total}} = \sqrt{E_1^2 + E_2^2 + E_3^2 + \cdots} \qquad 26.11$$

The error in the product of two quantities (x_1 and x_2) which have known errors (E_1 and E_2) is given by equation 26.12.

$$E_{\text{product}} = \sqrt{x_1^2 E_2^2 + x_2^2 E_1^2} \qquad 26.12$$

Example 26.14

The sides of a perfectly square rectangular section were determined to be $(1204.77 \pm .09)$ feet and $(765.31 \pm .04)$ feet respectively. What is the probable error in the area?

Solution

From equation 26.12,

$$E_{\text{area}} = \sqrt{(1204.77)^2(0.4)^2 + (765.31)^2(0.9)^2}$$
$$= 84.06\,\text{ft}^2$$

RESERVED FOR FUTURE USE

RESERVED FOR FUTURE USE

27

CONTRACT LAW, BUSINESS ASSOCIATIONS, AND ETHICS

PART 1: Contract Law

1 CONTRACTS

In its simplest sense, a contract is an agreement between two parties that is enforceable in a court of law. It is a promise by one person to another to do or not to do something in return for something else. *Contract law* is the enforcement of such promissory obligations, administered primarily by the state. The law differs in detail from state to state, but the basic procedures of contract law are fairly standard.

Contracts need not necessarily be in writing to be legally binding, as long as there is an implied intent to enter into a contractual agreement that is recognized by law. Such an agreement must meet several basic requirements:

- It must be voluntary for all parties concerned.

- It must be an exchange of goods or services.

- All parties must have legal capacity.

- The purpose of the contract must be legal.

In order to be enforceable, a contract requires several prominent features to be present. These are: offer, acceptance, and consideration. These three features usually develop from a promise.

2 PROMISE

A promise is an expression, no matter how expressed, made by one person that leads another person reasonably to expect a particular act from the promisor. Such an expression is a promise whether enforceable by law or not. It is a guarantee by one person that something shall or shall not happen.

3 OFFER

For a promise to be considered an offer in the legal sense, it must be clear, definite, and specific, with no room for ambiguity or misunderstanding. An offer is an act by a person that gives another person the legal power to create a contract. It is the basis of any contract, and must be explicitly made to a person or group of persons, and clear as to its terms and intent. When an offer is accepted by another person, it becomes a contract.

4 ACCEPTANCE

An acceptance is the agreement by a person to accept an offer made by another. Until such an agreement is made, no contract legally exists. As opposed to an offer, which must be explicit, an acceptance can be implied by the performance of the act requested by the offer. Both offer and acceptance must be voluntary acts; a contract cannot be forced on anyone.

5 CONSIDERATION

A contract must be supported by consideration. Consideration for a promise may be an act or the forbearance of an act, or the promise of such an act, or the creation, modification, or destruction of a legal relationship. Consideration may be given to the promisor or to some other person.

Consideration is a necessary element of any contract; a promise is not enforceable unless it is an agreement to exchange things of value. For example, if Smith and Jones enter into a contract in which Smith promises to pay Jones a sum of money for the performance of a service, both the money and the service are considerations.

It does not matter to the courts if the exchange is based on equal value or not, the promise to make the exchange is the legally binding contract. In addition, a contract cannot be made on a consideration unless the consideration is to be performed in the future—past acts or payments are not legally binding.

6 ORAL AND WRITTEN CONTRACTS

Not all contracts must be in writing to be enforceable. Many of our daily contractual agreements are not written agreements, such as short-term services or repairs. Sometimes a memorandum signed by one party is sufficient to create an enforceable contract, but interpretation is tricky and varies from state to state.

The *statute of frauds* states explicitly what types of contracts must be in writing. It originated in England common law with the passing in 1676 of *An Act of Prevention of Frauds and Perjuries*. The intention of the act was to prevent fraud resulting from perjury when a contract was not in writing. Each state now has its own version, but in general they require the same types of contracts to be in writing. These are:

- Promises by an administrator or executor of an estate to cover the debts of the estate from his own property.

- Any promise by one person to pay for the debts, damages, or misconduct of another person. This must be a direct promise to become a creditor. For example, if John says to Lynn, "Loan Sam $5.00 and I will pay you back if Sam doesn't," this promise is covered under the statute of frauds, and so must be in writing. But if John had said, "Give Sam $5.00 worth of goods and I will pay you for it," this would be a direct promise to pay money to Lynn. Because it is not a promise to cover the debts of another person, it does not come under the statute of frauds, so it would not need to be in writing to be enforceable. In general, if the primary purpose is for the benefit of the promisor, then it is usually a direct rather than a collateral promise (as is a co-signed loan), and it therefore does not come under the statute of frauds.

- An agreement not to be fulfilled within a year or a lifetime. This really means agreements that cannot or might not be performed within a year. If there is a possibility that full performance may be obtained within a year, the statute of frauds does not apply. For example, if John promised to support Bill for life, it would not come under the statute of frauds because John might die within a year, whereas if he promised to support Bill for two years, it would come under the statute of frauds because it cannot be completed until two years have elapsed.

- A contract for the sale or lease of real property.

- A contract for the sale of goods of substantial value (usually over $500).

The statute of frauds does not specify that if any of the above types of contracts are not in writing that they are illegal. But if they are disputed, they might not be enforceable in a court of law. Other types of oral contracts might still be enforceable because they do not come under the statute of frauds.

7 LEGAL CAPACITY

Both parties to a contract must have the legal capacity to enter into a contract. Capacity is the ability to take recognized legal action. It is the responsibility of each contracting party to be sure that the other party has legal capacity to contract.

Those persons without legal capacity are minors, mentally incompetent persons, and drunken or drugged persons. An agreement made by persons classified as such results in a voidable contract. This is for the protection of the less capable; those contracting with such parties might still be held responsible for their contractual obligations.

8 LEGALITY

You cannot create a legally binding contract to do something that is illegal or against public policy. Since there is no clear definition of public policy, each case is judged individually as to the intent of the contract.

9 THE INTERPRETATION OF CONTRACTS

Contracts are intended to provide a surety that one person upholds a legal promise to another. For this reason, the terms of the contract should convey the real intention of both parties toward one another. If one party does not perform as promised, the contract provides the other party legal recourse to receive compensation.

There are guidelines established that the courts apply in the interpretation of contracts when there is a dispute. These are not legally binding rules, but ways to fairly and consistently determine the actual facts of the agreement. In any contract dispute, the primary role of the court is first to determine what the parties had contracted for in the first place before any ruling can be affected concerning the completion of each party's contractual obligations.

If the agreement is in writing, then the written terms will take priority over any other evidence introduced. This it the *parol evidence rule.* This rule states very generally that if there is a written agreement, then any testimony outside that written agreement does not have to be considered. This rule only applies when the written agreement is clearly intended as the complete agreement. In most contracts there is a *merger clause*, which states explicitly that the contract is the complete agreement. If there is a contradiction between the written terms of the contract and an oral agreement made between the parties, the written terms will prevail.

If there is ambiguity in the wording of a written agreement, it will be interpreted in the light of the plain, usual, or literal meaning of the words. If, however, the words have a special meaning due to local usage or customs of a trade, or if there are technical words that have special meaning within the context of the contract, they will be interpreted as they apply. This relates to the parol evidence rule in that oral testimony may be allowed to explain ambiguity in a written document.

If there is some doubt about the meaning of a word in a contract, it is up to the contracting parties to agree on the interpretation of the wording within the context of the contract, and their agreement will be accepted by the court. If there is a dispute between the contracting parties as to the meaning of a word, the interpretation will be left up to the court, which is bound to judge in favor of the interpretation most commonly held, or most liable to make the contract valid.

In some cases, an ambiguity might create two possible interpretations of the terms of a contract. If one interpretation would render the contract void or such that it could not possibly be completed, it would be surrendered in favor of an interpretation that would make the contract valid and performable.

When a contract is disputed, and an interpretation must be made by the court before a decision can be reached regarding the fulfillment of the contract, the court will invariably interpret the contract in the favor of the contracting party. The party who prepared the contract, in many cases a business or insurance company, is at a disadvanage in a contract dispute. This is only in regards to the terms of the agreement, and does not necessarily affect the final decision regarding the satisfactory completion of the contract.

If a satisfactory interpretation of the terms of the contract cannot be determined, the contract will be considered unenforceable. It is always to the benefit of both parties that all terms of a contract are in writing and that there is no misunderstanding between them.

10 THE MODIFICATION OF A CONTRACT

A contract may be modified after it is agreed on if both parties agree to the modifications. Generally, if new terms are created, there must be new consideration. If not, then it is probably the case that the original contract is being rescinded by mutual agreement and a new contract is being formed.

The modification of a contract that is required to be in writing under the statute of frauds is also required to be in writing. An oral modification of such a contract is not binding, and, in effect, becomes a separate agreement. If the agreement made by the oral modification is disputed, it can be brought to court as a quasi-contract, but may not be allowed in the settlement of the original contract.

11 THE DISCHARGE OF A CONTRACT

A contract is said to be discharged when the agreement has been performed to the satisfaction of both parties. That is, both parties agree that they have each done what was originally required of them under the terms of the contract. When the contract is fully performed, it is discharged, and all rights and duties created under the contract are dissolved.

At any time, a contract may be discharged or altered if the contracting parties agree. A contract required to be in writing by the statute of frauds must be released by a written agreement.

A contract is discharged if it becomes impossible to perform it due to circumstances outside the control of the contracting parties. This could be due to several things. If the object of the contract were a piece of personal property, such as a house, and that property were destroyed by a natural disaster, the contract would no longer be binding. If the contract were for personal services that only a certain person could perform and that person were to die or otherwise be prevented from performing the contract, it would be discharged.

Extreme difficulty of executing the contract does not discharge it, even if it becomes more costly than originally anticipated. For example, if Jones has contracted to build a house for Brown for a certain price, and due to poor soil conditions, strikes, or difficulty in obtaining materials, the construction costs are higher than either Jones or Brown expected, Jones is still held to the original contract. If, however, there were contingency clauses in the contract to allow for modifications based on these unexpected happenings, it would be possible to modify the contract.

If the terms of the contract become impossible to complete, but there is still satisfactory completion, a contract is discharged if both parties agree. For example, if different materials than specified in the contract are substituted, but are considered of equal quality, that would be satisfactory completion. Or, if there has been a willful departure from the terms, as in the case where a house would be unsafe if built according to original terms, the contract could be discharged.

If an existing law changes, making the contract illegal, the contract is discharged. If only part of the contract is made illegal by changing legislation, the rest of the contract could remain binding. If a contract is made in a state where the contract would be considered void due to state laws, it will be considered void in other states, even if that state's laws were such that the contract would be legal. Some contracts may be dissolved by actions of the court, such as bankruptcy, public acts, or a declaration of war.

12 BREACH OF CONTRACT

The above describes situations where a contract is discharged by agreement of the contracting parties or by the court. However, sometimes a contract is discharged because one party fails to perform his part of the contract. This is known as *breach of contract*. A breach may be partial or total.

A *total breach* is sometimes called a *material breach*, and it means that the injured party has not received a substantial part of the performance of the contract. The injured party is released from all duties on his part of the contract and may then sue for damages. A *partial breach* involves only the non-performance of a small part of the contract, and the injured party is only allowed to sue for damages stemming from that part of the agreement, and is not released from his part of the contract. There is no definitive way of deciding whether a breach is partial or total; each case must be decided individually by the court.

If one party repudiates the contract, that is, announces that he does not intend to complete his part of the contract prior to the time he is required by the contract to perform his part, it is known as *anticipatory breach*. The injured party is then allowed to sue for total breach and is released from his obligations under the contract.

However, if a party wanted to bring suit for breach of contract because of the lack of satisfactory performance from the other party, the plaintiff would have to be in a situation to be able to perform his part of the contract, but be waiting for some action on the part of the other party. This is known as a *tender of performance*—the plaintiff is being kept from performing his part of the contract because of the neglect of the other party. In this case, the plaintiff may decide to rescind the contract and sue only for the restitution of any monies dispensed by him in the performance of the contract.

The plaintiff may also sue for *damages*. Damages are the losses incurred by the plaintiff as a result of the non-performance of the contract.

- *General or compensatory damages* are awarded to make up for the injury that was sustained.

- *Special damages* are awarded for the direct financial loss due to the breach.

- *Nominal damages* are awarded when responsibility has been established but the amount of injury is so slight as to be inconsequential.

- *Punitive or exemplary damages* are awarded usually in tort cases as a punishment for the defendant.

- *Consequential damages* provide compensation for losses incurred by the plaintiff not directly related to the contract in question but as an indirect result of the breach.

In a breach of contract suit, punitive damages cannot be claimed. Nor can the plaintiff ask for compensation for mental anguish caused by the breach.

Damages may be money damages, either the actual cost of damages suffered by the plaintiff due to the breach of contract, or a reasonable amount when an exact amount cannot be determined.

If money damages are inadequate compensation, the plaintiff may ask for *specific performance*, in which case the court might order the defendant to perform the contractual obligation. Specific performance is an equitable remedy available to enforce the conveyance of land and in certain contracts relating to personal property where money damages will not do justice. For example, a buyer might want certain articles for sentimental or personal reasons, and substituting a similar article will not satisfy him. Specific performance cannot be applied in cases where the performance of the contract would entail hardship or injustice, or in cases of personal services, where the requirement of performance would constitute forced labor.

Liquidated damages are damages that are specified in the contract to insure that if the contract becomes disputed, the amount of damages are agreed upon in advance. If the damages specified in this way are deemed at the time of the lawsuit to be unreasonable, they are not enforceable. In general, a clause in the contract that specifies damages ahead of time is only a guideline; the case will be judged in light of what actual damages have been incurred by the plaintiff.

The plaintiff has the responsibility to minimize his damages if he has the opportunity. If the plaintiff foresees a breach, he cannot act in such a way as to increase the damages that he could sue for. If the court determines that he could have lessened his damages through prudent action, the judgment will probably be less than compensatory.

There is another way to settle breach of contract disputes without going to court. This is *arbitration*. If the parties agree, either ahead of time as a clause of the contract or at the time of the dispute, they can submit their claim to the American Arbitration Association. Decisions of the arbitrators can be as binding as the judgments of a court, and such decisions have the additional advantage of being faster and cheaper for all parties involved. However, a plaintiff who has a strong position may be at a disadvantage in an arbitration case.

When a breach of contract suit has been decided in court or by arbitration, the contract is discharged, and the duties and responsibilities are no longer binding.

13 TORT

A *tort* is a civil wrong committed against a person or his property, business, emotional well-being, or reputation that causes damage. It is a breach of the rights of an individual to be secure in his person and property and to be free from undue harrassment. For an invasion of personal rights to be a tort and a valid cause for legal action, there must be resulting damages.

The difference between a tort and a crime, in a very general sense, is that a crime is a wrong against society, something that threatens the peace and safety of the community. A *criminal lawsuit* is then brought by the state against a defendant. A crime may also be a tort in that it results in personal damages. The victim of the crime may also bring a tort suit against the defendant to recover damages.

Tort law is primarily state law, and is predominantly case law rather than statutory. The decision in a tort case is based on three conditions:

- If a person's rights have been infringed upon.

- If it was a result of negligence or actual intent on the part of the plaintiff.

- If the defendant suffered damages as a direct result.

Tort law is concerned with compensation, not punishment. Compensation is usually a certain sum of money. The amount of the judgment stems directly from the amount of damage suffered as a direct result of the tort. Nominal damages may be awarded to the plaintiff as justification for the wrong suffered in cases where the actual damages do not warrant a substantial award. Special damages may be awarded if the defendant suffered other losses as a result of the original damage. Other types of damages may be awarded in special cases. For example, if the tort was the result of malicious intent, punitive damages may be awarded. But the damages in a tort case are primarily meant to be compensatory.

14 MISREPRESENTATION

Misrepresentation is a false statement by a person of a material fact that he knows to be false, with the intent to deceive another person. If that person then relies on this information and acts on it in such a way as to cause damages to himself, he can claim damages in a tort case.

Most commonly, misrepresentation is a factor in a breach of contract suit, but such a wrong can be treated as either a tort or as a breach of contract, depending on the type of damages requested by the defendant. For example, if you bought a car and it was stated in the contract that it had recently had a brake job and that the brakes were therefore in top condition, and you had an accident due to brake failure, you could sue for breach of contract or for personal injury in a tort case.

Sometimes misrepresentation does not involve a contractual agreement, and is therefore clearly a tort. In these cases, the reliance of the plaintiff on the misrepresented facts must be proven beyond a reasonable doubt and there must be an intent to deceive on the part of the defendant. Merely bragging or teasing does not constitute misrepresentation.

15 NEGLIGENCE

Negligence is not using proper care in a situation, resulting in damages to property or injury to persons. Proper care and safety can be a very subjective judgment, but in general terms it is that diligence exercised by a reasonably prudent person. Any damages suffered as the result of another's negligence can be compensated for in a tort case.

A tort of negligence requires three things:

- The plaintiff must have a duty to the defendant to provide safety.

- This duty must have been neglected by the defendant.

- The plaintiff must have suffered damages as a result.

The negligence must be the direct cause of the damages. However, merely proving negligence on the part of the defendant is not always sufficient to recover damages, because the defendant has several options that he can exercise as a defense. The most common of these is *contributory negligence*. This is where the defendant proves that the plaintiff was equally as negligent, or might have prevented the damages if the plaintiff had taken due care.

In some cases, the defendant can prove that the plaintiff willingly took an unnecessary risk and that he was aware of, thereby absolving the defendant of responsibility for any damages suffered. For example, if a person were to climb a scaffolding after being warned that it was dangerous, he could not sue for damages if he fell and suffered injuries. He has voluntarily assumed the risk, and thus no one else can be held liable.

PART 2: Forms of Business Associations

16 INTRODUCTION

Although most people work for others rather than being self-employed, it is not uncommon for the surveyor to at one time or another consider establishing a consulting practice. In general, there are three types of business ownership: sole proprietorships, partnerships, and corporations. Each form of ownership has its own benefits and disadvantages.

17 SOLE PROPRIETORSHIP

Ths sole proprietorship is probably the most common. It means the ownership and operation of a business is by an individual. The owner is solely responsible for the operation of the business, even if he hires others to assist him. He is also solely responsible for the debts of the business.

There is no separation between the property of the business and the owner's personal property. His personal property may be attached to pay the debts of his business. Similarly, all the profits of the business are solely his and he may do with them as he pleases. He need not account to anyone. He may also dissolve the business whenever he pleases, although he will still be personally liable for any debts incurred by his business.

An individual may do business under his own name or under an assumed (fictitious) name. Statutes governing the use of *fictitious names* vary from state to state, but generally require some form of public notice of the name of the business and the real name of the owner. He may also be required to file a statement with the County Clerk and display a copy of this statement at his place of business.

18 PARTNERSHIPS

A partnership is a business association of two or more persons operating for profit. Each *general partner* shares in the control of the business, the financial profits and losses, and the responsibilities and liabilities of the partnership. A *limited partner* makes a financial investment in a business, but does not share in the operation of the business, and is liable only for the amount of his investment.

Usually, a partnership is established by a contractual agreement. Although not required by law to be in writing, it is rare to enter into a business enterprise without a written contract. Most partnership agreements are fairly detailed and are put in writing for the protection of all partners. The laws governing partnerships are based on the *Uniform Partnership Act*, which has been adopted in its entirety by most states.

A contract establishing a partnership must meet all the requirements of any contract. In particular, it must be voluntary and all parties must be legally competent. The contract should be specific as to the financial investment and liability of each partner, the division of profits and losses, the management obligations of all partners, the accountability of the business's legal and financial dealings, and provisions for the dissolution of the partnership.

The original partnership contract can be amended at any time. If stated in the contract, a partnership interest may be transferred, allowing one partner to sell his interest in the business to another person. However, the original partner is still responsible for his share of the debts outstanding at the time of the transfer. If all parties want to redefine the distribution of duties or financial responsibilities, a new contract should be created.

A partnership is a legal entity and can do business under the assumed name of the partnership—it can acquire and dispose of real and personal property, and it can sue and be sued. Therefore, a certificate of assumed name must be filed, as with a sole proprietorship. In addition, some states require that a certificate of partnership be filed stating the real names of the partners, along with the name and purpose of the business. These certificates are in addition to the contract between the partners.

All partners have equal rights in managing the business, unless otherwise stated in the partnership agreement. All partners have the right to inspect the financial records or other business documents at any time. Partners do not receive a salary for their work; their compensation is a share of the profits as outlined in the partnership contract. The acquisition and disbursement of real property and personal property belonging to the business must be approved by all partners. In some cases, a majority of the partners can manage the firm business even if there is an objecting partner.

The business may be represented by any of the partners. Each partner has the power to bind the other partners (the partnership) when acting within the scope of the business of the partnership. The scope of the partnership business depends upon the nature of the business. For example, a movie theater is not in the business of lending money, so if one partner in a movie theater lends money in the name of the business, without the consent of the other partners, the others will not necessarily be bound by the loan agreement.

Notice given to one partner or knowledge obtained by one partner as to matters relating to the business usually serves to bind the partnership. In other words, all partners need not have actual knowledge; the knowledge of one binds the partnership. It is the responsibility of each partner to keep the others informed.

When a partner is acting within the scope of the partnership business and commits a civil wrong (tort), all of the partners may be held personally liable. The partnership is also liable for the misappropriation of property received by a partner within the scope of the partnership business from a third party (e.g., a partner damages the property of another while that property is being held by the partnerhsip). Partners are personally liable for all debts incurred by the partnership.

Whereas general partners are personally liable for partnership debts which exceed partnership assets, the limited partner has only limited liability. The limited partner invests a certain amount of money in the business, and he is not financially liable for any amount beyond this. His profits are also limited, and are usually specified in the partnership agreement. His participation in the running of the business is limited. Although he has the right to see the financial records in order to protect his own investment, he generally does not get involved in the management of the business. If he does, it could be construed that he is actually acting as a general partner and he could be held liable by creditors.

A partnership exists only as long as the partners choose; it may be dissolved at any time. The partnership dissolves upon the death of a partner or the withdrawal of a partner. The other partners may then create a new partnership agreement, but this will not negate the outstanding obligations of any partner under the terms of the original partnership contract.

19 CORPORATIONS

A corporation is a legal entity possessing many of the legal powers of individuals. For example, a corporation may hold title to property and may sue or be sued in its own name. In this way it is like a business partnerhsip.

But unlike a partnerhip, a corporation exists independently of the people who own and manage it. Its major advantage is, in fact, the separation of management and ownership. The owners of a corporation are shareholders whose liabilities extend only to the limits of their financial investments. The corporation is managed by professionals hired by the owners and paid a salary by the corporation.

Another advantage of a corporation is that the life of the corporation does not depend on the continuing participation of specific partners. The ownership of a corporation is easily transferable through stock and does not disrupt the conduct of business. The operation of the corporation is handled by paid employees, and the turnover of staff does not affect the ownership of the corporation.

Private corporations are established by individuals to do business for personal gain. They are also called *stock corporations* because they issue stock to shareholders, who in turn own the corporation, have a right of control over the business, and share in the financial profit. A large number of corporations are owned by the person or persons who established the corporation because they hold a majority of the stock. These are often referred to as *closed corporations* or *closely-held corporations*.

There are also *non-stock corporations*, known as membership corporations, set up as non-profit businesses. Shareholders do not receive financial profit nor do they have any right of control. An example of a memberhsip corporation is a public TV station whose members contribute money, but receive no dividends in return for their investments.

PART 3: Professional Ethics

20 DEALING WITH CLIENTS

Although a typical set of ethical codes is included as an appendix, most of the client codes can be summarized in the following items.

- The surveyor should protect his client's interest. This protection goes beyond normal business relationships and transcends the legal requirements of the surveyor-client contract.

- Confidential client information is kept confidential and remains the property of the client.

- The surveyor should avoid conflicts of interest and should inform his client of any business connections or interests that might influence the surveyor's judgment.

- The surveyor should recognize his own limitations. He should use associates and other experts when the requirements exceed his abilities.

- The surveyor should not accept discounts, allowances, commissions, or any other indirect compensation in connection with any work or recommendations. The surveyor's sole source of income is the fee paid by his client.

- If the surveyor's recommendations are questioned or are not accepted, he will present the consequences clearly to the client.

- The surveyor will not be bound by what the client wants in instances where such a plan would be unsuccessful.

- The surveyor admits freely and openly any errors to his client.

21 DEALING WITH OTHER SURVEYORS

Surveyors should try to protect the surveying profession as a whole, to strengthen it, and to enhance its public stature. The following items will guide the surveyor.

- The surveyor should not attempt to injure the professional reputation, business, or employment position of another surveyor.

- The surveyor should not review someone else's work while the other surveyor is still employed, unless the other surveyor is made aware of the review.

- The surveyor should not try to replace another surveyor once the other surveyor has received employment.

- The surveyor should not use the advantages of a salaried position to compete unfairly with other surveyors who would have to charge more for the same services.

22 DEALING WITH THE PUBLIC

The relationship between a surveyor and the public is essentially straightforward. Responsibilities to the public demand that the surveyor place service to mankind above personal gain. Furthermore, proper ethical behavior requires that a surveyor avoid association with projects that are contrary to public health and welfare, or are of questionable legal character.

Supplement to Chapter 27

TYPICAL CODE OF ETHICS

It should be considered unprofessional and inconsistent with honorable and dignified bearing for any professional practitioner:

1. To act for his client, or employer, in professional matters otherwise than as a faithful agent or trustee, or to accept any remuneration other than his stated recompense for services rendered.

2. To attempt to injure falsely or maliciously, directly or indirectly, the professional reputation, prospects, or business of anyone.

3. To attempt to supplant another fellow professional practitioner after definite steps have been taken toward his employment.

4. To compete with another fellow practitioner for employment by the use of unethical practices.

5. To review the work of another fellow practitioner for the same client, except with the knowledge of such practitioner or unless the connection of such practitioner with the work has terminated.

6. To attempt to obtain or render technical services or assistance without fair and just compensation commensurate with the services rendered.

7. To advertise in self-laudatory language or any other manner derogatory to the dignity of the profession.

8. To attempt to practice in any professional field in which the registrant is not proficient.

PROFESSIONAL PUBLICATIONS, INC. • P.O. Box 199, San Carlos, CA 94070

RESERVED FOR FUTURE USE

PROFESSIONAL PUBLICATIONS, INC. • P.O. Box 199, San Carlos, CA 94070

RESERVED FOR FUTURE USE

28 ECONOMIC ANALYSIS

Nomenclature

A	annual amount or annuity	$
B	present worth of all benefits	$
BV_j	book value at the end of the jth year	$
C	cost, or present worth of all costs	$
d	declining balance depreciation rate	decimal
D_j	depreciation in year j	$
D.R.	present worth of after-tax depreciation recovery	$
e	natural logarithm base (2.718)	–
EUAC	equivalent uniform annual cost	$
f	federal income tax rate	decimal
F	future amount or future worth	$
G	uniform gradient amount	$
l	effective rate per period (usually per year)	decimal
k	number of compounding periods per year	–
n	number of compounding periods, or life of asset	–
P	present worth or present value	$
P_t	present worth after taxes	$
ROR	rate of return	decimal
ROI	return on investment	$
r	nominal rate per year (rate per annum)	decimal
s	state income tax rate	decimal
S_n	expected salvage value in year n	$
t	composite tax rate, or time	decimal
ϕ	nominal rate per period	decimal

1 EQUIVALENCE

Business decision makers using engineering economics are concerned with the timing of a project's cash flows as well as with the total profitability of that project. In this situation, a method is required to compare projects involving receipts and disbursements occurring at different times.

By way of illustration, consider $100 placed in a bank account which pays 5% effective annual interest at the end of each year. After the first year, the account will have grown to $105. After the second year, the account will have grown to $110.25.

Assume that you will have no need for money during the next two years and that any money received would immediately go into your 5% account. Then, which of the following options would be more desirable?

option a: $100 now

option b: $105 to be delivered in one year

option c: $110.25 to be delivered in two years

In light of the previous illustration, none of the options is superior under the assumptions given. If the first option is chosen, you will immediately place $100 into a 5% account, and in two years the account will have grown to $110.25. In fact, the account will contain $110.25 at the end of two years regardless of the option chosen. Therefore, these alternatives are said to be *equivalent*.

2 CASH FLOW DIAGRAMS

Although they are not always necessary in simple problems (and they are often unwieldly in very complex problems), *cash flow diagrams* may be drawn to help visualize and simplify problems having diverse receipts and disbursements.

The conventions below are used to standardize cash flow diagrams.

- The horizontal (time) axis is marked off in equal increments, one per period, up to the duration or horizon of the project.

- All disbursements and receipts (cash flows) are assumed to take place at the end of the year in which they occur. This is known as the *year-end convention*. The exception to the year-end convention is any initial cost (purchase cost) which occurs at $t = 0$.

- Two or more transfers in the same year are placed end-to-end, and these may be combined.

- Expenses incurred before $t = 0$ are called *sunk costs*. Sunk costs are not relevant to the problem.

Example 28.1

A mechanical device will cost $20,000 when purchased. Maintenance will cost $1,000 each year. The device will generate revenues of $5,000 each year for 5 years after which the salvage value is expected to be $7,000. Draw and simplify the cash flow diagram.

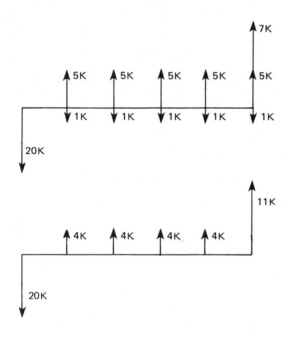

Figure 28.1   Example 28.1

3 TYPICAL PROBLEM FORMAT

With the exception of some investment and rate of return problems, the typical problem involving economic analysis will have the following characteristics:

- An interest rate will be given.

- Two or more alternatives will be competing for funding.

- Each alternative will have its own cash flows.

- It is necessary to select the best alternative.

Example 28.2

Investment A costs $10,000 today and pays back $11,500 two years from now. Investment B costs $8,000 today and pays back $4,500 each year for two years. If an interest rate of 5% is used, which alternative is superior?

The solution to this example is not difficult, but it will be postponed until methods of calculating equivalence have been covered.

4 CALCULATING EQUIVALENCE

It was previously illusrated that $100 now is equivalent at 5% to $105 in one year. The equivalence of any present amount, P, at $t = 0$ to any future amount, F, at $t = n$ is called the *future worth* and can be calculated from equation 28.1.

$$F = P(1 + i)^n \qquad 28.1$$

The factor $(1 + i)^n$ is known as the *compound amount factor* and has been tabulated at the end of this chapter for various combinations of i and n. Rather than actually writing the formula for the compound amount factor, the convention is to use the standard functional notation $(F/P, i\%, n)$. Thus,

$$F = P(F/P, i\%, n) \qquad 28.2$$

Similarly, the equivalence of any future amount to any present amount is called the *present worth* and can be calculated from

$$P = F(1 + i)^{-n} = F(P/F, i\%, n) \qquad 28.3$$

The factor $(1 + i)^{-n}$ is known as the *present worth factor*, with functional notation $(P/F, i\%, n)$. Tabulated values are also given for this factor at the end of this chapter.

Example 28.3

How much should you put into a 10% savings account in order to have $10,000 in 5 years?

This problem could also be stated: What is the equivalent present worth of $10,000 5 years from now if money is worth 10%?

$$P = F(1 + i)^{-n} = 10,000(1 + .10)^{-5} = 6,209$$

The factor .6209 would usually be obtained from the tables.

A cash flow which repeats regularly each year is known as an *annual amount*. Although the equivalent value for each of the n annual amounts could be calculated and then summed, it is much easier to use one of the *uniform series factors*, as illustrated in example 28.4.

Example 28.4

Maintenance costs for a machine are $250 each year. What is the present worth of these maintenance costs over a 12 year period if the interest rate is 8%?

Notice that

$$(P/A, 8\%, 12) = (P/F, 8\%, 1) + (P/F, 8\%, 2)$$
$$+ \cdots + (P/F, 8\%, 12)$$

Then

$$P = A(P/A, i\%, n) = -250(7.5361)$$
$$= -1{,}884$$

A common complication involves a uniformly increasing cash flow. Such an increasing cash flow should be handled with the *uniform gradient factor*, $(P/G, i\%, n)$. The uniform gradient factor finds the present worth of a uniformly increasing cash flow which starts in year 2 (not year 1) as shown in example 28.5.

Example 28.5

Maintenance on an old machine is $100 this year but is expected to increase by $25 each year thereafter. What is the present worth of 5 years of maintenance? Use an interest rate of 10%.

In this problem, the cash flow must be broken down into parts. Notice that the 5-year gradient factor is used even though there are only 4 non-zero gradient cash flows.

$$P = A(P/A, 10\%, 5) + G(P/G, 10\%, 5)$$
$$= -100(3.7908) - 25(6.8618) = -551$$

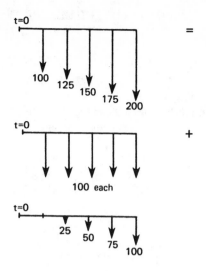

Figure 28.2 Example 28.5

Various combinations of the compounding and discounting factors are possible. For instance, the annual cash flow that would be equivalent to a uniform gradient may be found from

$$A = G(P/G, i\%, n)(A/P, i\%, n) \qquad 28.4$$

Formulas for all of the compounding and discounting factors are contained in table 28.1. Normally, it will not be necessary to calculate factors from the formulas. The tables at the end of this chapter are adequate for solving most problems.

Table 28.1
Discount Factors for Discrete Compounding

factor name	converts	symbol	formula
single payment compound amount	P to F	$(F/P, i\%, n)$	$(1+i)^n$
present worth	F to P	$(P/F, i\%, n)$	$(1+i)^{-n}$
uniform series	F to A	$(A/F, i\%, n)$	$\frac{i}{(1+i)^n - 1}$
sinking fund capital recovery	P to A	$(A/P, i\%, n)$	$\frac{i(1+i)^n}{(1+i)^n - 1}$
compound amount	A to F	$(F/A, i\%, n)$	$\frac{(1+i)^n - 1}{i}$
equal series	A to P	$(P/A, i\%, n)$	$\frac{(1+i)^n - 1}{i(1+i)^n}$
present worth uniform gradient	G to P	$(P/G, i\%, n)$	$\frac{(1+i)^n - 1}{i^2(1+i)^n} - \frac{n}{i(1+i)^n}$

5 THE MEANING OF "PRESENT WORTH" AND "i"

It is clear that $100 invested in a 5% bank account will allow you to remove $105 one year from now. If this investment is made, you will clearly receive a *return on investment* (ROI) of $5. The cash flow diagram and the present worth of the two transactions are

$$P = -100 + 105(P/F, 5\%, 1)$$
$$= -100 + 105(.9524) = 0$$

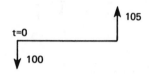

Figure 28.3

Notice that the present worth is zero even though you did receive a 5% return on your investment.

However, if you are offered $120 for the use of $100 over a one-year period, the cash flow diagram and present worth (at 5%) would be

$$P = -100 + 120(P/F, 5\%, 1)$$
$$= -100 + 120(.9524) = 14.29$$

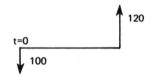

Figure 28.4

Therefore, it appears that the present worth of an alternative is equal to the equivalent value at $t = 0$ of the increase in return above that which you would be able to earn in an investment offering $i\%$ per period. In the above case, $14.29 is the present worth of ($20 − $5), the difference in the two *ROI*'s.

Alternatively, the actual earned interest rate, called *rate of return* (*ROR*), can be defined as the rate which makes the present worth of the alternative zero.

The *present worth* is also the amount that you would have to be given to dissuade you from making an investment, since placing the initial investment amount along with the present worth into a bank account earning $i\%$ will yield the same eventual *ROI*. Relating this to the

previous paragraphs, you could be dissuaded against investing $100 in an alternative which would return $120 in one year by a $t = 0$ payment of $14.29. Clearly, ($100 + $14.29) invested at $t = 0$ will also yield $120 in one year at 5%.

The selection of the interest rate is difficult in economic analysis problems. Usually it is taken as the average rate of return that an individual or business organization has realized in past investments. Fortunately, an interest rate is usually given. A company may not know what effective interest rate to use in an economic analysis. In such a case, the company can establish a minimum acceptable return on its investment. This *minimum attractive rate of return* (*MARR*) should be used as the effective interest rate i in economic analyses.

It should be obvious that alternatives with negative present worths are undesirable, and that alternatives with positive present worths are desirable because they increase the average earning power of invested capital.

6 CHOICE BETWEEN ALTERNATIVES

A variety of methods exists for selecting a superior alternative from among a group of proposals. Each method has its own merits and applications.

The *Present Worth Method* has already been implied. When two or more alternatives are capable of performing the same functions, the superior alternative will have the largest present worth. Ths method is suitable for ranking the desirability of alternatives. The present worth method is restricted to evaluating alternatives that are mutually exclusive and which have the same lives.

Returning to example 28.2, the present worth of each alternative should be found in order to determine which alternative is superior.

Example 28.2, continued

$$P(\mathbf{A}) = -10,000 + 11,500(P/F, 5\%, 2) = 431$$
$$P(\mathbf{B}) = -8,000 + 4,500(P/A, 5\%, 2) = 367$$

Alternative **A** is superior and should be chosen.

The present worth of a project with an infinite life is known as the *capitalized cost*. Capitalized cost is the amount of money at $t = 0$ needed to perpetually support the project on the earned interest only. Capitalized cost is a positive number when expenses exceed income.

$$\frac{\text{Capitalized}}{\text{Cost}} = \frac{\text{Initial}}{\text{Cost}} + \frac{\text{Annual Costs}}{i} \qquad 28.5$$

If disbursements occur irregularly instead of annually, the capitalized cost must be calculated from the equivalent uniform annual cost instead of the actual annual expenses.

$$\text{Capitalized} \atop \text{Cost} = \text{Inital} \atop \text{Cost} + \frac{\text{EUAC}}{i} \qquad 28.6$$

In comparing two alternatives, each of which is infinitely lived, the superior alternative will have the lowest capitalized cost.

Alternatives which accomplish the same purpose but which have unequal lives must be compared by the *Annual Cost Method*. The annual cost method assumes that each alternative will be replaced by an identical twin at the end of its useful life (infinite renewal). This method, which may also be used to rank alternatives according to their desirability, is also called the *Annual Return Method* and *the Capital Recovery Method*.

Restrictions are that the alternatives must be mutually exclusive and infinitely renewed up to the duration of the longest-lived alternative. The calculated annual cost is known as the *Equivalent Uniform Annual Cost, EUAC*. Cost is a positive number when expenses exceed income.

Example 28.6

Which of the following alternatives is superior over a 30 year period if the interest rate is 7%?

	A	B
type	brick	wood
life	30 years	10 years
cost	$1,800	$450
maintenance	$5/year	$20/year

$$\text{EUAC}(A) = 1,800(A/P, 7\%, 30) + 5 = 150$$
$$\text{EUAC}(B) = 450(A/P, 7\%, 10) + 20 = 84$$

Alternative **B** is superior since its annual cost of operation is the lowest. It is assumed that three wood facilities, each with a life of 10 years and a cost of $450, will be built to span the 30 year period.

The *Benefit-Cost Ratio Method* is often used in municipal project evaluations where benefits and costs accrue to different segments of the community. With this method, the present worth of all benefits (regardless of the beneficiary) is divided by the present worth of all costs. The project is considered acceptable if the ratio exceeds *one*.

When the benefit-cost ratio method is used, disbursements by the initiators or sponsors are *costs*. Disbursements by the users of the project are known as *disbenefits*. It is often difficult to determine whether a cash flow is a cost or a disbenefit (whether to place it in the numerator or denominator of the benefit-cost ratio calculation).

Regardless of where the cash flow is placed, an acceptable project will always have a benefit-cost ratio greater than one, although the actual numerical result will depend on the placement. For this reason, the benefit-cost ratio method should not be used to rank competing projects.

The benefit-cost ratio method may be used to rank alternative proposals only if an *incremental analysis* is used. First, determine that the ratio is greater than one for each alternative. Then, calculate the ratio of benefits to costs

$$\frac{(B_2 - B_1)}{(C_2 - C_1)}$$

for each possible pair of alternatives. If the ratio exceeds one, alternative 2 is superior to alternative 1. Otherwise, alternative 1 is superior.

Perhaps no method of analysis is less understood than the *Rate of Return* (ROR) *Method*. As was stated previously, the ROR is the interest rate that would yield identical profits if all money were invested at that rate. The present worth of any such investment is zero.

The ROR is defined as the interest rate that will discount all cash flows to a total present worth equal to the initial required investment. This definition is used to determine the ROR of an alternative. The ROR should not be used to rank or compare alternatives unless an incremental analysis is used. The advantage of the ROR method is that no knowledge of an interest rate is required.

To find the ROR of an alternative, proceed as follows:

Step 1: Set up the problem as if to calculate the present worth.

Step 2: Arbitrarily select a reasonable value for i. Calculate the present worth.

Step 3: Choose another value of i (not too close to the original value) and again solve for the present worth.

Step 4: Interpolate or extrapolate the value of i which gives a zero present worth.

Step 5: For increased accuracy, repeat steps (2) and (3) with two more values that straddle the value found in step (4).

A common, although incorrect, method of calculating the ROR involves dividing the annual receipts or returns by the initial investment. However this technique ignores such items as salvage, depreciation, taxes, and the time value of money. This technique also fails when the annual returns vary.

Example 28.7

What is the return on invested capital if $1,000 is invested now with $500 being returned in year 4 and $1,000 being returned in year 8?

First, set up the problem as a present worth calculation.

$$P = -1{,}000 + 500(P/F, i\%, 4) + 1{,}000(P/F, i\%, 8)$$

Arbitrarily select $i = 5\%$. The present worth is then found to be $88.15. Next, take a higher value of i to reduce the present worth. If $i = 10\%$, the present worth is $-\$192$. The ROR is found from simple interpolation to be approximately 6.6%.

7 TREATMENT OF SALVAGE VALUE IN REPLACEMENT STUDIES

An investigation into the retirement of an existing process or piece of equipment is known as a *replacement study*. Replacement studies are similar in most respects to other alternative comparison problems: an interest rate is given, two alternatives exist, and one of the previously mentioned methods of comparing alternatives is used to choose the superior alternative.

In replacement studies, the existing process or piece of equipment is known as the *defender*. The new process or piece of equipment being considered for purchase is known as the *challenger*.

Because most defenders still have some market value when they are retired, the problem of what to do with the salvage arises. It seems logical to use the salvage value of the defender to reduce the initial purchase cost of the challenger. This is consistent with what would actually happen if the defender were to be retired.

By convention, however, the salvage value is subtracted from the defender's present value. This does not seem logical, but it is done to keep all costs and benefits related to the defender with the defender. In this case, the salvage value is treated as an opportunity cost which would be incurred if the defender is not retired.

If the defender and the challenger have the same lives and a present worth study is used to choose the superior alternative, the placement of the salvage value will have no effect on the net difference between present worths for the challenger and defender. Although the values of the two present worths will be different depending on the placement, the difference in present worths will be the same.

If the defender and the challenger have different lives, an annual cost comparison must be made. Since the salvage value would be 'spread over' a different number of years depending on its placement, it is important to abide by the conventions listed in this section.

There are a number of ways to handle salvage value. The best way is to think of the EUAC of the defender as the cost of keeping the defender from now until next year. In addition to the usual operating and maintenance costs, that cost would include an opportunity interest cost incurred by not selling the defender and also a drop in the salvage value if the defender is kept for one additional year. Specifically,

$$
\begin{aligned}
\text{EUAC(defender)} = {}& \text{maintenance costs} \\
& + i(\text{current salvage value}) \\
& + (\text{current salvage} - \text{next} \\
& \quad \text{year's salvage})
\end{aligned}
\qquad 28.7
$$

It is important in retirement studies not to double count the salvage value. That is, it would be incorrect to add the salvage value to the defender and at the same time subtract it from the challenger.

8 BASIC INCOME TAX CONSIDERATIONS

Assume that an organization pay $f\%$ of its profits to the federal government as income taxes. If the organization also pays a state income tax of $s\%$, and if state taxes paid are recognized by the federal government as expenses, then the composite tax rate is

$$t = s + f - sf \qquad 28.8$$

The basic principles used to incorporate taxation into economic analyses are listed below.

a. Initial purchase cost is unaffected by income taxes.

b. Salvage value is unaffected by income taxes.

c. Deductible expenses, such as operating costs, maintenance costs, and interest payments, are reduced by $t\%$ (e.g., multiplied by the quantity $(1-t)$).

d. Revenues are reduced by $t\%$ (e.g., multiplied by the quantity $(1-t)$).

e. Depreciation is multiplied by t and added to the appropriate year's cash flow, increasing that year's present worth.

Income taxes and depreciation have no bearing on municipal or governmental projects since municipalities, states, and the U.S. Government pay no taxes.

Example 28.8

A corporation which pays 53% of its revenue in income taxes invests $10,000 in a project which will result in $3,000 anual revenue for 8 years. If the annual expenses are $700, salvage after 8 years is $500, and 9% interest is used, what is the after-tax present worth? Disregard depreciation.

$$P_t = -10,000 + 3000(P/A, 9\%, 8)(1 - .53)$$
$$- 700(P/A, 9\%, 8)(1 - .53)$$
$$+ 500(P/F, 9\%, 8)$$
$$= -3,766$$

It is interesting that the alternative evaluated in example 28.8 is undesirable if income taxes are considered but is desirable if income taxes are omitted.

9 DEPRECIATION

Although depreciation calculations may be considered independently in examination questions, it is important to recognize that depreciation has no effect on economic analysis calculations unless income taxes are also considered.

Generally, tax regulations do not allow the cost of equipment to be treated as a deductible expense in the year of purchase.[1] Rather, portions of the cost may be allocated to each of the years of the item's economic life (which may be different from the actual useful life). Each year, the book value (which is initially equal to the purchase price) is reduced by the depreciation in that year. Theoretically, the book value of an item will equal the market value at any time within the economic life of that item.

Since tax regulations allow the depreciation in any year to be handled as if it were an actual operating expense, and since operating expenses are deductible from the income base prior to taxation, the after-tax profits will be increased. If D is the depreciation, the net result to the after-tax cash flow will be the addition of tD.

The present worth of all depreciation over the economic life of the item is called the *depreciation recovery*. Although originally established to do so, depreciation recovery can never fully replace an item at the end of its life.

---

[1] The IRS tax regulations allow depreciation on almost all forms of *property* except land. The following types of property are distinguished; *real* (e.g., buildings used for business). *residential* (e.g., buildings used as rental property), and *personal* (e.g., equipment used for business). Personal property does *not* include items for personal use, despite its name. *Tangible* personal property is distinguished from *intangible property* (e.g., goodwill, copyrights, patents, trademarks, franchises, and agreements not to compete).

Depreciation is often confused with amortization and depletion. While depreciation spreads the cost of a fixed asset over a number of years, *amortization* spreads the cost of an intangible asset (e.g., a patent) over some basis such as time or expected units of production.

Depletion is another artificial deductible operating expense designed to compensate mining organizations for decreasing mineral reserves. Since original and remaining quantities of minerals are seldom known accurately, the *depletion allowance* is calculated as a fixed percentage of the organization's gross income. These percentages are usually in the 10%–20% range and apply to such mineral deposits as oil, natural gas, coal, uranium, and most metal ores.

There are three common methods of calculating depreciation. The book value of an asset depreciated with the *Straight Line* (SL) *Method* (also known as the *Fixed Percentage Method*) decreases linearly from the initial purchase at $t = 0$ to the estimated salvage at $t = n$. The depreciated amount is the same each year. The quantity $(C - S_n)$ in equation 28.9 is known as the *depreciation base*.

$$D_j = \frac{(C - S_n)}{n} \qquad 28.9$$

Double Declining Balance (DDB) depreciation is independent of salvage value.[2] Furthermore, the book value never stops decreasing, although the depreciation decreases in magnitude. Usually, any remaining book value is written off in the last year of the asset's estimated life. Unlike any of the other depreciation methods, DDB depends on accumulated depreciation.

$$D_j = \frac{2\left(C - \sum_{i=1}^{j-1} D_i\right)}{n} \qquad 28.10$$

In *Sum-of-the-Year's-Digits* (SOYD) depreciation, the digits from 1 to n inclusive, are summed. The total, T, can also be calculated from

$$T = \frac{1}{2}n(n + 1) \qquad 28.11$$

The depreciation can be found from

$$D_j = \frac{(C - S_n)(n - j + 1)}{T} \qquad 28.12$$

Example 28.9

An asset is purchased for $9,000. Its estimated economic life is 10 years, after which it will be sold for

---

[2] Double declining balance depreciation is a particular form of *declining balance depreciation*, as defined by the IRS tax regulations. Declining balance depreciation also includes 125% declining balance and 150% declining balance depreciations which can be calculated by substituting 1.25 and 1.50, respectively, for the 2 in equation 2.10.

$1,000. Find the depreciation in the first three years using SL, DDB, SOYD, and Sinking Fund at 6%.

SL: $\quad D = \frac{(9,000-1,000)}{10} \qquad = 800$ each year

DB: $\quad D_1 = \frac{2(9,000)}{10} \qquad\quad = 1,800$ in year 1

$\qquad\quad D_2 = \frac{2(9,000-1,800)}{10} \quad = 1,440$ in year 2

$\qquad\quad D_3 = \frac{2(9,000-3,240)}{10} \quad = 1,152$ in year 3

SOYD: $T = \frac{1}{2}(10)(11) = 55$

$\qquad\quad D_1 = \left(\frac{10}{55}\right)(9,000 - 1,000) = 1455$ in year 1

$\qquad\quad D_2 = \left(\frac{9}{55}\right)(8,000) \qquad = 1309$ in year 2

$\qquad\quad D_3 = \left(\frac{8}{55}\right)(8,000) \qquad = 1164$ in year 3

Example 28.10

For the asset described in example 28.9, calculate the book value during the first three years if SOYD depreciation is used.

The book value at the beginning of year 1 is $9,000. Then,

$$BV_1 = 9,000 - 1,455 = 7,545$$
$$BV_2 = 7,545 - 1,309 = 6,236$$
$$BV_3 = 6,236 - 1,164 = 5,072$$

Example 28.11

For the asset described in example 28.9, calculate the after-tax depreciation recovery with SL and SOYD depreciation methods. Use 6% interest with 48% income taxes.

SL: $\qquad D.R. = .48(800)(P/A, 6\%, 10) = 2,826$

SOYD: $\qquad$ The depreciation series can be thought of as a constant 1,454 term with a negative 145 gradient.

$$D.R. = .48(1,454)(P/A, 6\%, 10)$$
$$- .48(145)(P/G, 6\%, 10)$$
$$= 3,076$$

10 RATE AND PERIOD CHANGES

All of the foregoing calculations were based on compounding once a year at an *effective interest rate*, i. However, some problems specify compounding more frequently than annually. In such cases, a *nominal interest rate*, r, will be given. The nominal rate does not include the effect of compounding and is not the same as the effective rate, i. A nominal rate may be used to calculate the effective rate by using equation 28.13 or 28.14.

$$i = \left(1 + \frac{r}{k}\right)^k - 1 \qquad\qquad 28.13$$
$$= (1 + \phi)^k - 1 \qquad\qquad 28.14$$

A problem may also specify an effective rate per period, ϕ, (e.g., per month). However, that will be a simple problem since compounding for n periods at an effective rate per period is not affected by the definition or length of the period.

The following rules may be used to determine which interest rate is given in a problem:

- Unless specifically qualified in the problem, the interest rate given is an annual rate.

- If the compounding is annually, the rate given is the effective rate. If compounding is other than annually, the rate given is the nominal rate.

- If the type of compounding is not specified, assume annual compounding.

In the case of continuous compounding, the appropriate discount factors may be calculated from the formulas in table 28.2.

Table 28.2
Discount Factors for Continuous Compounding

(F/P)	e^{rn}
(P/F)	e^{-rn}
(A/F)	$(e^r - 1)/(e^{rn} - 1)$
(F/A)	$(e^{rn} - 1)/(e^r - 1)$
(A/P)	$(e^r - 1)/(1 - e^{-rn})$
(P/A)	$(1 - e^{-rn})/(e^r - 1)$

Example 28.12

A savings and loan offers $5\frac{1}{4}\%$ compounded daily. What is the annual effective rate?

method 1: $\quad r = .0525, \; k = 365 - 1$

$$i = \left(1 + \frac{.0525}{365}\right)^{365} = .0539$$

method 2: $\quad$ Assume daily compounding is the same as continuous compounding.

$$i = (F/P) - 1$$
$$= e^{.0525} - 1 = .0539$$

11 CONSUMER LOANS

Many consumer loans cannot be handled by the equivalence formulas presented up to this point. Many different arrangements can be made between lender and borrower. Four of the most common consumer loan arrangements are presented below. Refer to a real estate or investment analysis book for more complex loans.

A. SIMPLE INTEREST

Interest due does not compound with a *simple interest* loan. The interest due is merely proportional to the length of time the principal is outstanding. Because of this, simple interest loans are seldom made for long periods (e.g., longer than one year).

Example 28.13

A $12,000 simple interest loan is taken out at 16% per annum. The loan matures in one year with no intermediate payments. How much will be due at the end of the year?

$$\text{Amount due} = (1 + .16)(\$12,000) = \$13,920$$

For loans less than one year, it is commonly assumed that a year consists of 12 months of 30 days each.

Example 28.14

$4000 is borrowed for 75 days at 16% per annum simple interest. How much will be due at the end of 75 days?

$$\text{Amount due} = \$4000 + (.16)\left(\frac{75}{360}\right)(4000) = \$4{,}133$$

B. LOANS WITH CONSTANT AMOUNT PAID TOWARDS PRINCIPAL

With this loan type, the payment is not the same each period. The amount paid towards the principal is constant, but the interest varies from period to period. The following special symbols are used.

BAL_j balance after the jth payment

LV total value loaned (cost minus down payment)

j payment or period number

N total number of payments to pay off the loan

PI_j jth interest payment

PP_j jth principal payment

PT_j jth total payment

ϕ effective rate per period (r/k)

The equations which govern this type of loan are

$$BAL_j = LV - (j)(PP) \qquad 28.15$$
$$PI_j = \phi(BAL_j) \qquad 28.16$$
$$PT_j = PP + PI_j \qquad 28.17$$

C. DIRECT REDUCTION LOANS

This is the typical 'interest paid on unpaid balance' loan. The amount of the periodic payment is constant, but the amounts paid towards the principal and interest both vary.

The same symbols are used with this type of loan as are listed above.

$$N = \frac{ln[\frac{-\phi(LV)}{PT} + 1]}{ln(1 + \phi)} \qquad 28.18$$

$$BAL_{j-1} = PT\left[\frac{1 - (1 + \phi)^{j-1-N}}{\phi}\right] \qquad 28.19$$

$$PI_j = \phi(BAL_j) \qquad 28.20$$
$$PP_j = PT - PI_j \qquad 28.21$$
$$BAL_j = BAL_{j-1} - PP_j \qquad 28.22$$

Example 28.15

A $45,000 loan is financed at 9.25% per annum. The monthly payment is $385. What are the amounts paid toward interest and principal in the 14th period? What is the remaining principal balance after the 14th payment has been made?

The effective rate per month is

$$\phi = \frac{r}{k} = \frac{.0925}{12}$$
$$= .007708$$

$$N = -\frac{ln\left[\frac{-(.007708)(45,000)}{385} + 1\right]}{ln(1 + .007708)} = \$301$$

$$BAL_{13} = 385\left[\frac{1 - (1 + .007708)^{14-1-301}}{.007708}\right]$$
$$= \$44{,}476.39$$

$$PI_{14} = (.007708)(\$44{,}476.39) = \$342.82$$
$$PP_{14} = \$385 - \$342.82 = \$42.18$$
$$BAL_{14} = \$44{,}476.39 - \$42.18 = \$44{,}434.20$$

Equation 28.18 calculates the number of payments necessary to pay off a loan. This equation can be solved with effort for the total periodic payment (PT) or the initial value of the loan (LV). It is easier, however, to use the $(A/P, i\%, n)$ factor find the payment and loan value.

$$PT = (LV)(A/P, i\%, n) \qquad 28.23$$

If the loan is repaid in yearly installments, then i is the effective annual rate. If the loan is paid off monthly, then i should be replaced by the effective rate per month (ϕ from equation 28.14). For monthly payments, n is the number of months in the payback period.

D. DIRECT REDUCTION LOAN WITH BALLOON PAYMENT

This type of loan has a constant periodic payment, but the duration of the loan is insufficient to completely pay back the principal. Therefore, all remaining unpaid principal must be paid back in a lump sum when the loan matures. This large payment is known as a *balloon payment*.

Equation 28.18 through 28.22 can also be used with this type of loan. The remaining balance after the last payments is the balloon payment. This balloon payment must be repaid along with the last regular payment calculated.

STANDARD CASH FLOW FACTORS

MULTIPLY   *BY*   *TO OBTAIN*

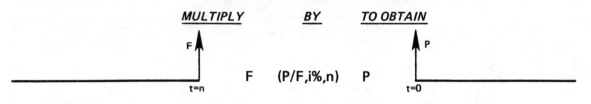

F   (P/F,i%,n)   P

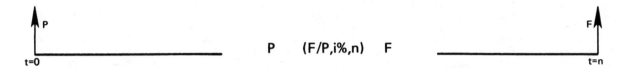

P   (F/P,i%,n)   F

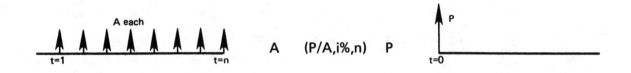

A   (P/A,i%,n)   P

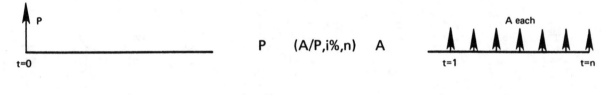

P   (A/P,i%,n)   A

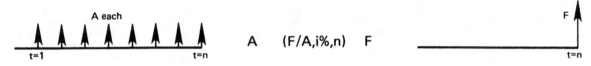

A   (F/A,i%,n)   F

F   (A/F,i%,n)   A

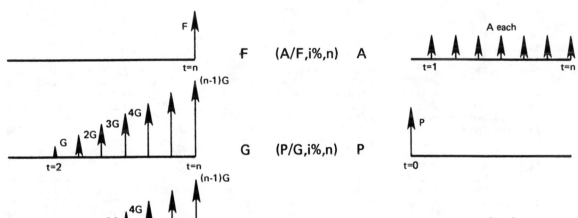

G   (P/G,i%,n)   P

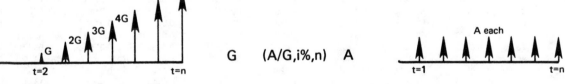

G   (A/G,i%,n)   A

PROFESSIONAL PUBLICATIONS, INC. ● P.O. Box 199, San Carlos, CA 94070

LAND SURVEYOR REFERENCE MANUAL

I = 0.50 %

N	(P/F)	(P/A)	(P/G)	(F/P)	(F/A)	(A/P)	(A/F)	(A/G)	N
1	.9950	0.9950	−0.0000	1.0050	1.0000	1.0050	1.0000	−0.0000	1
2	.9901	1.9851	0.9901	1.0100	2.0050	0.5038	0.4988	0.4988	2
3	.9851	2.9702	2.9604	1.0151	3.0150	0.3367	0.3317	0.9967	3
4	.9802	3.9505	5.9011	1.0202	4.0301	0.2531	0.2481	1.4938	4
5	.9754	4.9259	9.8026	1.0253	5.0503	0.2030	0.1980	1.9900	5
6	.9705	5.8964	14.6552	1.0304	6.0755	0.1696	0.1646	2.4855	6
7	.9657	6.8621	20.4493	1.0355	7.1059	0.1457	0.1407	2.9801	7
8	.9609	7.8230	27.1755	1.0407	8.1414	0.1278	0.1228	3.4738	8
9	.9561	8.7791	34.8244	1.0459	9.1821	0.1139	0.1089	3.9668	9
10	.9513	9.7304	43.3865	1.0511	10.2280	0.1028	0.0978	4.4589	10
11	.9466	10.6770	52.8526	1.0564	11.2792	0.0937	0.0887	4.9501	11
12	.9419	11.6189	63.2136	1.0617	12.3356	0.0861	0.0811	5.4406	12
13	.9372	12.5562	74.4602	1.0670	13.3972	0.0796	0.0746	5.9302	13
14	.9326	13.4887	86.5835	1.0723	14.4642	0.0741	0.0691	6.4190	14
15	.9279	14.4166	99.5743	1.0777	15.5365	0.0694	0.0644	6.9069	15
16	.9233	15.3399	113.4238	1.0831	16.6142	0.0652	0.0602	7.3940	16
17	.9187	16.2586	128.1231	1.0885	17.6973	0.0615	0.0565	7.8803	17
18	.9141	17.1728	143.6634	1.0939	18.7858	0.0582	0.0532	8.3658	18
19	.9096	18.0824	160.0360	1.0994	19.8797	0.0553	0.0503	8.8504	19
20	.9051	18.9874	177.2322	1.1049	20.9791	0.0527	0.0477	9.3342	20
21	.9006	19.8880	195.2434	1.1104	22.0840	0.0503	0.0453	9.8172	21
22	.8961	20.7841	214.0611	1.1160	23.1944	0.0481	0.0431	10.2993	22
23	.8916	21.6757	233.6768	1.1216	24.3104	0.0461	0.0411	10.7806	23
24	.8872	22.5629	254.0820	1.1272	25.4320	0.0443	0.0393	11.2611	24
25	.8828	23.4456	275.2686	1.1328	26.5591	0.0427	0.0377	11.7407	25
26	.8784	24.3240	297.2281	1.1385	27.6919	0.0411	0.0361	12.2195	26
27	.8740	25.1980	319.9523	1.1442	28.8304	0.0397	0.0347	12.6975	27
28	.8697	26.0677	343.4332	1.1499	29.9745	0.0384	0.0334	13.1747	28
29	.8653	26.9330	367.6625	1.1556	31.1244	0.0371	0.0321	13.6510	29
30	.8610	27.7941	392.6324	1.1614	32.2800	0.0360	0.0310	14.1265	30
31	.8567	28.6508	418.3348	1.1672	33.4414	0.0349	0.0299	14.6012	31
32	.8525	29.5033	444.7618	1.1730	34.6086	0.0339	0.0289	15.0750	32
33	.8482	30.3515	471.9055	1.1789	35.7817	0.0329	0.0279	15.5480	33
34	.8440	31.1955	499.7583	1.1848	36.9606	0.0321	0.0271	16.0202	34
35	.8398	32.0354	528.3123	1.1907	38.1454	0.0312	0.0262	16.4915	35
36	.8356	32.8710	557.5598	1.1967	39.3361	0.0304	0.0254	16.9621	36
37	.8315	33.7025	587.4934	1.2027	40.5328	0.0297	0.0247	17.4317	37
38	.8274	34.5299	618.1054	1.2087	41.7354	0.0290	0.0240	17.9006	38
39	.8232	35.3531	649.3883	1.2147	42.9441	0.0283	0.0233	18.3686	39
40	.8191	36.1722	681.3347	1.2208	44.1588	0.0276	0.0226	18.8359	40
41	.8151	36.9873	713.9372	1.2269	45.3796	0.0270	0.0220	19.3022	41
42	.8110	37.7983	747.1886	1.2330	46.6065	0.0265	0.0215	19.7678	42
43	.8070	38.6053	781.0815	1.2392	47.8396	0.0259	0.0209	20.2325	43
44	.8030	39.4082	815.6087	1.2454	49.0788	0.0254	0.0204	20.6964	44
45	.7990	40.2072	850.7631	1.2516	50.3242	0.0249	0.0199	21.1595	45
46	.7950	41.0022	886.5376	1.2579	51.5758	0.0244	0.0194	21.6217	46
47	.7910	41.7932	922.9252	1.2642	52.8337	0.0239	0.0189	22.0831	47
48	.7871	42.5803	959.9188	1.2705	54.0978	0.0235	0.0185	22.5437	48
49	.7832	43.3635	997.5116	1.2768	55.3683	0.0231	0.0181	23.0035	49
50	.7793	44.1428	1035.6966	1.2832	56.6452	0.0227	0.0177	23.4624	50
51	.7754	44.9182	1074.4670	1.2896	57.9284	0.0223	0.0173	23.9205	51
52	.7716	45.6897	1113.8162	1.2961	59.2180	0.0219	0.0169	24.3778	52
53	.7677	46.4575	1153.7372	1.3026	60.5141	0.0215	0.0165	24.8343	53
54	.7639	47.2214	1194.2236	1.3091	61.8167	0.0212	0.0162	25.2899	54
55	.7601	47.9814	1235.2686	1.3156	63.1258	0.0208	0.0158	25.7447	55
60	.7414	51.7256	1448.6458	1.3489	69.7700	0.0193	0.0143	28.0064	60
65	.7231	55.3775	1675.0272	1.3829	76.5821	0.0181	0.0131	30.2475	65
70	.7053	58.9394	1913.6427	1.4178	83.5661	0.0170	0.0120	32.4680	70
75	.6879	62.4136	2163.7525	1.4536	90.7265	0.0160	0.0110	34.6679	75
80	.6710	65.8023	2424.6455	1.4903	98.0677	0.0152	0.0102	36.8474	80
85	.6545	69.1075	2695.6389	1.5280	105.5943	0.0145	0.0095	39.0065	85
90	.6383	72.3313	2976.0769	1.5666	113.3109	0.0138	0.0088	41.1451	90
95	.6226	75.4757	3265.3298	1.6061	121.2224	0.0132	0.0082	43.2633	95
100	.6073	78.5426	3562.7934	1.6467	129.3337	0.0127	0.0077	45.3613	100

PROFESSIONAL PUBLICATIONS, INC. • P.O. Box 199, San Carlos, CA 94070

I = 0.75 %

N	(P/F)	(P/A)	(P/G)	(F/P)	(F/A)	(A/P)	(A/F)	(A/G)	N
1	.9926	0.9926	-0.0000	1.0075	1.0000	1.0075	1.0000	-0.0000	1
2	.9852	1.9777	0.9852	1.0151	2.0075	0.5056	0.4981	0.4981	2
3	.9778	2.9556	2.9408	1.0227	3.0226	0.3383	0.3308	0.9950	3
4	.9706	3.9261	5.8525	1.0303	4.0452	0.2547	0.2472	1.4907	4
5	.9633	4.8894	9.7058	1.0381	5.0756	0.2045	0.1970	1.9851	5
6	.9562	5.8456	14.4866	1.0459	6.1136	0.1711	0.1636	2.4782	6
7	.9490	6.7946	20.1808	1.0537	7.1595	0.1472	0.1397	2.9701	7
8	.9420	7.7366	26.7747	1.0616	8.2132	0.1293	0.1218	3.4608	8
9	.9350	8.6716	34.2544	1.0696	9.2748	0.1153	0.1078	3.9502	9
10	.9280	9.5996	42.6064	1.0776	10.3443	0.1042	0.0967	4.4384	10
11	.9211	10.5207	51.8174	1.0857	11.4219	0.0951	0.0876	4.9253	11
12	.9142	11.4349	61.8740	1.0938	12.5076	0.0875	0.0800	5.4110	12
13	.9074	12.3423	72.7632	1.1020	13.6014	0.0810	0.0735	5.8954	13
14	.9007	13.2430	84.4720	1.1103	14.7034	0.0755	0.0680	6.3786	14
15	.8940	14.1370	96.9876	1.1186	15.8137	0.0707	0.0632	6.8606	15
16	.8873	15.0243	110.2973	1.1270	16.9323	0.0666	0.0591	7.3413	16
17	.8807	15.9050	124.3887	1.1354	18.0593	0.0629	0.0554	7.8207	17
18	.8742	16.7792	139.2494	1.1440	19.1947	0.0596	0.0521	8.2989	18
19	.8676	17.6468	154.8671	1.1525	20.3387	0.0567	0.0492	8.7759	19
20	.8612	18.5080	171.2297	1.1612	21.4912	0.0540	0.0465	9.2516	20
21	.8548	19.3628	188.3253	1.1699	22.6524	0.0516	0.0441	9.7261	21
22	.8484	20.2112	206.1420	1.1787	23.8223	0.0495	0.0420	10.1994	22
23	.8421	21.0533	224.6682	1.1875	25.0010	0.0475	0.0400	10.6714	23
24	.8358	21.8891	243.8923	1.1964	26.1885	0.0457	0.0382	11.1422	24
25	.8296	22.7188	263.8029	1.2054	27.3849	0.0440	0.0365	11.6117	25
26	.8234	23.5422	284.3888	1.2144	28.5903	0.0425	0.0350	12.0800	26
27	.8173	24.3595	305.6387	1.2235	29.8047	0.0411	0.0336	12.5470	27
28	.8112	25.1707	327.5416	1.2327	31.0282	0.0397	0.0322	13.0128	28
29	.8052	25.9759	350.0867	1.2420	32.2609	0.0385	0.0310	13.4774	29
30	.7992	26.7751	373.2631	1.2513	33.5029	0.0373	0.0298	13.9407	30
31	.7932	27.5683	397.0602	1.2607	34.7542	0.0363	0.0288	14.4028	31
32	.7873	28.3557	421.4675	1.2701	36.0148	0.0353	0.0278	14.8636	32
33	.7815	29.1371	446.4746	1.2796	37.2849	0.0343	0.0268	15.3232	33
34	.7757	29.9128	472.0712	1.2892	38.5646	0.0334	0.0259	15.7816	34
35	.7699	30.6827	498.2471	1.2989	39.8538	0.0326	0.0251	16.2387	35
36	.7641	31.4468	524.9924	1.3086	41.1527	0.0318	0.0243	16.6946	36
37	.7585	32.2053	552.2969	1.3185	42.4614	0.0311	0.0236	17.1493	37
38	.7528	32.9581	580.1511	1.3283	43.7798	0.0303	0.0228	17.6027	38
39	.7472	33.7053	608.5451	1.3383	45.1082	0.0297	0.0222	18.0549	39
40	.7416	34.4469	637.4693	1.3483	46.4465	0.0290	0.0215	18.5058	40
41	.7361	35.1831	666.9144	1.3585	47.7948	0.0284	0.0209	18.9556	41
42	.7306	35.9137	696.8709	1.3686	49.1533	0.0278	0.0203	19.4040	42
43	.7252	36.6389	727.3297	1.3789	50.5219	0.0273	0.0198	19.8513	43
44	.7198	37.3587	758.2815	1.3893	51.9009	0.0268	0.0193	20.2973	44
45	.7145	38.0732	789.7173	1.3997	53.2901	0.0263	0.0188	20.7421	45
46	.7091	38.7823	821.6283	1.4102	54.6898	0.0258	0.0183	21.1856	46
47	.7039	39.4862	854.0056	1.4207	56.1000	0.0253	0.0178	21.6280	47
48	.6986	40.1848	886.8404	1.4314	57.5207	0.0249	0.0174	22.0691	48
49	.6934	40.8782	920.1243	1.4421	58.9521	0.0245	0.0170	22.5089	49
50	.6883	41.5664	953.8486	1.4530	60.3943	0.0241	0.0166	22.9476	50
51	.6831	42.2496	988.0050	1.4639	61.8472	0.0237	0.0162	23.3850	51
52	.6780	42.9276	1022.5852	1.4748	63.3111	0.0233	0.0158	23.8211	52
53	.6730	43.6006	1057.5810	1.4859	64.7859	0.0229	0.0154	24.2561	53
54	.6680	44.2686	1092.9842	1.4970	66.2718	0.0226	0.0151	24.6898	54
55	.6630	44.9316	1128.7869	1.5083	67.7688	0.0223	0.0148	25.1223	55
60	.6387	48.1734	1313.5189	1.5657	75.4241	0.0208	0.0133	27.2665	60
65	.6153	51.2963	1507.0910	1.6253	83.3709	0.0195	0.0120	29.3801	65
70	.5927	54.3046	1708.6065	1.6872	91.6201	0.0184	0.0109	31.4634	70
75	.5710	57.2027	1917.2225	1.7514	100.1833	0.0175	0.0100	33.5163	75
80	.5500	59.9944	2132.1472	1.8180	109.0725	0.0167	0.0092	35.5391	80
85	.5299	62.6838	2352.6375	1.8873	118.3001	0.0160	0.0085	37.5318	85
90	.5104	65.2746	2577.9961	1.9591	127.8790	0.0153	0.0078	39.4946	90
95	.4917	67.7704	2807.5694	2.0337	137.8225	0.0148	0.0073	41.4277	95
100	.4737	70.1746	3040.7453	2.1111	148.1445	0.0143	0.0068	43.3311	100

I = 1.00 %

N	(P/F)	(P/A)	(P/G)	(F/P)	(F/A)	(A/P)	(A/F)	(A/G)	N
1	.9901	0.9901	-0.0000	1.0100	1.0000	1.0100	1.0000	-0.0000	1
2	.9803	1.9704	0.9803	1.0201	2.0100	0.5075	0.4975	0.4975	2
3	.9706	2.9410	2.9215	1.0303	3.0301	0.3400	0.3300	0.9934	3
4	.9610	3.9020	5.8044	1.0406	4.0604	0.2563	0.2463	1.4876	4
5	.9515	4.8534	9.6103	1.0510	5.1010	0.2060	0.1960	1.9801	5
6	.9420	5.7955	14.3205	1.0615	6.1520	0.1725	0.1625	2.4710	6
7	.9327	6.7282	19.9168	1.0721	7.2135	0.1486	0.1386	2.9602	7
8	.9235	7.6517	26.3812	1.0829	8.2857	0.1307	0.1207	3.4478	8
9	.9143	8.5660	33.6959	1.0937	9.3685	0.1167	0.1067	3.9337	9
10	.9053	9.4713	41.8435	1.1046	10.4622	0.1056	0.0956	4.4179	10
11	.8963	10.3676	50.8067	1.1157	11.5668	0.0965	0.0865	4.9005	11
12	.8874	11.2551	60.5687	1.1268	12.6825	0.0888	0.0788	5.3815	12
13	.8787	12.1337	71.1126	1.1381	13.8093	0.0824	0.0724	5.8607	13
14	.8700	13.0037	82.4221	1.1495	14.9474	0.0769	0.0669	6.3384	14
15	.8613	13.8651	94.4810	1.1610	16.0969	0.0721	0.0621	6.8143	15
16	.8528	14.7179	107.2734	1.1726	17.2579	0.0679	0.0579	7.2886	16
17	.8444	15.5623	120.7834	1.1843	18.4304	0.0643	0.0543	7.7613	17
18	.8360	16.3983	134.9957	1.1961	19.6147	0.0610	0.0510	8.2323	18
19	.8277	17.2260	149.8950	1.2081	20.8109	0.0581	0.0481	8.7017	19
20	.8195	18.0456	165.4664	1.2202	22.0190	0.0554	0.0454	9.1694	20
21	.8114	18.8570	181.6950	1.2324	23.2392	0.0530	0.0430	9.6354	21
22	.8034	19.6604	198.5663	1.2447	24.4716	0.0509	0.0409	10.0998	22
23	.7954	20.4558	216.0660	1.2572	25.7163	0.0489	0.0389	10.5626	23
24	.7876	21.2434	234.1800	1.2697	26.9735	0.0471	0.0371	11.0237	24
25	.7798	22.0232	252.8945	1.2824	28.2432	0.0454	0.0354	11.4831	25
26	.7720	22.7952	272.1957	1.2953	29.5256	0.0439	0.0339	11.9409	26
27	.7644	23.5596	292.0702	1.3082	30.8209	0.0424	0.0324	12.3971	27
28	.7568	24.3164	312.5047	1.3213	32.1291	0.0411	0.0311	12.8516	28
29	.7493	25.0658	333.4863	1.3345	33.4504	0.0399	0.0299	13.3044	29
30	.7419	25.8077	355.0021	1.3478	34.7849	0.0387	0.0287	13.7557	30
31	.7346	26.5423	377.0394	1.3613	36.1327	0.0377	0.0277	14.2052	31
32	.7273	27.2696	399.5858	1.3749	37.4941	0.0367	0.0267	14.6532	32
33	.7201	27.9897	422.6291	1.3887	38.8690	0.0357	0.0257	15.0995	33
34	.7130	28.7027	446.1572	1.4026	40.2577	0.0348	0.0248	15.5441	34
35	.7059	29.4086	470.1583	1.4166	41.6603	0.0340	0.0240	15.9871	35
36	.6989	30.1075	494.6207	1.4308	43.0769	0.0332	0.0232	16.4285	36
37	.6920	30.7995	519.5329	1.4451	44.5076	0.0325	0.0225	16.8682	37
38	.6852	31.4847	544.8835	1.4595	45.9527	0.0318	0.0218	17.3063	38
39	.6784	32.1630	570.6616	1.4741	47.4123	0.0311	0.0211	17.7428	39
40	.6717	32.8347	596.8561	1.4889	48.8864	0.0305	0.0205	18.1776	40
41	.6650	33.4997	623.4562	1.5038	50.3752	0.0299	0.0199	18.6108	41
42	.6584	34.1581	650.4514	1.5188	51.8790	0.0293	0.0193	19.0424	42
43	.6519	34.8100	677.8312	1.5340	53.3978	0.0287	0.0187	19.4723	43
44	.6454	35.4555	705.5853	1.5493	54.9318	0.0282	0.0182	19.9006	44
45	.6391	36.0945	733.7037	1.5648	56.4811	0.0277	0.0177	20.3273	45
46	.6327	36.7272	762.1765	1.5805	58.0459	0.0272	0.0172	20.7524	46
47	.6265	37.3537	790.9938	1.5963	59.6263	0.0268	0.0168	21.1758	47
48	.6203	37.9740	820.1460	1.6122	61.2226	0.0263	0.0163	21.5976	48
49	.6141	38.5881	849.6237	1.6283	62.8348	0.0259	0.0159	22.0178	49
50	.6080	39.1961	879.4176	1.6446	64.4632	0.0255	0.0155	22.4363	50
51	.6020	39.7981	909.5186	1.6611	66.1078	0.0251	0.0151	22.8533	51
52	.5961	40.3942	939.9175	1.6777	67.7689	0.0248	0.0148	23.2686	52
53	.5902	40.9844	970.6057	1.6945	69.4466	0.0244	0.0144	23.6823	53
54	.5843	41.5687	1001.5743	1.7114	71.1410	0.0241	0.0141	24.0945	54
55	.5785	42.1472	1032.8148	1.7285	72.8525	0.0237	0.0137	24.5049	55
60	.5504	44.9550	1192.8061	1.8167	81.6697	0.0222	0.0122	26.5333	60
65	.5237	47.6266	1358.3903	1.9094	90.9366	0.0210	0.0110	28.5217	65
70	.4983	50.1685	1528.6474	2.0068	100.6763	0.0199	0.0099	30.4703	70
75	.4741	52.5871	1702.7340	2.1091	110.9128	0.0190	0.0090	32.3793	75
80	.4511	54.8882	1879.8771	2.2167	121.6715	0.0182	0.0082	34.2492	80
85	.4292	57.0777	2059.3701	2.3298	132.9790	0.0175	0.0075	36.0801	85
90	.4084	59.1609	2240.5675	2.4486	144.8633	0.0169	0.0069	37.8724	90
95	.3886	61.1430	2422.8811	2.5735	157.3538	0.0164	0.0064	39.6265	95
100	.3697	63.0289	2605.7758	2.7048	170.4814	0.0159	0.0059	41.3426	100

I = 1.50 %

N	(P/F)	(P/A)	(P/G)	(F/P)	(F/A)	(A/P)	(A/F)	(A/G)	N
1	.9852	0.9852	-0.0000	1.0150	1.0000	1.0150	1.0000	-0.0000	1
2	.9707	1.9559	0.9707	1.0302	2.0150	0.5113	0.4963	0.4963	2
3	.9563	2.9122	2.8833	1.0457	3.0452	0.3434	0.3284	0.9901	3
4	.9422	3.8544	5.7098	1.0614	4.0909	0.2594	0.2444	1.4814	4
5	.9283	4.7826	9.4229	1.0773	5.1523	0.2091	0.1941	1.9702	5
6	.9145	5.6972	13.9956	1.0934	6.2296	0.1755	0.1605	2.4566	6
7	.9010	6.5982	19.4018	1.1098	7.3230	0.1516	0.1366	2.9405	7
8	.8877	7.4859	25.6157	1.1265	8.4328	0.1336	0.1186	3.4219	8
9	.8746	8.3605	32.6125	1.1434	9.5593	0.1196	0.1046	3.9008	9
10	.8617	9.2222	40.3675	1.1605	10.7027	0.1084	0.0934	4.3772	10
11	.8489	10.0711	48.8568	1.1779	11.8633	0.0993	0.0843	4.8512	11
12	.8364	10.9075	58.0571	1.1956	13.0412	0.0917	0.0767	5.3227	12
13	.8240	11.7315	67.9454	1.2136	14.2368	0.0852	0.0702	5.7917	13
14	.8118	12.5434	78.4994	1.2318	15.4504	0.0797	0.0647	6.2582	14
15	.7999	13.3432	89.6974	1.2502	16.6821	0.0749	0.0599	6.7223	15
16	.7880	14.1313	101.5178	1.2690	17.9324	0.0708	0.0558	7.1839	16
17	.7764	14.9076	113.9400	1.2880	19.2014	0.0671	0.0521	7.6431	17
18	.7649	15.6726	126.9435	1.3073	20.4894	0.0638	0.0488	8.0997	18
19	.7536	16.4262	140.5084	1.3270	21.7967	0.0609	0.0459	8.5539	19
20	.7425	17.1686	154.6154	1.3469	23.1237	0.0582	0.0432	9.0057	20
21	.7315	17.9001	169.2453	1.3671	24.4705	0.0559	0.0409	9.4550	21
22	.7207	18.6208	184.3798	1.3876	25.8376	0.0537	0.0387	9.9018	22
23	.7100	19.3309	200.0006	1.4084	27.2251	0.0517	0.0367	10.3462	23
24	.6995	20.0304	216.0901	1.4295	28.6335	0.0499	0.0349	10.7881	24
25	.6892	20.7196	232.6310	1.4509	30.0630	0.0483	0.0333	11.2276	25
26	.6790	21.3986	249.6065	1.4727	31.5140	0.0467	0.0317	11.6646	26
27	.6690	22.0676	267.0002	1.4948	32.9867	0.0453	0.0303	12.0992	27
28	.6591	22.7267	284.7958	1.5172	34.4815	0.0440	0.0290	12.5313	28
29	.6494	23.3761	302.9779	1.5400	35.9987	0.0428	0.0278	12.9610	29
30	.6398	24.0158	321.5310	1.5631	37.5387	0.0416	0.0266	13.3883	30
31	.6303	24.6461	340.4402	1.5865	39.1018	0.0406	0.0256	13.8131	31
32	.6210	25.2671	359.6910	1.6103	40.6883	0.0396	0.0246	14.2355	32
33	.6118	25.8790	379.2691	1.6345	42.2986	0.0386	0.0236	14.6555	33
34	.6028	26.4817	399.1607	1.6590	43.9331	0.0378	0.0228	15.0731	34
35	.5939	27.0756	419.3521	1.6839	45.5921	0.0369	0.0219	15.4882	35
36	.5851	27.6607	439.8303	1.7091	47.2760	0.0362	0.0212	15.9009	36
37	.5764	28.2371	460.5822	1.7348	48.9851	0.0354	0.0204	16.3112	37
38	.5679	28.8051	481.5954	1.7608	50.7199	0.0347	0.0197	16.7191	38
39	.5595	29.3646	502.8576	1.7872	52.4807	0.0341	0.0191	17.1246	39
40	.5513	29.9158	524.3568	1.8140	54.2679	0.0334	0.0184	17.5277	40
41	.5431	30.4590	546.0814	1.8412	56.0819	0.0328	0.0178	17.9284	41
42	.5351	30.9941	568.0201	1.8688	57.9231	0.0323	0.0173	18.3267	42
43	.5272	31.5212	590.1617	1.8969	59.7920	0.0317	0.0167	18.7227	43
44	.5194	32.0406	612.4955	1.9253	61.6889	0.0312	0.0162	19.1162	44
45	.5117	32.5523	635.0110	1.9542	63.6142	0.0307	0.0157	19.5074	45
46	.5042	33.0565	657.6979	1.9835	65.5684	0.0303	0.0153	19.8962	46
47	.4967	33.5532	680.5462	2.0133	67.5519	0.0298	0.0148	20.2826	47
48	.4894	34.0426	703.5462	2.0435	69.5652	0.0294	0.0144	20.6667	48
49	.4821	34.5247	726.6884	2.0741	71.6087	0.0290	0.0140	21.0484	49
50	.4750	34.9997	749.9636	2.1052	73.6828	0.0286	0.0136	21.4277	50
51	.4680	35.4677	773.3629	2.1368	75.7881	0.0282	0.0132	21.8047	51
52	.4611	35.9287	796.8774	2.1689	77.9249	0.0278	0.0128	22.1794	52
53	.4543	36.3830	820.4986	2.2014	80.0938	0.0275	0.0125	22.5517	53
54	.4475	36.8305	844.2184	2.2344	82.2952	0.0272	0.0122	22.9217	54
55	.4409	37.2715	868.0285	2.2679	84.5296	0.0268	0.0118	23.2894	55
60	.4093	39.3803	988.1674	2.4432	96.2147	0.0254	0.0104	25.0930	60
65	.3799	41.3378	1109.4752	2.6320	108.8028	0.0242	0.0092	26.8393	65
70	.3527	43.1549	1231.1658	2.8355	122.3638	0.0232	0.0082	28.5290	70
75	.3274	44.8416	1352.5600	3.0546	136.9728	0.0223	0.0073	30.1631	75
80	.3039	46.4073	1473.0741	3.2907	152.7109	0.0215	0.0065	31.7423	80
85	.2821	47.8607	1592.2095	3.5450	169.6652	0.0209	0.0059	33.2676	85
90	.2619	49.2099	1709.5439	3.8189	187.9299	0.0203	0.0053	34.7399	90
95	.2431	50.4622	1824.7224	4.1141	207.6061	0.0198	0.0048	36.1602	95
100	.2256	51.6247	1937.4506	4.4320	228.8030	0.0194	0.0044	37.5295	100

PROFESSIONAL PUBLICATIONS, INC. • P.O. Box 199, San Carlos, CA   94070

I = 2.00 %

N	(P/F)	(P/A)	(P/G)	(F/P)	(F/A)	(A/P)	(A/F)	(A/G)	N
1	.9804	0.9804	-0.0000	1.0200	1.0000	1.0200	1.0000	-0.0000	1
2	.9612	1.9416	0.9612	1.0404	2.0200	0.5150	0.4950	0.4950	2
3	.9423	2.8839	2.8458	1.0612	3.0604	0.3468	0.3268	0.9868	3
4	.9238	3.8077	5.6173	1.0824	4.1216	0.2626	0.2426	1.4752	4
5	.9057	4.7135	9.2403	1.1041	5.2040	0.2122	0.1922	1.9604	5
6	.8880	5.6014	13.6801	1.1262	6.3081	0.1785	0.1585	2.4423	6
7	.8706	6.4720	18.9035	1.1487	7.4343	0.1545	0.1345	2.9208	7
8	.8535	7.3255	24.8779	1.1717	8.5830	0.1365	0.1165	3.3961	8
9	.8368	8.1622	31.5720	1.1951	9.7546	0.1225	0.1025	3.8681	9
10	.8203	8.9826	38.9551	1.2190	10.9497	0.1113	0.0913	4.3367	10
11	.8043	9.7868	46.9977	1.2434	12.1687	0.1022	0.0822	4.8021	11
12	.7885	10.5753	55.6712	1.2682	13.4121	0.0946	0.0746	5.2642	12
13	.7730	11.3484	64.9475	1.2936	14.6803	0.0881	0.0681	5.7231	13
14	.7579	12.1062	74.7999	1.3195	15.9739	0.0826	0.0626	6.1786	14
15	.7430	12.8493	85.2021	1.3459	17.2934	0.0778	0.0578	6.6309	15
16	.7284	13.5777	96.1288	1.3728	18.6393	0.0737	0.0537	7.0799	16
17	.7142	14.2919	107.5554	1.4002	20.0121	0.0700	0.0500	7.5256	17
18	.7002	14.9920	119.4581	1.4282	21.4123	0.0667	0.0467	7.9681	18
19	.6864	15.6785	131.8139	1.4568	22.8406	0.0638	0.0438	8.4073	19
20	.6730	16.3514	144.6003	1.4859	24.2974	0.0612	0.0412	8.8433	20
21	.6598	17.0112	157.7959	1.5157	25.7833	0.0588	0.0388	9.2760	21
22	.6468	17.6580	171.3795	1.5460	27.2990	0.0566	0.0366	9.7055	22
23	.6342	18.2922	185.3309	1.5769	28.8450	0.0547	0.0347	10.1317	23
24	.6217	18.9139	199.6305	1.6084	30.4219	0.0529	0.0329	10.5547	24
25	.6095	19.5235	214.2592	1.6406	32.0303	0.0512	0.0312	10.9745	25
26	.5976	20.1210	229.1987	1.6734	33.6709	0.0497	0.0297	11.3910	26
27	.5859	20.7069	244.4311	1.7069	35.3443	0.0483	0.0283	11.8043	27
28	.5744	21.2813	259.9392	1.7410	37.0512	0.0470	0.0270	12.2145	28
29	.5631	21.8444	275.7064	1.7758	38.7922	0.0458	0.0258	12.6214	29
30	.5521	22.3965	291.7164	1.8114	40.5681	0.0446	0.0246	13.0251	30
31	.5412	22.9377	307.9538	1.8476	42.3794	0.0436	0.0236	13.4257	31
32	.5306	23.4683	324.4035	1.8845	44.2270	0.0426	0.0226	13.8230	32
33	.5202	23.9886	341.0508	1.9222	46.1116	0.0417	0.0217	14.2172	33
34	.5100	24.4986	357.8817	1.9607	48.0338	0.0408	0.0208	14.6083	34
35	.5000	24.9986	374.8826	1.9999	49.9945	0.0400	0.0200	14.9961	35
36	.4902	25.4888	392.0405	2.0399	51.9944	0.0392	0.0192	15.3809	36
37	.4806	25.9695	409.3424	2.0807	54.0343	0.0385	0.0185	15.7625	37
38	.4712	26.4406	426.7764	2.1223	56.1149	0.0378	0.0178	16.1409	38
39	.4619	26.9026	444.3304	2.1647	58.2372	0.0372	0.0172	16.5163	39
40	.4529	27.3555	461.9931	2.2080	60.4020	0.0366	0.0166	16.8885	40
41	.4440	27.7995	479.7535	2.2522	62.6100	0.0360	0.0160	17.2576	41
42	.4353	28.2348	497.6010	2.2972	64.8622	0.0354	0.0154	17.6237	42
43	.4268	28.6616	515.5253	2.3432	67.1595	0.0349	0.0149	17.9866	43
44	.4184	29.0800	533.5165	2.3901	69.5027	0.0344	0.0144	18.3465	44
45	.4102	29.4902	551.5652	2.4379	71.8927	0.0339	0.0139	18.7034	45
46	.4022	29.8923	569.6621	2.4866	74.3306	0.0335	0.0135	19.0571	46
47	.3943	30.2866	587.7985	2.5363	76.8172	0.0330	0.0130	19.4079	47
48	.3865	30.6731	605.9657	2.5871	79.3535	0.0326	0.0126	19.7556	48
49	.3790	31.0521	624.1557	2.6388	81.9406	0.0322	0.0122	20.1003	49
50	.3715	31.4236	642.3606	2.6916	84.5794	0.0318	0.0118	20.4420	50
51	.3642	31.7878	660.5727	2.7454	87.2710	0.0315	0.0115	20.7807	51
52	.3571	32.1449	678.7849	2.8003	90.0164	0.0311	0.0111	21.1164	52
53	.3501	32.4950	696.9900	2.8563	92.8167	0.0308	0.0108	21.4491	53
54	.3432	32.8383	715.1815	2.9135	95.6731	0.0305	0.0105	21.7789	54
55	.3365	33.1748	733.3527	2.9717	98.5865	0.0301	0.0101	22.1057	55
60	.3048	34.7609	823.6975	3.2810	114.0515	0.0288	0.0088	23.6961	60
65	.2761	36.1975	912.7085	3.6225	131.1262	0.0276	0.0076	25.2147	65
70	.2500	37.4986	999.8343	3.9996	149.9779	0.0267	0.0067	26.6632	70
75	.2265	38.6771	1084.6393	4.4158	170.7918	0.0259	0.0059	28.0434	75
80	.2051	39.7445	1166.7868	4.8754	193.7720	0.0252	0.0052	29.3572	80
85	.1858	40.7113	1246.0241	5.3829	219.1439	0.0246	0.0046	30.6064	85
90	.1683	41.5869	1322.1701	5.9431	247.1567	0.0240	0.0040	31.7929	90
95	.1524	42.3800	1395.1033	6.5617	278.0850	0.0236	0.0036	32.9189	95
100	.1380	43.0984	1464.7527	7.2446	312.2323	0.0232	0.0032	33.9863	100

I = 3.00 %

N	(P/F)	(P/A)	(P/G)	(F/P)	(F/A)	(A/P)	(A/F)	(A/G)	N
1	.9709	0.9709	-0.0000	1.0300	1.0000	1.0300	1.0000	-0.0000	1
2	.9426	1.9135	0.9426	1.0609	2.0300	0.5226	0.4926	0.4926	2
3	.9151	2.8286	2.7729	1.0927	3.0909	0.3535	0.3235	0.9803	3
4	.8885	3.7171	5.4383	1.1255	4.1836	0.2690	0.2390	1.4631	4
5	.8626	4.5797	8.8888	1.1593	5.3091	0.2184	0.1884	1.9409	5
6	.8375	5.4172	13.0762	1.1941	6.4684	0.1846	0.1546	2.4138	6
7	.8131	6.2303	17.9547	1.2299	7.6625	0.1605	0.1305	2.8819	7
8	.7894	7.0197	23.4806	1.2668	8.8923	0.1425	0.1125	3.3450	8
9	.7664	7.7861	29.6119	1.3048	10.1591	0.1284	0.0984	3.8032	9
10	.7441	8.5302	36.3088	1.3439	11.4639	0.1172	0.0872	4.2565	10
11	.7224	9.2526	43.5330	1.3842	12.8078	0.1081	0.0781	4.7049	11
12	.7014	9.9540	51.2482	1.4258	14.1920	0.1005	0.0705	5.1485	12
13	.6810	10.6350	59.4196	1.4685	15.6178	0.0940	0.0640	5.5872	13
14	.6611	11.2961	68.0141	1.5126	17.0863	0.0885	0.0585	6.0210	14
15	.6419	11.9379	77.0002	1.5580	18.5989	0.0838	0.0538	6.4500	15
16	.6232	12.5611	86.3477	1.6047	20.1569	0.0796	0.0496	6.8742	16
17	.6050	13.1661	96.0280	1.6528	21.7616	0.0760	0.0460	7.2936	17
18	.5874	13.7535	106.0137	1.7024	23.4144	0.0727	0.0427	7.7081	18
19	.5703	14.3238	116.2788	1.7535	25.1169	0.0698	0.0398	8.1179	19
20	.5537	14.8775	126.7987	1.8061	26.8704	0.0672	0.0372	8.5229	20
21	.5375	15.4150	137.5496	1.8603	28.6765	0.0649	0.0349	8.9231	21
22	.5219	15.9369	148.5094	1.9161	30.5368	0.0627	0.0327	9.3186	22
23	.5067	16.4436	159.6566	1.9736	32.4529	0.0608	0.0308	9.7093	23
24	.4919	16.9355	170.9711	2.0328	34.4265	0.0590	0.0290	10.0954	24
25	.4776	17.4131	182.4336	2.0938	36.4593	0.0574	0.0274	10.4768	25
26	.4637	17.8768	194.0260	2.1566	38.5530	0.0559	0.0259	10.8535	26
27	.4502	18.3270	205.7309	2.2213	40.7096	0.0546	0.0246	11.2255	27
28	.4371	18.7641	217.5320	2.2879	42.9309	0.0533	0.0233	11.5930	28
29	.4243	19.1885	229.4137	2.3566	45.2189	0.0521	0.0221	11.9558	29
30	.4120	19.6004	241.3613	2.4273	47.5754	0.0510	0.0210	12.3141	30
31	.4000	20.0004	253.3609	2.5001	50.0027	0.0500	0.0200	12.6678	31
32	.3883	20.3888	265.3993	2.5751	52.5028	0.0490	0.0190	13.0169	32
33	.3770	20.7658	277.4642	2.6523	55.0778	0.0482	0.0182	13.3616	33
34	.3660	21.1318	289.5437	2.7319	57.7302	0.0473	0.0173	13.7018	34
35	.3554	21.4872	301.6267	2.8139	60.4621	0.0465	0.0165	14.0375	35
36	.3450	21.8323	313.7028	2.8983	63.2759	0.0458	0.0158	14.3688	36
37	.3350	22.1672	325.7622	2.9852	66.1742	0.0451	0.0151	14.6957	37
38	.3252	22.4925	337.7956	3.0748	69.1594	0.0445	0.0145	15.0182	38
39	.3158	22.8082	349.7942	3.1670	72.2342	0.0438	0.0138	15.3363	39
40	.3066	23.1148	361.7499	3.2620	75.4013	0.0433	0.0133	15.6502	40
41	.2976	23.4124	373.6551	3.3599	78.6633	0.0427	0.0127	15.9597	41
42	.2890	23.7014	385.5024	3.4607	82.0232	0.0422	0.0122	16.2650	42
43	.2805	23.9819	397.2852	3.5645	85.4839	0.0417	0.0117	16.5660	43
44	.2724	24.2543	408.9972	3.6715	89.0484	0.0412	0.0112	16.8629	44
45	.2644	24.5187	420.6325	3.7816	92.7199	0.0408	0.0108	17.1556	45
46	.2567	24.7754	432.1856	3.8950	96.5015	0.0404	0.0104	17.4441	46
47	.2493	25.0247	443.6515	4.0119	100.3965	0.0400	0.0100	17.7285	47
48	.2420	25.2667	455.0255	4.1323	104.4084	0.0396	0.0096	18.0089	48
49	.2350	25.5017	466.3031	4.2562	108.5406	0.0392	0.0092	18.2852	49
50	.2281	25.7298	477.4803	4.3839	112.7969	0.0389	0.0089	18.5575	50
51	.2215	25.9512	488.5535	4.5154	117.1808	0.0385	0.0085	18.8258	51
52	.2150	26.1662	499.5191	4.6509	121.6962	0.0382	0.0082	19.0902	52
53	.2088	26.3750	510.3742	4.7904	126.3471	0.0379	0.0079	19.3507	53
54	.2027	26.5777	521.1157	4.9341	131.1375	0.0376	0.0076	19.6073	54
55	.1968	26.7744	531.7411	5.0821	136.0716	0.0373	0.0073	19.8600	55
60	.1697	27.6756	583.0526	5.8916	163.0534	0.0361	0.0061	21.0674	60
65	.1464	28.4529	631.2010	6.8300	194.3328	0.0351	0.0051	22.1841	65
70	.1263	29.1234	676.0869	7.9178	230.5941	0.0343	0.0043	23.2145	70
75	.1089	29.7018	717.6978	9.1789	272.6309	0.0337	0.0037	24.1634	75
80	.0940	30.2008	756.0865	10.6409	321.3630	0.0331	0.0031	25.0353	80
85	.0811	30.6312	791.3529	12.3357	377.8570	0.0326	0.0026	25.8349	85
90	.0699	31.0024	823.6302	14.3005	443.3489	0.0323	0.0023	26.5667	90
95	.0603	31.3227	853.0742	16.5782	519.2720	0.0319	0.0019	27.2351	95
100	.0520	31.5989	879.8540	19.2186	607.2877	0.0316	0.0016	27.8444	100

LAND SURVEYOR REFERENCE MANUAL

I = 4.00 %

N	(P/F)	(P/A)	(P/G)	(F/P)	(F/A)	(A/P)	(A/F)	(A/G)	N
1	.9615	0.9615	-0.0000	1.0400	1.0000	1.0400	1.0000	-0.0000	1
2	.9246	1.8861	0.9246	1.0816	2.0400	0.5302	0.4902	0.4902	2
3	.8890	2.7751	2.7025	1.1249	3.1216	0.3603	0.3203	0.9739	3
4	.8548	3.6299	5.2670	1.1699	4.2465	0.2755	0.2355	1.4510	4
5	.8219	4.4518	8.5547	1.2167	5.4163	0.2246	0.1846	1.9216	5
6	.7903	5.2421	12.5062	1.2653	6.6330	0.1908	0.1508	2.3857	6
7	.7599	6.0021	17.0657	1.3159	7.8983	0.1666	0.1266	2.8433	7
8	.7307	6.7327	22.1806	1.3686	9.2142	0.1485	0.1085	3.2944	8
9	.7026	7.4353	27.8013	1.4233	10.5828	0.1345	0.0945	3.7391	9
10	.6756	8.1109	33.8814	1.4802	12.0061	0.1233	0.0833	4.1773	10
11	.6496	8.7605	40.3772	1.5395	13.4864	0.1141	0.0741	4.6090	11
12	.6246	9.3851	47.2477	1.6010	15.0258	0.1066	0.0666	5.0343	12
13	.6006	9.9856	54.4546	1.6651	16.6268	0.1001	0.0601	5.4533	13
14	.5775	10.5631	61.9618	1.7317	18.2919	0.0947	0.0547	5.8659	14
15	.5553	11.1184	69.7355	1.8009	20.0236	0.0899	0.0499	6.2721	15
16	.5339	11.6523	77.7441	1.8730	21.8245	0.0858	0.0458	6.6720	16
17	.5134	12.1657	85.9581	1.9479	23.6975	0.0822	0.0422	7.0656	17
18	.4936	12.6593	94.3498	2.0258	25.6454	0.0790	0.0390	7.4530	18
19	.4746	13.1339	102.8933	2.1068	27.6712	0.0761	0.0361	7.8342	19
20	.4564	13.5903	111.5647	2.1911	29.7781	0.0736	0.0336	8.2091	20
21	.4388	14.0292	120.3414	2.2788	31.9692	0.0713	0.0313	8.5779	21
22	.4220	14.4511	129.2024	2.3699	34.2480	0.0692	0.0292	8.9407	22
23	.4057	14.8568	138.1284	2.4647	36.6179	0.0673	0.0273	9.2973	23
24	.3901	15.2470	147.1012	2.5633	39.0826	0.0656	0.0256	9.6479	24
25	.3751	15.6221	156.1040	2.6658	41.6459	0.0640	0.0240	9.9925	25
26	.3607	15.9828	165.1212	2.7725	44.3117	0.0626	0.0226	10.3312	26
27	.3468	16.3296	174.1385	2.8834	47.0842	0.0612	0.0212	10.6640	27
28	.3335	16.6631	183.1424	2.9987	49.9676	0.0600	0.0200	10.9909	28
29	.3207	16.9837	192.1206	3.1187	52.9663	0.0589	0.0189	11.3120	29
30	.3083	17.2920	201.0618	3.2434	56.0849	0.0578	0.0178	11.6274	30
31	.2965	17.5885	209.9556	3.3731	59.3283	0.0569	0.0169	11.9371	31
32	.2851	17.8736	218.7924	3.5081	62.7015	0.0559	0.0159	12.2411	32
33	.2741	18.1476	227.5634	3.6484	66.2095	0.0551	0.0151	12.5396	33
34	.2636	18.4112	236.2607	3.7943	69.8579	0.0543	0.0143	12.8324	34
35	.2534	18.6646	244.8768	3.9461	73.6522	0.0536	0.0136	13.1198	35
36	.2437	18.9083	253.4052	4.1039	77.5983	0.0529	0.0129	13.4018	36
37	.2343	19.1426	261.8399	4.2681	81.7022	0.0522	0.0122	13.6784	37
38	.2253	19.3679	270.1754	4.4388	85.9703	0.0516	0.0116	13.9497	38
39	.2166	19.5845	278.4070	4.6164	90.4091	0.0511	0.0111	14.2157	39
40	.2083	19.7928	286.5303	4.8010	95.0255	0.0505	0.0105	14.4765	40
41	.2003	19.9931	294.5414	4.9931	99.8265	0.0500	0.0100	14.7322	41
42	.1926	20.1856	302.4370	5.1928	104.8196	0.0495	0.0095	14.9828	42
43	.1852	20.3708	310.2141	5.4005	110.0124	0.0491	0.0091	15.2284	43
44	.1780	20.5488	317.8700	5.6165	115.4129	0.0487	0.0087	15.4690	44
45	.1712	20.7200	325.4028	5.8412	121.0294	0.0483	0.0083	15.7047	45
46	.1646	20.8847	332.8104	6.0748	126.8706	0.0479	0.0079	15.9356	46
47	.1583	21.0429	340.0914	6.3178	132.9454	0.0475	0.0075	16.1618	47
48	.1522	21.1951	347.2446	6.5705	139.2632	0.0472	0.0072	16.3832	48
49	.1463	21.3415	354.2689	6.8333	145.8337	0.0469	0.0069	16.6000	49
50	.1407	21.4822	361.1638	7.1067	152.6671	0.0466	0.0066	16.8122	50
51	.1353	21.6175	367.9289	7.3910	159.7738	0.0463	0.0063	17.0200	51
52	.1301	21.7476	374.5638	7.6866	167.1647	0.0460	0.0060	17.2232	52
53	.1251	21.8727	381.0686	7.9941	174.8513	0.0457	0.0057	17.4221	53
54	.1203	21.9930	387.4436	8.3138	182.8454	0.0455	0.0055	17.6167	54
55	.1157	22.1086	393.6890	8.6464	191.1592	0.0452	0.0052	17.8070	55
60	.0951	22.6235	422.9966	10.5196	237.9907	0.0442	0.0042	18.6972	60
65	.0781	23.0467	449.2014	12.7987	294.9684	0.0434	0.0034	19.4909	65
70	.0642	23.3945	472.4789	15.5716	364.2905	0.0427	0.0027	20.1961	70
75	.0528	23.6804	493.0408	18.9453	448.6314	0.0422	0.0022	20.8206	75
80	.0434	23.9154	511.1161	23.0498	551.2450	0.0418	0.0018	21.3718	80
85	.0357	24.1085	526.9384	28.0436	676.0901	0.0415	0.0015	21.8569	85
90	.0293	24.2673	540.7369	34.1193	827.9833	0.0412	0.0012	22.2826	90
95	.0241	24.3978	552.7307	41.5114	1012.7846	0.0410	0.0010	22.6550	95
100	.0198	24.5050	563.1249	50.5049	1237.6237	0.0408	0.0008	22.9800	100

I = 5.00 %

N	(P/F)	(P/A)	(P/G)	(F/P)	(F/A)	(A/P)	(A/F)	(A/G)	N
1	.9524	0.9524	-0.0000	1.0500	1.0000	1.0500	1.0000	-0.0000	1
2	.9070	1.8594	0.9070	1.1025	2.0500	0.5378	0.4878	0.4878	2
3	.8638	2.7232	2.6347	1.1576	3.1525	0.3672	0.3172	0.9675	3
4	.8227	3.5460	5.1028	1.2155	4.3101	0.2820	0.2320	1.4391	4
5	.7835	4.3295	8.2369	1.2763	5.5256	0.2310	0.1810	1.9025	5
6	.7462	5.0757	11.9680	1.3401	6.8019	0.1970	0.1470	2.3579	6
7	.7107	5.7864	16.2321	1.4071	8.1420	0.1728	0.1228	2.8052	7
8	.6768	6.4632	20.9700	1.4775	9.5491	0.1547	0.1047	3.2445	8
9	.6446	7.1078	26.1268	1.5513	11.0266	0.1407	0.0907	3.6758	9
10	.6139	7.7217	31.6520	1.6289	12.5779	0.1295	0.0795	4.0991	10
11	.5847	8.3064	37.4988	1.7103	14.2068	0.1204	0.0704	4.5144	11
12	.5568	8.8633	43.6241	1.7959	15.9171	0.1128	0.0628	4.9219	12
13	.5303	9.3936	49.9879	1.8856	17.7130	0.1065	0.0565	5.3215	13
14	.5051	9.8986	56.5538	1.9799	19.5986	0.1010	0.0510	5.7133	14
15	.4810	10.3797	63.2880	2.0789	21.5786	0.0963	0.0463	6.0973	15
16	.4581	10.8378	70.1597	2.1829	23.6575	0.0923	0.0423	6.4736	16
17	.4363	11.2741	77.1405	2.2920	25.8404	0.0887	0.0387	6.8423	17
18	.4155	11.6896	84.2043	2.4066	28.1324	0.0855	0.0355	7.2034	18
19	.3957	12.0853	91.3275	2.5270	30.5390	0.0827	0.0327	7.5569	19
20	.3769	12.4622	98.4884	2.6533	33.0660	0.0802	0.0302	7.9030	20
21	.3589	12.8212	105.6673	2.7860	35.7193	0.0780	0.0280	8.2416	21
22	.3418	13.1630	112.8461	2.9253	38.5052	0.0760	0.0260	8.5730	22
23	.3256	13.4886	120.0087	3.0715	41.4305	0.0741	0.0241	8.8971	23
24	.3101	13.7986	127.1402	3.2251	44.5020	0.0725	0.0225	9.2140	24
25	.2953	14.0939	134.2275	3.3864	47.7271	0.0710	0.0210	9.5238	25
26	.2812	14.3752	141.2585	3.5557	51.1135	0.0696	0.0196	9.8266	26
27	.2678	14.6430	148.2226	3.7335	54.6691	0.0683	0.0183	10.1224	27
28	.2551	14.8981	155.1101	3.9201	58.4026	0.0671	0.0171	10.4114	28
29	.2429	15.1411	161.9126	4.1161	62.3227	0.0660	0.0160	10.6936	29
30	.2314	15.3725	168.6226	4.3219	66.4388	0.0651	0.0151	10.9691	30
31	.2204	15.5928	175.2333	4.5380	70.7608	0.0641	0.0141	11.2381	31
32	.2099	15.8027	181.7392	4.7649	75.2988	0.0633	0.0133	11.5005	32
33	.1999	16.0025	188.1351	5.0032	80.0638	0.0625	0.0125	11.7566	33
34	.1904	16.1929	194.4168	5.2533	85.0670	0.0618	0.0118	12.0063	34
35	.1813	16.3742	200.5807	5.5160	90.3203	0.0611	0.0111	12.2498	35
36	.1727	16.5469	206.6237	5.7918	95.8363	0.0604	0.0104	12.4872	36
37	.1644	16.7113	212.5434	6.0814	101.6281	0.0598	0.0098	12.7186	37
38	.1566	16.8679	218.3378	6.3855	107.7095	0.0593	0.0093	12.9440	38
39	.1491	17.0170	224.0054	6.7048	114.0950	0.0588	0.0088	13.1636	39
40	.1420	17.1591	229.5452	7.0400	120.7998	0.0583	0.0083	13.3775	40
41	.1353	17.2944	234.9564	7.3920	127.8398	0.0578	0.0078	13.5857	41
42	.1288	17.4232	240.2389	7.7616	135.2318	0.0574	0.0074	13.7884	42
43	.1227	17.5459	245.3925	8.1497	142.9933	0.0570	0.0070	13.9857	43
44	.1169	17.6628	250.4175	8.5572	151.1430	0.0566	0.0066	14.1777	44
45	.1113	17.7741	255.3145	8.9850	159.7002	0.0563	0.0063	14.3644	45
46	.1060	17.8801	260.0844	9.4343	168.6852	0.0559	0.0059	14.5461	46
47	.1009	17.9810	264.7281	9.9060	178.1194	0.0556	0.0056	14.7226	47
48	.0961	18.0772	269.2467	10.4013	188.0254	0.0553	0.0053	14.8943	48
49	.0916	18.1687	273.6418	10.9213	198.4267	0.0550	0.0050	15.0611	49
50	.0872	18.2559	277.9148	11.4674	209.3480	0.0548	0.0048	15.2233	50
51	.0831	18.3390	282.0673	12.0408	220.8154	0.0545	0.0045	15.3808	51
52	.0791	18.4181	286.1013	12.6428	232.8562	0.0543	0.0043	15.5337	52
53	.0753	18.4934	290.0184	13.2749	245.4990	0.0541	0.0041	15.6823	53
54	.0717	18.5651	293.8208	13.9387	258.7739	0.0539	0.0039	15.8265	54
55	.0683	18.6335	297.5104	14.6356	272.7126	0.0537	0.0037	15.9664	55
60	.0535	18.9293	314.3432	18.6792	353.5837	0.0528	0.0028	16.6062	60
65	.0419	19.1611	328.6910	23.8399	456.7980	0.0522	0.0022	17.1541	65
70	.0329	19.3427	340.8409	30.4264	588.5285	0.0517	0.0017	17.6212	70
75	.0258	19.4850	351.0721	38.8327	756.6537	0.0513	0.0013	18.0176	75
80	.0202	19.5965	359.6460	49.5614	971.2288	0.0510	0.0010	18.3526	80
85	.0158	19.6838	366.8007	63.2544	1245.0871	0.0508	0.0008	18.6346	85
90	.0124	19.7523	372.7488	80.7304	1594.6073	0.0506	0.0006	18.8712	90
95	.0097	19.8059	377.6774	103.0347	2040.6935	0.0505	0.0005	19.0689	95
100	.0076	19.8479	381.7492	131.5013	2610.0252	0.0504	0.0004	19.2337	100

I = 6.00 %

N	(P/F)	(P/A)	(P/G)	(F/P)	(F/A)	(A/P)	(A/F)	(A/G)	N
1	.9434	0.9434	-0.0000	1.0600	1.0000	1.0600	1.0000	-0.0000	1
2	.8900	1.8334	0.8900	1.1236	2.0600	0.5454	0.4854	0.4854	2
3	.8396	2.6730	2.5692	1.1910	3.1836	0.3741	0.3141	0.9612	3
4	.7921	3.4651	4.9455	1.2625	4.3746	0.2886	0.2286	1.4272	4
5	.7473	4.2124	7.9345	1.3382	5.6371	0.2374	0.1774	1.8836	5
6	.7050	4.9173	11.4594	1.4185	6.9753	0.2034	0.1434	2.3304	6
7	.6651	5.5824	15.4497	1.5036	8.3938	0.1791	0.1191	2.7676	7
8	.6274	6.2098	19.8416	1.5938	9.8975	0.1610	0.1010	3.1952	8
9	.5919	6.8017	24.5768	1.6895	11.4913	0.1470	0.0870	3.6133	9
10	.5584	7.3601	29.6023	1.7908	13.1808	0.1359	0.0759	4.0220	10
11	.5268	7.8869	34.8702	1.8983	14.9716	0.1268	0.0668	4.4213	11
12	.4970	8.3838	40.3369	2.0122	16.8699	0.1193	0.0593	4.8113	12
13	.4688	8.8527	45.9629	2.1329	18.8821	0.1130	0.0530	5.1920	13
14	.4423	9.2950	51.7128	2.2609	21.0151	0.1076	0.0476	5.5635	14
15	.4173	9.7122	57.5546	2.3966	23.2760	0.1030	0.0430	5.9260	15
16	.3936	10.1059	63.4592	2.5404	25.6725	0.0990	0.0390	6.2794	16
17	.3714	10.4773	69.4011	2.6928	28.2129	0.0954	0.0354	6.6240	17
18	.3503	10.8276	75.3569	2.8543	30.9057	0.0924	0.0324	6.9597	18
19	.3305	11.1581	81.3062	3.0256	33.7600	0.0896	0.0296	7.2867	19
20	.3118	11.4699	87.2304	3.2071	36.7856	0.0872	0.0272	7.6051	20
21	.2942	11.7641	93.1136	3.3996	39.9927	0.0850	0.0250	7.9151	21
22	.2775	12.0416	98.9412	3.6035	43.3923	0.0830	0.0230	8.2166	22
23	.2618	12.3034	104.7007	3.8197	46.9958	0.0813	0.0213	8.5099	23
24	.2470	12.5504	110.3812	4.0489	50.8156	0.0797	0.0197	8.7951	24
25	.2330	12.7834	115.9732	4.2919	54.8645	0.0782	0.0182	9.0722	25
26	.2198	13.0032	121.4684	4.5494	59.1564	0.0769	0.0169	9.3414	26
27	.2074	13.2105	126.8600	4.8223	63.7058	0.0757	0.0157	9.6029	27
28	.1956	13.4062	132.1420	5.1117	68.5281	0.0746	0.0146	9.8568	28
29	.1846	13.5907	137.3096	5.4184	73.6398	0.0736	0.0136	10.1032	29
30	.1741	13.7648	142.3588	5.7435	79.0582	0.0726	0.0126	10.3422	30
31	.1643	13.9291	147.2864	6.0881	84.8017	0.0718	0.0118	10.5740	31
32	.1550	14.0840	152.0901	6.4534	90.8898	0.0710	0.0110	10.7988	32
33	.1462	14.2302	156.7681	6.8406	97.3432	0.0703	0.0103	11.0166	33
34	.1379	14.3681	161.3192	7.2510	104.1838	0.0696	0.0096	11.2276	34
35	.1301	14.4982	165.7427	7.6861	111.4348	0.0690	0.0090	11.4319	35
36	.1227	14.6210	170.0387	8.1473	119.1209	0.0684	0.0084	11.6298	36
37	.1158	14.7368	174.2072	8.6361	127.2681	0.0679	0.0079	11.8213	37
38	.1092	14.8460	178.2490	9.1543	135.9042	0.0674	0.0074	12.0065	38
39	.1031	14.9491	182.1652	9.7035	145.0585	0.0669	0.0069	12.1857	39
40	.0972	15.0463	185.9568	10.2857	154.7620	0.0665	0.0065	12.3590	40
41	.0917	15.1380	189.6256	10.9029	165.0477	0.0661	0.0061	12.5264	41
42	.0865	15.2245	193.1732	11.5570	175.9505	0.0657	0.0057	12.6883	42
43	.0816	15.3062	196.6017	12.2505	187.5076	0.0653	0.0053	12.8446	43
44	.0770	15.3832	199.9130	12.9855	199.7580	0.0650	0.0050	12.9956	44
45	.0727	15.4558	203.1096	13.7646	212.7435	0.0647	0.0047	13.1413	45
46	.0685	15.5244	206.1938	14.5905	226.5081	0.0644	0.0044	13.2819	46
47	.0647	15.5890	209.1681	15.4659	241.0986	0.0641	0.0041	13.4177	47
48	.0610	15.6500	212.0351	16.3939	256.5645	0.0639	0.0039	13.5485	48
49	.0575	15.7076	214.7972	17.3775	272.9584	0.0637	0.0037	13.6748	49
50	.0543	15.7619	217.4574	18.4202	290.3359	0.0634	0.0034	13.7964	50
51	.0512	15.8131	220.0181	19.5254	308.7561	0.0632	0.0032	13.9137	51
52	.0483	15.8614	222.4823	20.6969	328.2814	0.0630	0.0030	14.0267	52
53	.0456	15.9070	224.8525	21.9387	348.9783	0.0629	0.0029	14.1355	53
54	.0430	15.9500	227.1316	23.2550	370.9170	0.0627	0.0027	14.2402	54
55	.0406	15.9905	229.3222	24.6503	394.1720	0.0625	0.0025	14.3411	55
60	.0303	16.1614	239.0428	32.9877	533.1282	0.0619	0.0019	14.7909	60
65	.0227	16.2891	246.9450	44.1450	719.0829	0.0614	0.0014	15.1601	65
70	.0169	16.3845	253.3271	59.0759	967.9322	0.0610	0.0010	15.4613	70
75	.0126	16.4558	258.4527	79.0569	1300.9487	0.0608	0.0008	15.7058	75
80	.0095	16.5091	262.5493	105.7960	1746.5999	0.0606	0.0006	15.9033	80
85	.0071	16.5489	265.8096	141.5789	2342.9817	0.0604	0.0004	16.0620	85
90	.0053	16.5787	268.3946	189.4645	3141.0752	0.0603	0.0003	16.1891	90
95	.0039	16.6009	270.4375	253.5463	4209.1042	0.0602	0.0002	16.2905	95
100	.0029	16.6175	272.0471	339.3021	5638.3681	0.0602	0.0002	16.3711	100

I = 7.00 %

N	(P/F)	(P/A)	(P/G)	(F/P)	(F/A)	(A/P)	(A/F)	(A/G)	N
1	.9346	0.9346	-0.0000	1.0700	1.0000	1.0700	1.0000	-0.0000	1
2	.8734	1.8080	0.8734	1.1449	2.0700	0.5531	0.4831	0.4831	2
3	.8163	2.6243	2.5060	1.2250	3.2149	0.3811	0.3111	0.9549	3
4	.7629	3.3872	4.7947	1.3108	4.4399	0.2952	0.2252	1.4155	4
5	.7130	4.1002	7.6467	1.4026	5.7507	0.2439	0.1739	1.8650	5
6	.6663	4.7665	10.9784	1.5007	7.1533	0.2098	0.1398	2.3032	6
7	.6227	5.3893	14.7149	1.6058	8.6540	0.1856	0.1156	2.7304	7
8	.5820	5.9713	18.7889	1.7182	10.2598	0.1675	0.0975	3.1465	8
9	.5439	6.5152	23.1404	1.8385	11.9780	0.1535	0.0835	3.5517	9
10	.5083	7.0236	27.7156	1.9672	13.8164	0.1424	0.0724	3.9461	10
11	.4751	7.4987	32.4665	2.1049	15.7836	0.1334	0.0634	4.3296	11
12	.4440	7.9427	37.3506	2.2522	17.8885	0.1259	0.0559	4.7025	12
13	.4150	8.3577	42.3302	2.4098	20.1406	0.1197	0.0497	5.0648	13
14	.3878	8.7455	47.3718	2.5785	22.5505	0.1143	0.0443	5.4167	14
15	.3624	9.1079	52.4461	2.7590	25.1290	0.1098	0.0398	5.7583	15
16	.3387	9.4466	57.5271	2.9522	27.8881	0.1059	0.0359	6.0897	16
17	.3166	9.7632	62.5923	3.1588	30.8402	0.1024	0.0324	6.4110	17
18	.2959	10.0591	67.6219	3.3799	33.9990	0.0994	0.0294	6.7225	18
19	.2765	10.3356	72.5991	3.6165	37.3790	0.0968	0.0268	7.0242	19
20	.2584	10.5940	77.5091	3.8697	40.9955	0.0944	0.0244	7.3163	20
21	.2415	10.8355	82.3393	4.1406	44.8652	0.0923	0.0223	7.5990	21
22	.2257	11.0612	87.0793	4.4304	49.0057	0.0904	0.0204	7.8725	22
23	.2109	11.2722	91.7201	4.7405	53.4361	0.0887	0.0187	8.1369	23
24	.1971	11.4693	96.2545	5.0724	58.1767	0.0872	0.0172	8.3923	24
25	.1842	11.6536	100.6765	5.4274	63.2490	0.0858	0.0158	8.6391	25
26	.1722	11.8258	104.9814	5.8074	68.6765	0.0846	0.0146	8.8773	26
27	.1609	11.9867	109.1656	6.2139	74.4838	0.0834	0.0134	9.1072	27
28	.1504	12.1371	113.2264	6.6488	80.6977	0.0824	0.0124	9.3289	28
29	.1406	12.2777	117.1622	7.1143	87.3465	0.0814	0.0114	9.5427	29
30	.1314	12.4090	120.9718	7.6123	94.4608	0.0806	0.0106	9.7487	30
31	.1228	12.5318	124.6550	8.1451	102.0730	0.0798	0.0098	9.9471	31
32	.1147	12.6466	128.2120	8.7153	110.2182	0.0791	0.0091	10.1381	32
33	.1072	12.7538	131.6435	9.3253	118.9334	0.0784	0.0084	10.3219	33
34	.1002	12.8540	134.9507	9.9781	128.2588	0.0778	0.0078	10.4987	34
35	.0937	12.9477	138.1353	10.6766	138.2369	0.0772	0.0072	10.6687	35
36	.0875	13.0352	141.1990	11.4239	148.9135	0.0767	0.0067	10.8321	36
37	.0818	13.1170	144.1441	12.2236	160.3374	0.0762	0.0062	10.9891	37
38	.0765	13.1935	146.9730	13.0793	172.5610	0.0758	0.0058	11.1398	38
39	.0715	13.2649	149.6883	13.9948	185.6403	0.0754	0.0054	11.2845	39
40	.0668	13.3317	152.2928	14.9745	199.6351	0.0750	0.0050	11.4233	40
41	.0624	13.3941	154.7892	16.0227	214.6096	0.0747	0.0047	11.5565	41
42	.0583	13.4524	157.1807	17.1443	230.6322	0.0743	0.0043	11.6842	42
43	.0545	13.5070	159.4702	18.3444	247.7765	0.0740	0.0040	11.8065	43
44	.0509	13.5579	161.6609	19.6285	266.1209	0.0738	0.0038	11.9237	44
45	.0476	13.6055	163.7559	21.0025	285.7493	0.0735	0.0035	12.0360	45
46	.0445	13.6500	165.7584	22.4726	306.7518	0.0733	0.0033	12.1435	46
47	.0416	13.6916	167.6714	24.0457	329.2244	0.0730	0.0030	12.2463	47
48	.0389	13.7305	169.4981	25.7289	353.2701	0.0728	0.0028	12.3447	48
49	.0363	13.7668	171.2417	27.5299	378.9990	0.0726	0.0026	12.4387	49
50	.0339	13.8007	172.9051	29.4570	406.5289	0.0725	0.0025	12.5287	50
51	.0317	13.8325	174.4915	31.5190	435.9860	0.0723	0.0023	12.6146	51
52	.0297	13.8621	176.0037	33.7253	467.5050	0.0721	0.0021	12.6967	52
53	.0277	13.8898	177.4447	36.0861	501.2303	0.0720	0.0020	12.7751	53
54	.0259	13.9157	178.8173	38.6122	537.3164	0.0719	0.0019	12.8500	54
55	.0242	13.9399	180.1243	41.3150	575.9286	0.0717	0.0017	12.9215	55
60	.0173	14.0392	185.7677	57.9464	813.5204	0.0712	0.0012	13.2321	60
65	.0123	14.1099	190.1452	81.2729	1146.7552	0.0709	0.0009	13.4760	65
70	.0088	14.1604	193.5185	113.9894	1614.1342	0.0706	0.0006	13.6662	70
75	.0063	14.1964	196.1035	159.8760	2269.6574	0.0704	0.0004	13.8136	75
80	.0045	14.2220	198.0748	224.2344	3189.0627	0.0703	0.0003	13.9273	80
85	.0032	14.2403	199.5717	314.5003	4478.5761	0.0702	0.0002	14.0146	85
90	.0023	14.2533	200.7042	441.1030	6287.1854	0.0702	0.0002	14.0812	90
95	.0016	14.2626	201.5581	618.6697	8823.8535	0.0701	0.0001	14.1319	95
100	.0012	14.2693	202.2001	867.7163	12381.6618	0.0701	0.0001	14.1703	100

I = 8.00 %

N	(P/F)	(P/A)	(P/G)	(F/P)	(F/A)	(A/P)	(A/F)	(A/G)	N
1	.9259	0.9259	-0.0000	1.0800	1.0000	1.0800	1.0000	-0.0000	1
2	.8573	1.7833	0.8573	1.1664	2.0800	0.5608	0.4808	0.4808	2
3	.7938	2.5771	2.4450	1.2597	3.2464	0.3880	0.3080	0.9487	3
4	.7350	3.3121	4.6501	1.3605	4.5061	0.3019	0.2219	1.4040	4
5	.6806	3.9927	7.3724	1.4693	5.8666	0.2505	0.1705	1.8465	5
6	.6302	4.6229	10.5233	1.5869	7.3359	0.2163	0.1363	2.2763	6
7	.5835	5.2064	14.0242	1.7138	8.9228	0.1921	0.1121	2.6937	7
8	.5403	5.7466	17.8061	1.8509	10.6366	0.1740	0.0940	3.0985	8
9	.5002	6.2469	21.8081	1.9990	12.4876	0.1601	0.0801	3.4910	9
10	.4632	6.7101	25.9768	2.1589	14.4866	0.1490	0.0690	3.8713	10
11	.4289	7.1390	30.2657	2.3316	16.6455	0.1401	0.0601	4.2395	11
12	.3971	7.5361	34.6339	2.5182	18.9771	0.1327	0.0527	4.5957	12
13	.3677	7.9038	39.0463	2.7196	21.4953	0.1265	0.0465	4.9402	13
14	.3405	8.2442	43.4723	2.9372	24.2149	0.1213	0.0413	5.2731	14
15	.3152	8.5595	47.8857	3.1722	27.1521	0.1168	0.0368	5.5945	15
16	.2919	8.8514	52.2640	3.4259	30.3243	0.1130	0.0330	5.9046	16
17	.2703	9.1216	56.5883	3.7000	33.7502	0.1096	0.0296	6.2037	17
18	.2502	9.3719	60.8426	3.9960	37.4502	0.1067	0.0267	6.4920	18
19	.2317	9.6036	65.0134	4.3157	41.4463	0.1041	0.0241	6.7697	19
20	.2145	9.8181	69.0898	4.6610	45.7620	0.1019	0.0219	7.0369	20
21	.1987	10.0168	73.0629	5.0338	50.4229	0.0998	0.0198	7.2940	21
22	.1839	10.2007	76.9257	5.4365	55.4568	0.0980	0.0180	7.5412	22
23	.1703	10.3711	80.6726	5.8715	60.8933	0.0964	0.0164	7.7786	23
24	.1577	10.5288	84.2997	6.3412	66.7648	0.0950	0.0150	8.0066	24
25	.1460	10.6748	87.8041	6.8485	73.1059	0.0937	0.0137	8.2254	25
26	.1352	10.8100	91.1842	7.3964	79.9544	0.0925	0.0125	8.4352	26
27	.1252	10.9352	94.4390	7.9881	87.3508	0.0914	0.0114	8.6363	27
28	.1159	11.0511	97.5687	8.6271	95.3388	0.0905	0.0105	8.8289	28
29	.1073	11.1584	100.5738	9.3173	103.9659	0.0896	0.0096	9.0133	29
30	.0994	11.2578	103.4558	10.0627	113.2832	0.0888	0.0088	9.1897	30
31	.0920	11.3498	106.2163	10.8677	123.3459	0.0881	0.0081	9.3584	31
32	.0852	11.4350	108.8575	11.7371	134.2135	0.0875	0.0075	9.5197	32
33	.0789	11.5139	111.3819	12.6760	145.9506	0.0869	0.0069	9.6737	33
34	.0730	11.5869	113.7924	13.6901	158.6267	0.0863	0.0063	9.8208	34
35	.0676	11.6546	116.0920	14.7853	172.3168	0.0858	0.0058	9.9611	35
36	.0626	11.7172	118.2839	15.9682	187.1021	0.0853	0.0053	10.0949	36
37	.0580	11.7752	120.3713	17.2456	203.0703	0.0849	0.0049	10.2225	37
38	.0537	11.8289	122.3579	18.6253	220.3159	0.0845	0.0045	10.3440	38
39	.0497	11.8786	124.2470	20.1153	238.9412	0.0842	0.0042	10.4597	39
40	.0460	11.9246	126.0422	21.7245	259.0565	0.0839	0.0039	10.5699	40
41	.0426	11.9672	127.7470	23.4625	280.7810	0.0836	0.0036	10.6747	41
42	.0395	12.0067	129.3651	25.3395	304.2435	0.0833	0.0033	10.7744	42
43	.0365	12.0432	130.8998	27.3666	329.5830	0.0830	0.0030	10.8692	43
44	.0338	12.0771	132.3547	29.5560	356.9496	0.0828	0.0028	10.9592	44
45	.0313	12.1084	133.7331	31.9204	386.5056	0.0826	0.0026	11.0447	45
46	.0290	12.1374	135.0384	34.4741	418.4261	0.0824	0.0024	11.1258	46
47	.0269	12.1643	136.2739	37.2320	452.9002	0.0822	0.0022	11.2028	47
48	.0249	12.1891	137.4428	40.2106	490.1322	0.0820	0.0020	11.2758	48
49	.0230	12.2122	138.5480	43.4274	530.3427	0.0819	0.0019	11.3451	49
50	.0213	12.2335	139.5928	46.9016	573.7702	0.0817	0.0017	11.4107	50
51	.0197	12.2532	140.5799	50.6537	620.6718	0.0816	0.0016	11.4729	51
52	.0183	12.2715	141.5121	54.7060	671.3255	0.0815	0.0015	11.5318	52
53	.0169	12.2884	142.3923	59.0825	726.0316	0.0814	0.0014	11.5875	53
54	.0157	12.3041	143.2229	63.8091	785.1141	0.0813	0.0013	11.6403	54
55	.0145	12.3186	144.0065	68.9139	848.9232	0.0812	0.0012	11.6902	55
60	.0099	12.3766	147.3000	101.2571	1253.2133	0.0808	0.0008	11.9015	60
65	.0067	12.4160	149.7387	148.7798	1847.2481	0.0805	0.0005	12.0602	65
70	.0046	12.4428	151.5326	218.6064	2720.0801	0.0804	0.0004	12.1783	70
75	.0031	12.4611	152.8448	321.2045	4002.5566	0.0802	0.0002	12.2658	75
80	.0021	12.4735	153.8001	471.9548	5886.9354	0.0802	0.0002	12.3301	80
85	.0014	12.4820	154.4925	693.4565	8655.7061	0.0801	0.0001	12.3772	85
90	.0010	12.4877	154.9925	1018.9151	12723.9386	0.0801	0.0001	12.4116	90
95	.0007	12.4917	155.3524	1497.1205	18701.5069	0.0801	0.0001	12.4365	95
100	.0005	12.4943	155.6107	2199.7613	27484.5157	0.0800	0.0000	12.4545	100

PROFESSIONAL PUBLICATIONS, INC. • P.O. Box 199, San Carlos, CA 94070

I = 9.00 %

N	(P/F)	(P/A)	(P/G)	(F/P)	(F/A)	(A/P)	(A/F)	(A/G)	N
1	.9174	0.9174	-0.0000	1.0900	1.0000	1.0900	1.0000	-0.0000	1
2	.8417	1.7591	0.8417	1.1881	2.0900	0.5685	0.4785	0.4785	2
3	.7722	2.5313	2.3860	1.2950	3.2781	0.3951	0.3051	0.9426	3
4	.7084	3.2397	4.5113	1.4116	4.5731	0.3087	0.2187	1.3925	4
5	.6499	3.8897	7.1110	1.5386	5.9847	0.2571	0.1671	1.8282	5
6	.5963	4.4859	10.0924	1.6771	7.5233	0.2229	0.1329	2.2498	6
7	.5470	5.0330	13.3746	1.8280	9.2004	0.1987	0.1087	2.6574	7
8	.5019	5.5348	16.8877	1.9926	11.0285	0.1807	0.0907	3.0512	8
9	.4604	5.9952	20.5711	2.1719	13.0210	0.1668	0.0768	3.4312	9
10	.4224	6.4177	24.3728	2.3674	15.1929	0.1558	0.0658	3.7978	10
11	.3875	6.8052	28.2481	2.5804	17.5603	0.1469	0.0569	4.1510	11
12	.3555	7.1607	32.1590	2.8127	20.1407	0.1397	0.0497	4.4910	12
13	.3262	7.4869	36.0731	3.0658	22.9534	0.1336	0.0436	4.8182	13
14	.2992	7.7862	39.9633	3.3417	26.0192	0.1284	0.0384	5.1326	14
15	.2745	8.0607	43.8069	3.6425	29.3609	0.1241	0.0341	5.4346	15
16	.2519	8.3126	47.5849	3.9703	33.0034	0.1203	0.0303	5.7245	16
17	.2311	8.5436	51.2821	4.3276	36.9737	0.1170	0.0270	6.0024	17
18	.2120	8.7556	54.8860	4.7171	41.3013	0.1142	0.0242	6.2687	18
19	.1945	8.9501	58.3868	5.1417	46.0185	0.1117	0.0217	6.5236	19
20	.1784	9.1285	61.7770	5.6044	51.1601	0.1095	0.0195	6.7674	20
21	.1637	9.2922	65.0509	6.1088	56.7645	0.1076	0.0176	7.0006	21
22	.1502	9.4424	68.2048	6.6586	62.8733	0.1059	0.0159	7.2232	22
23	.1378	9.5802	71.2359	7.2579	69.5319	0.1044	0.0144	7.4357	23
24	.1264	9.7066	74.1433	7.9111	76.7898	0.1030	0.0130	7.6384	24
25	.1160	9.8226	76.9265	8.6231	84.7009	0.1018	0.0118	7.8316	25
26	.1064	9.9290	79.5863	9.3992	93.3240	0.1007	0.0107	8.0156	26
27	.0976	10.0266	82.1241	10.2451	102.7231	0.0997	0.0097	8.1906	27
28	.0895	10.1161	84.5419	11.1671	112.9682	0.0989	0.0089	8.3571	28
29	.0822	10.1983	86.8422	12.1722	124.1354	0.0981	0.0081	8.5154	29
30	.0754	10.2737	89.0280	13.2677	136.3075	0.0973	0.0073	8.6657	30
31	.0691	10.3428	91.1024	14.4618	149.5752	0.0967	0.0067	8.8083	31
32	.0634	10.4062	93.0690	15.7633	164.0370	0.0961	0.0061	8.9436	32
33	.0582	10.4644	94.9314	17.1820	179.8003	0.0956	0.0056	9.0718	33
34	.0534	10.5178	96.6935	18.7284	196.9823	0.0951	0.0051	9.1933	34
35	.0490	10.5668	98.3590	20.4140	215.7108	0.0946	0.0046	9.3083	35
36	.0449	10.6118	99.9319	22.2512	236.1247	0.0942	0.0042	9.4171	36
37	.0412	10.6530	101.4162	24.2538	258.3759	0.0939	0.0039	9.5200	37
38	.0378	10.6908	102.8158	26.4367	282.6298	0.0935	0.0035	9.6172	38
39	.0347	10.7255	104.1345	28.8160	309.0665	0.0932	0.0032	9.7090	39
40	.0318	10.7574	105.3762	31.4094	337.8824	0.0930	0.0030	9.7957	40
41	.0292	10.7866	106.5445	34.2363	369.2919	0.0927	0.0027	9.8775	41
42	.0268	10.8134	107.6432	37.3175	403.5281	0.0925	0.0025	9.9546	42
43	.0246	10.8380	108.6758	40.6761	440.8457	0.0923	0.0023	10.0273	43
44	.0226	10.8605	109.6456	44.3370	481.5218	0.0921	0.0021	10.0958	44
45	.0207	10.8812	110.5561	48.3273	525.8587	0.0919	0.0019	10.1603	45
46	.0190	10.9002	111.4103	52.6767	574.1860	0.0917	0.0017	10.2210	46
47	.0174	10.9176	112.2115	57.4176	626.8628	0.0916	0.0016	10.2780	47
48	.0160	10.9336	112.9625	62.5852	684.2804	0.0915	0.0015	10.3317	48
49	.0147	10.9482	113.6661	68.2179	746.8656	0.0913	0.0013	10.3821	49
50	.0134	10.9617	114.3251	74.3575	815.0836	0.0912	0.0012	10.4295	50
51	.0123	10.9740	114.9420	81.0497	889.4411	0.0911	0.0011	10.4740	51
52	.0113	10.9853	115.5193	88.3442	970.4908	0.0910	0.0010	10.5158	52
53	.0104	10.9957	116.0593	96.2951	1058.8349	0.0909	0.0009	10.5549	53
54	.0095	11.0053	116.5642	104.9617	1155.1301	0.0909	0.0009	10.5917	54
55	.0087	11.0140	117.0362	114.4083	1260.0918	0.0908	0.0008	10.6261	55
60	.0057	11.0480	118.9683	176.0313	1944.7921	0.0905	0.0005	10.7683	60
65	.0037	11.0701	120.3344	270.8460	2998.2885	0.0903	0.0003	10.8702	65
70	.0024	11.0844	121.2942	416.7301	4619.2232	0.0902	0.0002	10.9427	70
75	.0016	11.0938	121.9646	641.1909	7113.2321	0.0901	0.0001	10.9940	75
80	.0010	11.0998	122.4306	986.5517	10950.5741	0.0901	0.0001	11.0299	80
85	.0007	11.1038	122.7533	1517.9320	16854.8003	0.0901	0.0001	11.0551	85
90	.0004	11.1064	122.9758	2335.5266	25939.1842	0.0900	0.0000	11.0726	90
95	.0003	11.1080	123.1287	3593.4971	39916.6350	0.0900	0.0000	11.0847	95
100	.0002	11.1091	123.2335	5529.0408	61422.6755	0.0900	0.0000	11.0930	100

PROFESSIONAL PUBLICATIONS, INC. • P.O. Box 199, San Carlos, CA 94070

LAND SURVEYOR REFERENCE MANUAL

I = 10.00 %

N	(P/F)	(P/A)	(P/G)	(F/P)	(F/A)	(A/P)	(A/F)	(A/G)	N
1	.9091	0.9091	− 0.0000	1.1000	1.0000	1.1000	1.0000	− 0.0000	1
2	.8264	1.7355	0.8264	1.2100	2.1000	0.5762	0.4762	0.4762	2
3	.7513	2.4869	2.3291	1.3310	3.3100	0.4021	0.3021	0.9366	3
4	.6830	3.1699	4.3781	1.4641	4.6410	0.3155	0.2155	1.3812	4
5	.6209	3.7908	6.8618	1.6105	6.1051	0.2638	0.1638	1.8101	5
6	.5645	4.3553	9.6842	1.7716	7.7156	0.2296	0.1296	2.2236	6
7	.5132	4.8684	12.7631	1.9487	9.4872	0.2054	0.1054	2.6216	7
8	.4665	5.3349	16.0287	2.1436	11.4359	0.1874	0.0874	3.0045	8
9	.4241	5.7590	19.4215	2.3579	13.5795	0.1736	0.0736	3.3724	9
10	.3855	6.1446	22.8913	2.5937	15.9374	0.1627	0.0627	3.7255	10
11	.3505	6.4951	26.3963	2.8531	18.5312	0.1540	0.0540	4.0641	11
12	.3186	6.8137	29.9012	3.1384	21.3843	0.1468	0.0468	4.3884	12
13	.2897	7.1034	33.3772	3.4523	24.5227	0.1408	0.0408	4.6988	13
14	.2633	7.3667	36.8005	3.7975	27.9750	0.1357	0.0357	4.9955	14
15	.2394	7.6061	40.1520	4.1772	31.7725	0.1315	0.0315	5.2789	15
16	.2176	7.8237	43.4164	4.5950	35.9497	0.1278	0.0278	5.5493	16
17	.1978	8.0216	46.5819	5.0545	40.5447	0.1247	0.0247	5.8071	17
18	.1799	8.2014	49.6395	5.5599	45.5992	0.1219	0.0219	6.0526	18
19	.1635	8.3649	52.5827	6.1159	51.1591	0.1195	0.0195	6.2861	19
20	.1486	8.5136	55.4069	6.7275	57.2750	0.1175	0.0175	6.5081	20
21	.1351	8.6487	58.1095	7.4002	64.0025	0.1156	0.0156	6.7189	21
22	.1228	8.7715	60.6893	8.1403	71.4027	0.1140	0.0140	6.9189	22
23	.1117	8.8832	63.1462	8.9543	79.5430	0.1126	0.0126	7.1085	23
24	.1015	8.9847	65.4813	9.8497	88.4973	0.1113	0.0113	7.2881	24
25	.0923	9.0770	67.6964	10.8347	98.3471	0.1102	0.0102	7.4580	25
26	.0839	9.1609	69.7940	11.9182	109.1818	0.1092	0.0092	7.6186	26
27	.0763	9.2372	71.7773	13.1100	121.0999	0.1083	0.0083	7.7704	27
28	.0693	9.3066	73.6495	14.4210	134.2099	0.1075	0.0075	7.9137	28
29	.0630	9.3696	75.4146	15.8631	148.6309	0.1067	0.0067	8.0489	29
30	.0573	9.4269	77.0766	17.4494	164.4940	0.1061	0.0061	8.1762	30
31	.0521	9.4790	78.6395	19.1943	181.9434	0.1055	0.0055	8.2962	31
32	.0474	9.5264	80.1078	21.1138	201.1378	0.1050	0.0050	8.4091	32
33	.0431	9.5694	81.4856	23.2252	222.2515	0.1045	0.0045	8.5152	33
34	.0391	9.6086	82.7773	25.5477	245.4767	0.1041	0.0041	8.6149	34
35	.0356	9.6442	83.9872	28.1024	271.0244	0.1037	0.0037	8.7086	35
36	.0323	9.6765	85.1194	30.9127	299.1268	0.1033	0.0033	8.7965	36
37	.0294	9.7059	86.1781	34.0039	330.0395	0.1030	0.0030	8.8789	37
38	.0267	9.7327	87.1673	37.4043	364.0434	0.1027	0.0027	8.9562	38
39	.0243	9.7570	88.0908	41.1448	401.4478	0.1025	0.0025	9.0285	39
40	.0221	9.7791	88.9525	45.2593	442.5926	0.1023	0.0023	9.0962	40
41	.0201	9.7991	89.7560	49.7852	487.8518	0.1020	0.0020	9.1596	41
42	.0183	9.8174	90.5047	54.7637	537.6370	0.1019	0.0019	9.2188	42
43	.0166	9.8340	91.2019	60.2401	592.4007	0.1017	0.0017	9.2741	43
44	.0151	9.8491	91.8508	66.2641	652.6408	0.1015	0.0015	9.3258	44
45	.0137	9.8628	92.4544	72.8905	718.9048	0.1014	0.0014	9.3740	45
46	.0125	9.8753	93.0157	80.1795	791.7953	0.1013	0.0013	9.4190	46
47	.0113	9.8866	93.5372	88.1975	871.9749	0.1011	0.0011	9.4610	47
48	.0103	9.8969	94.0217	97.0172	960.1723	0.1010	0.0010	9.5001	48
49	.0094	9.9063	94.4715	106.7190	1057.1896	0.1009	0.0009	9.5365	49
50	.0085	9.9148	94.8889	117.3909	1163.9085	0.1009	0.0009	9.5704	50
51	.0077	9.9226	95.2761	129.1299	1281.2994	0.1008	0.0008	9.6020	51
52	.0070	9.9296	95.6351	142.0429	1410.4293	0.1007	0.0007	9.6313	52
53	.0064	9.9360	95.9679	156.2472	1552.4723	0.1006	0.0006	9.6586	53
54	.0058	9.9418	96.2763	171.8719	1708.7195	0.1006	0.0006	9.6840	54
55	.0053	9.9471	96.5619	189.0591	1880.5914	0.1005	0.0005	9.7075	55
60	.0033	9.9672	97.7010	304.4816	3034.8164	0.1003	0.0003	9.8023	60
65	.0020	9.9796	98.4705	490.3707	4893.7073	0.1002	0.0002	9.8672	65
70	.0013	9.9873	98.9870	789.7470	7887.4696	0.1001	0.0001	9.9113	70
75	.0008	9.9921	99.3317	1271.8954	12708.9537	0.1001	0.0001	9.9410	75
80	.0005	9.9951	99.5606	2048.4002	20474.0021	0.1000	0.0000	9.9609	80
85	.0003	9.9970	99.7120	3298.9690	32979.6903	0.1000	0.0000	9.9742	85
90	.0002	9.9981	99.8118	5313.0226	53120.2261	0.1000	0.0000	9.9831	90
95	.0001	9.9988	99.8773	8556.6760	85556.7605	0.1000	0.0000	9.9889	95
100	.0001	9.9993	99.9202	13780.6123	137796.1234	0.1000	0.0000	9.9927	100

I = 12.00 %

N	(P/F)	(P/A)	(P/G)	(F/P)	(F/A)	(A/P)	(A/F)	(A/G)	N
1	.8929	0.8929	-0.0000	1.1200	1.0000	1.1200	1.0000	-0.0000	1
2	.7972	1.6901	0.7972	1.2544	2.1200	0.5917	0.4717	0.4717	2
3	.7118	2.4018	2.2208	1.4049	3.3744	0.4163	0.2963	0.9246	3
4	.6355	3.0373	4.1273	1.5735	4.7793	0.3292	0.2092	1.3589	4
5	.5674	3.6048	6.3970	1.7623	6.3528	0.2774	0.1574	1.7746	5
6	.5066	4.1114	8.9302	1.9738	8.1152	0.2432	0.1232	2.1720	6
7	.4523	4.5638	11.6443	2.2107	10.0890	0.2191	0.0991	2.5515	7
8	.4039	4.9676	14.4714	2.4760	12.2997	0.2013	0.0813	2.9131	8
9	.3606	5.3282	17.3563	2.7731	14.7757	0.1877	0.0677	3.2574	9
10	.3220	5.6502	20.2541	3.1058	17.5487	0.1770	0.0570	3.5847	10
11	.2875	5.9377	23.1288	3.4785	20.6546	0.1684	0.0484	3.8953	11
12	.2567	6.1944	25.9523	3.8960	24.1331	0.1614	0.0414	4.1897	12
13	.2292	6.4235	28.7024	4.3635	28.0291	0.1557	0.0357	4.4683	13
14	.2046	6.6282	31.3624	4.8871	32.3926	0.1509	0.0309	4.7317	14
15	.1827	6.8109	33.9202	5.4736	37.2797	0.1468	0.0268	4.9803	15
16	.1631	6.9740	36.3670	6.1304	42.7533	0.1434	0.0234	5.2147	16
17	.1456	7.1196	38.6973	6.8660	48.8837	0.1405	0.0205	5.4353	17
18	.1300	7.2497	40.9080	7.6900	55.7497	0.1379	0.0179	5.6427	18
19	.1161	7.3658	42.9979	8.6128	63.4397	0.1358	0.0158	5.8375	19
20	.1037	7.4694	44.9676	9.6463	72.0524	0.1339	0.0139	6.0202	20
21	.0926	7.5620	46.8188	10.8038	81.6987	0.1322	0.0122	6.1913	21
22	.0826	7.6446	48.5543	12.1003	92.5026	0.1308	0.0108	6.3514	22
23	.0738	7.7184	50.1776	13.5523	104.6029	0.1296	0.0096	6.5010	23
24	.0659	7.7843	51.6929	15.1786	118.1552	0.1285	0.0085	6.6406	24
25	.0588	7.8431	53.1046	17.0001	133.3339	0.1275	0.0075	6.7708	25
26	.0525	7.8957	54.4177	19.0401	150.3339	0.1267	0.0067	6.8921	26
27	.0469	7.9426	55.6369	21.3249	169.3740	0.1259	0.0059	7.0049	27
28	.0419	7.9844	56.7674	23.8839	190.6989	0.1252	0.0052	7.1098	28
29	.0374	8.0218	57.8141	26.7499	214.5828	0.1247	0.0047	7.2071	29
30	.0334	8.0552	58.7821	29.9599	241.3327	0.1241	0.0041	7.2974	30
31	.0298	8.0850	59.6761	33.5551	271.2926	0.1237	0.0037	7.3811	31
32	.0266	8.1116	60.5010	37.5817	304.8477	0.1233	0.0033	7.4586	32
33	.0238	8.1354	61.2612	42.0915	342.4294	0.1229	0.0029	7.5302	33
34	.0212	8.1566	61.9612	47.1425	384.5210	0.1226	0.0026	7.5965	34
35	.0189	8.1755	62.6052	52.7996	431.6635	0.1223	0.0023	7.6577	35
36	.0169	8.1924	63.1970	59.1356	484.4631	0.1221	0.0021	7.7141	36
37	.0151	8.2075	63.7406	66.2318	543.5987	0.1218	0.0018	7.7661	37
38	.0135	8.2210	64.2394	74.1797	609.8305	0.1216	0.0016	7.8141	38
39	.0120	8.2330	64.6967	83.0812	684.0102	0.1215	0.0015	7.8582	39
40	.0107	8.2438	65.1159	93.0510	767.0914	0.1213	0.0013	7.8988	40
41	.0096	8.2534	65.4997	104.2171	860.1424	0.1212	0.0012	7.9361	41
42	.0086	8.2619	65.8509	116.7231	964.3595	0.1210	0.0010	7.9704	42
43	.0076	8.2696	66.1722	130.7299	1081.0826	0.1209	0.0009	8.0019	43
44	.0068	8.2764	66.4659	146.4175	1211.8125	0.1208	0.0008	8.0308	44
45	.0061	8.2825	66.7342	163.9876	1358.2300	0.1207	0.0007	8.0572	45
46	.0054	8.2880	66.9792	183.6661	1522.2176	0.1207	0.0007	8.0815	46
47	.0049	8.2928	67.2028	205.7061	1705.8838	0.1206	0.0006	8.1037	47
48	.0043	8.2972	67.4068	230.3908	1911.5898	0.1205	0.0005	8.1241	48
49	.0039	8.3010	67.5929	258.0377	2141.9806	0.1205	0.0005	8.1427	49
50	.0035	8.3045	67.7624	289.0022	2400.0182	0.1204	0.0004	8.1597	50
51	.0031	8.3076	67.9169	323.6825	2689.0204	0.1204	0.0004	8.1753	51
52	.0028	8.3103	68.0576	362.5243	3012.7029	0.1203	0.0003	8.1895	52
53	.0025	8.3128	68.1856	406.0273	3375.2272	0.1203	0.0003	8.2025	53
54	.0022	8.3150	68.3022	454.7505	3781.2545	0.1203	0.0003	8.2143	54
55	.0020	8.3170	68.4082	509.3206	4236.0050	0.1202	0.0002	8.2251	55
60	.0011	8.3240	68.8100	897.5969	7471.6411	0.1201	0.0001	8.2664	60
65	.0006	8.3281	69.0581	1581.8725	13173.9374	0.1201	0.0001	8.2922	65
70	.0004	8.3303	69.2103	2787.7998	23223.3319	0.1200	0.0000	8.3082	70
75	.0002	8.3316	69.3031	4913.0558	40933.7987	0.1200	0.0000	8.3181	75
80	.0001	8.3324	69.3594	8658.4831	72145.6925	0.1200	0.0000	8.3241	80
85	.0001	8.3328	69.3935	15259.2057	127151.7140	0.1200	0.0000	8.3278	85
90	.0000	8.3330	69.4140	26891.9342	224091.1185	0.1200	0.0000	8.3300	90
95	.0000	8.3332	69.4263	47392.7766	394931.4719	0.1200	0.0000	8.3313	95
100	.0000	8.3332	69.4336	83522.2657	696010.5477	0.1200	0.0000	8.3321	100

LAND SURVEYOR REFERENCE MANUAL

I = 15.00 %

N	(P/F)	(P/A)	(P/G)	(F/P)	(F/A)	(A/P)	(A/F)	(A/G)	N
1	.8696	0.8696	-0.0000	1.1500	1.0000	1.1500	1.0000	-0.0000	1
2	.7561	1.6257	0.7561	1.3225	2.1500	0.6151	0.4651	0.4651	2
3	.6575	2.2832	2.0712	1.5209	3.4725	0.4380	0.2880	0.9071	3
4	.5718	2.8550	3.7864	1.7490	4.9934	0.3503	0.2003	1.3263	4
5	.4972	3.3522	5.7751	2.0114	6.7424	0.2983	0.1483	1.7228	5
6	.4323	3.7845	7.9368	2.3131	8.7537	0.2642	0.1142	2.0972	6
7	.3759	4.1604	10.1924	2.6600	11.0668	0.2404	0.0904	2.4498	7
8	.3269	4.4873	12.4807	3.0590	13.7268	0.2229	0.0729	2.7813	8
9	.2843	4.7716	14.7548	3.5179	16.7858	0.2096	0.0596	3.0922	9
10	.2472	5.0188	16.9795	4.0456	20.3037	0.1993	0.0493	3.3832	10
11	.2149	5.2337	19.1289	4.6524	24.3493	0.1911	0.0411	3.6549	11
12	.1869	5.4206	21.1849	5.3503	29.0017	0.1845	0.0345	3.9082	12
13	.1625	5.5831	23.1352	6.1528	34.3519	0.1791	0.0291	4.1438	13
14	.1413	5.7245	24.9725	7.0757	40.5047	0.1747	0.0247	4.3624	14
15	.1229	5.8474	26.6930	8.1371	47.5804	0.1710	0.0210	4.5650	15
16	.1069	5.9542	28.2960	9.3576	55.7175	0.1679	0.0179	4.7522	16
17	.0929	6.0472	29.7828	10.7613	65.0751	0.1654	0.0154	4.9251	17
18	.0808	6.1280	31.1565	12.3755	75.8364	0.1632	0.0132	5.0843	18
19	.0703	6.1982	32.4213	14.2318	88.2118	0.1613	0.0113	5.2307	19
20	.0611	6.2593	33.5822	16.3665	102.4436	0.1598	0.0098	5.3651	20
21	.0531	6.3125	34.6448	18.8215	118.8101	0.1584	0.0084	5.4883	21
22	.0462	6.3587	35.6150	21.6447	137.6316	0.1573	0.0073	5.6010	22
23	.0402	6.3988	36.4988	24.8915	159.2764	0.1563	0.0063	5.7040	23
24	.0349	6.4338	37.3023	28.6252	184.1678	0.1554	0.0054	5.7979	24
25	.0304	6.4641	38.0314	32.9190	212.7930	0.1547	0.0047	5.8834	25
26	.0264	6.4906	38.6918	37.8568	245.7120	0.1541	0.0041	5.9612	26
27	.0230	6.5135	39.2890	43.5353	283.5688	0.1535	0.0035	6.0319	27
28	.0200	6.5335	39.8283	50.0656	327.1041	0.1531	0.0031	6.0960	28
29	.0174	6.5509	40.3146	57.5755	377.1697	0.1527	0.0027	6.1541	29
30	.0151	6.5660	40.7526	66.2118	434.7451	0.1523	0.0023	6.2066	30
31	.0131	6.5791	41.1466	76.1435	500.9569	0.1520	0.0020	6.2541	31
32	.0114	6.5905	41.5006	87.5651	577.1005	0.1517	0.0017	6.2970	32
33	.0099	6.6005	41.8184	100.6998	664.6655	0.1515	0.0015	6.3357	33
34	.0086	6.6091	42.1033	115.8048	765.3654	0.1513	0.0013	6.3705	34
35	.0075	6.6166	42.3586	133.1755	881:1702	0.1511	0.0011	6.4019	35
36	.0065	6.6231	42.5872	153.1519	1014.3457	0.1510	0.0010	6.4301	36
37	.0057	6.6288	42.7916	176.1246	1167.4975	0.1509	0.0009	6.4554	37
38	.0049	6.6338	42.9743	202.5433	1343.6222	0.1507	0.0007	6.4781	38
39	.0043	6.6380	43.1374	232.9248	1546.1655	0.1506	0.0006	6.4985	39
40	.0037	6.6418	43.2830	267.8635	1779.0903	0.1506	0.0006	6.5168	40
41	.0032	6.6450	43.4128	308.0431	2046.9539	0.1505	0.0005	6.5331	41
42	.0028	6.6478	43.5286	354.2495	2354.9969	0.1504	0.0004	6.5478	42
43	.0025	6.6503	43.6317	407.3870	2709.2465	0.1504	0.0004	6.5609	43
44	.0021	6.6524	43.7235	468.4950	3116.6334	0.1503	0.0003	6.5725	44
45	.0019	6.6543	43.8051	538.7693	3585.1285	0.1503	0.0003	6.5830	45
46	.0016	6.6559	43.8778	619.5847	4123.8977	0.1502	0.0002	6.5923	46
47	.0014	6.6573	43.9423	712.5224	4743.4824	0.1502	0.0002	6.6006	47
48	.0012	6.6585	43.9997	819.4007	5456.0047	0.1502	0.0002	6.6080	48
49	.0011	6.6596	44.0506	942.3108	6275.4055	0.1502	0.0002	6.6146	49
50	.0009	6.6605	44.0958	1083.6574	7217.7163	0.1501	0.0001	6.6205	50
51	.0008	6.6613	44.1360	1246.2061	8301.3737	0.1501	0.0001	6.6257	51
52	.0007	6.6620	44.1715	1433.1370	9547.5798	0.1501	0.0001	6.6304	52
53	.0006	6.6626	44.2031	1648.1075	10980.7167	0.1501	0.0001	6.6345	53
54	.0005	6.6631	44.2311	1895.3236	12628.8243	0.1501	0.0001	6.6382	54
55	.0005	6.6636	44.2558	2179.6222	14524.1479	0.1501	0.0001	6.6414	55
60	.0002	6.6651	44.3431	4383.9987	29219.9916	0.1500	0.0000	6.6530	60
65	.0001	6.6659	44.3903	8817.7874	58778.5826	0.1500	0.0000	6.6593	65
70	.0001	6.6663	44.4156	17735.7200	118231.4669	0.1500	0.0000	6.6627	70
75	.0000	6.6665	44.4292	35672.8680	237812.4532	0.1500	0.0000	6.6646	75
80	.0000	6.6666	44.4364	71750.8794	478332.5293	0.1500	0.0000	6.6656	80
85	.0000	6.6666	44.4402	144316.6470	962104.3133	0.1500	0.0000	6.6661	85
90	.0000	6.6666	44.4422	290272.3252	1935142.1680	0.1500	0.0000	6.6664	90
95	.0000	6.6667	44.4433	583841.3276	3892268.8509	0.1500	0.0000	6.6665	95
100	.0000	6.6667	44.4438	1174313.4507	7828749.6713	0.1500	0.0000	6.6666	100

I = 20.00 %

N	(P/F)	(P/A)	(P/G)	(F/P)	(F/A)	(A/P)	(A/F)	(A/G)	N
1	.8333	0.8333	-0.0000	1.2000	1.0000	1.2000	1.0000	-0.0000	1
2	.6944	1.5278	0.6944	1.4400	2.2000	0.6545	0.4545	0.4545	2
3	.5787	2.1065	1.8519	1.7280	3.6400	0.4747	0.2747	0.8791	3
4	.4823	2.5887	3.2986	2.0736	5.3680	0.3863	0.1863	1.2742	4
5	.4019	2.9906	4.9061	2.4883	7.4416	0.3344	0.1344	1.6405	5
6	.3349	3.3255	6.5806	2.9860	9.9299	0.3007	0.1007	1.9788	6
7	.2791	3.6046	8.2551	3.5832	12.9159	0.2774	0.0774	2.2902	7
8	.2326	3.8372	9.8831	4.2998	16.4991	0.2606	0.0606	2.5756	8
9	.1938	4.0310	11.4335	5.1598	20.7989	0.2481	0.0481	2.8364	9
10	.1615	4.1925	12.8871	6.1917	25.9587	0.2385	0.0385	3.0739	10
11	.1346	4.3271	14.2330	7.4301	32.1504	0.2311	0.0311	3.2893	11
12	.1122	4.4392	15.4667	8.9161	39.5805	0.2253	0.0253	3.4841	12
13	.0935	4.5327	16.5883	10.6993	48.4966	0.2206	0.0206	3.6597	13
14	.0779	4.6106	17.6008	12.8392	59.1959	0.2169	0.0169	3.8175	14
15	.0649	4.6755	18.5095	15.4070	72.0351	0.2139	0.0139	3.9588	15
16	.0541	4.7296	19.3208	18.4884	87.4421	0.2114	0.0114	4.0851	16
17	.0451	4.7746	20.0419	22.1861	105.9306	0.2094	0.0094	4.1976	17
18	.0376	4.8122	20.6805	26.6233	128.1167	0.2078	0.0078	4.2975	18
19	.0313	4.8435	21.2439	31.9480	154.7400	0.2065	0.0065	4.3861	19
20	.0261	4.8696	21.7395	38.3376	186.6880	0.2054	0.0054	4.4643	20
21	.0217	4.8913	22.1742	46.0051	225.0256	0.2044	0.0044	4.5334	21
22	.0181	4.9094	22.5546	55.2061	271.0307	0.2037	0.0037	4.5941	22
23	.0151	4.9245	22.8867	66.2474	326.2369	0.2031	0.0031	4.6475	23
24	.0126	4.9371	23.1760	79.4968	392.4842	0.2025	0.0025	4.6943	24
25	.0105	4.9476	23.4276	95.3962	471.9811	0.2021	0.0021	4.7352	25
26	.0087	4.9563	23.6460	114.4755	567.3773	0.2018	0.0018	4.7709	26
27	.0073	4.9636	23.8353	137.3706	681.8528	0.2015	0.0015	4.8020	27
28	.0061	4.9697	23.9991	164.8447	819.2233	0.2012	0.0012	4.8291	28
29	.0051	4.9747	24.1406	197.8136	984.0680	0.2010	0.0010	4.8527	29
30	.0042	4.9789	24.2628	237.3763	1181.8816	0.2008	0.0008	4.8731	30
31	.0035	4.9824	24.3681	284.8516	1419.2579	0.2007	0.0007	4.8908	31
32	.0029	4.9854	24.4588	341.8219	1704.1095	0.2006	0.0006	4.9061	32
33	.0024	4.9878	24.5368	410.1863	2045.9314	0.2005	0.0005	4.9194	33
34	.0020	4.9898	24.6038	492.2235	2456.1176	0.2004	0.0004	4.9308	34
35	.0017	4.9915	24.6614	590.6682	2948.3411	0.2003	0.0003	4.9406	35
36	.0014	4.9929	24.7108	708.8019	3539.0094	0.2003	0.0003	4.9491	36
37	.0012	4.9941	24.7531	850.5622	4247.8112	0.2002	0.0002	4.9564	37
38	.0010	4.9951	24.7894	1020.6747	5098.3735	0.2002	0.0002	4.9627	38
39	.0008	4.9959	24.8204	1224.8096	6119.0482	0.2002	0.0002	4.9681	39
40	.0007	4.9966	24.8469	1469.7716	7343.8578	0.2001	0.0001	4.9728	40
41	.0006	4.9972	24.8696	1763.7259	8813.6294	0.2001	0.0001	4.9767	41
42	.0005	4.9976	24.8890	2116.4711	10577.3553	0.2001	0.0001	4.9801	42
43	.0004	4.9980	24.9055	2539.7653	12693.8263	0.2001	0.0001	4.9831	43
44	.0003	4.9984	24.9196	3047.7183	15233.5916	0.2001	0.0001	4.9856	44
45	.0003	4.9986	24.9316	3657.2620	18281.3099	0.2001	0.0001	4.9877	45
46	.0002	4.9989	24.9419	4388.7144	21938.5719	0.2000	0.0000	4.9895	46
47	.0002	4.9991	24.9506	5266.4573	26327.2863	0.2000	0.0000	4.9911	47
48	.0002	4.9992	24.9581	6319.7487	31593.7436	0.2000	0.0000	4.9924	48
49	.0001	4.9993	24.9644	7583.6985	37913.4923	0.2000	0.0000	4.9935	49
50	.0001	4.9995	24.9698	9100.4382	45497.1908	0.2000	0.0000	4.9945	50
51	.0001	4.9995	24.9744	10920.5258	54597.6289	0.2000	0.0000	4.9953	51
52	.0001	4.9996	24.9783	13104.6309	65518.1547	0.2000	0.0000	4.9960	52
53	.0001	4.9997	24.9816	15725.5571	78622.7856	0.2000	0.0000	4.9966	53
54	.0001	4.9997	24.9844	18870.6685	94348.3427	0.2000	0.0000	4.9971	54
55	.0000	4.9998	24.9868	22644.8023	113219.0113	0.2000	0.0000	4.9976	55
60	.0000	4.9999	24.9942	56347.5144	281732.5718	0.2000	0.0000	4.9989	60
65	.0000	5.0000	24.9975	140210.6469	701048.2346	0.2000	0.0000	4.9995	65
70	.0000	5.0000	24.9989	348888.9569	1744439.7847	0.2000	0.0000	4.9998	70
75	.0000	5.0000	24.9995	868147.3693	4340731.8466	0.2000	0.0000	4.9999	75

I = 25.00 %

N	(P/F)	(P/A)	(P/G)	(F/P)	(F/A)	(A/P)	(A/F)	(A/G)	N
1	.8000	0.8000	0.0	1.2500	1.0000	1.2500	1.0000	0.0	1
2	.6400	1.4400	0.6400	1.5625	2.2500	0.6944	0.4444	0.4444	2
3	.5120	1.9520	1.6640	1.9531	3.8125	0.5123	0.2623	0.8525	3
4	.4096	2.3616	2.8928	2.4414	5.7656	0.4234	0.1734	1.2249	4
5	.3277	2.6893	4.2035	3.0518	8.2070	0.3718	0.1218	1.5631	5
6	.2621	2.9514	5.5142	3.8147	11.2588	0.3388	0.0888	1.8683	6
7	.2097	3.1611	6.7725	4.7684	15.0735	0.3163	0.0663	2.1424	7
8	.1678	3.3289	7.9469	5.9605	19.8419	0.3004	0.0504	2.3872	8
9	.1342	3.4631	9.0207	7.4506	25.8023	0.2888	0.0388	2.6048	9
10	.1074	3.5705	9.9870	9.3132	33.2529	0.2801	0.0301	2.7971	10
11	.0859	3.6564	10.8460	11.6415	42.5661	0.2735	0.0235	2.9663	11
12	.0687	3.7251	11.6020	14.5519	54.2077	0.2684	0.0184	3.1145	12
13	.0550	3.7801	12.2617	18.1899	68.7596	0.2645	0.0145	3.2437	13
14	.0440	3.8241	12.8334	22.7374	86.9495	0.2615	0.0115	3.3559	14
15	.0352	3.8593	13.3260	28.4217	109.6868	0.2591	0.0091	3.4530	15
16	.0281	3.8874	13.7482	35.5271	138.1085	0.2572	0.0072	3.5366	16
17	.0225	3.9099	14.1085	44.4089	173.6357	0.2558	0.0058	3.6084	17
18	.0180	3.9279	14.4147	55.5112	218.0446	0.2546	0.0046	3.6698	18
19	.0144	3.9424	14.6741	69.3889	273.5558	0.2537	0.0037	3.7222	19
20	.0115	3.9539	14.8932	86.7362	342.9447	0.2529	0.0029	3.7667	20
21	.0092	3.9631	15.0777	108.4202	429.6809	0.2523	0.0023	3.8045	21
22	.0074	3.9705	15.2326	135.5253	538.1011	0.2519	0.0019	3.8365	22
23	.0059	3.9764	15.3625	169.4066	673.6264	0.2515	0.0015	3.8634	23
24	.0047	3.9811	15.4711	211.7582	843.0329	0.2512	0.0012	3.8861	24
25	.0038	3.9849	15.5618	264.6978	1054.7912	0.2509	0.0009	3.9052	25
26	.0030	3.9879	15.6373	330.8722	1319.4890	0.2508	0.0008	3.9212	26
27	.0024	3.9903	15.7002	413.5903	1650.3612	0.2506	0.0006	3.9346	27
28	.0019	3.9923	15.7524	516.9879	2063.9515	0.2505	0.0005	3.9457	28
29	.0015	3.9938	15.7957	646.2349	2580.9394	0.2504	0.0004	3.9551	29
30	.0012	3.9950	15.8316	807.7936	3227.1743	0.2503	0.0003	3.9628	30
31	.0010	3.9960	15.8614	1009.7420	4034.9678	0.2502	0.0002	3.9693	31
32	.0008	3.9968	15.8859	1262.1774	5044.7098	0.2502	0.0002	3.9746	32
33	.0006	3.9975	15.9062	1577.7218	6306.8872	0.2502	0.0002	3.9791	33
34	.0005	3.9980	15.9229	1972.1523	7884.6091	0.2501	0.0001	3.9828	34
35	.0004	3.9984	15.9367	2465.1903	9856.7613	0.2501	0.0001	3.9858	35
36	.0003	3.9987	15.9481	3081.4879	12321.9516	0.2501	0.0001	3.9883	36
37	.0003	3.9990	15.9574	3851.8599	15403.4396	0.2501	0.0001	3.9904	37
38	.0002	3.9992	15.9651	4814.8249	19255.2994	0.2501	0.0001	3.9921	38
39	.0002	3.9993	15.9714	6018.5311	24070.1243	0.2500	0.0000	3.9935	39
40	.0001	3.9995	15.9766	7523.1638	30088.6554	0.2500	0.0000	3.9947	40
41	.0001	3.9996	15.9809	9403.9548	37611.8192	0.2500	0.0000	3.9956	41
42	.0001	3.9997	15.9843	11754.9435	47015.7740	0.2500	0.0000	3.9964	42
43	.0001	3.9997	15.9872	14693.6794	58770.7175	0.2500	0.0000	3.9971	43
44	.0001	3.9998	15.9895	18367.0992	73464.3969	0.2500	0.0000	3.9976	44
45	.0000	3.9998	15.9915	22958.8740	91831.4962	0.2500	0.0000	3.9980	45
46	.0000	3.9999	15.9930	28698.5925	114790.3702	0.2500	0.0000	3.9984	46
47	.0000	3.9999	15.9943	35873.2407	143488.9627	0.2500	0.0000	3.9987	47
48	.0000	3.9999	15.9954	44841.5509	179362.2034	0.2500	0.0000	3.9989	48
49	.0000	3.9999	15.9962	56051.9386	224203.7543	0.2500	0.0000	3.9991	49
50	.0000	3.9999	15.9969	70064.9232	280255.6929	0.2500	0.0000	3.9993	50
51	.0000	4.0000	15.9975	87581.1540	350320.6161	0.2500	0.0000	3.9994	51
52	.0000	4.0000	15.9980	109476.4425	437901.7701	0.2500	0.0000	3.9995	52
53	.0000	4.0000	15.9983	136845.5532	547378.2126	0.2500	0.0000	3.9996	53
54	.0000	4.0000	15.9986	171056.9414	684223.7658	0.2500	0.0000	3.9997	54
55	.0000	4.0000	15.9989	213821.1768	855280.7072	0.2500	0.0000	3.9997	55
60	.0000	4.0000	15.9996	652530.4468	2610117.7872	0.2500	0.0000	3.9999	60

PROFESSIONAL PUBLICATIONS, INC. • P.O. Box 199, San Carlos, CA 94070

PRACTICE PROBLEMS FOR CHAPTER 28

1. How much will be accumulated at 6% if $1000 is invested for 10 years?

2. What is the present worth at 6% of $2000 which becomes available in 4 years?

3. How much will it take to accumulate $2000 in 20 years at 6%?

4. What year-end annual amount over 7 years at 6% is equivalent to $500 invested now?

5. $50 is invested at the end of each year for 10 years. What will be the accumulated amount at the end of 10 years at 6%?

6. How much should you deposit at 6% at the start of each year for 10 years in order to empty the fund by drawing out $200 at the end of each year for 10 years?

7. You need $2000 on the date of your last deposit. How much should be deposited at the start of each year for 5 years at 6%?

8. How much will be accumulated at 6% in 10 years if you deposit 3 payments of $100 every other year for 4 years, with the first payment occurring at $t = 0$?

9. $500 is compounded monthly at a 6% annual rate. How much will you have in 5 years?

10. What is the rate of return of an $80 investment that pays back $120 in 7 years?

11. A new machine will cost $17000 and will have a value of $14000 in 5 years. Special tooling will cost $5000, and it will have a resale value of $2500 after 5 years. Maintenance will be $200 per year. What will be the average cost of ownership during the next five years if interest is 6%?

12. An old highway bridge may be strengthened at a cost of $9000 or it may be replaced for $40,000. The present salvage value of the old bridge is $13,000. It is estimated that the reinforced bridge will last for 20 years with an annual cost of $500 and will have a salvage value of $10,000 at the end of 20 years. The estimated salvage of the new bridge after 25 years is $15,000. The maintenance for the new bridge will be $100 annually. Which is the best alternative at 8%?

13. A firm expects to receive $32,000 each year for 15 years from the sale of a product. It will require an initial investment of $150,000. Expenses will run $7530 per year. Salvage is zero, and straight-line depreciation is used. The tax rate is 48%. What is the after-tax rate of return?

14. A public works project has initial costs of $1,000,000; benefits of $1,500,000; and disbenefits of $300,000. (a) What is the benefit/cost ratio? (b) What is the excess of benefits over costs?

15. A speculator in land pays $14,000 for property that he expects to hold for 10 years. $1000 is spent in renovation, and a monthly rent of $75 is collected from the tenants. Taxes are $150 per year, and maintenance costs are $250. What must be the sales price in 10 years to realize a 10% rate of return? Use the year-end convention.

16. What is the effective interest rate for a payment plan of 30 equal payments of $89.30 per month when a lump sum of $2000 would have been an outright purchase?

17. An apartment complex is purchased for $500,000. What is the depreciation in each of the first 3 years if the salvage value is $100,000 in 25 years? Use (a) straight line, (b) sum-of-the-years' digits, and (c) double declining balance.

18. Equipment is purchased for $12,000 which is expected to be sold after 10 years for $2000. The estimated maintenance is $1000 the first year but is expected to increase $200 each year thereafter. Using 10%, find the (a) present worth and (b) annual cost.

Appendix A: Sines, Cosines, and Tangents

	0° sine	cosine	tangent	1° sine	cosine	tangent	2° sine	cosine	tangent	
0	0.0000000	1.0000000	0.0000000	0.0174524	0.9998478	0.0174551	0.0348995	0.9993908	0.0349208	60
1	0.0002909	1.0000000	0.0002909	0.0177433	0.9998427	0.0177460	0.0351902	0.9993806	0.0352120	59
2	0.0005818	0.9999999	0.0005818	0.0180341	0.9998373	0.0180370	0.0354809	0.9993705	0.0355033	58
3	0.0008727	0.9999996	0.0008727	0.0183249	0.9998322	0.0183280	0.0357716	0.9993600	0.0357945	57
4	0.0011636	0.9999993	0.0011636	0.0186158	0.9998267	0.0186190	0.0360623	0.9993495	0.0360858	56
5	0.0014544	0.9999990	0.0014544	0.0189066	0.9998213	0.0189100	0.0363530	0.9993390	0.0363771	55
6	0.0017453	0.9999985	0.0017453	0.0191974	0.9998158	0.0192010	0.0366437	0.9993284	0.0366683	54
7	0.0020362	0.9999980	0.0020362	0.0194883	0.9998102	0.0194920	0.0369344	0.9993177	0.0369596	53
8	0.0023271	0.9999974	0.0023271	0.0197791	0.9998044	0.0197830	0.0372251	0.9993070	0.0372509	52
9	0.0026180	0.9999966	0.0026180	0.0200699	0.9997986	0.0200740	0.0375158	0.9992961	0.0375422	51
10	0.0029089	0.9999958	0.0029089	0.0203608	0.9997928	0.0203650	0.0378065	0.9992852	0.0378335	50
11	0.0031998	0.9999949	0.0031998	0.0206516	0.9997869	0.0206560	0.0380971	0.9992741	0.0381248	49
12	0.0034907	0.9999939	0.0034907	0.0209424	0.9997807	0.0209470	0.0383878	0.9992629	0.0384161	48
13	0.0037815	0.9999929	0.0037816	0.0212332	0.9997746	0.0212380	0.0386785	0.9992517	0.0387074	47
14	0.0040724	0.9999917	0.0040725	0.0215241	0.9997684	0.0215291	0.0389692	0.9992404	0.0389988	46
15	0.0043633	0.9999906	0.0043634	0.0218149	0.9997621	0.0218201	0.0392598	0.9992291	0.0392901	45
16	0.0046542	0.9999892	0.0046542	0.0221057	0.9997556	0.0221111	0.0395505	0.9992176	0.0395814	44
17	0.0049451	0.9999879	0.0049451	0.0223965	0.9997492	0.0224021	0.0398411	0.9992060	0.0398728	43
18	0.0052360	0.9999863	0.0052360	0.0226873	0.9997426	0.0226932	0.0401318	0.9991944	0.0401641	42
19	0.0055268	0.9999848	0.0055269	0.0229781	0.9997360	0.0229842	0.0404224	0.9991827	0.0404555	41
20	0.0058177	0.9999831	0.0058178	0.0232690	0.9997293	0.0232753	0.0407131	0.9991709	0.0407469	40
21	0.0061086	0.9999813	0.0061087	0.0235598	0.9997225	0.0235663	0.0410037	0.9991589	0.0410383	39
22	0.0063995	0.9999796	0.0063996	0.0238506	0.9997156	0.0238574	0.0412944	0.9991470	0.0413296	38
23	0.0066904	0.9999776	0.0066905	0.0241414	0.9997086	0.0241484	0.0415850	0.9991350	0.0416210	37
24	0.0069813	0.9999757	0.0069814	0.0244322	0.9997015	0.0244395	0.0418757	0.9991229	0.0419124	36
25	0.0072721	0.9999736	0.0072723	0.0247230	0.9996944	0.0247305	0.0421663	0.9991106	0.0422038	35
26	0.0075630	0.9999714	0.0075632	0.0250138	0.9996872	0.0250216	0.0424569	0.9990984	0.0424952	34
27	0.0078539	0.9999692	0.0078541	0.0253046	0.9996798	0.0253127	0.0427475	0.9990859	0.0427867	33
28	0.0081448	0.9999669	0.0081450	0.0255954	0.9996724	0.0256038	0.0430382	0.9990735	0.0430781	32
29	0.0084357	0.9999645	0.0084360	0.0258862	0.9996650	0.0258948	0.0433288	0.9990609	0.0433695	31
30	0.0087265	0.9999620	0.0087269	0.0261769	0.9996574	0.0261859	0.0436194	0.9990483	0.0436609	30
31	0.0090174	0.9999593	0.0090178	0.0264677	0.9996498	0.0264770	0.0439100	0.9990355	0.0439524	29
32	0.0093083	0.9999568	0.0093087	0.0267585	0.9996419	0.0267681	0.0442006	0.9990228	0.0442438	28
33	0.0095992	0.9999539	0.0095996	0.0270493	0.9996342	0.0270592	0.0444912	0.9990098	0.0445353	27
34	0.0098900	0.9999512	0.0098905	0.0273401	0.9996262	0.0273503	0.0447818	0.9989969	0.0448268	26
35	0.0101809	0.9999482	0.0101814	0.0276309	0.9996182	0.0276414	0.0450724	0.9989838	0.0451182	25
36	0.0104718	0.9999452	0.0104724	0.0279216	0.9996101	0.0279325	0.0453630	0.9989706	0.0454097	24
37	0.0107627	0.9999421	0.0107633	0.0282124	0.9996020	0.0282236	0.0456536	0.9989573	0.0457012	23
38	0.0110535	0.9999390	0.0110542	0.0285032	0.9995937	0.0285148	0.0459442	0.9989441	0.0459927	22
39	0.0113444	0.9999357	0.0113451	0.0287940	0.9995854	0.0288059	0.0462347	0.9989307	0.0462842	21
40	0.0116353	0.9999323	0.0116361	0.0290847	0.9995770	0.0290970	0.0465253	0.9989172	0.0465757	20
41	0.0119261	0.9999290	0.0119270	0.0293755	0.9995685	0.0293882	0.0468159	0.9989035	0.0468673	19
42	0.0122170	0.9999254	0.0122179	0.0296662	0.9995599	0.0296793	0.0471065	0.9988900	0.0471588	18
43	0.0125079	0.9999218	0.0125088	0.0299570	0.9995512	0.0299705	0.0473970	0.9988762	0.0474503	17
44	0.0127987	0.9999180	0.0127998	0.0302478	0.9995425	0.0302616	0.0476876	0.9988623	0.0477419	16
45	0.0130896	0.9999143	0.0130907	0.0305385	0.9995337	0.0305528	0.0479781	0.9988483	0.0480334	15
46	0.0133805	0.9999106	0.0133817	0.0308293	0.9995248	0.0308439	0.0482687	0.9988344	0.0483250	14
47	0.0136713	0.9999065	0.0136726	0.0311200	0.9995157	0.0311351	0.0485592	0.9988204	0.0486166	13
48	0.0139622	0.9999025	0.0139635	0.0314108	0.9995065	0.0314263	0.0488498	0.9988061	0.0489082	12
49	0.0142530	0.9998986	0.0142545	0.0317015	0.9994974	0.0317174	0.0491403	0.9987918	0.0491997	11
50	0.0145439	0.9998944	0.0145454	0.0319922	0.9994882	0.0320086	0.0494308	0.9987776	0.0494913	10
51	0.0148348	0.9998901	0.0148364	0.0322830	0.9994789	0.0322998	0.0497214	0.9987633	0.0497829	9
52	0.0151256	0.9998857	0.0151273	0.0325737	0.9994694	0.0325910	0.0500119	0.9987487	0.0500746	8
53	0.0154165	0.9998812	0.0154183	0.0328644	0.9994599	0.0328822	0.0503024	0.9987341	0.0503662	7
54	0.0157073	0.9998766	0.0157093	0.0331552	0.9994502	0.0331734	0.0505929	0.9987194	0.0506578	6
55	0.0159982	0.9998720	0.0160002	0.0334459	0.9994405	0.0334646	0.0508835	0.9987047	0.0509495	5
56	0.0162890	0.9998673	0.0162912	0.0337366	0.9994308	0.0337558	0.0511740	0.9986897	0.0512411	4
57	0.0165799	0.9998626	0.0165821	0.0340273	0.9994209	0.0340471	0.0514645	0.9986749	0.0515328	3
58	0.0168707	0.9998577	0.0168731	0.0343181	0.9994110	0.0343383	0.0517550	0.9986599	0.0518244	2
59	0.0171616	0.9998527	0.0171641	0.0346088	0.9994009	0.0346295	0.0520455	0.9986448	0.0521161	1
60	0.0174524	0.9998478	0.0174551	0.0348995	0.9993908	0.0349208	0.0523360	0.9986296	0.0524078	0
	cosine	sine 89°	cotangent	cosine	sine 88°	cotangent	cosine	sine 87°	cotangent	

PROFESSIONAL PUBLICATIONS, INC. ● P.O. Box 199, San Carlos, CA 94070

LAND SURVEYOR REFERENCE MANUAL

	3°			4°			5°			
	sine	cosine	tangent	sine	cosine	tangent	sine	cosine	tangent	
0	0.0523360	0.9986296	0.0524078	0.0697565	0.9975641	0.0699268	0.0871557	0.9961947	0.0874887	60
1	0.0526264	0.9986144	0.0526995	0.0700467	0.9975437	0.0702191	0.0874455	0.9961693	0.0877818	59
2	0.0529169	0.9985990	0.0529912	0.0703368	0.9975234	0.0705115	0.0877353	0.9961439	0.0880749	58
3	0.0532074	0.9985835	0.0532829	0.0706270	0.9975028	0.0708038	0.0880251	0.9961184	0.0883681	57
4	0.0534979	0.9985680	0.0535746	0.0709171	0.9974822	0.0710961	0.0883148	0.9960926	0.0886612	56
5	0.0537883	0.9985524	0.0538663	0.0712073	0.9974615	0.0713885	0.0886046	0.9960669	0.0889544	55
6	0.0540788	0.9985366	0.0541581	0.0714974	0.9974409	0.0716809	0.0888943	0.9960410	0.0892476	54
7	0.0543693	0.9985209	0.0544498	0.0717876	0.9974199	0.0719733	0.0891840	0.9960153	0.0895408	53
8	0.0546597	0.9985051	0.0547416	0.0720777	0.9973991	0.0722657	0.0894738	0.9959892	0.0898341	52
9	0.0549502	0.9984891	0.0550333	0.0723679	0.9973780	0.0725581	0.0897635	0.9959632	0.0901273	51
10	0.0552406	0.9984730	0.0553251	0.0726580	0.9973570	0.0728505	0.0900532	0.9959369	0.0904206	50
11	0.0555311	0.9984570	0.0556169	0.0729481	0.9973358	0.0731430	0.0903429	0.9959108	0.0907138	49
12	0.0558215	0.9984409	0.0559087	0.0732382	0.9973145	0.0734354	0.0906326	0.9958845	0.0910071	48
13	0.0561119	0.9984245	0.0562005	0.0735283	0.9972931	0.0737279	0.0909223	0.9958580	0.0913004	47
14	0.0564024	0.9984081	0.0564923	0.0738184	0.9972718	0.0740203	0.0912119	0.9958315	0.0915938	46
15	0.0566928	0.9983917	0.0567841	0.0741085	0.9972503	0.0743128	0.0915016	0.9958050	0.0918871	45
16	0.0569832	0.9983752	0.0570759	0.0743986	0.9972286	0.0746053	0.0917913	0.9957784	0.0921804	44
17	0.0572736	0.9983585	0.0573678	0.0746887	0.9972069	0.0748979	0.0920809	0.9957515	0.0924738	43
18	0.0575640	0.9983417	0.0576596	0.0749787	0.9971851	0.0751904	0.0923706	0.9957248	0.0927672	42
19	0.0578544	0.9983250	0.0579515	0.0752688	0.9971634	0.0754829	0.0926602	0.9956979	0.0930606	41
20	0.0581448	0.9983082	0.0582434	0.0755589	0.9971414	0.0757755	0.0929499	0.9956707	0.0933540	40
21	0.0584352	0.9982913	0.0585352	0.0758489	0.9971194	0.0760680	0.0932395	0.9956437	0.0936474	39
22	0.0587256	0.9982742	0.0588271	0.0761390	0.9970973	0.0763606	0.0935291	0.9956166	0.0939409	38
23	0.0590160	0.9982570	0.0591190	0.0764290	0.9970750	0.0766532	0.0938187	0.9955893	0.0942343	37
24	0.0593064	0.9982399	0.0594109	0.0767190	0.9970527	0.0769458	0.0941083	0.9955620	0.0945278	36
25	0.0595968	0.9982225	0.0597029	0.0770091	0.9970305	0.0772384	0.0943979	0.9955346	0.0948213	35
26	0.0598871	0.9982051	0.0599948	0.0772991	0.9970080	0.0775311	0.0946875	0.9955071	0.0951148	34
27	0.0601775	0.9981877	0.0602867	0.0775891	0.9969855	0.0778237	0.0949771	0.9954795	0.0954084	33
28	0.0604678	0.9981702	0.0605787	0.0778791	0.9969628	0.0781164	0.0952666	0.9954517	0.0957019	32
29	0.0607582	0.9981525	0.0608706	0.0781691	0.9969403	0.0784090	0.0955562	0.9954241	0.0959955	31
30	0.0610485	0.9981349	0.0611626	0.0784591	0.9969174	0.0787017	0.0958458	0.9953963	0.0962890	30
31	0.0613389	0.9981171	0.0614546	0.0787491	0.9968945	0.0789944	0.0961353	0.9953683	0.0965826	29
32	0.0616292	0.9980991	0.0617466	0.0790391	0.9968715	0.0792871	0.0964248	0.9953403	0.0968763	28
33	0.0619196	0.9980811	0.0620386	0.0793290	0.9968486	0.0795798	0.0967144	0.9953123	0.0971699	27
34	0.0622099	0.9980632	0.0623306	0.0796190	0.9968254	0.0798726	0.0970039	0.9952840	0.0974635	26
35	0.0625002	0.9980450	0.0626226	0.0799090	0.9968022	0.0801653	0.0972934	0.9952557	0.0977572	25
36	0.0627905	0.9980267	0.0629147	0.0801989	0.9967790	0.0804581	0.0975829	0.9952273	0.0980509	24
37	0.0630808	0.9980085	0.0632067	0.0804889	0.9967555	0.0807509	0.0978724	0.9951990	0.0983446	23
38	0.0633711	0.9979901	0.0634988	0.0807788	0.9967321	0.0810437	0.0981619	0.9951705	0.0986383	22
39	0.0636614	0.9979716	0.0637908	0.0810687	0.9967086	0.0813365	0.0984514	0.9951419	0.0989320	21
40	0.0639517	0.9979530	0.0640829	0.0813587	0.9966848	0.0816293	0.0987408	0.9951133	0.0992257	20
41	0.0642420	0.9979343	0.0643750	0.0816486	0.9966613	0.0819221	0.0990303	0.9950845	0.0995195	19
42	0.0645323	0.9979156	0.0646671	0.0819385	0.9966375	0.0822150	0.0993198	0.9950556	0.0998133	18
43	0.0648226	0.9978969	0.0649592	0.0822284	0.9966136	0.0825078	0.0996092	0.9950266	0.1001071	17
44	0.0651129	0.9978780	0.0652513	0.0825183	0.9965895	0.0828007	0.0998986	0.9949976	0.1004009	16
45	0.0654031	0.9978589	0.0655435	0.0828082	0.9965655	0.0830936	0.1001881	0.9949685	0.1006947	15
46	0.0656934	0.9978399	0.0658356	0.0830981	0.9965413	0.0833865	0.1004775	0.9949394	0.1009885	14
47	0.0659837	0.9978207	0.0661278	0.0833880	0.9965172	0.0836794	0.1007669	0.9949101	0.1012824	13
48	0.0662739	0.9978015	0.0664199	0.0836779	0.9964929	0.0839723	0.1010563	0.9948807	0.1015763	12
49	0.0665642	0.9977821	0.0667121	0.0839677	0.9964685	0.0842653	0.1013457	0.9948513	0.1018702	11
50	0.0668544	0.9977629	0.0670043	0.0842576	0.9964441	0.0845582	0.1016351	0.9948218	0.1021641	10
51	0.0671446	0.9977433	0.0672965	0.0845474	0.9964195	0.0848512	0.1019245	0.9947922	0.1024580	9
52	0.0674348	0.9977237	0.0675887	0.0848373	0.9963949	0.0851442	0.1022138	0.9947625	0.1027520	8
53	0.0677251	0.9977041	0.0678809	0.0851271	0.9963701	0.0854372	0.1025032	0.9947327	0.1030460	7
54	0.0680153	0.9976842	0.0681732	0.0854169	0.9963453	0.0857302	0.1027925	0.9947028	0.1033399	6
55	0.0683055	0.9976644	0.0684654	0.0857067	0.9963204	0.0860233	0.1030819	0.9946729	0.1036339	5
56	0.0685957	0.9976446	0.0687577	0.0859966	0.9962954	0.0863163	0.1033712	0.9946428	0.1039280	4
57	0.0688859	0.9976246	0.0690499	0.0862864	0.9962704	0.0866094	0.1036605	0.9946127	0.1042220	3
58	0.0691761	0.9976045	0.0693422	0.0865762	0.9962454	0.0869025	0.1039499	0.9945826	0.1045161	2
59	0.0694663	0.9975843	0.0696345	0.0868660	0.9962201	0.0871956	0.1042392	0.9945523	0.1048102	1
60	0.0697565	0.9975641	0.0699268	0.0871557	0.9961947	0.0874887	0.1045285	0.9945219	0.1051042	0
	cosine	sine	cotangent	cosine	sine	cotangent	cosine	sine	cotangent	
		86°			85°			84°		

PROFESSIONAL PUBLICATIONS, INC. ● P.O. Box 199, San Carlos, CA 94070

	6° sine	6° cosine	6° tangent	7° sine	7° cosine	7° tangent	8° sine	8° cosine	8° tangent	
0	0.1045285	0.9945219	0.1051042	0.1218693	0.9925462	0.1227846	0.1391731	0.9902681	0.1405408	60
1	0.1048178	0.9944914	0.1053984	0.1221581	0.9925107	0.1230798	0.1394612	0.9902275	0.1408375	59
2	0.1051070	0.9944609	0.1056925	0.1224468	0.9924752	0.1233751	0.1397492	0.9901870	0.1411341	58
3	0.1053963	0.9944302	0.1059866	0.1227355	0.9924394	0.1236705	0.1400372	0.9901463	0.1414309	57
4	0.1056856	0.9943996	0.1062808	0.1230241	0.9924037	0.1239658	0.1403252	0.9901056	0.1417276	56
5	0.1059748	0.9943689	0.1065750	0.1233128	0.9923678	0.1242612	0.1406132	0.9900646	0.1420243	55
6	0.1062641	0.9943380	0.1068692	0.1236015	0.9923319	0.1245566	0.1409013	0.9900236	0.1423211	54
7	0.1065533	0.9943070	0.1071634	0.1238901	0.9922960	0.1248520	0.1411892	0.9899827	0.1426179	53
8	0.1068425	0.9942761	0.1074576	0.1241788	0.9922599	0.1251474	0.1414772	0.9899416	0.1429147	52
9	0.1071318	0.9942449	0.1077519	0.1244674	0.9922237	0.1254429	0.1417651	0.9899004	0.1432115	51
10	0.1074210	0.9942137	0.1080462	0.1247560	0.9921875	0.1257384	0.1420531	0.9898591	0.1435084	50
11	0.1077102	0.9941824	0.1083404	0.1250446	0.9921511	0.1260339	0.1423410	0.9898177	0.1438053	49
12	0.1079994	0.9941510	0.1086348	0.1253332	0.9921148	0.1263294	0.1426289	0.9897763	0.1441022	48
13	0.1082885	0.9941195	0.1089291	0.1256218	0.9920782	0.1266249	0.1429168	0.9897347	0.1443991	47
14	0.1085777	0.9940879	0.1092234	0.1259104	0.9920416	0.1269205	0.1432047	0.9896931	0.1446961	46
15	0.1088669	0.9940564	0.1095178	0.1261990	0.9920049	0.1272161	0.1434926	0.9896514	0.1449931	45
16	0.1091560	0.9940247	0.1098122	0.1264875	0.9919682	0.1275117	0.1437805	0.9896097	0.1452901	44
17	0.1094452	0.9939930	0.1101066	0.1267761	0.9919314	0.1278073	0.1440684	0.9895678	0.1455872	43
18	0.1097343	0.9939610	0.1104010	0.1270646	0.9918945	0.1281030	0.1443562	0.9895259	0.1458842	42
19	0.1100234	0.9939290	0.1106955	0.1273531	0.9918574	0.1283986	0.1446440	0.9894838	0.1461813	41
20	0.1103126	0.9938970	0.1109899	0.1276416	0.9918204	0.1286943	0.1449319	0.9894417	0.1464784	40
21	0.1106017	0.9938648	0.1112844	0.1279301	0.9917832	0.1289900	0.1452197	0.9893995	0.1467756	39
22	0.1108908	0.9938326	0.1115789	0.1282186	0.9917459	0.1292858	0.1455075	0.9893571	0.1470728	38
23	0.1111799	0.9938004	0.1118734	0.1285071	0.9917086	0.1295815	0.1457953	0.9893149	0.1473699	37
24	0.1114689	0.9937680	0.1121680	0.1287956	0.9916711	0.1298773	0.1460830	0.9892724	0.1476672	36
25	0.1117580	0.9937355	0.1124625	0.1290841	0.9916337	0.1301731	0.1463708	0.9892299	0.1479644	35
26	0.1120471	0.9937029	0.1127571	0.1293725	0.9915961	0.1304689	0.1466586	0.9891872	0.1482617	34
27	0.1123361	0.9936703	0.1130517	0.1296609	0.9915584	0.1307648	0.1469463	0.9891446	0.1485590	33
28	0.1126252	0.9936376	0.1133463	0.1299494	0.9915206	0.1310607	0.1472340	0.9891017	0.1488563	32
29	0.1129142	0.9936047	0.1136409	0.1302378	0.9914828	0.1313566	0.1475217	0.9890589	0.1491536	31
30	0.1132032	0.9935719	0.1139356	0.1305262	0.9914449	0.1316525	0.1478094	0.9890159	0.1494510	30
31	0.1134922	0.9935389	0.1142303	0.1308146	0.9914069	0.1319485	0.1480971	0.9889729	0.1497484	29
32	0.1137812	0.9935059	0.1145250	0.1311030	0.9913688	0.1322444	0.1483848	0.9889296	0.1500458	28
33	0.1140702	0.9934727	0.1148197	0.1313913	0.9913305	0.1325404	0.1486724	0.9888866	0.1503433	27
34	0.1143592	0.9934395	0.1151144	0.1316797	0.9912924	0.1328364	0.1489601	0.9888433	0.1506407	26
35	0.1146482	0.9934062	0.1154092	0.1319681	0.9912540	0.1331324	0.1492477	0.9887998	0.1509383	25
36	0.1149372	0.9933728	0.1157039	0.1322564	0.9912156	0.1334285	0.1495354	0.9887564	0.1512358	24
37	0.1152261	0.9933392	0.1159988	0.1325447	0.9911771	0.1337246	0.1498230	0.9887129	0.1515333	23
38	0.1155151	0.9933058	0.1162935	0.1328330	0.9911385	0.1340207	0.1501106	0.9886693	0.1518309	22
39	0.1158040	0.9932721	0.1165884	0.1331213	0.9910998	0.1343168	0.1503981	0.9886255	0.1521285	21
40	0.1160929	0.9932384	0.1168832	0.1334096	0.9910610	0.1346129	0.1506857	0.9885818	0.1524262	20
41	0.1163818	0.9932045	0.1171781	0.1336979	0.9910222	0.1349091	0.1509733	0.9885378	0.1527238	19
42	0.1166707	0.9931707	0.1174730	0.1339862	0.9909832	0.1352053	0.1512608	0.9884940	0.1530215	18
43	0.1169596	0.9931366	0.1177679	0.1342744	0.9909442	0.1355015	0.1515484	0.9884499	0.1533192	17
44	0.1172485	0.9931026	0.1180629	0.1345627	0.9909051	0.1357978	0.1518359	0.9884058	0.1536169	16
45	0.1175374	0.9930685	0.1183578	0.1348509	0.9908659	0.1360940	0.1521234	0.9883615	0.1539147	15
46	0.1178263	0.9930342	0.1186528	0.1351392	0.9908267	0.1363903	0.1524109	0.9883172	0.1542125	14
47	0.1181151	0.9929999	0.1189478	0.1354274	0.9907874	0.1366866	0.1526984	0.9882728	0.1545103	13
48	0.1184040	0.9929656	0.1192428	0.1357156	0.9907479	0.1369830	0.1529858	0.9882285	0.1548082	12
49	0.1186928	0.9929311	0.1195378	0.1360038	0.9907083	0.1372793	0.1532733	0.9881838	0.1551061	11
50	0.1189816	0.9928965	0.1198329	0.1362919	0.9906688	0.1375757	0.1535607	0.9881393	0.1554039	10
51	0.1192704	0.9928619	0.1201279	0.1365801	0.9906291	0.1378721	0.1538482	0.9880946	0.1557019	9
52	0.1195593	0.9928271	0.1204230	0.1368683	0.9905893	0.1381686	0.1541356	0.9880497	0.1559998	8
53	0.1198480	0.9927922	0.1207182	0.1371564	0.9905494	0.1384650	0.1544230	0.9880049	0.1562978	7
54	0.1201368	0.9927573	0.1210133	0.1374446	0.9905095	0.1387615	0.1547104	0.9879599	0.1565958	6
55	0.1204256	0.9927224	0.1213085	0.1377327	0.9904695	0.1390580	0.1549978	0.9879149	0.1568938	5
56	0.1207144	0.9926874	0.1216036	0.1380208	0.9904294	0.1393545	0.1552851	0.9878697	0.1571919	4
57	0.1210031	0.9926521	0.1218988	0.1383089	0.9903892	0.1396510	0.1555725	0.9878245	0.1574900	3
58	0.1212919	0.9926169	0.1221940	0.1385970	0.9903489	0.1399476	0.1558598	0.9877791	0.1577881	2
59	0.1215806	0.9925816	0.1224893	0.1388851	0.9903085	0.1402442	0.1561472	0.9877338	0.1580863	1
60	0.1218693	0.9925462	0.1227846	0.1391731	0.9902681	0.1405408	0.1564345	0.9876884	0.1583844	0
	cosine	sine 83°	cotangent	cosine	sine 82°	cotangent	cosine	sine 81°	cotangent	

		9°			10°			11°		
	sine	cosine	tangent	sine	cosine	tangent	sine	cosine	tangent	
0	0.1564345	0.9876884	0.1583844	0.1736482	0.9848078	0.1763270	0.1908090	0.9816272	0.1943803	60
1	0.1567218	0.9876429	0.1586826	0.1739346	0.9847572	0.1766269	0.1910945	0.9815717	0.1946822	59
2	0.1570091	0.9875972	0.1589809	0.1742211	0.9847066	0.1769269	0.1913801	0.9815159	0.1949842	58
3	0.1572963	0.9875515	0.1592791	0.1745075	0.9846559	0.1772269	0.1916656	0.9814603	0.1952861	57
4	0.1575836	0.9875057	0.1595774	0.1747939	0.9846051	0.1775270	0.1919511	0.9814045	0.1955881	56
5	0.1578708	0.9874598	0.1598757	0.1750803	0.9845542	0.1778270	0.1922365	0.9813486	0.1958901	55
6	0.1581581	0.9874138	0.1601741	0.1753667	0.9845032	0.1781271	0.1925220	0.9812927	0.1961922	54
7	0.1584453	0.9873677	0.1604724	0.1756531	0.9844521	0.1784273	0.1928074	0.9812366	0.1964943	53
8	0.1587325	0.9873216	0.1607708	0.1759395	0.9844010	0.1787274	0.1930928	0.9811805	0.1967964	52
9	0.1590197	0.9872755	0.1610692	0.1762258	0.9843498	0.1790276	0.1933782	0.9811243	0.1970986	51
10	0.1593069	0.9872291	0.1613677	0.1765121	0.9842985	0.1793278	0.1936636	0.9810680	0.1974008	50
11	0.1595940	0.9871828	0.1616662	0.1767984	0.9842471	0.1796281	0.1939490	0.9810116	0.1977031	49
12	0.1598812	0.9871364	0.1619646	0.1770847	0.9841956	0.1799284	0.1942343	0.9809552	0.1980053	48
13	0.1601683	0.9870898	0.1622632	0.1773710	0.9841441	0.1802287	0.1945197	0.9808987	0.1983076	47
14	0.1604555	0.9870431	0.1625617	0.1776573	0.9840924	0.1805291	0.1948050	0.9808421	0.1986100	46
15	0.1607426	0.9869964	0.1628603	0.1779436	0.9840408	0.1808295	0.1950903	0.9807854	0.1989124	45
16	0.1610297	0.9869496	0.1631589	0.1782298	0.9839890	0.1811299	0.1953756	0.9807286	0.1992148	44
17	0.1613168	0.9869028	0.1634576	0.1785160	0.9839370	0.1814303	0.1956609	0.9806716	0.1995172	43
18	0.1616038	0.9868557	0.1637563	0.1788022	0.9838850	0.1817308	0.1959462	0.9806147	0.1998197	42
19	0.1618909	0.9868087	0.1640550	0.1790884	0.9838330	0.1820313	0.1962314	0.9805576	0.2001223	41
20	0.1621779	0.9867615	0.1643537	0.1793746	0.9837809	0.1823318	0.1965166	0.9805005	0.2004248	40
21	0.1624650	0.9867144	0.1646525	0.1796608	0.9837287	0.1826324	0.1968018	0.9804432	0.2007274	39
22	0.1627520	0.9866670	0.1649513	0.1799469	0.9836763	0.1829330	0.1970870	0.9803861	0.2010300	38
23	0.1630390	0.9866197	0.1652501	0.1802330	0.9836240	0.1832336	0.1973722	0.9803287	0.2013327	37
24	0.1633260	0.9865721	0.1655489	0.1805191	0.9835715	0.1835343	0.1976573	0.9802712	0.2016354	36
25	0.1636129	0.9865246	0.1658478	0.1808053	0.9835190	0.1838351	0.1979425	0.9802136	0.2019381	35
26	0.1638999	0.9864770	0.1661467	0.1810913	0.9834663	0.1841358	0.1982276	0.9801560	0.2022409	34
27	0.1641869	0.9864293	0.1664456	0.1813774	0.9834136	0.1844365	0.1985127	0.9800984	0.2025436	33
28	0.1644738	0.9863815	0.1667446	0.1816635	0.9833608	0.1847373	0.1987978	0.9800406	0.2028465	32
29	0.1647607	0.9863335	0.1670436	0.1819495	0.9833079	0.1850382	0.1990829	0.9799827	0.2031494	31
30	0.1650476	0.9862856	0.1673426	0.1822355	0.9832549	0.1853391	0.1993679	0.9799247	0.2034523	30
31	0.1653345	0.9862376	0.1676417	0.1825215	0.9832019	0.1856399	0.1996530	0.9798667	0.2037553	29
32	0.1656214	0.9861894	0.1679407	0.1828075	0.9831487	0.1859409	0.1999380	0.9798086	0.2040582	28
33	0.1659082	0.9861413	0.1682398	0.1830935	0.9830955	0.1862418	0.2002230	0.9797504	0.2043612	27
34	0.1661951	0.9860929	0.1685390	0.1833795	0.9830422	0.1865428	0.2005080	0.9796922	0.2046643	26
35	0.1664819	0.9860445	0.1688381	0.1836654	0.9829888	0.1868439	0.2007930	0.9796337	0.2049674	25
36	0.1667687	0.9859961	0.1691373	0.1839514	0.9829354	0.1871449	0.2010779	0.9795753	0.2052705	24
37	0.1670556	0.9859475	0.1694366	0.1842373	0.9828818	0.1874460	0.2013629	0.9795167	0.2055737	23
38	0.1673423	0.9858989	0.1697358	0.1845232	0.9828283	0.1877471	0.2016478	0.9794582	0.2058769	22
39	0.1676291	0.9858502	0.1700351	0.1848091	0.9827744	0.1880483	0.2019327	0.9793994	0.2061801	21
40	0.1679159	0.9858013	0.1703344	0.1850949	0.9827207	0.1883495	0.2022176	0.9793407	0.2064834	20
41	0.1682026	0.9857525	0.1706337	0.1853808	0.9826668	0.1886507	0.2025025	0.9792818	0.2067867	19
42	0.1684894	0.9857035	0.1709331	0.1856666	0.9826128	0.1889520	0.2027873	0.9792228	0.2070900	18
43	0.1687761	0.9856545	0.1712325	0.1859524	0,9825588	0.1892532	0.2030721	0.9791638	0.2073934	17
44	0.1690628	0.9856054	0.1715320	0.1862383	0.9825047	0.1895546	0.2033570	0.9791047	0.2076969	16
45	0.1693495	0.9855561	0.1718314	0.1865240	0.9824504	0.1898559	0.2036418	0.9790455	0.2080003	15
46	0.1696362	0.9855068	0.1721309	0.1868098	0.9823961	0.1901573	0.2039265	0.9789863	0.2083038	14
47	0.1699229	0.9854574	0.1724304	0.1870956	0.9823418	0.1904587	0.2042113	0.9789268	0.2086073	13
48	0.1702095	0.9854079	0.1727300	0.1873813	0.9822872	0.1907602	0.2044961	0.9788674	0.2089109	12
49	0.1704961	0.9853583	0.1730296	0.1876670	0.9822327	0.1910617	0.2047808	0.9788079	0.2092145	11
50	0.1707828	0.9853087	0.1733292	0.1879528	0.9821781	0.1913632	0.2050655	0.9787483	0.2095181	10
51	0.1710694	0.9852590	0.1736289	0.1882385	0.9821233	0.1916648	0.2053502	0.9786885	0.2098218	9
52	0.1713560	0.9852092	0.1739285	0.1885241	0.9820686	0.1919664	0.2056349	0.9786289	0.2101255	8
53	0.1716425	0.9851594	0.1742282	0.1888098	0.9820137	0.1922680	0.2059195	0.9785689	0.2104293	7
54	0.1719291	0.9851093	0.1745279	0.1890954	0.9819587	0.1925696	0.2062042	0.9785091	0.2107330	6
55	0.1722157	0.9850593	0.1748277	0.1893811	0.9819037	0.1928713	0.2064888	0.9784490	0.2110369	5
56	0.1725022	0.9850091	0.1751275	0.1896667	0.9818486	0.1931731	0.2067734	0.9783888	0.2113407	4
57	0.1727887	0.9849590	0.1754273	0.1899523	0.9817933	0.1934748	0.2070580	0.9783287	0.2116446	3
58	0.1730752	0.9849086	0.1757272	0.1902379	0.9817380	0.1937766	0.2073426	0.9782684	0.2119486	2
59	0.1733617	0.9848583	0.1760270	0.1905235	0.9816827	0.1940785	0.2076271	0.9782081	0.2122525	1
60	0.1736482	0.9848078	0.1763270	0.1908090	0.9816272	0.1943803	0.2079117	0.9781476	0.2125566	0
	cosine	sine	cotangent	cosine	sine	cotangent	cosine	sine	cotangent	
		80°			79°			78°		

PROFESSIONAL PUBLICATIONS, INC. ● P.O. Box 199, San Carlos, CA 94070

	12° sine	12° cosine	12° tangent	13° sine	13° cosine	13° tangent	14° sine	14° cosine	14° tangent	
0	0.2079117	0.9781476	0.2125566	0.2249510	0.9743701	0.2308682	0.2419219	0.9702958	0.2493280	60
1	0.2081962	0.9780871	0.2128606	0.2252345	0.9743046	0.2311746	0.2422041	0.9702253	0.2496370	59
2	0.2084807	0.9780264	0.2131647	0.2255179	0.9742391	0.2314811	0.2424864	0.9701549	0.2499460	58
3	0.2087652	0.9779659	0.2134688	0.2258013	0.9741735	0.2317875	0.2427686	0.9700842	0.2502551	57
4	0.2090497	0.9779051	0.2137730	0.2260846	0.9741077	0.2320941	0.2430507	0.9700136	0.2505642	56
5	0.2093341	0.9778442	0.2140772	0.2263680	0.9740419	0.2324006	0.2433329	0.9699429	0.2508734	55
6	0.2096186	0.9777833	0.2143814	0.2266513	0.9739760	0.2327073	0.2436150	0.9698721	0.2511826	54
7	0.2099030	0.9777223	0.2146857	0.2269346	0.9739100	0.2330140	0.2438972	0.9698011	0.2514919	53
8	0.2101874	0.9776611	0.2149900	0.2272179	0.9738440	0.2333207	0.2441792	0.9697302	0.2518012	52
9	0.2104718	0.9775999	0.2152944	0.2275012	0.9737779	0.2336274	0.2444613	0.9696591	0.2521105	51
10	0.2107561	0.9775387	0.2155988	0.2277844	0.9737117	0.2339342	0.2447434	0.9695880	0.2524199	50
11	0.2110405	0.9774774	0.2159032	0.2280677	0.9736453	0.2342410	0.2450254	0.9695168	0.2527294	49
12	0.2113248	0.9774159	0.2162077	0.2283509	0.9735789	0.2345479	0.2453074	0.9694453	0.2530389	48
13	0.2116091	0.9773544	0.2165121	0.2286341	0.9735125	0.2348548	0.2455894	0.9693739	0.2533485	47
14	0.2118934	0.9772928	0.2168167	0.2289173	0.9734460	0.2351618	0.2458714	0.9693025	0.2536580	46
15	0.2121777	0.9772311	0.2171213	0.2292004	0.9733794	0.2354687	0.2461533	0.9692309	0.2539676	45
16	0.2124619	0.9771694	0.2174259	0.2294835	0.9733126	0.2357758	0.2464352	0.9691594	0.2542773	44
17	0.2127462	0.9771076	0.2177306	0.2297667	0.9732458	0.2360829	0.2467171	0.9690877	0.2545870	43
18	0.2130304	0.9770456	0.2180353	0.2300498	0.9731789	0.2363900	0.2469990	0.9690158	0.2548968	42
19	0.2133146	0.9769837	0.2183400	0.2303328	0.9731120	0.2366971	0.2472809	0.9689439	0.2552066	41
20	0.2135988	0.9769215	0.2186448	0.2306159	0.9730449	0.2370043	0.2475627	0.9688718	0.2555165	40
21	0.2138830	0.9768593	0.2189496	0.2308989	0.9729778	0.2373116	0.2478446	0.9687998	0.2558264	39
22	0.2141671	0.9767971	0.2192544	0.2311819	0.9729105	0.2376189	0.2481263	0.9687277	0.2561363	38
23	0.2144512	0.9767348	0.2195593	0.2314649	0.9728432	0.2379262	0.2484082	0.9686555	0.2564463	37
24	0.2147353	0.9766724	0.2198642	0.2317479	0.9727759	0.2382336	0.2486899	0.9685832	0.2567563	36
25	0.2150194	0.9766098	0.2201692	0.2320309	0.9727085	0.2385410	0.2489716	0.9685107	0.2570665	35
26	0.2153035	0.9765472	0.2204742	0.2323138	0.9726410	0.2388484	0.2492534	0.9684384	0.2573766	34
27	0.2155876	0.9764846	0.2207793	0.2325967	0.9725733	0.2391560	0.2495351	0.9683658	0.2576868	33
28	0.2158716	0.9764218	0.2210844	0.2328796	0.9725056	0.2394635	0.2498167	0.9682931	0.2579970	32
29	0.2161556	0.9763590	0.2213895	0.2331625	0.9724379	0.2397711	0.2500984	0.9682205	0.2583073	31
30	0.2164396	0.9762959	0.2216947	0.2334453	0.9723700	0.2400787	0.2503800	0.9681476	0.2586176	30
31	0.2167236	0.9762330	0.2219999	0.2337282	0.9723021	0.2403864	0.2506616	0.9680748	0.2589279	29
32	0.2170076	0.9761699	0.2223051	0.2340110	0.9722340	0.2406942	0.2509432	0.9680018	0.2592383	28
33	0.2172915	0.9761068	0.2226104	0.2342938	0.9721659	0.2410019	0.2512248	0.9679289	0.2595488	27
34	0.2175754	0.9760436	0.2229157	0.2345766	0.9720977	0.2413097	0.2515063	0.9678557	0.2598593	26
35	0.2178594	0.9759802	0.2232211	0.2348594	0.9720294	0.2416175	0.2517878	0.9677825	0.2601699	25
36	0.2181432	0.9759168	0.2235265	0.2351421	0.9719610	0.2419255	0.2520694	0.9677092	0.2604805	24
37	0.2184271	0.9758533	0.2238319	0.2354249	0.9718926	0.2422334	0.2523508	0.9676358	0.2607911	23
38	0.2187110	0.9757897	0.2241374	0.2357076	0.9718240	0.2425414	0.2526323	0.9675624	0.2611018	22
39	0.2189948	0.9757261	0.2244429	0.2359902	0.9717555	0.2428494	0.2529137	0.9674889	0.2614126	21
40	0.2192786	0.9756623	0.2247485	0.2362729	0.9716868	0.2431575	0.2531952	0.9674153	0.2617233	20
41	0.2195624	0.9755985	0.2250541	0.2365555	0.9716180	0.2434656	0.2534766	0.9673416	0.2620342	19
42	0.2198462	0.9755346	0.2253597	0.2368381	0.9715491	0.2437737	0.2537580	0.9672678	0.2623451	18
43	0.2201300	0.9754706	0.2256654	0.2371207	0.9714803	0.2440819	0.2540393	0.9671939	0.2626560	17
44	0.2204137	0.9754065	0.2259712	0.2374033	0.9714112	0.2443902	0.2543206	0.9671199	0.2629670	16
45	0.2206974	0.9753424	0.2262769	0.2376859	0.9713421	0.2446984	0.2546020	0.9670459	0.2632780	15
46	0.2209811	0.9752781	0.2265827	0.2379684	0.9712729	0.2450068	0.2548832	0.9669719	0.2635891	14
47	0.2212648	0.9752138	0.2268885	0.2382510	0.9712036	0.2453152	0.2551645	0.9668977	0.2639002	13
48	0.2215485	0.9751495	0.2271944	0.2385335	0.9711343	0.2456236	0.2554458	0.9668234	0.2642114	12
49	0.2218321	0.9750849	0.2275003	0.2388159	0.9710649	0.2459320	0.2557270	0.9667490	0.2645226	11
50	0.2221158	0.9750203	0.2278063	0.2390984	0.9709954	0.2462405	0.2560082	0.9666747	0.2648338	10
51	0.2223994	0.9749557	0.2281123	0.2393809	0.9709257	0.2465491	0.2562894	0.9666001	0.2651452	9
52	0.2226830	0.9748909	0.2284184	0.2396633	0.9708561	0.2468577	0.2565705	0.9665256	0.2654565	8
53	0.2229666	0.9748261	0.2287245	0.2399457	0.9707863	0.2471663	0.2568517	0.9664509	0.2657680	7
54	0.2232501	0.9747613	0.2290305	0.2402280	0.9707165	0.2474750	0.2571328	0.9663761	0.2660794	6
55	0.2235337	0.9746963	0.2293367	0.2405104	0.9706466	0.2477837	0.2574139	0.9663013	0.2663909	5
56	0.2238172	0.9746312	0.2296429	0.2407927	0.9705766	0.2480924	0.2576950	0.9662263	0.2667025	4
57	0.2241007	0.9745660	0.2299492	0.2410751	0.9705066	0.2484013	0.2579760	0.9661513	0.2670141	3
58	0.2243841	0.9745008	0.2302555	0.2413574	0.9704364	0.2487101	0.2582570	0.9660763	0.2673257	2
59	0.2246676	0.9744356	0.2305618	0.2416396	0.9703661	0.2490190	0.2585381	0.9660011	0.2676375	1
60	0.2249510	0.9743701	0.2308682	0.2419219	0.9702958	0.2493280	0.2588191	0.9659258	0.2679492	0
	cosine	sine 77°	cotangent	cosine	sine 76°	cotangent	cosine	sine 75°	cotangent	

PROFESSIONAL PUBLICATIONS, INC. • P.O. Box 199, San Carlos, CA 94070

	15° sine	cosine	tangent	16° sine	cosine	tangent	17° sine	cosine	tangent	
0	0.2588191	0.9659258	0.2679492	0.2756374	0.9612617	0.2867454	0.2923717	0.9563047	0.3057307	60
1	0.2591000	0.9658505	0.2682610	0.2759169	0.9611815	0.2870602	0.2926499	0.9562197	0.3060488	59
2	0.2593810	0.9657751	0.2685728	0.2761966	0.9611012	0.2873751	0.2929280	0.9561346	0.3063669	58
3	0.2596619	0.9656996	0.2688847	0.2764761	0.9610208	0.2876900	0.2932061	0.9560492	0.3066852	57
4	0.2599428	0.9656241	0.2691967	0.2767557	0.9609404	0.2880051	0.2934842	0.9559640	0.3070034	56
5	0.2602237	0.9655484	0.2695087	0.2770352	0.9608598	0.2883201	0.2937623	0.9558786	0.3073218	55
6	0.2605045	0.9654727	0.2698207	0.2773147	0.9607791	0.2886352	0.2940403	0.9557931	0.3076401	54
7	0.2607853	0.9653969	0.2701328	0.2775941	0.9606985	0.2889503	0.2943183	0.9557074	0.3079586	53
8	0.2610661	0.9653209	0.2704449	0.2778736	0.9606177	0.2892655	0.2945963	0.9556218	0.3082771	52
9	0.2613470	0.9652449	0.2707571	0.2781530	0.9605368	0.2895807	0.2948743	0.9555361	0.3085957	51
10	0.2616277	0.9651689	0.2710694	0.2784324	0.9604559	0.2898960	0.2951522	0.9554502	0.3089143	50
11	0.2619085	0.9650928	0.2713816	0.2787118	0.9603749	0.2902114	0.2954302	0.9553644	0.3092330	49
12	0.2621892	0.9650165	0.2716940	0.2789911	0.9602937	0.2905269	0.2957081	0.9552784	0.3095517	48
13	0.2624699	0.9649402	0.2720064	0.2792704	0.9602125	0.2908423	0.2959859	0.9551924	0.3098705	47
14	0.2627506	0.9648639	0.2723188	0.2795497	0.9601312	0.2911578	0.2962638	0.9551061	0.3101894	46
15	0.2630312	0.9647874	0.2726313	0.2798290	0.9600498	0.2914734	0.2965416	0.9550200	0.3105083	45
16	0.2633119	0.9647108	0.2729438	0.2801083	0.9599685	0.2917890	0.2968194	0.9549338	0.3108272	44
17	0.2635925	0.9646342	0.2732564	0.2803875	0.9598869	0.2921047	0.2970971	0.9548473	0.3111463	43
18	0.2638730	0.9645575	0.2735690	0.2806667	0.9598054	0.2924204	0.2973749	0.9547608	0.3114653	42
19	0.2641536	0.9644806	0.2738817	0.2809459	0.9597237	0.2927363	0.2976526	0.9546743	0.3117844	41
20	0.2644342	0.9644039	0.2741944	0.2812251	0.9596419	0.2930521	0.2979303	0.9545876	0.3121036	40
21	0.2647147	0.9643268	0.2745073	0.2815042	0.9595600	0.2933680	0.2982080	0.9545010	0.3124229	39
22	0.2649952	0.9642498	0.2748201	0.2817833	0.9594781	0.2936839	0.2984856	0.9544142	0.3127422	38
23	0.2652757	0.9641726	0.2751330	0.2820624	0.9593961	0.2939999	0.2987632	0.9543273	0.3130616	37
24	0.2655561	0.9640955	0.2754459	0.2823415	0.9593140	0.2943160	0.2990408	0.9542404	0.3133810	36
25	0.2658366	0.9640182	0.2757589	0.2826205	0.9592319	0.2946321	0.2993184	0.9541533	0.3137005	35
26	0.2661169	0.9639409	0.2760718	0.2828995	0.9591496	0.2949483	0.2995959	0.9540662	0.3140200	34
27	0.2663973	0.9638633	0.2763849	0.2831785	0.9590672	0.2952645	0.2998734	0.9539791	0.3143396	33
28	0.2666777	0.9637858	0.2766981	0.2834575	0.9589849	0.2955807	0.3001509	0.9538918	0.3146593	32
29	0.2669581	0.9637082	0.2770113	0.2837364	0.9589023	0.2958971	0.3004284	0.9538044	0.3149790	31
30	0.2672384	0.9636305	0.2773245	0.2840154	0.9588197	0.2962135	0.3007058	0.9537169	0.3152988	30
31	0.2675187	0.9635527	0.2776378	0.2842942	0.9587371	0.2965299	0.3009832	0.9536295	0.3156186	29
32	0.2677990	0.9634749	0.2779512	0.2845731	0.9586543	0.2968464	0.3012606	0.9535419	0.3159385	28
33	0.2680792	0.9633969	0.2782646	0.2848520	0.9585715	0.2971630	0.3015379	0.9534541	0.3162585	27
34	0.2683594	0.9633189	0.2785780	0.2851308	0.9584887	0.2974796	0.3018153	0.9533665	0.3165785	26
35	0.2686397	0.9632407	0.2788915	0.2854096	0.9584056	0.2977962	0.3020926	0.9532786	0.3168986	25
36	0.2689198	0.9631626	0.2792050	0.2856884	0.9583227	0.2981129	0.3023699	0.9531907	0.3172187	24
37	0.2692000	0.9630844	0.2795186	0.2859671	0.9582395	0.2984297	0.3026472	0.9531026	0.3175389	23
38	0.2694801	0.9630060	0.2798322	0.2862459	0.9581562	0.2987465	0.3029244	0.9530146	0.3178592	22
39	0.2697602	0.9629276	0.2801459	0.2865246	0.9580730	0.2990634	0.3032016	0.9529265	0.3181794	21
40	0.2700403	0.9628491	0.2804597	0.2868032	0.9579896	0.2993803	0.3034787	0.9528382	0.3184997	20
41	0.2703204	0.9627705	0.2807735	0.2870819	0.9579061	0.2996973	0.3037559	0.9527499	0.3188202	19
42	0.2706004	0.9626918	0.2810873	0.2873605	0.9578226	0.3000144	0.3040331	0.9526615	0.3191407	18
43	0.2708805	0.9626130	0.2814012	0.2876392	0.9577389	0.3003315	0.3043102	0.9525731	0.3194613	17
44	0.2711605	0.9625342	0.2817152	0.2879177	0.9576551	0.3006487	0.3045872	0.9524844	0.3197819	16
45	0.2714404	0.9624552	0.2820292	0.2881963	0.9575714	0.3009658	0.3048643	0.9523958	0.3201025	15
46	0.2717204	0.9623762	0.2823432	0.2884748	0.9574876	0.3012831	0.3051413	0.9523071	0.3204232	14
47	0.2720004	0.9622972	0.2826574	0.2887533	0.9574036	0.3016004	0.3054183	0.9522183	0.3207440	13
48	0.2722803	0.9622181	0.2829715	0.2890318	0.9573196	0.3019178	0.3056953	0.9521294	0.3210649	12
49	0.2725601	0.9621388	0.2832857	0.2893103	0.9572355	0.3022352	0.3059723	0.9520405	0.3213858	11
50	0.2728400	0.9620594	0.2836000	0.2895887	0.9571513	0.3025527	0.3062492	0.9519514	0.3217067	10
51	0.2731199	0.9619800	0.2839143	0.2898671	0.9570670	0.3028703	0.3065261	0.9518623	0.3220278	9
52	0.2733997	0.9619005	0.2842286	0.2901455	0.9569826	0.3031879	0.3068030	0.9517731	0.3223488	8
53	0.2736795	0.9618210	0.2845430	0.2904239	0.9568982	0.3035055	0.3070798	0.9516838	0.3226700	7
54	0.2739592	0.9617413	0.2848575	0.2907022	0.9568136	0.3038232	0.3073566	0.9515945	0.3229912	6
55	0.2742390	0.9616616	0.2851720	0.2909805	0.9567291	0.3041409	0.3076334	0.9515050	0.3233124	5
56	0.2745187	0.9615817	0.2854866	0.2912588	0.9566444	0.3044588	0.3079102	0.9514155	0.3236338	4
57	0.2747984	0.9615019	0.2858012	0.2915371	0.9565595	0.3047768	0.3081869	0.9513259	0.3239551	3
58	0.2750781	0.9614219	0.2861159	0.2918153	0.9564747	0.3050946	0.3084637	0.9512362	0.3242766	2
59	0.2753578	0.9613419	0.2864306	0.2920935	0.9563898	0.3054126	0.3087403	0.9511464	0.3245981	1
60	0.2756374	0.9612617	0.2867454	0.2923717	0.9563047	0.3057307	0.3090170	0.9510565	0.3249198	0
	cosine	sine	cotangent	cosine	sine	cotangent	cosine	sine	cotangent	
		74°			73°			72°		

PROFESSIONAL PUBLICATIONS, INC. • P.O. Box 199, San Carlos, CA 94070

	sine	18° cosine	tangent	sine	19° cosine	tangent	sine	20° cosine	tangent	
0	0.3090170	0.9510565	0.3249198	0.3255682	0.9455186	0.3443276	0.3420202	0.9396927	0.3639702	60
1	0.3092936	0.9509667	0.3252413	0.3258432	0.9454238	0.3446530	0.3422935	0.9395931	0.3642997	59
2	0.3095703	0.9508766	0.3255630	0.3261182	0.9453291	0.3449785	0.3425668	0.9394935	0.3646292	58
3	0.3098468	0.9507865	0.3258848	0.3263932	0.9452342	0.3453040	0.3428400	0.9393938	0.3649588	57
4	0.3101234	0.9506963	0.3262066	0.3266681	0.9451391	0.3456297	0.3431133	0.9392941	0.3652885	56
5	0.3103999	0.9506061	0.3265285	0.3269430	0.9450440	0.3459553	0.3433865	0.9391942	0.3656182	55
6	0.3106764	0.9505157	0.3268504	0.3272179	0.9449489	0.3462811	0.3436597	0.9390943	0.3659480	54
7	0.3109529	0.9504253	0.3271724	0.3274928	0.9448537	0.3466069	0.3439328	0.9389943	0.3662779	53
8	0.3112294	0.9503348	0.3274944	0.3277676	0.9447584	0.3469327	0.3442060	0.9388942	0.3666078	52
9	0.3115058	0.9502442	0.3278165	0.3280424	0.9446630	0.3472587	0.3444791	0.9387940	0.3669379	51
10	0.3117822	0.9501536	0.3281387	0.3283171	0.9445676	0.3475846	0.3447522	0.9386938	0.3672680	50
11	0.3120586	0.9500629	0.3284609	0.3285919	0.9444721	0.3479107	0.3450252	0.9385934	0.3675981	49
12	0.3123349	0.9499720	0.3287833	0.3288667	0.9443764	0.3482369	0.3452982	0.9384931	0.3679284	48
13	0.3126113	0.9498812	0.3291056	0.3291414	0.9442807	0.3485630	0.3455712	0.9383925	0.3682587	47
14	0.3128875	0.9497902	0.3294280	0.3294160	0.9441850	0.3488893	0.3458441	0.9382920	0.3685890	46
15	0.3131638	0.9496992	0.3297505	0.3296907	0.9440891	0.3492156	0.3461171	0.9381914	0.3689195	45
16	0.3134401	0.9496080	0.3300731	0.3299653	0.9439932	0.3495420	0.3463900	0.9380907	0.3692500	44
17	0.3137163	0.9495168	0.3303957	0.3302398	0.9438971	0.3498685	0.3466628	0.9379899	0.3695805	43
18	0.3139925	0.9494255	0.3307184	0.3305144	0.9438010	0.3501950	0.3469356	0.9378890	0.3699112	42
19	0.3142686	0.9493341	0.3310412	0.3307890	0.9437048	0.3505217	0.3472085	0.9377880	0.3702420	41
20	0.3145448	0.9492427	0.3313639	0.3310634	0.9436085	0.3508483	0.3474812	0.9376870	0.3705727	40
21	0.3148209	0.9491512	0.3316868	0.3313379	0.9435123	0.3511750	0.3477540	0.9375859	0.3709036	39
22	0.3150970	0.9490595	0.3320097	0.3316123	0.9434158	0.3515018	0.3480267	0.9374847	0.3712345	38
23	0.3153730	0.9489678	0.3323327	0.3318867	0.9433193	0.3518286	0.3482994	0.9373834	0.3715656	37
24	0.3156491	0.9488761	0.3326557	0.3321611	0.9432227	0.3521556	0.3485720	0.9372820	0.3718967	36
25	0.3159250	0.9487842	0.3329788	0.3324355	0.9431261	0.3524825	0.3488447	0.9371806	0.3722278	35
26	0.3162010	0.9486922	0.3333020	0.3327098	0.9430293	0.3528096	0.3491173	0.9370791	0.3725590	34
27	0.3164770	0.9486002	0.3336253	0.3329841	0.9429325	0.3531367	0.3493899	0.9369775	0.3728903	33
28	0.3167529	0.9485081	0.3339485	0.3332584	0.9428356	0.3534640	0.3496624	0.9368758	0.3732217	32
29	0.3170288	0.9484160	0.3342719	0.3335326	0.9427386	0.3537912	0.3499349	0.9367741	0.3735532	31
30	0.3173046	0.9483237	0.3345953	0.3338069	0.9426416	0.3541186	0.3502074	0.9366722	0.3738847	30
31	0.3175805	0.9482313	0.3349188	0.3340811	0.9425444	0.3544460	0.3504798	0.9365703	0.3742163	29
32	0.3178563	0.9481389	0.3352424	0.3343552	0.9424472	0.3547734	0.3507523	0.9364683	0.3745479	28
33	0.3181321	0.9480465	0.3355659	0.3346294	0.9423499	0.3551010	0.3510247	0.9363663	0.3748797	27
34	0.3184079	0.9479538	0.3358896	0.3349034	0.9422525	0.3554285	0.3512971	0.9362642	0.3752115	26
35	0.3186836	0.9478612	0.3362134	0.3351775	0.9421551	0.3557562	0.3515694	0.9361619	0.3755433	25
36	0.3189593	0.9477685	0.3365372	0.3354516	0.9420575	0.3560840	0.3518417	0.9360596	0.3758753	24
37	0.3192350	0.9476756	0.3368610	0.3357256	0.9419599	0.3564118	0.3521139	0.9359572	0.3762073	23
38	0.3195107	0.9475828	0.3371850	0.3359996	0.9418622	0.3567397	0.3523862	0.9358547	0.3765394	22
39	0.3197863	0.9474898	0.3375089	0.3362736	0.9417644	0.3570676	0.3526584	0.9357522	0.3768716	21
40	0.3200619	0.9473967	0.3378330	0.3365475	0.9416665	0.3573956	0.3529305	0.9356496	0.3772038	20
41	0.3203374	0.9473035	0.3381571	0.3368214	0.9415686	0.3577237	0.3532027	0.9355469	0.3775361	19
42	0.3206130	0.9472104	0.3384813	0.3370953	0.9414706	0.3580518	0.3534749	0.9354441	0.3778685	18
43	0.3208885	0.9471170	0.3388056	0.3373691	0.9413725	0.3583801	0.3537470	0.9353412	0.3782010	17
44	0.3211640	0.9470236	0.3391299	0.3376429	0.9412743	0.3587084	0.3540190	0.9352382	0.3785335	16
45	0.3214395	0.9469302	0.3394543	0.3379167	0.9411761	0.3590367	0.3542911	0.9351352	0.3788661	15
46	0.3217149	0.9468366	0.3397787	0.3381905	0.9410777	0.3593651	0.3545631	0.9350322	0.3791988	14
47	0.3219903	0.9467430	0.3401032	0.3384642	0.9409794	0.3596936	0.3548350	0.9349290	0.3795315	13
48	0.3222657	0.9466493	0.3404278	0.3387379	0.9408808	0.3600222	0.3551069	0.9348257	0.3798643	12
49	0.3225411	0.9465556	0.3407524	0.3390116	0.9407822	0.3603508	0.3553789	0.9347223	0.3801973	11
50	0.3228164	0.9464616	0.3410771	0.3392853	0.9406835	0.3606795	0.3556508	0.9346189	0.3805302	10
51	0.3230917	0.9463677	0.3414018	0.3395589	0.9405849	0.3610082	0.3559226	0.9345154	0.3808633	9
52	0.3233669	0.9462736	0.3417267	0.3398325	0.9404861	0.3613371	0.3561944	0.9344119	0.3811964	8
53	0.3236422	0.9461795	0.3420516	0.3401060	0.9403871	0.3616660	0.3564662	0.9343083	0.3815296	7
54	0.3239174	0.9460855	0.3423765	0.3403795	0.9402881	0.3619949	0.3567380	0.9342046	0.3818628	6
55	0.3241926	0.9459912	0.3427015	0.3406531	0.9401891	0.3623240	0.3570097	0.9341007	0.3821962	5
56	0.3244678	0.9458968	0.3430266	0.3409265	0.9400900	0.3626530	0.3572814	0.9339969	0.3825296	4
57	0.3247429	0.9458024	0.3433518	0.3412000	0.9399908	0.3629823	0.3575531	0.9338928	0.3828631	3
58	0.3250180	0.9457079	0.3436770	0.3414734	0.9398915	0.3633115	0.3578248	0.9337888	0.3831967	2
59	0.3252931	0.9456133	0.3440023	0.3417468	0.9397920	0.3636409	0.3580964	0.9336847	0.3835303	1
60	0.3255682	0.9455186	0.3443276	0.3420202	0.9396927	0.3639702	0.3583679	0.9335805	0.3838640	0
	cosine	sine 71°	cotangent	cosine	sine 70°	cotangent	cosine	sine 69°	cotangent	

PROFESSIONAL PUBLICATIONS, INC. • P.O. Box 199, San Carlos, CA 94070

	21° sine	21° cosine	21° tangent	22° sine	22° cosine	22° tangent	23° sine	23° cosine	23° tangent	
0	0.3583679	0.9335805	0.3838640	0.3746066	0.9271839	0.4040262	0.3907311	0.9205049	0.4244748	60
1	0.3586395	0.9334762	0.3841978	0.3748763	0.9270748	0.4043646	0.3909989	0.9203912	0.4248181	59
2	0.3589110	0.9333718	0.3845317	0.3751460	0.9269658	0.4047031	0.3912666	0.9202774	0.4251616	58
3	0.3591825	0.9332674	0.3848656	0.3754156	0.9268566	0.4050417	0.3915343	0.9201636	0.4255051	57
4	0.3594540	0.9331628	0.3851996	0.3756852	0.9267474	0.4053804	0.3918019	0.9200496	0.4258487	56
5	0.3597254	0.9330583	0.3855337	0.3759547	0.9266381	0.4057191	0.3920696	0.9199356	0.4261924	55
6	0.3599968	0.9329536	0.3858679	0.3762243	0.9265286	0.4060579	0.3923371	0.9198216	0.4265361	54
7	0.3602682	0.9328488	0.3862021	0.3764938	0.9264192	0.4063968	0.3926047	0.9197074	0.4268800	53
8	0.3605395	0.9327439	0.3865364	0.3767633	0.9263096	0.4067357	0.3928722	0.9195932	0.4272239	52
9	0.3608108	0.9326390	0.3868708	0.3770327	0.9262000	0.4070748	0.3931397	0.9194788	0.4275679	51
10	0.3610821	0.9325340	0.3872053	0.3773021	0.9260902	0.4074140	0.3934071	0.9193644	0.4279121	50
11	0.3613534	0.9324290	0.3875398	0.3775714	0.9259804	0.4077531	0.3936745	0.9192499	0.4282563	49
12	0.3616246	0.9323239	0.3878744	0.3778408	0.9258706	0.4080925	0.3939419	0.9191354	0.4286005	48
13	0.3618958	0.9322186	0.3882092	0.3781101	0.9257606	0.4084318	0.3942093	0.9190208	0.4289449	47
14	0.3621669	0.9321133	0.3885439	0.3783794	0.9256506	0.4087713	0.3944766	0.9189060	0.4292894	46
15	0.3624381	0.9320078	0.3888788	0.3786486	0.9255406	0.4091108	0.3947439	0.9187912	0.4296339	45
16	0.3627091	0.9319025	0.3892136	0.3789178	0.9254303	0.4094504	0.3950111	0.9186764	0.4299785	44
17	0.3629802	0.9317969	0.3895486	0.3791870	0.9253201	0.4097901	0.3952784	0.9185614	0.4303233	43
18	0.3632512	0.9316913	0.3898837	0.3794562	0.9252098	0.4101299	0.3955455	0.9184464	0.4306680	42
19	0.3635222	0.9315855	0.3902189	0.3797253	0.9250994	0.4104698	0.3958127	0.9183313	0.4310130	41
20	0.3637932	0.9314798	0.3905541	0.3799944	0.9249888	0.4108097	0.3960798	0.9182162	0.4313579	40
21	0.3640642	0.9313740	0.3908893	0.3802634	0.9248782	0.4111497	0.3963469	0.9181009	0.4317029	39
22	0.3643351	0.9312680	0.3912247	0.3805325	0.9247676	0.4114898	0.3966139	0.9179856	0.4320481	38
23	0.3646059	0.9311620	0.3915601	0.3808014	0.9246569	0.4118300	0.3968810	0.9178702	0.4323933	37
24	0.3648768	0.9310558	0.3918957	0.3810704	0.9245461	0.4121703	0.3971479	0.9177547	0.4327386	36
25	0.3651476	0.9309497	0.3922313	0.3813393	0.9244352	0.4125106	0.3974149	0.9176391	0.4330841	35
26	0.3654184	0.9308434	0.3925670	0.3816082	0.9243242	0.4128510	0.3976818	0.9175234	0.4334296	34
27	0.3656892	0.9307370	0.3929028	0.3818771	0.9242132	0.4131915	0.3979487	0.9174078	0.4337751	33
28	0.3659599	0.9306307	0.3932386	0.3821459	0.9241021	0.4135321	0.3982155	0.9172919	0.4341208	32
29	0.3662306	0.9305241	0.3935745	0.3824147	0.9239909	0.4138728	0.3984823	0.9171761	0.4344665	31
30	0.3665012	0.9304176	0.3939105	0.3826835	0.9238796	0.4142136	0.3987491	0.9170600	0.4348124	30
31	0.3667719	0.9303109	0.3942465	0.3829522	0.9237682	0.4145544	0.3990158	0.9169441	0.4351582	29
32	0.3670425	0.9302043	0.3945826	0.3832209	0.9236567	0.4148953	0.3992826	0.9168279	0.4355044	28
33	0.3673131	0.9300974	0.3949189	0.3834895	0.9235453	0.4152363	0.3995492	0.9167118	0.4358504	27
34	0.3675836	0.9299906	0.3952552	0.3837582	0.9234337	0.4155774	0.3998159	0.9165955	0.4361967	26
35	0.3678541	0.9298836	0.3955916	0.3840268	0.9233220	0.4159186	0.4000825	0.9164792	0.4365429	25
36	0.3681246	0.9297765	0.3959280	0.3842953	0.9232102	0.4162599	0.4003491	0.9163628	0.4368892	24
37	0.3683950	0.9296694	0.3962645	0.3845639	0.9230984	0.4166012	0.4006155	0.9162462	0.4372357	23
38	0.3686654	0.9295622	0.3966011	0.3848324	0.9229866	0.4169425	0.4008821	0.9161297	0.4375822	22
39	0.3689358	0.9294550	0.3969378	0.3851008	0.9228745	0.4172841	0.4011486	0.9160131	0.4379289	21
40	0.3692062	0.9293476	0.3972746	0.3853693	0.9227625	0.4176257	0.4014150	0.9158964	0.4382756	20
41	0.3694765	0.9292402	0.3976114	0.3856377	0.9226503	0.4179673	0.4016814	0.9157795	0.4386224	19
42	0.3697468	0.9291326	0.3979484	0.3859061	0.9225381	0.4183091	0.4019478	0.9156626	0.4389693	18
43	0.3700170	0.9290251	0.3982853	0.3861744	0.9224258	0.4186509	0.4022141	0.9155456	0.4393163	17
44	0.3702873	0.9289173	0.3986224	0.3864427	0.9223135	0.4189928	0.4024804	0.9154286	0.4396634	16
45	0.3705574	0.9288096	0.3989595	0.3867109	0.9222010	0.4193347	0.4027467	0.9153115	0.4400105	15
46	0.3708276	0.9287018	0.3992967	0.3869793	0.9220885	0.4196769	0.4030129	0.9151943	0.4403577	14
47	0.3710977	0.9285938	0.3996341	0.3872474	0.9219759	0.4200190	0.4032791	0.9150770	0.4407051	13
48	0.3713678	0.9284859	0.3999714	0.3875156	0.9218632	0.4203613	0.4035453	0.9149597	0.4410526	12
49	0.3716379	0.9283777	0.4003089	0.3877837	0.9217504	0.4207036	0.4038115	0.9148423	0.4414001	11
50	0.3719079	0.9282697	0.4006464	0.3880519	0.9216376	0.4210460	0.4040776	0.9147247	0.4417477	10
51	0.3721780	0.9281614	0.4009841	0.3883199	0.9215246	0.4213885	0.4043436	0.9146072	0.4420953	9
52	0.3724479	0.9280531	0.4013218	0.3885880	0.9214116	0.4217312	0.4046097	0.9144895	0.4424432	8
53	0.3727179	0.9279447	0.4016596	0.3888560	0.9212986	0.4220738	0.4048756	0.9143717	0.4427911	7
54	0.3729878	0.9278363	0.4019974	0.3891239	0.9211854	0.4224165	0.4051416	0.9142540	0.4431390	6
55	0.3732577	0.9277278	0.4023354	0.3893919	0.9210722	0.4227594	0.4054075	0.9141361	0.4434871	5
56	0.3735275	0.9276192	0.4026734	0.3896598	0.9209589	0.4231023	0.4056734	0.9140182	0.4438351	4
57	0.3737974	0.9275104	0.4030115	0.3899277	0.9208456	0.4234453	0.4059393	0.9139001	0.4441834	3
58	0.3740671	0.9274017	0.4033496	0.3901955	0.9207321	0.4237884	0.4062051	0.9137819	0.4445318	2
59	0.3743369	0.9272928	0.4036879	0.3904634	0.9206185	0.4241316	0.4064709	0.9136638	0.4448802	1
60	0.3746066	0.9271839	0.4040262	0.3907311	0.9205049	0.4244748	0.4067367	0.9135455	0.4452287	0
	cosine	sine 68°	cotangent	cosine	sine 67°	cotangent	cosine	sine 66°	cotangent	

PROFESSIONAL PUBLICATIONS, INC. • P.O. Box 199, San Carlos, CA 94070

		24°			25°			26°		
	sine	cosine	tangent	sine	cosine	tangent	sine	cosine	tangent	
0	0.4067367	0.9135455	0.4452287	0.4226183	0.9063079	0.4663076	0.4383712	0.8987941	0.4877326	60
1	0.4070024	0.9134271	0.4455773	0.4228819	0.9061849	0.4666618	0.4386326	0.8986665	0.4880927	59
2	0.4072681	0.9133087	0.4459260	0.4231455	0.9060618	0.4670161	0.4388940	0.8985389	0.4884529	58
3	0.4075337	0.9131902	0.4462748	0.4234090	0.9059387	0.4673705	0.4391553	0.8984112	0.4888133	57
	0.4077994	0.9130716	0.4466236	0.4236725	0.9058154	0.4677250	0.4394166	0.8982834	0.4891737	56
5	0.4080649	0.9129530	0.4469726	0.4239360	0.9056922	0.4680796	0.4396779	0.8981556	0.4895343	55
6	0.4083305	0.9128342	0.4473216	0.4241995	0.9055688	0.4684343	0.4399392	0.8980277	0.4898949	54
7	0.4085960	0.9127155	0.4476707	0.4244628	0.9054453	0.4687890	0.4402004	0.8978995	0.4902557	53
8	0.4088615	0.9125965	0.4480200	0.4247262	0.9053220	0.4691438	0.4404615	0.8977715	0.4906165	52
9	0.4091269	0.9124776	0.4483693	0.4249895	0.9051983	0.4694988	0.4407226	0.8976434	0.4909774	51
10	0.4093923	0.9123585	0.4487187	0.4252528	0.9050747	0.4698539	0.4409838	0.8975151	0.4913386	50
11	0.4096577	0.9122394	0.4490682	0.4255161	0.9049509	0.4702090	0.4412448	0.8973868	0.4916997	49
12	0.4099230	0.9121202	0.4494178	0.4257793	0.9048272	0.4705642	0.4415059	0.8972584	0.4920610	48
13	0.4101884	0.9120008	0.4497676	0.4260425	0.9047031	0.4709197	0.4417669	0.8971300	0.4924224	47
14	0.4104536	0.9118815	0.4501173	0.4263056	0.9045792	0.4712751	0.4420278	0.8970013	0.4927839	46
15	0.4107189	0.9117621	0.4504671	0.4265687	0.9044552	0.4716306	0.4422887	0.8968729	0.4931453	45
16	0.4109841	0.9116426	0.4508171	0.4268318	0.9043310	0.4719863	0.4425496	0.8967441	0.4935071	44
17	0.4112492	0.9115230	0.4511671	0.4270949	0.9042069	0.4723420	0.4428104	0.8966154	0.4938689	43
18	0.4115144	0.9114034	0.4515173	0.4273579	0.9040826	0.4726978	0.4430712	0.8964865	0.4942307	42
19	0.4117795	0.9112836	0.4518676	0.4276209	0.9039583	0.4730538	0.4433320	0.8963575	0.4945928	41
20	0.4120446	0.9111638	0.4522179	0.4278838	0.9038339	0.4734098	0.4435927	0.8962285	0.4949549	40
21	0.4123096	0.9110438	0.4525683	0.4281467	0.9037093	0.4737659	0.4438533	0.8960994	0.4953171	39
22	0.4125746	0.9109239	0.4529188	0.4284095	0.9035847	0.4741222	0.4441140	0.8959703	0.4956794	38
23	0.4128395	0.9108038	0.4532694	0.4286723	0.9034601	0.4744784	0.4443746	0.8958411	0.4960418	37
24	0.4131044	0.9106837	0.4536201	0.4289351	0.9033353	0.4748349	0.4446352	0.8957118	0.4964043	36
25	0.4133693	0.9105635	0.4539709	0.4291979	0.9032105	0.4751914	0.4448957	0.8955824	0.4967669	35
26	0.4136342	0.9104432	0.4543217	0.4294606	0.9030856	0.4755481	0.4451562	0.8954530	0.4971297	34
27	0.4138990	0.9103229	0.4546728	0.4297233	0.9029607	0.4759048	0.4454167	0.8953234	0.4974926	33
28	0.4141638	0.9102024	0.4550238	0.4299860	0.9028357	0.4762616	0.4456771	0.8951939	0.4978554	32
29	0.4144285	0.9100819	0.4553750	0.4302485	0.9027105	0.4766185	0.4459375	0.8950641	0.4982184	31
30	0.4146933	0.9099613	0.4557262	0.4305111	0.9025853	0.4769755	0.4461978	0.8949344	0.4985816	30
31	0.4149579	0.9098406	0.4560776	0.4307736	0.9024600	0.4773327	0.4464582	0.8948046	0.4989449	29
32	0.4152226	0.9097199	0.4564290	0.4310361	0.9023347	0.4776898	0.4467184	0.8946747	0.4993082	28
33	0.4154872	0.9095991	0.4567806	0.4312986	0.9022093	0.4780471	0.4469786	0.8945447	0.4996716	27
34	0.4157518	0.9094782	0.4571322	0.4315610	0.9020838	0.4784046	0.4472389	0.8944145	0.5000353	26
35	0.4160163	0.9093572	0.4574839	0.4318234	0.9019582	0.4787621	0.4474990	0.8942844	0.5003990	25
36	0.4162808	0.9092362	0.4578357	0.4320858	0.9018325	0.4791197	0.4477591	0.8941542	0.5007627	24
37	0.4165453	0.9091150	0.4581876	0.4323481	0.9017068	0.4794775	0.4480192	0.8940240	0.5011266	23
38	0.4168097	0.9089938	0.4585396	0.4326103	0.9015810	0.4798352	0.4482792	0.8938936	0.5014905	22
39	0.4170741	0.9088725	0.4588917	0.4328726	0.9014552	0.4801932	0.4485393	0.8937632	0.5018547	21
40	0.4173385	0.9087511	0.4592440	0.4331348	0.9013292	0.4805512	0.4487992	0.8936327	0.5022188	20
41	0.4176027	0.9086298	0.4595962	0.4333970	0.9012032	0.4809093	0.4490591	0.8935021	0.5025831	19
42	0.4178671	0.9085082	0.4599486	0.4336591	0.9010770	0.4812675	0.4493190	0.8933714	0.5029476	18
43	0.4181314	0.9083866	0.4603011	0.4339212	0.9009508	0.4816258	0.4495789	0.8932407	0.5033122	17
44	0.4183956	0.9082649	0.4606537	0.4341832	0.9008246	0.4819842	0.4498387	0.8931099	0.5036768	16
45	0.4186597	0.9081432	0.4610063	0.4344453	0.9006982	0.4823428	0.4500985	0.8929790	0.5040415	15
46	0.4189239	0.9080214	0.4613591	0.4347073	0.9005719	0.4827014	0.4503582	0.8928481	0.5044063	14
47	0.4191880	0.9078994	0.4617119	0.4349692	0.9004453	0.4830601	0.4506179	0.8927170	0.5047712	13
48	0.4194521	0.9077775	0.4620649	0.4352311	0.9003188	0.4834189	0.4508775	0.8925858	0.5051364	12
49	0.4197161	0.9076555	0.4624179	0.4354930	0.9001921	0.4837779	0.4511372	0.8924546	0.5055015	11
50	0.4199801	0.9075333	0.4627710	0.4357548	0.9000655	0.4841368	0.4513968	0.8923234	0.5058668	10
51	0.4202442	0.9074111	0.4631243	0.4360166	0.8999386	0.4844960	0.4516563	0.8921921	0.5062321	9
52	0.4205081	0.9072888	0.4634776	0.4362784	0.8998118	0.4848551	0.4519158	0.8920606	0.5065976	8
53	0.4207720	0.9071665	0.4638311	0.4365401	0.8996848	0.4852145	0.4521753	0.8919291	0.5069633	7
54	0.4210358	0.9070441	0.4641845	0.4368018	0.8995578	0.4855739	0.4524347	0.8917975	0.5073290	6
55	0.4212996	0.9069216	0.4645381	0.4370634	0.8994308	0.4859334	0.4526941	0.8916659	0.5076947	5
56	0.4215634	0.9067989	0.4648919	0.4373251	0.8993035	0.4862931	0.4529534	0.8915341	0.5080606	4
57	0.4218272	0.9066762	0.4652457	0.4375867	0.8991763	0.4866528	0.4532128	0.8914024	0.5084267	3
58	0.4220909	0.9065535	0.4655996	0.4378482	0.8990489	0.4870126	0.4534721	0.8912705	0.5087929	2
59	0.4223546	0.9064307	0.4659536	0.4381097	0.8989215	0.4873726	0.4537313	0.8911386	0.5091591	1
60	0.4226183	0.9063079	0.4663076	0.4383712	0.8987941	0.4877326	0.4539905	0.8910066	0.5095254	0
	cosine	sine	cotangent	cosine	sine	cotangent	cosine	sine	cotangent	
		65°			64°			63°		

PROFESSIONAL PUBLICATIONS, INC. ● P.O. Box 199, San Carlos, CA 94070

	27° sine	cosine	tangent	28° sine	cosine	tangent	29° sine	cosine	tangent	
0	0.4539905	0.8910066	0.5095254	0.4694716	0.8829476	0.5317094	0.4848097	0.8746197	0.5543091	60
1	0.4542497	0.8908744	0.5098919	0.4697284	0.8828110	0.5320826	0.4850640	0.8744787	0.5546893	59
2	0.4545088	0.8907423	0.5102585	0.4699852	0.8826744	0.5324559	0.4853183	0.8743375	0.5550698	58
3	0.4547679	0.8906100	0.5106252	0.4702419	0.8825376	0.5328293	0.4855727	0.8741964	0.5554504	57
4	0.4550270	0.8904777	0.5109919	0.4704987	0.8824008	0.5332029	0.4858269	0.8740551	0.5558310	56
5	0.4552860	0.8903453	0.5113589	0.4707553	0.8822639	0.5335765	0.4860812	0.8739136	0.5562120	55
6	0.4555449	0.8902128	0.5117258	0.4710119	0.8821269	0.5339503	0.4863354	0.8737722	0.5565929	54
7	0.4558039	0.8900803	0.5120929	0.4712685	0.8819898	0.5343242	0.4865895	0.8736308	0.5569739	53
8	0.4560627	0.8899477	0.5124602	0.4715250	0.8818527	0.5346982	0.4868436	0.8734892	0.5573551	52
9	0.4563216	0.8898150	0.5128275	0.4717816	0.8817155	0.5350723	0.4870977	0.8733475	0.5577365	51
10	0.4565804	0.8896822	0.5131950	0.4720380	0.8815783	0.5354465	0.4873517	0.8732058	0.5581178	50
11	0.4568392	0.8895494	0.5135625	0.4722944	0.8814409	0.5358209	0.4876057	0.8730640	0.5584994	49
12	0.4570979	0.8894164	0.5139302	0.4725508	0.8813035	0.5361953	0.4878597	0.8729221	0.5588812	48
13	0.4573567	0.8892835	0.5142980	0.4728071	0.8811660	0.5365699	0.4881136	0.8727801	0.5592630	47
14	0.4576153	0.8891503	0.5146658	0.4730634	0.8810284	0.5369446	0.4883674	0.8726382	0.5596448	46
15	0.4578739	0.8890172	0.5150338	0.4733197	0.8808908	0.5373194	0.4886212	0.8724960	0.5600269	45
16	0.4581325	0.8888840	0.5154019	0.4735759	0.8807531	0.5376943	0.4888750	0.8723539	0.5604090	44
17	0.4583911	0.8887507	0.5157701	0.4738320	0.8806153	0.5380693	0.4891288	0.8722116	0.5607914	43
18	0.4586496	0.8886173	0.5161385	0.4740882	0.8804774	0.5384445	0.4893824	0.8720694	0.5611737	42
19	0.4589081	0.8884838	0.5165070	0.4743443	0.8803394	0.5388197	0.4896361	0.8719269	0.5615564	41
20	0.4591665	0.8883502	0.5168755	0.4746004	0.8802014	0.5391953	0.4898897	0.8717844	0.5619391	40
21	0.4594249	0.8882167	0.5172441	0.4748564	0.8800634	0.5395707	0.4901433	0.8716419	0.5623218	39
22	0.4596832	0.8880830	0.5176129	0.4751124	0.8799252	0.5399463	0.4903969	0.8714992	0.5627049	38
23	0.4599415	0.8879492	0.5179818	0.4753683	0.8797870	0.5403221	0.4906504	0.8713566	0.5630879	37
24	0.4601998	0.8878154	0.5183507	0.4756242	0.8796486	0.5406980	0.4909037	0.8712139	0.5634710	36
25	0.4604580	0.8876815	0.5187199	0.4758801	0.8795102	0.5410739	0.4911572	0.8710710	0.5638543	35
26	0.4607162	0.8875475	0.5190890	0.4761359	0.8793718	0.5414501	0.4914105	0.8709282	0.5642377	34
27	0.4609744	0.8874134	0.5194584	0.4763917	0.8792332	0.5418263	0.4916639	0.8707851	0.5646213	33
28	0.4612325	0.8872793	0.5198279	0.4766474	0.8790946	0.5422027	0.4919171	0.8706421	0.5650050	32
29	0.4614906	0.8871452	0.5201974	0.4769031	0.8789560	0.5425791	0.4921704	0.8704990	0.5653889	31
30	0.4617486	0.8870108	0.5205671	0.4771588	0.8788172	0.5429557	0.4924236	0.8703557	0.5657728	30
31	0.4620066	0.8868765	0.5209368	0.4774144	0.8786783	0.5433325	0.4926767	0.8702124	0.5661568	29
32	0.4622646	0.8867422	0.5213066	0.4776699	0.8785394	0.5437092	0.4929298	0.8700691	0.5665410	28
33	0.4625225	0.8866076	0.5216767	0.4779255	0.8784004	0.5440862	0.4931830	0.8699256	0.5669254	27
34	0.4627804	0.8864729	0.5220469	0.4781810	0.8782613	0.5444632	0.4934359	0.8697821	0.5673098	26
35	0.4630382	0.8863384	0.5224170	0.4784365	0.8781223	0.5448404	0.4936890	0.8696386	0.5676944	25
36	0.4632961	0.8862035	0.5227874	0.4786919	0.8779831	0.5452176	0.4939419	0.8694950	0.5680791	24
37	0.4635538	0.8860688	0.5231578	0.4789473	0.8778437	0.5455952	0.4941948	0.8693512	0.5684639	23
38	0.4638115	0.8859339	0.5235283	0.4792026	0.8777044	0.5459726	0.4944476	0.8692074	0.5688489	22
39	0.4640692	0.8857989	0.5238991	0.4794579	0.8775649	0.5463503	0.4947004	0.8690636	0.5692338	21
40	0.4643269	0.8856640	0.5242698	0.4797131	0.8774254	0.5467281	0.4949532	0.8689196	0.5696191	20
41	0.4645845	0.8855289	0.5246407	0.4799683	0.8772860	0.5471059	0.4952060	0.8687757	0.5700045	19
42	0.4648421	0.8853937	0.5250117	0.4802235	0.8771461	0.5474841	0.4954587	0.8686315	0.5703900	18
43	0.4650996	0.8852584	0.5253829	0.4804786	0.8770065	0.5478621	0.4957114	0.8684874	0.5707755	17
44	0.4653571	0.8851231	0.5257541	0.4807338	0.8768666	0.5482404	0.4959640	0.8683432	0.5711612	16
45	0.4656146	0.8849877	0.5261255	0.4809888	0.8767269	0.5486187	0.4962165	0.8681989	0.5715470	15
46	0.4658719	0.8848522	0.5264969	0.4812438	0.8765868	0.5489973	0.4964691	0.8680545	0.5719330	14
47	0.4661293	0.8847166	0.5268685	0.4814987	0.8764468	0.5493759	0.4967215	0.8679100	0.5723191	13
48	0.4663866	0.8845810	0.5272402	0.4817537	0.8763067	0.5497547	0.4969740	0.8677655	0.5727054	12
49	0.4666440	0.8844453	0.5276120	0.4820086	0.8761666	0.5501335	0.4972264	0.8676209	0.5730918	11
50	0.4669012	0.8843095	0.5279840	0.4822634	0.8760262	0.5505126	0.4974788	0.8674762	0.5734783	10
51	0.4671584	0.8841737	0.5283560	0.4825182	0.8758860	0.5508916	0.4977310	0.8673315	0.5738648	9
52	0.4674156	0.8840377	0.5287281	0.4827730	0.8757456	0.5512708	0.4979833	0.8671866	0.5742516	8
53	0.4676727	0.8839017	0.5291004	0.4830277	0.8756051	0.5516502	0.4982356	0.8670417	0.5746386	7
54	0.4679298	0.8837656	0.5294727	0.4832824	0.8754646	0.5520297	0.4984877	0.8668968	0.5750254	6
55	0.4681869	0.8836295	0.5298452	0.4835370	0.8753240	0.5524092	0.4987399	0.8667518	0.5754126	5
56	0.4684439	0.8834933	0.5302178	0.4837917	0.8751832	0.5527890	0.4989920	0.8666066	0.5757999	4
57	0.4687009	0.8833569	0.5305906	0.4840462	0.8750424	0.5531688	0.4992440	0.8664615	0.5761873	3
58	0.4689578	0.8832206	0.5309634	0.4843008	0.8749017	0.5535488	0.4994961	0.8663162	0.5765748	2
59	0.4692147	0.8830841	0.5313364	0.4845552	0.8747607	0.5539289	0.4997481	0.8661709	0.5769624	1
60	0.4694716	0.8829476	0.5317094	0.4848097	0.8746197	0.5543091	0.5000000	0.8660254	0.5773503	0
	cosine	sine	cotangent	cosine	sine	cotangent	cosine	sine	cotangent	
		62°			61°			60°		

PROFESSIONAL PUBLICATIONS, INC. • P.O. Box 199, San Carlos, CA   94070

	30° sine	30° cosine	30° tangent	31° sine	31° cosine	31° tangent	32° sine	32° cosine	32° tangent	
0	0.5000000	0.8660254	0.5773503	0.5150381	0.8571673	0.6008606	0.5299193	0.8480481	0.6248693	60
1	0.5002519	0.8658800	0.5777382	0.5152874	0.8570175	0.6012566	0.5301659	0.8478940	0.6252738	59
2	0.5005038	0.8657344	0.5781263	0.5155367	0.8568676	0.6016526	0.5304126	0.8477397	0.6256785	58
3	0.5007555	0.8655887	0.5785144	0.5157859	0.8567175	0.6020489	0.5306591	0.8475854	0.6260834	57
4	0.5010073	0.8654431	0.5789027	0.5160351	0.8565675	0.6024454	0.5309057	0.8474310	0.6264884	56
5	0.5012590	0.8652973	0.5792911	0.5162843	0.8564172	0.6028420	0.5311521	0.8472766	0.6268935	55
6	0.5015107	0.8651514	0.5796797	0.5165333	0.8562671	0.6032385	0.5313985	0.8471220	0.6272987	54
7	0.5017624	0.8650056	0.5800684	0.5167824	0.8561168	0.6036354	0.5316450	0.8469674	0.6277042	53
8	0.5020140	0.8648595	0.5804573	0.5170314	0.8559664	0.6040323	0.5318914	0.8468126	0.6281099	52
9	0.5022655	0.8647135	0.5808462	0.5172804	0.8558160	0.6044294	0.5321377	0.8466579	0.6285155	51
10	0.5025170	0.8645674	0.5812352	0.5175292	0.8556655	0.6048266	0.5323839	0.8465030	0.6289214	50
11	0.5027685	0.8644211	0.5816245	0.5177782	0.8555150	0.6052240	0.5326301	0.8463482	0.6293275	49
12	0.5030200	0.8642749	0.5820139	0.5180270	0.8553643	0.6056215	0.5328763	0.8461931	0.6297337	48
13	0.5032713	0.8641284	0.5824034	0.5182759	0.8552136	0.6060192	0.5331224	0.8460382	0.6301399	47
14	0.5035228	0.8639820	0.5827931	0.5185246	0.8550627	0.6064170	0.5333685	0.8458830	0.6305464	46
15	0.5037740	0.8638356	0.5831828	0.5187733	0.8549119	0.6068150	0.5336145	0.8457279	0.6309530	45
16	0.5040253	0.8636889	0.5835727	0.5190219	0.8547610	0.6072129	0.5338606	0.8455725	0.6313599	44
17	0.5042765	0.8635423	0.5839626	0.5192705	0.8546100	0.6076111	0.5341065	0.8454172	0.6317668	43
18	0.5045276	0.8633956	0.5843528	0.5195191	0.8544589	0.6080095	0.5343524	0.8452619	0.6321737	42
19	0.5047788	0.8632488	0.5847431	0.5197677	0.8543077	0.6084081	0.5345982	0.8451064	0.6325809	41
20	0.5050299	0.8631019	0.5851336	0.5200161	0.8541565	0.6088066	0.5348440	0.8449509	0.6329883	40
21	0.5052809	0.8629550	0.5855241	0.5202646	0.8540052	0.6092054	0.5350898	0.8447952	0.6333959	39
22	0.5055319	0.8628079	0.5859148	0.5205130	0.8538538	0.6096044	0.5353355	0.8446395	0.6338035	38
23	0.5057828	0.8626609	0.5863055	0.5207613	0.8537024	0.6100034	0.5355812	0.8444838	0.6342113	37
24	0.5060338	0.8625137	0.5866966	0.5210096	0.8535509	0.6104026	0.5358269	0.8443280	0.6346194	36
25	0.5062847	0.8623664	0.5870876	0.5212579	0.8533992	0.6108020	0.5360724	0.8441721	0.6350275	35
26	0.5065355	0.8622192	0.5874788	0.5215061	0.8532475	0.6112015	0.5363179	0.8440161	0.6354357	34
27	0.5067863	0.8620718	0.5878702	0.5217544	0.8530958	0.6116012	0.5365634	0.8438601	0.6358441	33
28	0.5070370	0.8619244	0.5882617	0.5220025	0.8529440	0.6120008	0.5368088	0.8437039	0.6362526	32
29	0.5072877	0.8617768	0.5886532	0.5222505	0.8527922	0.6124007	0.5370543	0.8435478	0.6366614	31
30	0.5075384	0.8616292	0.5890450	0.5224986	0.8526402	0.6128008	0.5372996	0.8433915	0.6370702	30
31	0.5077890	0.8614815	0.5894369	0.5227466	0.8524882	0.6132010	0.5375450	0.8432352	0.6374793	29
32	0.5080396	0.8613338	0.5898290	0.5229946	0.8523360	0.6136014	0.5377902	0.8430787	0.6378884	28
33	0.5082901	0.8611860	0.5902210	0.5232425	0.8521839	0.6140019	0.5380354	0.8429223	0.6382977	27
34	0.5085406	0.8610380	0.5906134	0.5234904	0.8520316	0.6144025	0.5382805	0.8427657	0.6387072	26
35	0.5087910	0.8608901	0.5910058	0.5237381	0.8518794	0.6148031	0.5385257	0.8426091	0.6391169	25
36	0.5090414	0.8607421	0.5913983	0.5239859	0.8517270	0.6152041	0.5387708	0.8424525	0.6395267	24
37	0.5092918	0.8605940	0.5917910	0.5242336	0.8515745	0.6156052	0.5390158	0.8422957	0.6399365	23
38	0.5095421	0.8604457	0.5921839	0.5244813	0.8514220	0.6160063	0.5392609	0.8421388	0.6403468	22
39	0.5097924	0.8602976	0.5925768	0.5247290	0.8512694	0.6164078	0.5395058	0.8419819	0.6407570	21
40	0.5100425	0.8601493	0.5929698	0.5249766	0.8511167	0.6168092	0.5397507	0.8418249	0.6411674	20
41	0.5102928	0.8600008	0.5933632	0.5252241	0.8509640	0.6172107	0.5399956	0.8416680	0.6415779	19
42	0.5105429	0.8598523	0.5937566	0.5254717	0.8508111	0.6176126	0.5402403	0.8415108	0.6419886	18
43	0.5107930	0.8597038	0.5941500	0.5257191	0.8506582	0.6180145	0.5404851	0.8413536	0.6423994	17
44	0.5110431	0.8595551	0.5945438	0.5259665	0.8505053	0.6184165	0.5407298	0.8411964	0.6428105	16
45	0.5112931	0.8594064	0.5949375	0.5262139	0.8503522	0.6188188	0.5409745	0.8410391	0.6432216	15
46	0.5115430	0.8592576	0.5953314	0.5264613	0.8501992	0.6192212	0.5412191	0.8408816	0.6436329	14
47	0.5117930	0.8591089	0.5957254	0.5267085	0.8500460	0.6196236	0.5414637	0.8407242	0.6440445	13
48	0.5120429	0.8589599	0.5961196	0.5269558	0.8498927	0.6200262	0.5417082	0.8405666	0.6444560	12
49	0.5122927	0.8588109	0.5965140	0.5272030	0.8497394	0.6204290	0.5419527	0.8404090	0.6448677	11
50	0.5125425	0.8586618	0.5969085	0.5274501	0.8495860	0.6208320	0.5421971	0.8402514	0.6452796	10
51	0.5127923	0.8585127	0.5973031	0.5276973	0.8494325	0.6212351	0.5424415	0.8400936	0.6456918	9
52	0.5130420	0.8583636	0.5976977	0.5279444	0.8492789	0.6216384	0.5426859	0.8399358	0.6461041	8
53	0.5132917	0.8582143	0.5980926	0.5281914	0.8491254	0.6220417	0.5429302	0.8397778	0.6465165	7
54	0.5135412	0.8580649	0.5984876	0.5284384	0.8489717	0.6224452	0.5431745	0.8396199	0.6469291	6
55	0.5137908	0.8579155	0.5988828	0.5286853	0.8488179	0.6228489	0.5434188	0.8394619	0.6473418	5
56	0.5140404	0.8577660	0.5992780	0.5289322	0.8486641	0.6232527	0.5436628	0.8393037	0.6477546	4
57	0.5142899	0.8576165	0.5996736	0.5291790	0.8485103	0.6236566	0.5439070	0.8391455	0.6481677	3
58	0.5145394	0.8574669	0.6000691	0.5294259	0.8483562	0.6240608	0.5441510	0.8389873	0.6485807	2
59	0.5147887	0.8573171	0.6004648	0.5296726	0.8482022	0.6244649	0.5443950	0.8388290	0.6489940	1
60	0.5150381	0.8571673	0.6008606	0.5299193	0.8480481	0.6248693	0.5446391	0.8386706	0.6494076	0
	cosine	sine 59°	cotangent	cosine	sine 58°	cotangent	cosine	sine 57°	cotangent	

	sine	33° cosine	tangent	sine	34° cosine	tangent	sine	35° cosine	tangent	
0	0.5446391	0.8386706	0.6494076	0.5591929	0.8290376	0.6745085	0.5735765	0.8191521	0.7002075	60
1	0.5448830	0.8385121	0.6498212	0.5594341	0.8288749	0.6749318	0.5738147	0.8189852	0.7006412	59
2	0.5451269	0.8383536	0.6502350	0.5596751	0.8287122	0.6753553	0.5740529	0.8188183	0.7010749	58
3	0.5453707	0.8381950	0.6506490	0.5599162	0.8285493	0.6757790	0.5742911	0.8186512	0.7015089	57
4	0.5456145	0.8380363	0.6510631	0.5601572	0.8283864	0.6762028	0.5745292	0.8184841	0.7019430	56
5	0.5458583	0.8378776	0.6514773	0.5603981	0.8282234	0.6766267	0.5747672	0.8183169	0.7023773	55
6	0.5461020	0.8377188	0.6518918	0.5606390	0.8280604	0.6770509	0.5750052	0.8181497	0.7028117	54
7	0.5463456	0.8375599	0.6523063	0.5608799	0.8278973	0.6774752	0.5752432	0.8179824	0.7032464	53
8	0.5465893	0.8374009	0.6527212	0.5611206	0.8277341	0.6778997	0.5754812	0.8178151	0.7036814	52
9	0.5468329	0.8372419	0.6531361	0.5613614	0.8275708	0.6783243	0.5757190	0.8176476	0.7041163	51
10	0.5470763	0.8370827	0.6535511	0.5616022	0.8274075	0.6787492	0.5759568	0.8174801	0.7045515	50
11	0.5473198	0.8369236	0.6539663	0.5618428	0.8272440	0.6791742	0.5761946	0.8173125	0.7049869	49
12	0.5475633	0.8367643	0.6543817	0.5620834	0.8270806	0.6795993	0.5764324	0.8171449	0.7054225	48
13	0.5478066	0.8366050	0.6547972	0.5623240	0.8269171	0.6800247	0.5766701	0.8169772	0.7058582	47
14	0.5480500	0.8364457	0.6552129	0.5625645	0.8267534	0.6804501	0.5769077	0.8168094	0.7062941	46
15	0.5482932	0.8362862	0.6556287	0.5628050	0.8265898	0.6808758	0.5771452	0.8166416	0.7067300	45
16	0.5485365	0.8361267	0.6560447	0.5630454	0.8264260	0.6813016	0.5773827	0.8164736	0.7071664	44
17	0.5487797	0.8359671	0.6564609	0.5632858	0.8262622	0.6817276	0.5776202	0.8163056	0.7076029	43
18	0.5490229	0.8358074	0.6568772	0.5635260	0.8260983	0.6821537	0.5778576	0.8161376	0.7080395	42
19	0.5492659	0.8356476	0.6572937	0.5637664	0.8259344	0.6825801	0.5780950	0.8159695	0.7084763	41
20	0.5495090	0.8354878	0.6577104	0.5640066	0.8257703	0.6830066	0.5783323	0.8158013	0.7089131	40
21	0.5497520	0.8353280	0.6581271	0.5642467	0.8256062	0.6834332	0.5785697	0.8156330	0.7093505	39
22	0.5499949	0.8351679	0.6585441	0.5644869	0.8254421	0.6838601	0.5788068	0.8154647	0.7097877	38
23	0.5502379	0.8350080	0.6589612	0.5647271	0.8252778	0.6842873	0.5790440	0.8152962	0.7102253	37
24	0.5504808	0.8348478	0.6593786	0.5649670	0.8251135	0.6847144	0.5792812	0.8151278	0.7106630	36
25	0.5507236	0.8346877	0.6597960	0.5652071	0.8249491	0.6851417	0.5795183	0.8149592	0.7111009	35
26	0.5509664	0.8345274	0.6602136	0.5654470	0.8247847	0.6855692	0.5797554	0.8147907	0.7115390	34
27	0.5512091	0.8343672	0.6606314	0.5656868	0.8246202	0.6859968	0.5799923	0.8146220	0.7119772	33
28	0.5514518	0.8342068	0.6610492	0.5659267	0.8244556	0.6864247	0.5802292	0.8144532	0.7124157	32
29	0.5516944	0.8340464	0.6614673	0.5661665	0.8242909	0.6868528	0.5804662	0.8142844	0.7128543	31
30	0.5519370	0.8338858	0.6618856	0.5664063	0.8241262	0.6872810	0.5807030	0.8141155	0.7132931	30
31	0.5521796	0.8337253	0.6623040	0.5666459	0.8239615	0.6877092	0.5809398	0.8139466	0.7137321	29
32	0.5524221	0.8335646	0.6627226	0.5668857	0.8237966	0.6881379	0.5811765	0.8137776	0.7141712	28
33	0.5526645	0.8334039	0.6631412	0.5671252	0.8236316	0.6885666	0.5814132	0.8136085	0.7146106	27
34	0.5529069	0.8332431	0.6635601	0.5673648	0.8234666	0.6889955	0.5816498	0.8134393	0.7150501	26
35	0.5531493	0.8330823	0.6639792	0.5676042	0.8233016	0.6894245	0.5818864	0.8132702	0.7154897	25
36	0.5533916	0.8329213	0.6643984	0.5678437	0.8231364	0.6898537	0.5821230	0.8131008	0.7159296	24
37	0.5536338	0.8327603	0.6648178	0.5680832	0.8229712	0.6902832	0.5823595	0.8129314	0.7163697	23
38	0.5538761	0.8325992	0.6652374	0.5683225	0.8228059	0.6907128	0.5825959	0.8127619	0.7168101	22
39	0.5541183	0.8324381	0.6656570	0.5685619	0.8226406	0.6911424	0.5828323	0.8125924	0.7172505	21
40	0.5543604	0.8322768	0.6660770	0.5688012	0.8224751	0.6915725	0.5830687	0.8124229	0.7176911	20
41	0.5546024	0.8321155	0.6664969	0.5690404	0.8223096	0.6920027	0.5833050	0.8122532	0.7181320	19
42	0.5548444	0.8319541	0.6669171	0.5692796	0.8221440	0.6924329	0.5835413	0.8120836	0.7185729	18
43	0.5550864	0.8317927	0.6673375	0.5695186	0.8219784	0.6928633	0.5837774	0.8119137	0.7190141	17
44	0.5553284	0.8316311	0.6677581	0.5697578	0.8218127	0.6932939	0.5840136	0.8117440	0.7194554	16
45	0.5555702	0.8314696	0.6681787	0.5699968	0.8216469	0.6937247	0.5842497	0.8115740	0.7198971	15
46	0.5558121	0.8313081	0.6685994	0.5702358	0.8214812	0.6941556	0.5844857	0.8114040	0.7203387	14
47	0.5560539	0.8311462	0.6690205	0.5704747	0.8213152	0.6945868	0.5847218	0.8112339	0.7207807	13
48	0.5562956	0.8309845	0.6694416	0.5707136	0.8211492	0.6950181	0.5849577	0.8110638	0.7212227	12
49	0.5565373	0.8308226	0.6698630	0.5709524	0.8209832	0.6954495	0.5851936	0.8108937	0.7216650	11
50	0.5567790	0.8306608	0.6702845	0.5711912	0.8208170	0.6958813	0.5854294	0.8107234	0.7221075	10
51	0.5570205	0.8304988	0.6707060	0.5714300	0.8206509	0.6963131	0.5856653	0.8105531	0.7225502	9
52	0.5572621	0.8303367	0.6711279	0.5716686	0.8204846	0.6967450	0.5859010	0.8103827	0.7229930	8
53	0.5575037	0.8301744	0.6715500	0.5719073	0.8203183	0.6971772	0.5861368	0.8102122	0.7234361	7
54	0.5577452	0.8300123	0.6719722	0.5721459	0.8201519	0.6976097	0.5863724	0.8100417	0.7238792	6
55	0.5579866	0.8298500	0.6723945	0.5723845	0.8199854	0.6980422	0.5866081	0.8098711	0.7243227	5
56	0.5582279	0.8296877	0.6728169	0.5726230	0.8198189	0.6984750	0.5868436	0.8097004	0.7247663	4
57	0.5584692	0.8295253	0.6732395	0.5728614	0.8196523	0.6989078	0.5870791	0.8095297	0.7252101	3
58	0.5587105	0.8293628	0.6736624	0.5730999	0.8194856	0.6993409	0.5873145	0.8093589	0.7256539	2
59	0.5589517	0.8292002	0.6740854	0.5733382	0.8193189	0.6997741	0.5875499	0.8091880	0.7260981	1
60	0.5591929	0.8290376	0.6745085	0.5735765	0.8191521	0.7002075	0.5877853	0.8090171	0.7265425	0
	cosine	sine 56°	cotangent	cosine	sine 55°	cotangent	cosine	sine 54°	cotangent	

PROFESSIONAL PUBLICATIONS, INC. ● P.O. Box 199, San Carlos, CA 94070

APPENDIX

A-13

	sine	36° cosine	tangent	sine	37° cosine	tangent	sine	38° cosine	tangent	
0	0.5877853	0.8090171	0.7265425	0.6018150	0.7986355	0.7535541	0.6156615	0.7880107	0.7812857	60
1	0.5880206	0.8088459	0.7269872	0.6020473	0.7984604	0.7540102	0.6158907	0.7878317	0.7817541	59
2	0.5882558	0.8086749	0.7274318	0.6022795	0.7982852	0.7544666	0.6161198	0.7876524	0.7822229	58
3	0.5884910	0.8085039	0.7278765	0.6025118	0.7981101	0.7549232	0.6163489	0.7874732	0.7826920	57
4	0.5887262	0.8083326	0.7283218	0.6027439	0.7979348	0.7553799	0.6165780	0.7872939	0.7831612	56
5	0.5889613	0.8081613	0.7287670	0.6029760	0.7977594	0.7558369	0.6168069	0.7871145	0.7836305	55
6	0.5891964	0.8079899	0.7292126	0.6032080	0.7975840	0.7562940	0.6170359	0.7869351	0.7841001	54
7	0.5894313	0.8078185	0.7296582	0.6034400	0.7974085	0.7567514	0.6172647	0.7867556	0.7845699	53
8	0.5896664	0.8076469	0.7301041	0.6036719	0.7972329	0.7572090	0.6174936	0.7865759	0.7850401	52
9	0.5899013	0.8074754	0.7305501	0.6039038	0.7970573	0.7576668	0.6177224	0.7863962	0.7855104	51
10	0.5901361	0.8073037	0.7309964	0.6041357	0.7968814	0.7581249	0.6179511	0.7862166	0.7859808	50
11	0.5903709	0.8071321	0.7314427	0.6043674	0.7967058	0.7585829	0.6181797	0.7860368	0.7864514	49
12	0.5906057	0.8069603	0.7318894	0.6045991	0.7965299	0.7590413	0.6184084	0.7858569	0.7869225	48
13	0.5908405	0.8067885	0.7323362	0.6048308	0.7963540	0.7594999	0.6186370	0.7856770	0.7873935	47
14	0.5910751	0.8066166	0.7327831	0.6050624	0.7961780	0.7599586	0.6188655	0.7854970	0.7878649	46
15	0.5913097	0.8064446	0.7332304	0.6052940	0.7960020	0.7604176	0.6190940	0.7853170	0.7883364	45
16	0.5915442	0.8062727	0.7336776	0.6055255	0.7958260	0.7608768	0.6193224	0.7851368	0.7888083	44
17	0.5917787	0.8061005	0.7341252	0.6057570	0.7956498	0.7613362	0.6195508	0.7849567	0.7892802	43
18	0.5920132	0.8059283	0.7345730	0.6059884	0.7954735	0.7617958	0.6197791	0.7847763	0.7897525	42
19	0.5922475	0.8057561	0.7350209	0.6062198	0.7952971	0.7622557	0.6200073	0.7845961	0.7902248	41
20	0.5924819	0.8055838	0.7354690	0.6064511	0.7951208	0.7627156	0.6202354	0.7844157	0.7906973	40
21	0.5927162	0.8054114	0.7359173	0.6066824	0.7949444	0.7631758	0.6204637	0.7842352	0.7911704	39
22	0.5929506	0.8052389	0.7363660	0.6069136	0.7947679	0.7636362	0.6206917	0.7840548	0.7916433	38
23	0.5931848	0.8050663	0.7368147	0.6071448	0.7945912	0.7640970	0.6209198	0.7838742	0.7921167	37
24	0.5934189	0.8048939	0.7372636	0.6073759	0.7944146	0.7645579	0.6211478	0.7836935	0.7925903	36
25	0.5936530	0.8047212	0.7377127	0.6076069	0.7942380	0.7650188	0.6213757	0.7835128	0.7930639	35
26	0.5938871	0.8045484	0.7381620	0.6078379	0.7940611	0.7654800	0.6216037	0.7833319	0.7935380	34
27	0.5941211	0.8043757	0.7386115	0.6080689	0.7938843	0.7659414	0.6218315	0.7831511	0.7940121	33
28	0.5943550	0.8042028	0.7390612	0.6082998	0.7937074	0.7664030	0.6220592	0.7829702	0.7944865	32
29	0.5945889	0.8040299	0.7395110	0.6085307	0.7935304	0.7668650	0.6222870	0.7827892	0.7949611	31
30	0.5948228	0.8038569	0.7399610	0.6087614	0.7933534	0.7673269	0.6225147	0.7826082	0.7954359	30
31	0.5950567	0.8036838	0.7404114	0.6089922	0.7931763	0.7677892	0.6227423	0.7824271	0.7959110	29
32	0.5952904	0.8035107	0.7408618	0.6092229	0.7929991	0.7682517	0.6229698	0.7822459	0.7963862	28
33	0.5955241	0.8033375	0.7413124	0.6094536	0.7928218	0.7687145	0.6231974	0.7820647	0.7968617	27
34	0.5957577	0.8031643	0.7417632	0.6096841	0.7926445	0.7691773	0.6234248	0.7818834	0.7973374	26
35	0.5959913	0.8029909	0.7422143	0.6099147	0.7924672	0.7696402	0.6236522	0.7817020	0.7978133	25
36	0.5962249	0.8028175	0.7426656	0.6101452	0.7922897	0.7701036	0.6238796	0.7815205	0.7982894	24
37	0.5964584	0.8026441	0.7431169	0.6103756	0.7921122	0.7705671	0.6241069	0.7813390	0.7987658	23
38	0.5966919	0.8024705	0.7435687	0.6106060	0.7919346	0.7710308	0.6243343	0.7811574	0.7992426	22
39	0.5969253	0.8022968	0.7440205	0.6108364	0.7917569	0.7714949	0.6245614	0.7809758	0.7997193	21
40	0.5971587	0.8021232	0.7444725	0.6110666	0.7915792	0.7719589	0.6247885	0.7807940	0.8001963	20
41	0.5973919	0.8019495	0.7449246	0.6112970	0.7914014	0.7724234	0.6250156	0.7806122	0.8006736	19
42	0.5976252	0.8017757	0.7453771	0.6115271	0.7912235	0.7728879	0.6252427	0.7804304	0.8011512	18
43	0.5978584	0.8016018	0.7458296	0.6117572	0.7910457	0.7733525	0.6254697	0.7802485	0.8016288	17
44	0.5980915	0.8014278	0.7462825	0.6119873	0.7908677	0.7738175	0.6256966	0.7800665	0.8021067	16
45	0.5983247	0.8012539	0.7467355	0.6122174	0.7906896	0.7742828	0.6259235	0.7798845	0.8025848	15
46	0.5985577	0.8010798	0.7471886	0.6124473	0.7905115	0.7747481	0.6261503	0.7797024	0.8030632	14
47	0.5987906	0.8009056	0.7476420	0.6126772	0.7903332	0.7752137	0.6263771	0.7795202	0.8035419	13
48	0.5990236	0.8007314	0.7480956	0.6129071	0.7901551	0.7756795	0.6266038	0.7793380	0.8040206	12
49	0.5992565	0.8005571	0.7485493	0.6131368	0.7899768	0.7761454	0.6268305	0.7791557	0.8044997	11
50	0.5994893	0.8003828	0.7490033	0.6133667	0.7897983	0.7766117	0.6270571	0.7789733	0.8049788	10
51	0.5997221	0.8002083	0.7494575	0.6135963	0.7896199	0.7770781	0.6272836	0.7787910	0.8054583	9
52	0.5999548	0.8000339	0.7499117	0.6138260	0.7894414	0.7775447	0.6275102	0.7786083	0.8059382	8
53	0.6001877	0.7998593	0.7503666	0.6140556	0.7892628	0.7780117	0.6277367	0.7784258	0.8064182	7
54	0.6004203	0.7996847	0.7508213	0.6142852	0.7890841	0.7784788	0.6279631	0.7782431	0.8068983	6
55	0.6006529	0.7995099	0.7512763	0.6145148	0.7889054	0.7789460	0.6281894	0.7780605	0.8073786	5
56	0.6008853	0.7993352	0.7517313	0.6147442	0.7887266	0.7794135	0.6284157	0.7778776	0.8078594	4
57	0.6011179	0.7991604	0.7521868	0.6149736	0.7885478	0.7798812	0.6286420	0.7776948	0.8083402	3
58	0.6013503	0.7989855	0.7526423	0.6152030	0.7883688	0.7803493	0.6288682	0.7775120	0.8088211	2
59	0.6015827	0.7988105	0.7530981	0.6154323	0.7881898	0.7808173	0.6290944	0.7773290	0.8093026	1
60	0.6018150	0.7986355	0.7535541	0.6156615	0.7880107	0.7812857	0.6293204	0.7771460	0.8097841	0
	cosine	sine 53°	cotangent	cosine	sine 52°	cotangent	cosine	sine 51°	cotangent	

	sine	39°\ncosine	tangent	sine	40°\ncosine	tangent	sine	41°\ncosine	tangent	
0	0.6293204	0.7771460	0.8097841	0.6427876	0.7660445	0.8390995	0.6560591	0.7547097	0.8692867	60
1	0.6295465	0.7769629	0.8102658	0.6430104	0.7658575	0.8395954	0.6562785	0.7545188	0.8697974	59
2	0.6297724	0.7767798	0.8107477	0.6432332	0.7656704	0.8400915	0.6564980	0.7543278	0.8703086	58
3	0.6299983	0.7765965	0.8112299	0.6434559	0.7654833	0.8405878	0.6567174	0.7541368	0.8708200	57
4	0.6302242	0.7764132	0.8117123	0.6436785	0.7652960	0.8410843	0.6569368	0.7539456	0.8713318	56
5	0.6304501	0.7762298	0.8121951	0.6439011	0.7651088	0.8415810	0.6571561	0.7537547	0.8718434	55
6	0.6306758	0.7760465	0.8126779	0.6441236	0.7649214	0.8420781	0.6573753	0.7535635	0.8723556	54
7	0.6309015	0.7758630	0.8131610	0.6443461	0.7647340	0.8425754	0.6575944	0.7533722	0.8728679	53
8	0.6311272	0.7756794	0.8136443	0.6445686	0.7645466	0.8430730	0.6578136	0.7531808	0.8733807	52
9	0.6313528	0.7754957	0.8141281	0.6447909	0.7643591	0.8435708	0.6580326	0.7529895	0.8738935	51
10	0.6315784	0.7753121	0.8146119	0.6450133	0.7641714	0.8440689	0.6582516	0.7527980	0.8744067	50
11	0.6318039	0.7751283	0.8150960	0.6452355	0.7639838	0.8445671	0.6584705	0.7526065	0.8749200	49
12	0.6320293	0.7749445	0.8155800	0.6454577	0.7637960	0.8450655	0.6586895	0.7524149	0.8754339	48
13	0.6322548	0.7747605	0.8160648	0.6456798	0.7636083	0.8455642	0.6589083	0.7522233	0.8759478	47
14	0.6324801	0.7745766	0.8165494	0.6459019	0.7634205	0.8460631	0.6591271	0.7520316	0.8764620	46
15	0.6327053	0.7743927	0.8170342	0.6461240	0.7632325	0.8465625	0.6593458	0.7518398	0.8769764	45
16	0.6329306	0.7742086	0.8175194	0.6463460	0.7630445	0.8470619	0.6595646	0.7516480	0.8774912	44
17	0.6331557	0.7740245	0.8180047	0.6465679	0.7628564	0.8475618	0.6597831	0.7514561	0.8780062	43
18	0.6333809	0.7738402	0.8184906	0.6467897	0.7626684	0.8480616	0.6600017	0.7512642	0.8785215	42
19	0.6336060	0.7736560	0.8189764	0.6470116	0.7624801	0.8485619	0.6602201	0.7510722	0.8790369	41
20	0.6338310	0.7734717	0.8194624	0.6472334	0.7622919	0.8490624	0.6604386	0.7508801	0.8795527	40
21	0.6340559	0.7732872	0.8199487	0.6474551	0.7621036	0.8495631	0.6606570	0.7506879	0.8800689	39
22	0.6342809	0.7731028	0.8204354	0.6476767	0.7619153	0.8500640	0.6608754	0.7504957	0.8805852	38
23	0.6345058	0.7729182	0.8209223	0.6478984	0.7617267	0.8505654	0.6610937	0.7503034	0.8811017	37
24	0.6347305	0.7727336	0.8214092	0.6481199	0.7615383	0.8510668	0.6613119	0.7501110	0.8816186	36
25	0.6349552	0.7725490	0.8218964	0.6483415	0.7613498	0.8515685	0.6615301	0.7499187	0.8821357	35
26	0.6351800	0.7723642	0.8223841	0.6485628	0.7611611	0.8520703	0.6617482	0.7497262	0.8826531	34
27	0.6354046	0.7721794	0.8228718	0.6487842	0.7609725	0.8525725	0.6619663	0.7495337	0.8831708	33
28	0.6356292	0.7719945	0.8233598	0.6490055	0.7607837	0.8530750	0.6621842	0.7493411	0.8836886	32
29	0.6358538	0.7718096	0.8238479	0.6492268	0.7605948	0.8535777	0.6624022	0.7491484	0.8842068	31
30	0.6360782	0.7716246	0.8243363	0.6494481	0.7604060	0.8540807	0.6626201	0.7489557	0.8847255	30
31	0.6363026	0.7714396	0.8248250	0.6496692	0.7602171	0.8545839	0.6628379	0.7487629	0.8852441	29
32	0.6365271	0.7712544	0.8253140	0.6498903	0.7600281	0.8550872	0.6630557	0.7485702	0.8857628	28
33	0.6367514	0.7710692	0.8258031	0.6501114	0.7598389	0.8555911	0.6632734	0.7483773	0.8862821	27
34	0.6369756	0.7708840	0.8262925	0.6503324	0.7596499	0.8560949	0.6634911	0.7481843	0.8868017	26
35	0.6371999	0.7706987	0.8267821	0.6505533	0.7594606	0.8565992	0.6637087	0.7479913	0.8873214	25
36	0.6374240	0.7705132	0.8272719	0.6507742	0.7592713	0.8571036	0.6639262	0.7477981	0.8878415	24
37	0.6376481	0.7703278	0.8277620	0.6509951	0.7590820	0.8576084	0.6641437	0.7476050	0.8883618	23
38	0.6378722	0.7701423	0.8282524	0.6512159	0.7588926	0.8581133	0.6643612	0.7474117	0.8888825	22
39	0.6380962	0.7699567	0.8287430	0.6514366	0.7587031	0.8586186	0.6645786	0.7472184	0.8894033	21
40	0.6383201	0.7697711	0.8292337	0.6516573	0.7585136	0.8591240	0.6647959	0.7470251	0.8899245	20
41	0.6385440	0.7695854	0.8297247	0.6518779	0.7583240	0.8596298	0.6650131	0.7468317	0.8904458	19
42	0.6387678	0.7693996	0.8302160	0.6520984	0.7581343	0.8601357	0.6652303	0.7466382	0.8909674	18
43	0.6389917	0.7692137	0.8307076	0.6523190	0.7579446	0.8606420	0.6654475	0.7464447	0.8914894	17
44	0.6392154	0.7690279	0.8311992	0.6525394	0.7577549	0.8611484	0.6656646	0.7462510	0.8920116	16
45	0.6394390	0.7688418	0.8316912	0.6527597	0.7575650	0.8616551	0.6658817	0.7460574	0.8925341	15
46	0.6396626	0.7686558	0.8321834	0.6529801	0.7573750	0.8621622	0.6660987	0.7458637	0.8930568	14
47	0.6398862	0.7684698	0.8326759	0.6532004	0.7571852	0.8626693	0.6663156	0.7456698	0.8935800	13
48	0.6401097	0.7682836	0.8331685	0.6534206	0.7569951	0.8631768	0.6665325	0.7454761	0.8941031	12
49	0.6403332	0.7680974	0.8336614	0.6536407	0.7568050	0.8636845	0.6667493	0.7452821	0.8946267	11
50	0.6405566	0.7679110	0.8341547	0.6538609	0.7566148	0.8641926	0.6669660	0.7450882	0.8951505	10
51	0.6407799	0.7677247	0.8346481	0.6540809	0.7564245	0.8647008	0.6671827	0.7448941	0.8956746	9
52	0.6410031	0.7675383	0.8351415	0.6543009	0.7562343	0.8652093	0.6673995	0.7447000	0.8961992	8
53	0.6412265	0.7673517	0.8356357	0.6545209	0.7560439	0.8657181	0.6676161	0.7445058	0.8967239	7
54	0.6414497	0.7671651	0.8361299	0.6547409	0.7558534	0.8662273	0.6678326	0.7443116	0.8972486	6
55	0.6416728	0.7669786	0.8366241	0.6549607	0.7556630	0.8667365	0.6680490	0.7441173	0.8977739	5
56	0.6418959	0.7667918	0.8371189	0.6551805	0.7554725	0.8672460	0.6682655	0.7439229	0.8982994	4
57	0.6421189	0.7666051	0.8376137	0.6554002	0.7552819	0.8677557	0.6684819	0.7437285	0.8988252	3
58	0.6423419	0.7664183	0.8381087	0.6556199	0.7550911	0.8682659	0.6686982	0.7435340	0.8993512	2
59	0.6425648	0.7662314	0.8386040	0.6558395	0.7549004	0.8687762	0.6689144	0.7433395	0.8998774	1
60	0.6427876	0.7660445	0.8390995	0.6560591	0.7547097	0.8692867	0.6691306	0.7431448	0.9004041	0
	cosine	sine\n50°	cotangent	cosine	sine\n49°	cotangent	cosine	sine\n48°	cotangent	

PROFESSIONAL PUBLICATIONS, INC. ● P.O. Box 199, San Carlos, CA 94070

	42° sine	cosine	tangent	43° sine	cosine	tangent	44° sine	cosine	tangent	
0	0.6691306	0.7431448	0.9004041	0.6819984	0.7313537	0.9325151	0.6946584	0.7193398	0.9656888	60
1	0.6693468	0.7429502	0.9009309	0.6822111	0.7311553	0.9330590	0.6948676	0.7191377	0.9662511	59
2	0.6695629	0.7427554	0.9014581	0.6824238	0.7309568	0.9336034	0.6950768	0.7189355	0.9668137	58
3	0.6697789	0.7425607	0.9019854	0.6826364	0.7307583	0.9341480	0.6952859	0.7187333	0.9673767	57
4	0.6699948	0.7423659	0.9025130	0.6828489	0.7305598	0.9346927	0.6954949	0.7185311	0.9679399	56
5	0.6702108	0.7421709	0.9030411	0.6830613	0.7303611	0.9352378	0.6957039	0.7183288	0.9685034	55
6	0.6704266	0.7419759	0.9035692	0.6832738	0.7301623	0.9357834	0.6959128	0.7181263	0.9690674	54
7	0.6706424	0.7417808	0.9040979	0.6834862	0.7299635	0.9363292	0.6961216	0.7179239	0.9696314	53
8	0.6708581	0.7415857	0.9046265	0.6836985	0.7297646	0.9368753	0.6963305	0.7177213	0.9701962	52
9	0.6710739	0.7413905	0.9051558	0.6839107	0.7295657	0.9374217	0.6965393	0.7175187	0.9707611	51
10	0.6712895	0.7411953	0.9056850	0.6841230	0.7293668	0.9379684	0.6967480	0.7173160	0.9713264	50
11	0.6715052	0.7409999	0.9062148	0.6843351	0.7291678	0.9385153	0.6969565	0.7171134	0.9718916	49
12	0.6717206	0.7408047	0.9067446	0.6845471	0.7289687	0.9390625	0.6971651	0.7169106	0.9724575	48
13	0.6719362	0.7406091	0.9072750	0.6847591	0.7287695	0.9396100	0.6973736	0.7167078	0.9730236	47
14	0.6721514	0.7404137	0.9078053	0.6849710	0.7285703	0.9401578	0.6975821	0.7165049	0.9735900	46
15	0.6723669	0.7402182	0.9083360	0.6851830	0.7283710	0.9407061	0.6977904	0.7163020	0.9741568	45
16	0.6725821	0.7400225	0.9088671	0.6853948	0.7281716	0.9412545	0.6979988	0.7160989	0.9747241	44
17	0.6727973	0.7398269	0.9093984	0.6856067	0.7279723	0.9418033	0.6982071	0.7158958	0.9752914	43
18	0.6730125	0.7396311	0.9099299	0.6858184	0.7277728	0.9423522	0.6984153	0.7156928	0.9758590	42
19	0.6732277	0.7394353	0.9104619	0.6860300	0.7275733	0.9429016	0.6986234	0.7154896	0.9764270	41
20	0.6734427	0.7392395	0.9109939	0.6862416	0.7273737	0.9434513	0.6988316	0.7152863	0.9769956	40
21	0.6736577	0.7390435	0.9115264	0.6864532	0.7271740	0.9440014	0.6990396	0.7150830	0.9775643	39
22	0.6738727	0.7388476	0.9120592	0.6866646	0.7269743	0.9445515	0.6992476	0.7148796	0.9781334	38
23	0.6740876	0.7386515	0.9125922	0.6868762	0.7267745	0.9451023	0.6994555	0.7146762	0.9787027	37
24	0.6743025	0.7384554	0.9131256	0.6870876	0.7265746	0.9456532	0.6996633	0.7144727	0.9792723	36
25	0.6745172	0.7382592	0.9136591	0.6872988	0.7263747	0.9462042	0.6998712	0.7142692	0.9798424	35
26	0.6747320	0.7380629	0.9141930	0.6875101	0.7261748	0.9467556	0.7000789	0.7140655	0.9804127	34
27	0.6749466	0.7378666	0.9147272	0.6877213	0.7259749	0.9473073	0.7002866	0.7138619	0.9809834	33
28	0.6751612	0.7376703	0.9152614	0.6879325	0.7257748	0.9478595	0.7004942	0.7136581	0.9815543	32
29	0.6753757	0.7374739	0.9157960	0.6881436	0.7255746	0.9484118	0.7007018	0.7134543	0.9821256	31
30	0.6755902	0.7372774	0.9163311	0.6883546	0.7253744	0.9489646	0.7009093	0.7132504	0.9826974	30
31	0.6758047	0.7370807	0.9168666	0.6885656	0.7251742	0.9495174	0.7011167	0.7130466	0.9832692	29
32	0.6760190	0.7368842	0.9174020	0.6887764	0.7249739	0.9500707	0.7013242	0.7128426	0.9838415	28
33	0.6762334	0.7366875	0.9179378	0.6889874	0.7247735	0.9506245	0.7015314	0.7126386	0.9844140	27
34	0.6764476	0.7364908	0.9184740	0.6891981	0.7245730	0.9511783	0.7017387	0.7124345	0.9849870	26
35	0.6766618	0.7362940	0.9190103	0.6894089	0.7243724	0.9517326	0.7019459	0.7122303	0.9855604	25
36	0.6768760	0.7360971	0.9195472	0.6896195	0.7241720	0.9522870	0.7021530	0.7120261	0.9861338	24
37	0.6770900	0.7359002	0.9200841	0.6898302	0.7239712	0.9528421	0.7023602	0.7118219	0.9867077	23
38	0.6773041	0.7357031	0.9206216	0.6900408	0.7237706	0.9533971	0.7025673	0.7116174	0.9872823	22
39	0.6775181	0.7355062	0.9211589	0.6902513	0.7235698	0.9539526	0.7027742	0.7114130	0.9878569	21
40	0.6777321	0.7353089	0.9216970	0.6904617	0.7233691	0.9545082	0.7029811	0.7112086	0.9884317	20
41	0.6779459	0.7351118	0.9222350	0.6906721	0.7231681	0.9550644	0.7031879	0.7110041	0.9890069	19
42	0.6781597	0.7349146	0.9227734	0.6908824	0.7229672	0.9556207	0.7033947	0.7107995	0.9895824	18
43	0.6783734	0.7347172	0.9233122	0.6910927	0.7227662	0.9561775	0.7036015	0.7105948	0.9901584	17
44	0.6785871	0.7345200	0.9238512	0.6913029	0.7225651	0.9567344	0.7038081	0.7103902	0.9907346	16
45	0.6788008	0.7343225	0.9243905	0.6915131	0.7223640	0.9572917	0.7040147	0.7101854	0.9913111	15
46	0.6790143	0.7341251	0.9249300	0.6917232	0.7221628	0.9578493	0.7042213	0.7099806	0.9918881	14
47	0.6792278	0.7339275	0.9254699	0.6919332	0.7219616	0.9584073	0.7044278	0.7097757	0.9924654	13
48	0.6794413	0.7337299	0.9260101	0.6921432	0.7217603	0.9589655	0.7046342	0.7095708	0.9930429	12
49	0.6796547	0.7335322	0.9265506	0.6923531	0.7215589	0.9595239	0.7048406	0.7093658	0.9936208	11
50	0.6798681	0.7333345	0.9270914	0.6925630	0.7213574	0.9600830	0.7050469	0.7091608	0.9941989	10
51	0.6800814	0.7331368	0.9276323	0.6927728	0.7211560	0.9606421	0.7052532	0.7089556	0.9947777	9
52	0.6802946	0.7329389	0.9281737	0.6929825	0.7209544	0.9612015	0.7054594	0.7087504	0.9953565	8
53	0.6805078	0.7327409	0.9287155	0.6931922	0.7207528	0.9617614	0.7056655	0.7085451	0.9959359	7
54	0.6807209	0.7325429	0.9292574	0.6934019	0.7205511	0.9623216	0.7058716	0.7083399	0.9965154	6
55	0.6809340	0.7323449	0.9297996	0.6936114	0.7203494	0.9628819	0.7060776	0.7081345	0.9970953	5
56	0.6811470	0.7321467	0.9303421	0.6938210	0.7201476	0.9634427	0.7062836	0.7079291	0.9976757	4
57	0.6813599	0.7319486	0.9308848	0.6940304	0.7199458	0.9640037	0.7064895	0.7077236	0.9982563	3
58	0.6815728	0.7317504	0.9314280	0.6942398	0.7197438	0.9645651	0.7066953	0.7075181	0.9988371	2
59	0.6817856	0.7315521	0.9319714	0.6944492	0.7195418	0.9651270	0.7069011	0.7073125	0.9994184	1
60	0.6819984	0.7313537	0.9325151	0.6946584	0.7193398	0.9656888	0.7071068	0.7071068	1.0000000	0
	cosine	sine 47°	cotangent	cosine	sine 46°	cotangent	cosine	sine 45°	cotangent	

PROFESSIONAL PUBLICATIONS, INC. • P.O. Box 199, San Carlos, CA 94070

Appendix B: Tangents and Externals for a 1° Curve

$$\Delta = \text{angle}$$

$$T = (5729.578)\tan\left(\frac{\Delta}{2}\right)$$

$$E = (5729.578)\left[\sec\left(\frac{\Delta}{2}\right) - 1\right]$$

For arc definition, calculate T and E for any degree of curve D by dividing both table values by D.

	Δ = 0°		Δ = 1°		Δ = 2°		Δ = 3°		Δ = 4°		Δ = 5°		Δ = 6°		
	T	E	T	E	T	E	T	E	T	E	T	E	T	E	
0	0.00	0.00	50.00	0.22	100.01	0.87	150.03	1.96	200.08	3.49	250.16	5.46	300.27	7.86	0
1	0.83	0.00	50.83	0.23	100.84	0.89	150.87	1.99	200.92	3.52	250.99	5.50	301.11	7.91	1
2	1.67	0.00	51.67	0.23	101.68	0.90	151.70	2.01	201.75	3.55	251.83	5.53	301.95	7.95	2
3	2.50	0.00	52.50	0.24	102.51	0.92	152.54	2.03	202.58	3.58	252.66	5.57	302.78	7.99	3
4	3.33	0.00	53.33	0.25	103.34	0.93	153.37	2.05	203.42	3.61	253.50	5.60	303.62	8.04	4
5	4.17	0.00	54.17	0.26	104.18	0.95	154.20	2.08	204.25	3.64	254.33	5.64	304.45	8.08	5
6	5.00	0.00	55.00	0.26	105.01	0.96	155.04	2.10	205.09	3.67	255.17	5.68	305.29	8.13	6
7	5.83	0.00	55.84	0.27	105.85	0.98	155.87	2.12	205.92	3.70	256.00	5.72	306.12	8.17	7
8	6.67	0.00	56.67	0.28	106.68	0.99	156.71	2.14	206.76	3.73	256.84	5.75	306.96	8.22	8
9	7.50	0.00	57.50	0.29	107.51	1.01	157.54	2.17	207.59	3.76	257.67	5.79	307.80	8.26	9
10	8.33	0.01	58.34	0.30	108.35	1.02	158.37	2.19	208.43	3.79	258.51	5.83	308.63	8.31	10
11	9.17	0.01	59.17	0.31	109.18	1.04	159.21	2.21	209.26	3.82	259.34	5.87	309.47	8.35	11
12	10.00	0.01	60.00	0.31	110.01	1.06	160.04	2.23	210.09	3.85	260.18	5.90	310.30	8.40	12
13	10.83	0.01	60.84	0.32	110.85	1.07	160.88	2.26	210.93	3.88	261.01	5.94	311.14	8.44	13
14	11.67	0.01	61.67	0.33	111.68	1.09	161.71	2.28	211.76	3.91	261.85	5.98	311.97	8.49	14
15	12.50	0.01	62.50	0.34	112.51	1.10	162.54	2.30	212.60	3.94	262.68	6.02	312.81	8.53	15
16	13.33	0.02	63.34	0.35	113.35	1.12	163.38	2.33	213.43	3.97	263.52	6.06	313.65	8.58	16
17	14.17	0.02	64.17	0.36	114.18	1.14	164.21	2.35	214.27	4.00	264.35	6.09	314.48	8.62	17
18	15.00	0.02	65.00	0.37	115.02	1.15	165.05	2.38	215.10	4.04	265.19	6.13	315.32	8.67	18
19	15.83	0.02	65.84	0.38	115.85	1.17	165.88	2.40	215.94	4.07	266.02	6.17	316.15	8.72	19
20	16.67	0.02	66.67	0.39	116.68	1.19	166.71	2.42	216.77	4.10	266.86	6.21	316.99	8.76	20
21	17.50	0.03	67.50	0.40	117.52	1.20	167.55	2.45	217.60	4.13	267.69	6.25	317.83	8.81	21
22	18.33	0.03	68.34	0.41	118.35	1.22	168.38	2.47	218.44	4.16	268.53	6.29	318.66	8.85	22
23	19.17	0.03	69.17	0.42	119.18	1.24	169.22	2.50	219.27	4.19	269.36	6.33	319.50	8.90	23
24	20.00	0.03	70.00	0.43	120.02	1.26	170.05	2.52	220.11	4.23	270.20	6.37	320.33	8.95	24
25	20.83	0.04	70.84	0.44	120.85	1.27	170.88	2.55	220.94	4.26	271.04	6.41	321.17	8.99	25
26	21.67	0.04	71.67	0.45	121.68	1.29	171.72	2.57	221.78	4.29	271.87	6.45	322.01	9.04	26
27	22.50	0.04	72.50	0.46	122.52	1.31	172.55	2.60	222.61	4.32	272.71	6.49	322.84	9.09	27
28	23.33	0.05	73.34	0.47	123.35	1.33	173.39	2.62	223.45	4.36	273.54	6.53	323.68	9.14	28
29	24.17	0.05	74.17	0.48	124.19	1.34	174.22	2.65	224.28	4.39	274.38	6.57	324.51	9.18	29
30	25.00	0.05	75.00	0.49	125.02	1.36	175.05	2.67	225.12	4.42	275.21	6.61	325.35	9.23	30
31	25.83	0.06	75.84	0.50	125.85	1.38	175.89	2.70	225.95	4.45	276.05	6.65	326.19	9.28	31
32	26.67	0.06	76.67	0.51	126.69	1.40	176.72	2.72	226.78	4.49	276.88	6.69	327.02	9.32	32
33	27.50	0.07	77.50	0.52	127.52	1.42	177.56	2.75	227.62	4.52	277.72	6.73	327.86	9.37	33
34	28.33	0.07	78.34	0.54	128.35	1.44	178.39	2.78	228.45	4.55	278.55	6.77	328.69	9.42	34
35	29.17	0.07	79.17	0.55	129.19	1.46	179.23	2.80	229.29	4.59	279.39	6.81	329.53	9.47	35
36	30.00	0.08	80.01	0.56	130.02	1.48	180.06	2.83	230.12	4.62	280.22	6.85	330.37	9.52	36
37	30.83	0.08	80.84	0.57	130.86	1.49	180.89	2.86	230.96	4.65	281.06	6.89	331.20	9.56	37
38	31.67	0.09	81.67	0.58	131.69	1.51	181.73	2.88	231.79	4.69	281.89	6.93	332.04	9.61	38
39	32.50	0.09	82.51	0.59	132.52	1.53	182.56	2.91	232.63	4.72	282.73	6.97	332.87	9.66	39
40	33.33	0.10	83.34	0.61	133.36	1.55	183.40	2.93	233.46	4.75	283.56	7.01	333.71	9.71	40
41	34.17	0.10	84.17	0.62	134.19	1.57	184.23	2.96	234.30	4.79	284.40	7.05	334.55	9.76	41
42	35.00	0.11	85.01	0.63	135.02	1.59	185.06	2.99	235.13	4.82	285.24	7.09	335.38	9.81	42
43	35.83	0.11	85.84	0.64	135.86	1.61	185.90	3.01	235.97	4.86	286.07	7.14	336.22	9.86	43
44	36.67	0.12	86.67	0.66	136.69	1.63	186.73	3.04	236.80	4.89	286.91	7.18	337.05	9.91	44
45	37.50	0.12	87.51	0.67	137.53	1.65	187.57	3.07	237.64	4.93	287.74	7.22	337.89	9.95	45
46	38.33	0.13	88.34	0.68	138.36	1.67	188.40	3.10	238.47	4.96	288.58	7.26	338.73	10.00	46
47	39.17	0.13	89.17	0.69	139.19	1.69	189.24	3.12	239.31	4.99	289.41	7.30	339.56	10.05	47
48	40.00	0.14	90.01	0.71	140.03	1.71	190.07	3.15	240.14	5.03	290.25	7.35	340.40	10.10	48
49	40.83	0.15	90.84	0.72	140.86	1.73	190.90	3.18	240.98	5.07	291.08	7.39	341.24	10.15	49
50	41.67	0.15	91.67	0.73	141.70	1.75	191.74	3.21	241.81	5.10	291.92	7.43	342.07	10.20	50
51	42.50	0.16	92.51	0.75	142.53	1.77	192.57	3.23	242.64	5.13	292.75	7.47	342.91	10.25	51
52	43.33	0.16	93.34	0.76	143.36	1.79	193.41	3.26	243.48	5.17	293.59	7.52	343.74	10.30	52
53	44.17	0.17	94.18	0.77	144.20	1.81	194.24	3.29	244.31	5.21	294.43	7.56	344.58	10.35	53
54	45.00	0.18	95.01	0.79	145.03	1.84	195.08	3.32	245.15	5.24	295.26	7.60	345.42	10.40	54
55	45.83	0.18	95.84	0.80	145.86	1.86	195.91	3.35	245.98	5.28	296.10	7.65	346.25	10.45	55
56	46.67	0.19	96.68	0.82	146.70	1.88	196.74	3.38	246.82	5.31	296.93	7.69	347.09	10.50	56
57	47.50	0.20	97.51	0.83	147.53	1.90	197.58	3.41	247.65	5.35	297.77	7.73	347.93	10.55	57
58	48.33	0.20	98.34	0.84	148.37	1.92	198.41	3.43	248.49	5.39	298.60	7.78	348.76	10.61	58
59	49.17	0.21	99.18	0.86	149.20	1.94	199.25	3.46	249.32	5.42	299.44	7.82	349.60	10.66	59

LAND SURVEYOR REFERENCE MANUAL

$$\Delta = \text{angle}$$
$$T = (5729.578)\tan\left(\frac{\Delta}{2}\right)$$
$$E = (5729.578)\left[\sec\left(\frac{\Delta}{2}\right) - 1\right]$$

For arc definition, calculate T and E for any degree of curve D by dividing both table values by D.

	Δ = 7°		Δ = 8°		Δ = 9°		Δ = 10°		Δ = 11°		Δ = 12°		Δ = 13°		
	T	E	T	E	T	E	T	E	T	E	T	E	T	E	
0	350.44	10.71	400.65	13.99	450.93	17.72	501.27	21.89	551.70	26.50	602.20	31.56	652.80	37.07	0
1	351.27	10.76	401.49	14.05	451.77	17.78	502.11	21.96	552.54	26.58	603.05	31.65	653.65	37.16	1
2	352.11	10.81	402.33	14.11	452.60	17.85	502.95	22.03	553.38	26.66	603.89	31.74	654.49	37.26	2
3	352.95	10.86	403.16	14.17	453.44	17.91	503.79	22.11	554.22	26.74	604.73	31.82	655.34	37.36	3
4	353.78	10.91	404.00	14.23	454.28	17.98	504.63	22.18	555.06	26.82	605.57	31.91	656.18	37.45	4
5	354.62	10.96	404.84	14.28	455.12	18.05	505.47	22.25	555.90	26.90	606.42	32.00	657.02	37.55	5
6	355.45	11.02	405.68	14.34	455.96	18.11	506.31	22.33	556.74	26.99	607.26	32.09	657.87	37.64	6
7	356.29	11.07	406.51	14.40	456.80	18.18	507.15	22.40	557.58	27.07	608.10	32.18	658.71	37.74	7
8	357.13	11.12	407.35	14.46	457.64	18.25	507.99	22.48	558.42	27.15	608.94	32.27	659.56	37.84	8
9	357.96	11.17	408.19	14.52	458.47	18.31	508.83	22.55	559.27	27.23	609.79	32.36	660.40	37.93	9
10	358.80	11.22	409.03	14.58	459.31	18.38	509.67	22.62	560.11	27.31	610.63	32.45	661.25	38.03	10
11	359.64	11.28	409.86	14.64	460.15	18.45	510.51	22.70	560.95	27.39	611.47	32.54	662.09	38.13	11
12	360.47	11.33	410.70	14.70	460.99	18.51	511.35	22.77	561.79	27.48	612.32	32.63	662.93	38.22	12
13	361.31	11.38	411.54	14.76	461.83	18.58	512.19	22.85	562.63	27.56	613.16	32.72	663.78	38.32	13
14	362.15	11.43	412.38	14.82	462.67	18.65	513.03	22.92	563.47	27.64	614.00	32.80	664.62	38.42	14
15	362.98	11.49	413.21	14.88	463.51	18.72	513.87	23.00	564.31	27.72	614.84	32.89	665.47	38.52	15
16	363.82	11.54	414.05	14.94	464.35	18.79	514.71	23.07	565.16	27.81	615.69	32.98	666.31	38.61	16
17	364.66	11.59	414.89	15.00	465.18	18.85	515.55	23.15	566.00	27.89	616.53	33.08	667.16	38.71	17
18	365.49	11.65	415.73	15.06	466.02	18.92	516.39	23.22	566.84	27.97	617.37	33.17	668.00	38.81	18
19	366.33	11.70	416.56	15.12	466.86	18.99	517.23	23.30	567.68	28.05	618.22	33.26	668.85	38.91	19
20	367.17	11.75	417.40	15.18	467.70	19.06	518.07	23.37	568.52	28.14	619.06	33.35	669.69	39.00	20
21	368.00	11.81	418.24	15.24	468.54	19.13	518.91	23.45	569.36	28.22	619.90	33.44	670.54	39.10	21
22	368.84	11.86	419.08	15.31	469.38	19.19	519.75	23.53	570.20	28.30	620.75	33.53	671.38	39.20	22
23	369.68	11.91	419.92	15.37	470.22	19.26	520.59	23.60	571.05	28.39	621.59	33.62	672.23	39.30	23
24	370.52	11.97	420.75	15.43	471.06	19.33	521.43	23.68	571.89	28.47	622.43	33.71	673.07	39.40	24
25	371.35	12.02	421.59	15.49	471.90	19.40	522.27	23.75	572.73	28.55	623.27	33.80	673.92	39.50	25
26	372.19	12.08	422.43	15.55	472.73	19.47	523.11	23.83	573.57	28.64	624.12	33.89	674.76	39.60	26
27	373.03	12.13	423.27	15.61	473.57	19.54	523.95	23.91	574.41	28.72	624.96	33.98	675.61	39.69	27
28	373.86	12.18	424.11	15.67	474.41	19.61	524.79	23.98	575.25	28.81	625.80	34.08	676.45	39.79	28
29	374.70	12.24	424.94	15.74	475.25	19.68	525.63	24.06	576.10	28.89	626.65	34.17	677.30	39.89	29
30	375.54	12.29	425.78	15.80	476.09	19.75	526.47	24.14	576.94	28.97	627.49	34.26	678.14	39.99	30
31	376.37	12.35	426.62	15.86	476.93	19.82	527.31	24.21	577.78	29.06	628.33	34.35	678.99	40.09	31
32	377.21	12.40	427.46	15.92	477.77	19.89	528.16	24.29	578.62	29.14	629.18	34.44	679.83	40.19	32
33	378.05	12.46	428.30	15.99	478.61	19.96	529.00	24.37	579.46	29.23	630.02	34.53	680.68	40.29	33
34	378.88	12.51	429.13	16.05	479.45	20.02	529.84	24.45	580.31	29.31	630.86	34.63	681.52	40.39	34
35	379.72	12.57	429.97	16.11	480.29	20.10	530.68	24.52	581.15	29.40	631.71	34.72	682.37	40.49	35
36	380.56	12.62	430.81	16.17	481.13	20.17	531.52	24.60	581.99	29.48	632.55	34.81	683.21	40.59	36
37	381.40	12.68	431.65	16.24	481.97	20.24	532.36	24.68	582.83	29.57	633.39	34.90	684.06	40.69	37
38	382.23	12.74	432.49	16.30	482.80	20.31	533.20	24.76	583.67	29.65	634.24	35.00	684.90	40.79	38
39	383.07	12.79	433.32	16.36	483.64	20.38	534.04	24.83	584.52	29.74	635.08	35.09	685.75	40.89	39
40	383.91	12.85	434.16	16.43	484.48	20.45	534.88	24.91	585.36	29.82	635.93	35.18	686.59	40.99	40
41	384.74	12.90	435.00	16.49	485.32	20.52	535.72	24.99	586.20	29.91	636.77	35.28	687.44	41.09	41
42	385.58	12.96	435.84	16.55	486.16	20.59	536.56	25.07	587.04	29.99	637.61	35.37	688.28	41.19	42
43	386.42	13.02	436.68	16.62	487.00	20.66	537.40	25.15	587.88	30.08	638.46	35.46	689.13	41.29	43
44	387.25	13.07	437.51	16.68	487.84	20.73	538.24	25.23	588.73	30.17	639.30	35.56	689.97	41.39	44
45	388.09	13.13	438.35	16.74	488.68	20.80	539.08	25.30	589.57	30.25	640.14	35.65	690.82	41.50	45
46	388.93	13.18	439.19	16.81	489.52	20.87	539.92	25.38	590.41	30.34	640.99	35.74	691.66	41.60	46
47	389.77	13.24	440.03	16.87	490.36	20.94	540.76	25.46	591.25	30.43	641.83	35.84	692.51	41.70	47
48	390.60	13.30	440.87	16.94	491.20	21.02	541.60	25.54	592.09	30.51	642.68	35.93	693.36	41.80	48
49	391.44	13.36	441.71	17.00	492.04	21.09	542.45	25.62	592.94	30.60	643.52	36.03	694.20	41.90	49
50	392.28	13.41	442.54	17.06	492.88	21.16	543.29	25.70	593.78	30.69	644.36	36.12	695.05	42.00	50
51	393.12	13.47	443.38	17.13	493.72	21.23	544.13	25.78	594.62	30.77	645.21	36.21	695.89	42.11	51
52	393.95	13.53	444.22	17.19	494.56	21.30	544.97	25.86	595.46	30.86	646.05	36.31	696.74	42.21	52
53	394.79	13.59	445.06	17.26	495.40	21.38	545.81	25.94	596.31	30.95	646.89	36.40	697.58	42.31	53
54	395.63	13.64	445.90	17.32	496.24	21.45	546.65	26.02	597.15	31.03	647.74	36.50	698.43	42.41	54
55	396.46	13.70	446.74	17.39	497.07	21.52	547.49	26.10	597.99	31.12	648.58	36.59	699.27	42.51	55
56	397.30	13.76	447.57	17.46	497.91	21.59	548.33	26.18	598.83	31.21	649.43	36.69	700.12	42.62	56
57	398.14	13.82	448.41	17.52	498.75	21.67	549.17	26.26	599.68	31.30	650.27	36.78	700.97	42.72	57
58	398.98	13.87	449.25	17.59	499.59	21.74	550.01	26.34	600.52	31.38	651.11	36.88	701.81	42.82	58
59	399.81	13.93	450.09	17.65	500.43	21.81	550.85	26.42	601.36	31.47	651.96	36.97	702.66	42.92	59

$$\Delta = \text{angle}$$
$$T = (5729.578)\tan\left(\frac{\Delta}{2}\right)$$
$$E = (5729.578)\left[\sec\left(\frac{\Delta}{2}\right) - 1\right]$$

For arc definition, calculate T and E for any degree of curve D by dividing both table values by D.

	$\Delta = 14°$ T	E	$\Delta = 15°$ T	E	$\Delta = 16°$ T	E	$\Delta = 17°$ T	E	$\Delta = 18°$ T	E	$\Delta = 19°$ T	E	$\Delta = 20°$ T	E	
0	703.50	43.03	754.31	49.44	805.24	56.31	856.29	63.63	907.48	71.42	958.80	79.67	1010.28	88.39	0
1	704.35	43.13	755.16	49.55	806.09	56.43	857.14	63.76	908.33	71.55	959.66	79.81	1011.14	88.54	1
2	705.20	43.23	756.01	49.66	806.94	56.55	858.00	63.88	909.18	71.69	960.52	79.95	1012.00	88.69	2
3	706.04	43.34	756.86	49.77	807.79	56.66	858.85	64.01	910.04	71.82	961.37	80.09	1012.86	88.84	3
4	706.89	43.44	757.70	49.88	808.64	56.78	859.70	64.14	910.89	71.96	962.23	80.24	1013.72	88.99	4
5	707.73	43.55	758.55	49.99	809.49	56.90	860.55	64.26	911.75	72.09	963.09	80.38	1014.58	89.14	5
6	708.58	43.65	759.40	50.11	810.34	57.02	861.40	64.39	912.60	72.22	963.94	80.52	1015.44	89.29	6
7	709.43	43.75	760.25	50.22	811.19	57.14	862.26	64.52	913.46	72.36	964.80	80.66	1016.29	89.44	7
8	710.27	43.86	761.10	50.33	812.04	57.26	863.11	64.64	914.31	72.49	965.66	80.81	1017.15	89.59	8
9	711.12	43.96	761.94	50.44	812.89	57.38	863.96	64.77	915.17	72.63	966.51	80.95	1018.01	89.74	9
10	711.96	44.07	762.79	50.55	813.74	57.50	864.81	64.90	916.02	72.76	967.37	81.09	1018.87	89.89	10
11	712.81	44.17	763.64	50.67	814.59	57.62	865.66	65.03	916.88	72.90	968.23	81.23	1019.73	90.04	11
12	713.66	44.27	764.49	50.78	815.44	57.74	866.52	65.15	917.73	73.03	969.09	81.38	1020.59	90.19	12
13	714.50	44.38	765.34	50.89	816.29	57.86	867.37	65.28	918.58	73.17	969.94	81.52	1021.45	90.34	13
14	715.35	44.48	766.19	51.00	817.14	57.98	868.22	65.41	919.44	73.30	970.80	81.66	1022.31	90.49	14
15	716.20	44.59	767.03	51.11	817.99	58.10	869.07	65.54	920.29	73.44	971.66	81.81	1023.17	90.64	15
16	717.04	44.69	767.88	51.23	818.84	58.22	869.93	65.66	921.15	73.57	972.51	81.95	1024.03	90.79	16
17	717.89	44.80	768.73	51.34	819.69	58.34	870.78	65.79	922.00	73.71	973.37	82.09	1024.89	90.94	17
18	718.73	44.90	769.58	51.45	820.54	58.46	871.63	65.92	922.86	73.85	974.23	82.24	1025.75	91.09	18
19	719.58	45.01	770.43	51.57	821.39	58.58	872.48	66.05	923.71	73.98	975.09	82.38	1026.61	91.25	19
20	720.43	45.11	771.28	51.68	822.24	58.70	873.34	66.18	924.57	74.12	975.94	82.52	1027.47	91.40	20
21	721.27	45.22	772.12	51.79	823.09	58.82	874.19	66.31	925.42	74.25	976.80	82.67	1028.33	91.55	21
22	722.12	45.33	772.97	51.91	823.94	58.94	875.04	66.43	926.28	74.39	977.66	82.81	1029.19	91.70	22
23	722.97	45.43	773.82	52.02	824.79	59.06	875.90	66.56	927.13	74.53	978.52	82.96	1030.05	91.85	23
24	723.81	45.54	774.67	52.13	825.64	59.18	876.75	66.69	927.99	74.66	979.37	83.10	1030.91	92.01	24
25	724.66	45.64	775.52	52.25	826.50	59.30	877.60	66.82	928.84	74.80	980.23	83.25	1031.77	92.16	25
26	725.51	45.75	776.37	52.36	827.35	59.43	878.45	66.95	929.70	74.94	981.09	83.39	1032.63	92.31	26
27	726.35	45.86	777.22	52.47	828.20	59.55	879.31	67.08	930.56	75.08	981.95	83.53	1033.49	92.46	27
28	727.20	45.96	778.06	52.59	829.05	59.67	880.16	67.21	931.41	75.21	982.81	83.68	1034.36	92.62	28
29	728.05	46.07	778.91	52.70	829.90	59.79	881.01	67.34	932.27	75.35	983.66	83.83	1035.22	92.77	29
30	728.89	46.18	779.76	52.82	830.75	59.91	881.87	67.47	933.12	75.49	984.52	83.97	1036.08	92.92	30
31	729.74	46.28	780.61	52.93	831.60	60.03	882.72	67.60	933.98	75.62	985.38	84.12	1036.94	93.08	31
32	730.59	46.39	781.46	53.05	832.45	60.16	883.57	67.73	934.83	75.76	986.24	84.26	1037.80	93.23	32
33	731.44	46.50	782.31	53.16	833.30	60.28	884.43	67.86	935.69	75.90	987.10	84.41	1038.66	93.38	33
34	732.28	46.61	783.16	53.28	834.15	60.40	885.28	67.99	936.54	76.04	987.95	84.55	1039.52	93.54	34
35	733.13	46.71	784.01	53.39	835.00	60.53	886.13	68.12	937.40	76.18	988.81	84.70	1040.38	93.69	35
36	733.98	46.82	784.85	53.51	835.86	60.65	886.99	68.25	938.25	76.31	989.67	84.84	1041.24	93.84	36
37	734.82	46.93	785.70	53.62	836.71	60.77	887.84	68.38	939.11	76.45	990.53	84.99	1042.10	94.00	37
38	735.67	47.04	786.55	53.74	837.56	60.89	888.69	68.51	939.97	76.59	991.39	85.14	1042.96	94.15	38
39	736.52	47.14	787.40	53.85	838.41	61.02	889.55	68.64	940.82	76.73	992.24	85.28	1043.82	94.31	39
40	737.36	47.25	788.25	53.97	839.26	61.14	890.40	68.77	941.68	76.87	993.10	85.43	1044.68	94.46	40
41	738.21	47.36	789.10	54.08	840.11	61.26	891.25	68.90	942.53	77.01	993.96	85.58	1045.55	94.62	41
42	739.06	47.47	789.95	54.20	840.96	61.39	892.11	69.03	943.39	77.15	994.82	85.72	1046.41	94.77	42
43	739.91	47.58	790.80	54.32	841.81	61.51	892.96	69.17	944.25	77.28	995.68	85.87	1047.27	94.93	43
44	740.75	47.69	791.65	54.43	842.66	61.63	893.81	69.30	945.10	77.42	996.54	86.02	1048.13	95.08	44
45	741.60	47.79	792.50	54.55	843.52	61.76	894.67	69.43	945.96	77.56	997.40	86.16	1048.99	95.23	45
46	742.45	47.90	793.35	54.66	844.37	61.88	895.52	69.56	946.81	77.70	998.25	86.31	1049.85	95.39	46
47	743.29	48.01	794.20	54.78	845.22	62.01	896.37	69.69	947.67	77.84	999.11	86.46	1050.71	95.54	47
48	744.14	48.12	795.04	54.90	846.07	62.13	897.23	69.83	948.53	77.98	999.97	86.61	1051.57	95.70	48
49	744.99	48.23	795.89	55.01	846.92	62.26	898.08	69.96	949.38	78.12	1000.83	86.75	1052.44	95.86	49
50	745.84	48.34	796.74	55.13	847.77	62.38	898.94	70.09	950.24	78.26	1001.69	86.90	1053.30	96.01	50
51	746.68	48.45	797.59	55.25	848.63	62.51	899.79	70.22	951.09	78.40	1002.55	87.05	1054.16	96.17	51
52	747.53	48.56	798.44	55.37	849.48	62.63	900.64	70.35	951.95	78.54	1003.41	87.20	1055.02	96.32	52
53	748.38	48.67	799.29	55.48	850.33	62.76	901.50	70.49	952.81	78.68	1004.27	87.35	1055.88	96.48	53
54	749.23	48.78	800.14	55.60	851.18	62.88	902.35	70.62	953.66	78.82	1005.12	87.49	1056.74	96.64	54
55	750.07	48.89	800.99	55.72	852.03	63.01	903.21	70.75	954.52	78.97	1005.98	87.64	1057.61	96.79	55
56	750.92	49.00	801.84	55.84	852.88	63.13	904.06	70.89	955.38	79.11	1006.84	87.79	1058.47	96.95	56
57	751.77	49.11	802.69	55.95	853.74	63.26	904.91	71.02	956.23	79.25	1007.70	87.94	1059.33	97.11	57
58	752.62	49.22	803.54	56.07	854.59	63.38	905.77	71.15	957.09	79.39	1008.56	88.09	1060.19	97.26	58
59	753.47	49.33	804.39	56.19	855.44	63.51	906.62	71.29	957.95	79.53	1009.42	88.24	1061.05	97.42	59

$$\Delta = \text{angle}$$
$$T = (5729.578)\tan\left(\frac{\Delta}{2}\right)$$
$$E = (5729.578)\left[\sec\left(\frac{\Delta}{2}\right) - 1\right]$$

For arc definition, calculate T and E for any degree of curve D by dividing both table values by D.

	$\Delta = 21°$		$\Delta = 22°$		$\Delta = 23°$		$\Delta = 24°$		$\Delta = 25°$		$\Delta = 26°$		$\Delta = 27°$		
	T	E	T	E	T	E	T	E	T	E	T	E	T	E	
0	1061.91	97.58	1113.72	107.24	1165.70	117.38	1217.86	128.00	1270.22	139.11	1322.78	150.71	1375.55	162.81	0
1	1062.78	97.73	1114.58	107.40	1166.56	117.55	1218.73	128.18	1271.09	139.30	1323.66	150.91	1376.43	163.01	1
2	1063.64	97.89	1115.45	107.57	1167.43	117.73	1219.60	128.36	1271.97	139.49	1324.53	151.11	1377.31	163.22	2
3	1064.50	98.05	1116.31	107.73	1168.30	117.90	1220.47	128.55	1272.84	139.68	1325.41	151.30	1378.19	163.42	3
4	1065.36	98.21	1117.18	107.90	1169.17	118.07	1221.34	128.73	1273.71	139.87	1326.29	151.50	1379.08	163.63	4
5	1066.22	98.36	1118.04	108.07	1170.04	118.25	1222.22	128.91	1274.59	140.06	1327.17	151.70	1379.96	163.84	5
6	1067.09	98.52	1118.91	108.23	1170.90	118.42	1223.09	129.09	1275.46	140.25	1328.04	151.90	1380.84	164.04	6
7	1067.95	98.68	1119.77	108.40	1171.77	118.59	1223.96	129.27	1276.34	140.44	1328.92	152.10	1381.72	164.25	7
8	1068.81	98.84	1120.64	108.56	1172.64	118.77	1224.83	129.45	1277.21	140.63	1329.80	152.30	1382.60	164.46	8
9	1069.67	99.00	1121.50	108.73	1173.51	118.94	1225.70	129.64	1278.09	140.82	1330.68	152.49	1383.48	164.66	9
10	1070.54	99.15	1122.37	108.90	1174.38	119.12	1226.57	129.82	1278.96	141.01	1331.56	152.69	1384.37	164.87	10
11	1071.40	99.31	1123.23	109.06	1175.25	119.29	1227.44	130.00	1279.84	141.20	1332.44	152.89	1385.25	165.08	11
12	1072.26	99.47	1124.10	109.23	1176.11	119.46	1228.32	130.18	1280.71	141.39	1333.31	153.09	1386.13	165.29	12
13	1073.12	99.63	1124.96	109.40	1176.98	119.64	1229.19	130.37	1281.59	141.58	1334.19	153.29	1387.01	165.49	13
14	1073.99	99.79	1125.83	109.56	1177.85	119.81	1230.06	130.55	1282.46	141.77	1335.07	153.49	1387.90	165.70	14
15	1074.85	99.95	1126.69	109.73	1178.72	119.99	1230.93	130.73	1283.34	141.96	1335.95	153.69	1388.78	165.91	15
16	1075.71	100.11	1127.56	109.90	1179.59	120.16	1231.80	130.92	1284.21	142.16	1336.83	153.89	1389.66	166.12	16
17	1076.57	100.27	1128.43	110.06	1180.46	120.34	1232.67	131.10	1285.09	142.35	1337.71	154.09	1390.54	166.33	17
18	1077.44	100.42	1129.29	110.23	1181.33	120.52	1233.55	131.28	1285.96	142.54	1338.59	154.29	1391.42	166.53	18
19	1078.30	100.58	1130.16	110.40	1182.19	120.69	1234.42	131.47	1286.84	142.73	1339.47	154.49	1392.31	166.74	19
20	1079.16	100.74	1131.02	110.57	1183.06	120.87	1235.29	131.65	1287.71	142.92	1340.34	154.69	1393.19	166.95	20
21	1080.03	100.90	1131.89	110.73	1183.93	121.04	1236.16	131.83	1288.59	143.12	1341.22	154.89	1394.07	167.16	21
22	1080.89	101.06	1132.76	110.90	1184.80	121.22	1237.03	132.02	1289.47	143.31	1342.10	155.09	1394.96	167.37	22
23	1081.75	101.22	1133.62	111.07	1185.67	121.39	1237.91	132.20	1290.34	143.50	1342.98	155.29	1395.84	167.58	23
24	1082.61	101.38	1134.49	111.24	1186.54	121.57	1238.78	132.39	1291.22	143.69	1343.86	155.49	1396.72	167.79	24
25	1083.48	101.54	1135.35	111.41	1187.41	121.75	1239.65	132.57	1292.09	143.88	1344.74	155.69	1397.60	167.99	25
26	1084.34	101.70	1136.22	111.57	1188.28	121.92	1240.52	132.76	1292.97	144.08	1345.62	155.89	1398.49	168.20	26
27	1085.20	101.87	1137.09	111.74	1189.15	122.10	1241.40	132.94	1293.84	144.27	1346.50	156.09	1399.37	168.41	27
28	1086.07	102.03	1137.95	111.91	1190.02	122.28	1242.27	133.13	1294.72	144.46	1347.38	156.29	1400.25	168.62	28
29	1086.93	102.19	1138.82	112.08	1190.88	122.45	1243.14	133.31	1295.60	144.66	1348.26	156.50	1401.14	168.83	29
30	1087.79	102.35	1139.68	112.25	1191.75	122.63	1244.01	133.50	1296.47	144.85	1349.14	156.70	1402.02	169.04	30
31	1088.66	102.51	1140.55	112.42	1192.62	122.81	1244.89	133.68	1297.35	145.04	1350.02	156.90	1402.90	169.25	31
32	1089.52	102.67	1141.42	112.59	1193.49	122.98	1245.76	133.87	1298.22	145.24	1350.90	157.10	1403.79	169.46	32
33	1090.38	102.83	1142.28	112.76	1194.36	123.16	1246.63	134.05	1299.10	145.43	1351.78	157.30	1404.67	169.67	33
34	1091.25	102.99	1143.15	112.93	1195.23	123.34	1247.50	134.24	1299.98	145.62	1352.66	157.50	1405.55	169.88	34
35	1092.11	103.15	1144.02	113.10	1196.10	123.52	1248.38	134.42	1300.85	145.82	1353.53	157.71	1406.44	170.09	35
36	1092.98	103.32	1144.88	113.27	1196.97	123.69	1249.25	134.61	1301.73	146.01	1354.41	157.91	1407.32	170.30	36
37	1093.84	103.48	1145.75	113.44	1197.84	123.87	1250.12	134.80	1302.60	146.21	1355.29	158.11	1408.20	170.51	37
38	1094.70	103.64	1146.62	113.61	1198.71	124.05	1251.00	134.98	1303.48	146.40	1356.17	158.31	1409.09	170.73	38
39	1095.57	103.80	1147.48	113.78	1199.58	124.23	1251.87	135.17	1304.36	146.59	1357.05	158.52	1409.97	170.94	39
40	1096.43	103.97	1148.35	113.95	1200.45	124.41	1252.74	135.35	1305.23	146.79	1357.93	158.72	1410.85	171.15	40
41	1097.29	104.13	1149.22	114.12	1201.32	124.59	1253.62	135.54	1306.11	146.98	1358.81	158.92	1411.74	171.36	41
42	1098.16	104.29	1150.08	114.29	1202.19	124.76	1254.49	135.73	1306.99	147.18	1359.70	159.13	1412.62	171.57	42
43	1099.02	104.46	1150.95	114.46	1203.06	124.94	1255.36	135.91	1307.86	147.37	1360.58	159.33	1413.51	171.78	43
44	1099.89	104.62	1151.82	114.63	1203.93	125.12	1256.24	136.10	1308.74	147.57	1361.46	159.53	1414.39	171.99	44
45	1100.75	104.78	1152.68	114.80	1204.80	125.30	1257.11	136.29	1309.62	147.76	1362.34	159.74	1415.27	172.21	45
46	1101.61	104.94	1153.55	114.97	1205.67	125.48	1257.98	136.48	1310.49	147.96	1363.22	159.94	1416.16	172.42	46
47	1102.48	105.10	1154.42	115.14	1206.54	125.66	1258.86	136.66	1311.37	148.16	1364.10	160.14	1417.04	172.63	47
48	1103.34	105.27	1155.29	115.31	1207.41	125.84	1259.73	136.85	1312.25	148.35	1364.98	160.35	1417.93	172.84	48
49	1104.21	105.43	1156.15	115.48	1208.28	126.02	1260.60	137.04	1313.13	148.55	1365.86	160.55	1418.81	173.06	49
50	1105.07	105.60	1157.02	115.66	1209.15	126.20	1261.48	137.23	1314.00	148.74	1366.74	160.76	1419.70	173.27	50
51	1105.94	105.76	1157.89	115.83	1210.02	126.38	1262.35	137.41	1314.88	148.94	1367.62	160.96	1420.58	173.48	51
52	1106.80	105.92	1158.76	116.00	1210.89	126.56	1263.22	137.60	1315.76	149.14	1368.50	161.17	1421.47	173.69	52
53	1107.66	106.09	1159.62	116.17	1211.76	126.74	1264.10	137.79	1316.63	149.33	1369.38	161.37	1422.35	173.91	53
54	1108.53	106.25	1160.49	116.34	1212.63	126.92	1264.97	137.98	1317.51	149.53	1370.26	161.57	1423.23	174.12	54
55	1109.39	106.41	1161.36	116.52	1213.51	127.10	1265.85	138.17	1318.39	149.73	1371.14	161.78	1424.12	174.33	55
56	1110.26	106.58	1162.22	116.69	1214.38	127.28	1266.72	138.36	1319.27	149.92	1372.02	161.99	1425.00	174.55	56
57	1111.12	106.74	1163.09	116.86	1215.25	127.46	1267.59	138.54	1320.14	150.12	1372.91	162.19	1425.89	174.76	57
58	1111.99	106.91	1163.96	117.03	1216.12	127.64	1268.47	138.73	1321.02	150.32	1373.79	162.40	1426.77	174.98	58
59	1112.85	107.07	1164.83	117.21	1216.99	127.82	1269.34	138.92	1321.90	150.51	1374.67	162.60	1427.66	175.19	59

$$\Delta = \text{angle}$$
$$T = (5729.578)\tan\left(\frac{\Delta}{2}\right)$$
$$E = (5729.578)\left[\sec\left(\frac{\Delta}{2}\right) - 1\right]$$

For arc definition, calculate T and E for any degree of curve D by dividing both table values by D.

	$\Delta=28°$ T	E	$\Delta=29°$ T	E	$\Delta=30°$ T	E	$\Delta=31°$ T	E	$\Delta=32°$ T	E	$\Delta=33°$ T	E	$\Delta=34°$ T	E	
0	1428.54	175.40	1481.77	188.51	1535.24	202.12	1588.95	216.25	1642.93	230.90	1697.18	246.08	1751.71	261.79	0
1	1429.43	175.62	1482.66	188.73	1536.13	202.35	1589.85	216.49	1643.83	231.15	1698.08	246.34	1752.62	262.06	1
2	1430.31	175.83	1483.55	188.95	1537.02	202.58	1590.75	216.73	1644.73	231.40	1698.99	246.59	1753.53	262.33	2
3	1431.20	176.05	1484.44	189.17	1537.92	202.81	1591.65	216.97	1645.64	231.64	1699.90	246.85	1754.44	262.59	3
4	1432.09	176.26	1485.33	189.40	1538.81	203.04	1592.54	217.21	1646.54	231.89	1700.80	247.11	1755.35	262.86	4
5	1432.97	176.48	1486.22	189.62	1539.70	203.27	1593.44	217.45	1647.44	232.14	1701.71	247.37	1756.26	263.13	5
6	1433.86	176.69	1487.11	189.84	1540.60	203.51	1594.34	217.69	1648.34	232.39	1702.62	247.63	1757.18	263.40	6
7	1434.74	176.91	1487.99	190.07	1541.49	203.74	1595.24	217.93	1649.24	232.64	1703.53	247.88	1758.09	263.66	7
8	1435.63	177.12	1488.88	190.29	1542.38	203.97	1596.13	218.17	1650.15	232.89	1704.43	248.14	1759.00	263.93	8
9	1436.51	177.34	1489.77	190.51	1543.28	204.20	1597.03	218.41	1651.05	233.14	1705.34	248.40	1759.91	264.20	9
10	1437.40	177.55	1490.66	190.74	1544.17	204.44	1597.93	218.65	1651.95	233.39	1706.25	248.66	1760.82	264.47	10
11	1438.28	177.77	1491.55	190.96	1545.06	204.67	1598.83	218.89	1652.85	233.64	1707.15	248.92	1761.74	264.73	11
12	1439.17	177.98	1492.44	191.19	1545.96	204.90	1599.73	219.14	1653.76	233.89	1708.06	249.18	1762.65	265.00	12
13	1440.06	178.20	1493.33	191.41	1546.85	205.13	1600.63	219.38	1654.66	234.14	1708.97	249.44	1763.56	265.27	13
14	1440.94	178.41	1494.22	191.64	1547.75	205.37	1601.52	219.62	1655.56	234.39	1709.88	249.70	1764.47	265.54	14
15	1441.83	178.63	1495.11	191.86	1548.64	205.60	1602.42	219.86	1656.47	234.64	1710.78	249.96	1765.39	265.81	15
16	1442.71	178.85	1496.00	192.08	1549.54	205.83	1603.32	220.10	1657.37	234.89	1711.69	250.22	1766.30	266.08	16
17	1443.60	179.06	1496.89	192.31	1550.43	206.07	1604.22	220.35	1658.27	235.15	1712.60	250.48	1767.21	266.35	17
18	1444.49	179.28	1497.78	192.53	1551.32	206.30	1605.12	220.59	1659.18	235.40	1713.51	250.74	1768.12	266.61	18
19	1445.37	179.50	1498.67	192.76	1552.22	206.54	1606.02	220.83	1660.08	235.65	1714.42	251.00	1769.04	266.88	19
20	1446.26	179.71	1499.56	192.98	1553.11	206.77	1606.92	221.07	1660.98	235.90	1715.32	251.26	1769.95	267.15	20
21	1447.15	179.93	1500.45	193.21	1554.01	207.00	1607.81	221.32	1661.89	236.15	1716.23	251.52	1770.86	267.42	21
22	1448.03	180.15	1501.35	193.44	1554.90	207.24	1608.71	221.56	1662.79	236.40	1717.14	251.78	1771.77	267.69	22
23	1448.92	180.36	1502.24	193.66	1555.80	207.47	1609.61	221.80	1663.69	236.66	1718.05	252.04	1772.69	267.96	23
24	1449.81	180.58	1503.13	193.89	1556.69	207.71	1610.51	222.04	1664.60	236.91	1718.96	252.30	1773.60	268.23	24
25	1450.69	180.80	1504.02	194.11	1557.59	207.94	1611.41	222.29	1665.50	237.16	1719.86	252.56	1774.51	268.50	25
26	1451.58	181.02	1504.91	194.34	1558.48	208.18	1612.31	222.53	1666.40	237.41	1720.77	252.82	1775.43	268.77	26
27	1452.47	181.24	1505.80	194.57	1559.38	208.41	1613.21	222.77	1667.31	237.66	1721.68	253.08	1776.34	269.04	27
28	1453.35	181.45	1506.69	194.79	1560.27	208.65	1614.11	223.02	1668.21	237.92	1722.59	253.35	1777.25	269.31	28
29	1454.24	181.67	1507.58	195.02	1561.17	208.88	1615.01	223.26	1669.12	238.17	1723.50	253.61	1778.17	269.58	29
30	1455.13	181.89	1508.47	195.25	1562.06	209.12	1615.91	223.51	1670.02	238.42	1724.41	253.87	1779.08	269.85	30
31	1456.01	182.11	1509.36	195.47	1562.96	209.35	1616.81	223.75	1670.92	238.68	1725.32	254.13	1779.99	270.13	31
32	1456.90	182.33	1510.25	195.70	1563.85	209.59	1617.71	224.00	1671.83	238.93	1726.22	254.39	1780.91	270.40	32
33	1457.79	182.55	1511.15	195.93	1564.75	209.82	1618.61	224.24	1672.73	239.18	1727.13	254.66	1781.82	270.67	33
34	1458.68	182.76	1512.04	196.16	1565.64	210.06	1619.51	224.49	1673.64	239.44	1728.04	254.92	1782.74	270.94	34
35	1459.56	182.98	1512.93	196.38	1566.54	210.30	1620.41	224.73	1674.54	239.69	1728.95	255.18	1783.65	271.21	35
36	1460.45	183.20	1513.82	196.61	1567.44	210.53	1621.31	224.97	1675.45	239.94	1729.86	255.44	1784.56	271.48	36
37	1461.34	183.42	1514.71	196.84	1568.33	210.77	1622.21	225.22	1676.35	240.20	1730.77	255.71	1785.48	271.76	37
38	1462.23	183.64	1515.60	197.07	1569.23	211.01	1623.11	225.47	1677.26	240.45	1731.68	255.97	1786.39	272.03	38
39	1463.11	183.86	1516.49	197.29	1570.12	211.24	1624.01	225.71	1678.16	240.71	1732.59	256.23	1787.31	272.30	39
40	1464.00	184.08	1517.39	197.52	1571.02	211.48	1624.91	225.96	1679.06	240.96	1733.50	256.50	1788.22	272.57	40
41	1464.89	184.30	1518.28	197.75	1571.91	211.72	1625.81	226.20	1679.97	241.21	1734.41	256.76	1789.14	272.84	41
42	1465.78	184.52	1519.17	197.98	1572.81	211.95	1626.71	226.45	1680.88	241.47	1735.32	257.02	1790.05	273.12	42
43	1466.67	184.74	1520.06	198.21	1573.71	212.19	1627.61	226.69	1681.78	241.72	1736.23	257.29	1790.97	273.39	43
44	1467.55	184.96	1520.95	198.44	1574.60	212.43	1628.51	226.94	1682.69	241.98	1737.14	257.55	1791.88	273.66	44
45	1468.44	185.18	1521.85	198.67	1575.50	212.67	1629.41	227.19	1683.59	242.23	1738.05	257.82	1792.80	273.94	45
46	1469.33	185.40	1522.74	198.90	1576.40	212.90	1630.31	227.43	1684.50	242.49	1738.96	258.08	1793.71	274.21	46
47	1470.22	185.62	1523.63	199.12	1577.29	213.14	1631.21	227.68	1685.40	242.74	1739.87	258.34	1794.63	274.48	47
48	1471.11	185.84	1524.52	199.35	1578.19	213.38	1632.11	227.93	1686.31	243.00	1740.78	258.61	1795.54	274.76	48
49	1471.99	186.06	1525.42	199.58	1579.09	213.62	1633.01	228.17	1687.21	243.26	1741.69	258.87	1796.46	275.03	49
50	1472.88	186.29	1526.31	199.81	1579.98	213.86	1633.92	228.42	1688.12	243.51	1742.60	259.14	1797.37	275.30	50
51	1473.77	186.51	1527.20	200.04	1580.88	214.09	1634.82	228.67	1689.02	243.77	1743.51	259.40	1798.29	275.58	51
52	1474.66	186.73	1528.09	200.27	1581.78	214.33	1635.72	228.92	1689.93	244.02	1744.42	259.67	1799.20	275.85	52
53	1475.55	186.95	1528.99	200.50	1582.67	214.57	1636.62	229.16	1690.84	244.28	1745.33	259.93	1800.12	276.13	53
54	1476.44	187.17	1529.88	200.73	1583.57	214.81	1637.52	229.41	1691.74	244.54	1746.24	260.20	1801.03	276.40	54
55	1477.32	187.39	1530.77	200.96	1584.47	215.05	1638.42	229.66	1692.65	244.79	1747.15	260.46	1801.95	276.68	55
56	1478.21	187.62	1531.66	201.19	1585.36	215.29	1639.32	229.91	1693.55	245.05	1748.06	260.73	1802.87	276.95	56
57	1479.10	187.84	1532.56	201.42	1586.26	215.53	1640.22	230.15	1694.46	245.31	1748.97	261.00	1803.78	277.23	57
58	1479.99	188.06	1533.45	201.66	1587.16	215.77	1641.13	230.40	1695.37	245.56	1749.89	261.26	1804.70	277.50	58
59	1480.88	188.28	1534.34	201.89	1588.06	216.01	1642.03	230.65	1696.27	245.82	1750.80	261.53	1805.61	277.78	59

$$\Delta = \text{angle}$$
$$T = (5729.578)\tan\left(\frac{\Delta}{2}\right)$$
$$E = (5729.578)\left[\sec\left(\frac{\Delta}{2}\right) - 1\right]$$

For arc definition, calculate T and E for any degree of curve D by dividing both table values by D.

	Δ = 35°		Δ = 36°		Δ = 37°		Δ = 38°		Δ = 39°		Δ = 40°		Δ = 41°		
	T	E	T	E	T	E	T	E	T	E	T	E	T	E	
0	1806.53	278.05	1861.65	294.86	1917.09	312.22	1972.85	330.14	2028.95	348.64	2085.40	367.71	2142.20	387.37	0
1	1807.45	278.33	1862.57	295.14	1918.02	312.51	1973.78	330.45	2029.89	348.95	2086.34	368.03	2143.15	387.71	1
2	1808.36	278.60	1863.50	295.43	1918.94	312.81	1974.72	330.75	2030.83	349.26	2087.28	368.36	2144.10	388.04	2
3	1809.28	278.88	1864.42	295.71	1919.87	313.10	1975.65	331.05	2031.76	349.58	2088.23	368.68	2145.05	388.37	3
4	1810.19	279.15	1865.34	296.00	1920.80	313.40	1976.58	331.36	2032.70	349.89	2089.17	369.00	2146.00	388.70	4
5	1811.11	279.43	1866.26	296.28	1921.72	313.69	1977.51	331.66	2033.64	350.20	2090.12	369.33	2146.95	389.04	5
6	1812.03	279.71	1867.18	296.57	1922.65	313.98	1978.45	331.97	2034.58	350.52	2091.06	369.65	2147.90	389.37	6
7	1812.94	279.98	1868.10	296.85	1923.58	314.28	1979.38	332.27	2035.52	350.83	2092.00	369.97	2148.85	389.71	7
8	1813.86	280.26	1869.03	297.14	1924.51	314.58	1980.31	332.57	2036.46	351.15	2092.95	370.30	2149.80	390.04	8
9	1814.78	280.54	1869.95	297.43	1925.43	314.87	1981.24	332.88	2037.39	351.46	2093.89	370.62	2150.75	390.37	9
10	1815.70	280.81	1870.87	297.71	1926.36	315.17	1982.18	333.18	2038.33	351.78	2094.84	370.95	2151.70	390.71	10
11	1816.61	281.09	1871.79	298.00	1927.29	315.46	1983.11	333.49	2039.27	352.09	2095.78	371.27	2152.66	391.04	11
12	1817.53	281.37	1872.71	298.28	1928.22	315.76	1984.04	333.80	2040.21	352.41	2096.73	371.60	2153.61	391.38	12
13	1818.45	281.65	1873.64	298.57	1929.14	316.05	1984.98	334.10	2041.15	352.72	2097.67	371.92	2154.56	391.71	13
14	1819.36	281.92	1874.56	298.86	1930.07	316.35	1985.91	334.41	2042.09	353.04	2098.62	372.25	2155.51	392.05	14
15	1820.28	282.20	1875.48	299.14	1931.00	316.65	1986.84	334.71	2043.03	353.35	2099.56	372.57	2156.46	392.38	15
16	1821.20	282.48	1876.41	299.43	1931.93	316.94	1987.78	335.02	2043.97	353.67	2100.51	372.90	2157.41	392.72	16
17	1822.12	282.76	1877.33	299.72	1932.86	317.24	1988.71	335.32	2044.91	353.98	2101.45	373.22	2158.36	393.05	17
18	1823.03	283.03	1878.25	300.01	1933.78	317.53	1989.65	335.63	2045.85	354.30	2102.40	373.55	2159.31	393.39	18
19	1823.95	283.31	1879.17	300.29	1934.71	317.83	1990.58	335.94	2046.79	354.61	2103.34	373.87	2160.27	393.72	19
20	1824.87	283.59	1880.10	300.58	1935.64	318.13	1991.51	336.24	2047.73	354.93	2104.29	374.20	2161.22	394.06	20
21	1825.79	283.87	1881.02	300.87	1936.57	318.43	1992.45	336.55	2048.67	355.25	2105.24	374.53	2162.17	394.39	21
22	1826.71	284.15	1881.94	301.16	1937.50	318.72	1993.38	336.86	2049.61	355.56	2106.18	374.85	2163.12	394.73	22
23	1827.62	284.43	1882.87	301.45	1938.43	319.02	1994.32	337.16	2050.55	355.88	2107.13	375.18	2164.07	395.07	23
24	1828.54	284.71	1883.79	301.73	1939.36	319.32	1995.25	337.47	2051.49	356.20	2108.07	375.50	2165.03	395.40	24
25	1829.46	284.99	1884.71	302.02	1940.29	319.62	1996.18	337.78	2052.43	356.51	2109.02	375.83	2165.98	395.74	25
26	1830.38	285.27	1885.64	302.31	1941.21	319.92	1997.12	338.09	2053.37	356.83	2109.97	376.16	2166.93	396.08	26
27	1831.30	285.54	1886.56	302.60	1942.14	320.21	1998.05	338.39	2054.31	357.15	2110.91	376.49	2167.88	396.41	27
28	1832.22	285.83	1887.48	302.89	1943.07	320.51	1998.99	338.70	2055.25	357.47	2111.86	376.81	2168.84	396.75	28
29	1833.13	286.10	1888.41	303.18	1944.00	320.81	1999.92	339.01	2056.19	357.78	2112.81	377.14	2169.79	397.09	29
30	1834.05	286.39	1889.33	303.47	1944.93	321.11	2000.86	339.32	2057.13	358.10	2113.75	377.47	2170.74	397.43	30
31	1834.97	286.67	1890.26	303.76	1945.86	321.41	2001.79	339.63	2058.07	358.42	2114.70	377.80	2171.70	397.76	31
32	1835.89	286.95	1891.18	304.05	1946.79	321.71	2002.73	339.93	2059.01	358.74	2115.65	378.12	2172.65	398.10	32
33	1836.81	287.23	1892.10	304.34	1947.72	322.01	2003.66	340.24	2059.95	359.06	2116.59	378.45	2173.60	398.44	33
34	1837.73	287.51	1893.03	304.63	1948.65	322.31	2004.60	340.55	2060.89	359.37	2117.54	378.78	2174.56	398.78	34
35	1838.65	287.79	1893.95	304.92	1949.58	322.60	2005.53	340.86	2061.83	359.69	2118.49	379.11	2175.51	399.12	35
36	1839.57	288.07	1894.88	305.21	1950.51	322.90	2006.47	341.17	2062.78	360.01	2119.44	379.44	2176.46	399.46	36
37	1840.49	288.35	1895.80	305.50	1951.44	323.20	2007.41	341.48	2063.72	360.33	2120.38	379.77	2177.42	399.79	37
38	1841.41	288.63	1896.73	305.79	1952.37	323.50	2008.34	341.79	2064.66	360.65	2121.33	380.10	2178.37	400.13	38
39	1842.32	288.91	1897.65	306.08	1953.30	323.80	2009.28	342.10	2065.60	360.97	2122.28	380.42	2179.32	400.47	39
40	1843.24	289.19	1898.58	306.37	1954.23	324.10	2010.21	342.41	2066.54	361.29	2123.23	380.75	2180.28	400.81	40
41	1844.16	289.48	1899.50	306.66	1955.16	324.40	2011.15	342.72	2067.48	361.61	2124.17	381.08	2181.23	401.15	41
42	1845.08	289.76	1900.43	306.95	1956.09	324.71	2012.08	343.03	2068.42	361.93	2125.12	381.41	2182.19	401.49	42
43	1846.00	290.04	1901.35	307.24	1957.02	325.01	2013.02	343.34	2069.37	362.25	2126.07	381.74	2183.14	401.83	43
44	1846.92	290.32	1902.28	307.53	1957.95	325.31	2013.96	343.65	2070.31	362.57	2127.02	382.07	2184.09	402.17	44
45	1847.84	290.60	1903.20	307.83	1958.88	325.61	2014.89	343.96	2071.25	362.89	2127.97	382.40	2185.05	402.51	45
46	1848.76	290.89	1904.13	308.12	1959.81	325.91	2015.83	344.27	2072.19	363.21	2128.91	382.73	2186.00	402.85	46
47	1849.68	291.17	1905.05	308.41	1960.74	326.21	2016.77	344.58	2073.14	363.53	2129.86	383.06	2186.96	403.19	47
48	1850.60	291.45	1905.98	308.70	1961.67	326.51	2017.70	344.89	2074.08	363.85	2130.81	383.39	2187.91	403.53	48
49	1851.52	291.73	1906.90	308.99	1962.60	326.81	2018.64	345.20	2075.02	364.17	2131.76	383.72	2188.87	403.87	49
50	1852.44	292.02	1907.83	309.29	1963.54	327.11	2019.58	345.51	2075.96	364.49	2132.71	384.05	2189.82	404.21	50
51	1853.36	292.30	1908.75	309.58	1964.47	327.42	2020.51	345.83	2076.91	364.81	2133.66	384.39	2190.78	404.55	51
52	1854.29	292.58	1909.68	309.87	1965.40	327.72	2021.45	346.14	2077.85	365.13	2134.61	384.72	2191.73	404.89	52
53	1855.21	292.87	1910.61	310.16	1966.33	328.02	2022.39	346.45	2078.79	365.46	2135.56	385.05	2192.69	405.24	53
54	1856.13	293.15	1911.53	310.46	1967.26	328.32	2023.32	346.76	2079.74	365.78	2136.50	385.38	2193.64	405.58	54
55	1857.05	293.44	1912.46	310.75	1968.19	328.63	2024.26	347.07	2080.68	366.10	2137.45	385.71	2194.60	405.92	55
56	1857.97	293.72	1913.38	311.04	1969.12	328.93	2025.20	347.39	2081.62	366.42	2138.40	386.04	2195.56	406.26	56
57	1858.89	294.00	1914.31	311.34	1970.06	329.23	2026.14	347.70	2082.57	366.74	2139.35	386.38	2196.51	406.60	57
58	1859.81	294.29	1915.24	311.63	1970.99	329.53	2027.07	348.01	2083.51	367.07	2140.30	386.71	2197.47	406.95	58
59	1860.73	294.57	1916.16	311.92	1971.92	329.84	2028.01	348.32	2084.45	367.39	2141.25	387.04	2198.42	407.29	59

APPENDIX

$$\Delta = \text{angle}$$
$$T = (5729.578)\tan\left(\frac{\Delta}{2}\right)$$
$$E = (5729.578)\left[\sec\left(\frac{\Delta}{2}\right) - 1\right]$$

For arc definition, calculate T and E for any degree of curve D by dividing both table values by D.

	Δ = 42° T	E	Δ = 43° T	E	Δ = 44° T	E	Δ = 45° T	E	
0	2199.38	407.63	2256.94	428.49	2314.90	449.97	2373.27	472.07	0
1	2200.34	407.97	2257.90	428.85	2315.87	450.33	2374.25	472.45	1
2	2201.29	408.32	2258.87	429.20	2316.84	450.70	2375.22	472.82	2
3	2202.25	408.66	2259.83	429.55	2317.81	451.06	2376.20	473.19	3
4	2203.20	409.00	2260.79	429.91	2318.78	451.42	2377.18	473.57	4
5	2204.16	409.35	2261.76	430.26	2319.75	451.79	2378.15	473.94	5
6	2205.12	409.69	2262.72	430.61	2320.72	452.15	2379.13	474.32	6
7	2206.07	410.03	2263.68	430.97	2321.69	452.52	2380.11	474.69	7
8	2207.03	410.38	2264.65	431.32	2322.66	452.88	2381.08	475.07	8
9	2207.99	410.72	2265.61	431.68	2323.63	453.25	2382.06	475.44	9
10	2208.95	411.06	2266.57	432.03	2324.60	453.61	2383.04	475.82	10
11	2209.90	411.41	2267.54	432.38	2325.57	453.98	2384.02	476.19	11
12	2210.86	411.75	2268.50	432.74	2326.54	454.34	2384.99	476.57	12
13	2211.82	412.10	2269.46	433.09	2327.51	454.71	2385.97	476.94	13
14	2212.78	412.44	2270.43	433.45	2328.48	455.07	2386.95	477.32	14
15	2213.73	412.79	2271.39	433.81	2329.45	455.44	2387.93	477.70	15
16	2214.69	413.13	2272.36	434.16	2330.42	455.80	2388.90	478.07	16
17	2215.65	413.48	2273.32	434.52	2331.40	456.17	2389.88	478.45	17
18	2216.61	413.83	2274.29	434.87	2332.37	456.54	2390.86	478.83	18
19	2217.56	414.17	2275.25	435.23	2333.34	456.90	2391.84	479.20	19
20	2218.52	414.52	2276.22	435.58	2334.31	457.27	2392.82	479.58	20
21	2219.48	414.86	2277.18	435.94	2335.28	457.63	2393.80	479.96	21
22	2220.44	415.21	2278.15	436.30	2336.25	458.00	2394.78	480.34	22
23	2221.40	415.56	2279.11	436.65	2337.23	458.37	2395.76	480.71	23
24	2222.36	415.90	2280.08	437.01	2338.20	458.74	2396.73	481.09	24
25	2223.32	416.25	2281.04	437.37	2339.17	459.10	2397.71	481.47	25
26	2224.27	416.60	2282.01	437.72	2340.14	459.47	2398.69	481.85	26
27	2225.23	416.94	2282.97	438.08	2341.11	459.84	2399.67	482.22	27
28	2226.19	417.29	2283.94	438.44	2342.09	460.21	2400.65	482.60	28
29	2227.15	417.64	2284.90	438.80	2343.06	460.57	2401.63	482.98	29
30	2228.11	417.99	2285.87	439.15	2344.03	460.94	2402.61	483.36	30
31	2229.07	418.33	2286.84	439.51	2345.01	461.31	2403.59	483.74	31
32	2230.03	418.68	2287.80	439.87	2345.98	461.68	2404.57	484.12	32
33	2230.99	419.03	2288.77	440.23	2346.95	462.05	2405.55	484.50	33
34	2231.95	419.38	2289.73	440.59	2347.92	462.42	2406.53	484.88	34
35	2232.91	419.73	2290.70	440.95	2348.90	462.79	2407.51	485.26	35
36	2233.87	420.08	2291.67	441.31	2349.87	463.16	2408.49	485.64	36
37	2234.83	420.42	2292.63	441.66	2350.84	463.53	2409.47	486.02	37
38	2235.79	420.77	2293.60	442.02	2351.82	463.89	2410.45	486.40	38
39	2236.75	421.12	2294.57	442.38	2352.79	464.27	2411.44	486.78	39
40	2237.71	421.47	2295.54	442.74	2353.77	464.64	2412.42	487.16	40
41	2238.67	421.82	2296.50	443.10	2354.74	465.00	2413.40	487.54	41
42	2239.63	422.17	2297.47	443.46	2355.71	465.38	2414.38	487.92	42
43	2240.59	422.52	2298.44	443.82	2356.69	465.75	2415.36	488.30	43
44	2241.55	422.87	2299.40	444.18	2357.66	466.12	2416.34	488.68	44
45	2242.51	423.22	2300.37	444.54	2358.64	466.49	2417.32	489.06	45
46	2243.47	423.57	2301.34	444.90	2359.61	466.86	2418.30	489.45	46
47	2244.44	423.92	2302.31	445.26	2360.59	467.23	2419.29	489.83	47
48	2245.40	424.27	2303.28	445.63	2361.56	467.60	2420.27	490.21	48
49	2246.36	424.62	2304.24	445.99	2362.54	467.97	2421.25	490.59	49
50	2247.32	424.97	2305.21	446.35	2363.51	468.34	2422.23	490.97	50
51	2248.28	425.32	2306.18	446.71	2364.49	468.72	2423.22	491.36	51
52	2249.24	425.68	2307.15	447.07	2365.46	469.09	2424.20	491.74	52
53	2250.21	426.03	2308.12	447.43	2366.44	469.46	2425.18	492.12	53
54	2251.17	426.38	2309.09	447.79	2367.41	469.83	2426.16	492.51	54
55	2252.13	426.73	2310.05	448.16	2368.39	470.21	2427.15	492.89	55
56	2253.09	427.08	2311.02	448.52	2369.36	470.58	2428.13	493.27	56
57	2254.05	427.44	2311.99	448.88	2370.34	470.95	2429.11	493.66	57
58	2255.02	427.79	2312.96	449.24	2371.32	471.32	2430.09	494.04	58
59	2255.98	428.14	2313.93	449.61	2372.29	471.70	2431.08	494.42	59

Appendix C: Radius When Degree of Curve is Known

$$R = \frac{5729.578}{D}$$

D	radius	D	radius	D	radius	D	radius	D	radius
0°00′		9°00′	636.62	18°00′	318.31	27°00′	212.21	36°00′	159.15
15′	22918.31	15′	619.41	15′	313.95	15′	210.26	15′	158.06
30′	11459.16	30′	603.11	30′	309.71	30′	208.35	30′	156.97
45′	7639.44	45′	587.65	45′	305.58	45′	206.47	45′	155.91
1°00′	5729.58	10°00′	572.96	19°00′	301.56	28°00′	204.63	37°00′	154.85
15′	4583.66	15′	558.98	15′	297.64	15′	202.82	15′	153.81
30′	3819.72	30′	545.67	30′	293.82	30′	201.04	30′	152.79
45′	3274.04	45′	532.98	45′	290.11	45′	199.29	45′	151.78
2°00′	2864.79	11°00′	520.87	20°00′	286.48	29°00′	197.57	38°00′	150.78
15′	2546.48	15′	509.30	15′	282.94	15′	195.88	15′	149.79
30′	2291.83	30′	498.22	30′	279.49	30′	194.22	30′	148.82
45′	2083.48	45′	487.62	45′	276.12	45′	192.59	45′	147.86
3°00′	1909.86	12°00′	477.46	21°00′	272.84	30°00′	190.99	39°00′	146.91
15′	1762.95	15′	467.72	15′	269.63	15′	189.41	15′	145.98
30′	1637.02	30′	458.37	30′	266.49	30′	187.86	30′	145.05
45′	1527.89	45′	449.38	45′	263.43	45′	186.33	45′	144.14
4°00′	1432.39	13°00′	440.74	22°00′	260.44	31°00′	184.83	40°00′	143.24
15′	1348.14	15′	432.42	15′	257.51	15′	183.35	15′	142.35
30′	1273.24	30′	424.41	30′	254.65	30′	181.89	30′	141.47
45′	1206.23	45′	416.70	45′	251.85	45′	180.46	45′	140.60
5°00′	1145.92	14°00′	409.26	23°00′	249.11	32°00′	179.05	41°00′	139.75
15′	1091.35	15′	402.08	15′	246.43	15′	177.66	15′	138.90
30′	1041.74	30′	395.14	30′	243.81	30′	176.29	30′	138.06
45′	996.45	45′	388.45	45′	241.25	45′	174.95	45′	137.24
6°00′	954.93	15°00′	381.97	24°00′	238.73	33°00′	173.62	42°00′	136.42
15′	916.73	15′	375.71	15′	236.27	15′	172.32	15′	135.61
30′	881.47	30′	369.65	30′	233.86	30′	171.03	30′	134.81
45′	848.83	45′	363.78	45′	231.50	45′	169.77	45′	134.03
7°00′	818.51	16°00′	358.10	25°00′	229.18	34°00′	168.52	43°00′	133.25
15′	790.29	15′	352.59	15′	226.91	15′	167.29	15′	132.48
30′	763.94	30′	347.25	30′	224.69	30′	166.07	30′	131.71
45′	739.30	45′	342.06	45′	222.51	45′	164.88	45′	130.96
8°00′	716.20	17°00′	337.03	26°00′	220.37	35°00′	163.70	44°00′	130.22
15′	694.49	15′	332.15	15′	218.27	15′	162.54	15′	129.48
30′	674.07	30′	327.40	30′	216.21	30′	161.40	30′	128.75
45′	654.81	45′	322.79	45′	214.19	45′	160.27	45′	128.04
9°00′	636.62	18°00′	318.31	27°00′	212.21	36°00′	159.15	45°00′	127.32
15′	619.41	15′	313.95	15′	210.26	15′	158.06	15′	126.62
30′	603.11	30′	309.71	30′	208.35	30′	156.97	30′	125.92
45′	587.65	45′	305.58	45′	206.47	45′	155.91	45′	125.24

PROFESSIONAL PUBLICATIONS, INC. ● P.O. Box 199, San Carlos, CA 94070

$$R = \frac{5729.578}{D}$$

D	radius	D	radius	D	radius	D	radius	D	radius
45°00′		54°00′	106.10	63°00′	90.95	72°00′	79.58	81°00′	70.74
15′	126.62	15′	105.61	15′	90.59	15′	79.30	15′	70.52
30′	125.92	30′	105.13	30′	90.23	30′	79.03	30′	70.30
45′	125.24	45′	104.65	45′	89.88	45′	78.76	45′	70.09
46°00′	124.56	55°00′	104.17	64°00′	89.52	73°00′	78.49	82°00′	69.87
15′	123.88	15′	103.70	15′	89.18	15′	78.22	15′	69.66
30′	123.22	30′	103.24	30′	88.83	30′	77.95	30′	69.45
45′	122.56	45′	102.77	45′	88.49	45′	77.69	45′	69.24
47°00′	121.91	56°00′	102.31	65°00′	88.15	74°00′	77.43	83°00′	69.03
15′	121.26	15′	101.86	15′	87.81	15′	77.17	15′	68.82
30′	120.62	30′	101.41	30′	87.47	30′	76.91	30′	68.62
45′	119.99	45′	100.96	45′	87.14	45′	76.65	45′	68.41
48°00′	119.37	57°00′	100.52	66°00′	86.81	75°00′	76.39	84°00′	68.21
15′	118.75	15′	100.08	15′	86.48	15′	76.14	15′	68.01
30′	118.14	30′	99.64	30′	86.16	30′	75.89	30′	67.81
45′	117.53	45′	99.21	45′	85.84	45′	75.64	45′	67.61
49°00′	116.93	58°00′	98.79	67°00′	85.52	76°00′	75.39	85°00′	67.41
15′	116.34	15′	98.36	15′	85.20	15′	75.14	15′	67.21
30′	115.75	30′	97.94	30′	84.88	30′	74.90	30′	67.01
45′	115.17	45′	97.52	45′	84.57	45′	74.65	45′	66.82
50°00′	114.59	59°00′	97.11	68°00′	84.26	77°00′	74.41	86°00′	66.62
15′	114.02	15′	96.70	15′	83.95	15′	74.17	15′	66.43
30′	113.46	30′	96.30	30′	83.64	30′	73.93	30′	66.24
45′	112.90	45′	95.89	45′	83.34	45′	73.69	45′	66.05
51°00′	112.34	60°00′	95.49	69°00′	83.04	78°00′	73.46	87°00′	65.86
15′	111.80	15′	95.10	15′	82.74	15′	73.22	15′	65.67
30′	111.25	30′	94.70	30′	82.44	30′	72.99	30′	65.48
45′	110.72	45′	94.31	45′	82.14	45′	72.76	45′	65.29
52°00′	110.18	61°00′	93.93	70°00′	81.85	79°00′	72.53	88°00′	65.11
15′	109.66	15′	93.54	15′	81.56	15′	72.30	15′	64.92
30′	109.13	30′	93.16	30′	81.27	30′	72.07	30′	64.74
45′	108.62	45′	92.79	45′	80.98	45′	71.84	45′	64.56
53°00′	108.11	62°00′	92.41	71°00′	80.70	80°00′	71.62	89°00′	64.38
15′	107.60	15′	92.04	15′	80.42	15′	71.40	15′	64.20
30′	107.09	30′	91.67	30′	80.13	30′	71.17	30′	64.02
45′	106.60	45′	91.31	45′	79.85	45′	70.95	45′	63.84
54°00′	106.10	63°00′	90.95	72°00′	79.58	81°00′	70.74	90°00′	63.66
15′	105.61	15′	90.59	15′	79.30	15′	70.52	15′	63.49
30′	105.13	30′	90.23	30′	79.03	30′	70.30	30′	63.31
45′	104.65	45′	89.88	45′	78.76	45′	70.09	45′	63.14

Appendix D: Chord Lengths of Circular Arcs (Arc Definition)

Degree of curve	Chords								Degree of curve
	For arcs of								
	100′	95′	90′	85′	80′	75′	70′	60′	
1°	100′	95′	90′	85′	80′	75′	70′	60′	1°
2°	100	95	90	85	80	75	70	60	2°
3°	99.99	94.99	89.99	85	80	75	70	60	3°
4°	99.98	94.98	89.98	84.99	79.99	74.99	70	60	4°
5°	99.97	94.97	89.98	84.98	79.98	74.99	69.99	59.99	5°
6°	99.95	94.96	89.97	84.97	79.98	74.98	69.98	59.99	6°
7°	99.94	94.95	89.96	84.96	79.96	74.97	69.98	59.99	7°
8°	99.92	94.93	89.94	84.95	79.96	74.97	69.97	59.98	8°
9°	99.90	94.91	89.93	84.94	79.95	74.96	69.96	59.98	9°
10°	99.87	94.89	89.91	84.92	79.94	74.95	69.96	59.97	10°

Degree of curve	For arc of								Degree of curve
	100′	60′	55′	50′	45′	40′	35′	30′	
11°	99.85′	59.97′	54.97′	49.98′	44.99′	39.99′	34.99′	30′	11°
12°	99.82	59.96	54.97	49.98	44.98	39.99	34.99	33	12°
13°	99.79	59.95	54.96	49.97	44.98	39.99	34.99	29.99	13°
14°	99.75	59.95	54.96	49.97	44.98	39.98	34.99	29.99	14°
15°	99.71	59.94	54.95	49.96	44.97	39.98	34.99	29.99	15°
16°	99.68	59.93	54.95	49.96	44.97	39.98	34.99	29.99	16°
17°	99.63	59.92	54.94	49.95	44.97	39.98	34.98	29.99	17°
18°	99.59	59.91	54.93	49.95	44.96	39.97	34.98	29.99	18°
19°	99.54	59.90	54.92	49.94	44.96	39.97	34.98	29.99	19°
20°	99.49	59.89	54.92	49.94	44.95	39.97	34.98	29.99	20°
21°	99.44	59.88	54.91	49.93	44.95	39.96	34.98	29.98	21°
22°	99.39	59.87	54.90	49.92	44.94	39.96	34.97	29.98	22°
23°	99.33	59.85	54.89	49.92	44.94	39.96	34.97	29.98	23°
24°	99.27	59.84	54.88	49.91	44.93	39.95	34.97	29.98	24°
25°	99.21	59.83	54.87	49.90	44.93	39.95	34.97	29.98	25°

PROFESSIONAL PUBLICATIONS, INC. ● P.O. Box 199, San Carlos, CA 94070

Appendix E: Areas Under The Standard Normal Curve (0 to Z)

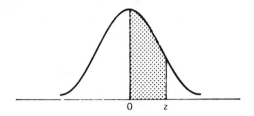

z	0	1	2	3	4	5	6	7	8	9
0.0	.0000	.0040	.0080	.0120	.0160	.0199	.0239	.0279	.0319	.0359
0.1	.0398	.0438	.0478	.0517	.0557	.0596	.0636	.0675	.0714	.0754
0.2	.0793	.0832	.0871	.0910	.0948	.0987	.1026	.1064	.1103	.1141
0.3	.1179	.1217	.1255	.1293	.1331	.1368	.1406	.1443	.1480	.1517
0.4	.1554	.1591	.1628	.1664	.1700	.1736	.1772	.1808	.1844	.1879
0.5	.1915	.1950	.1985	.2019	.2054	.2088	.2123	.2157	.2190	.2224
0.6	.2258	.2291	.2324	.2357	.2389	.2422	.2454	.2486	.2518	.2549
0.7	.2580	.2612	.2642	.2673	.2704	.2734	.2764	.2794	.2823	.2852
0.8	.2881	.2910	.2939	.2967	.2996	.3023	.3051	.3078	.3106	.3133
0.9	.3159	.3186	.3212	.3238	.3264	.3289	.3315	.3340	.3365	.3389
1.0	.3413	.3438	.3461	.3485	.3508	.3531	.3554	.3577	.3599	.3621
1.1	.3643	.3665	.3686	.3708	.3729	.3749	.3770	.3790	.3810	.3830
1.2	.3849	.3869	.3888	.3907	.3925	.3944	.3962	.3980	.3997	.4015
1.3	.4032	.4049	.4066	.4082	.4099	.4115	.4131	.4147	.4162	.4177
1.4	.4192	.4207	.4222	.4236	.4251	.4265	.4279	.4292	.4306	.4319
1.5	.4332	.4345	.4357	.4370	.4382	.4394	.4406	.4418	.4429	.4441
1.6	.4452	.4463	.4474	.4484	.4495	.4505	.4515	.4525	.4535	.4545
1.7	.4554	.4564	.4573	.4582	.4591	.4599	.4608	.4616	.4625	.4633
1.8	.4641	.4649	.4656	.4664	.4671	.4678	.4686	.4693	.4699	.4706
1.9	.4713	.4719	.4726	.4732	.4738	.4744	.4750	.4756	.4761	.4767
2.0	.4772	.4778	.4783	.4788	.4793	.4798	.4803	.4808	.4812	.4817
2.1	.4821	.4826	.4830	.4834	.4838	.4842	.4846	.4850	.4854	.4857
2.2	.4861	.4864	.4868	.4871	.4875	.4878	.4881	.4884	.4887	.4890
2.3	.4893	.4896	.4898	.4901	.4904	.4906	.4909	.4911	.4913	.4916
2.4	.4918	.4920	.4922	.4925	.4927	.4929	.4931	.4932	.4934	.4936
2.5	.4938	.4940	.4941	.4943	.4945	.4946	.4948	.4949	.4951	.4952
2.6	.4953	.4955	.4956	.4957	.4959	.4960	.4961	.4962	.4963	.4964
2.7	.4965	.4966	.4967	.4968	.4969	.4970	.4971	.4972	.4973	.4974
2.8	.4974	.4975	.4976	.4977	.4977	.4978	.4979	.4979	.4980	.4981
2.9	.4981	.4982	.4982	.4983	.4984	.4984	.4985	.4985	.4986	.4986
3.0	.4987	.4987	.4987	.4988	.4988	.4989	.4989	.4989	.4990	.4990
3.1	.4990	.4991	.4991	.4991	.4992	.4992	.4992	.4992	.4993	.4993
3.2	.4993	.4993	.4994	.4994	.4994	.4994	.4994	.4995	.4995	.4995
3.3	.4995	.4995	.4995	.4996	.4996	.4996	.4996	.4996	.4996	.4997
3.4	.4997	.4997	.4997	.4997	.4997	.4997	.4997	.4997	.4997	.4998
3.5	.4998	.4998	.4998	.4998	.4998	.4998	.4998	.4998	.4998	.4998
3.6	.4998	.4998	.4999	.4999	.4999	.4999	.4999	.4999	.4999	.4999
3.7	.4999	.4999	.4999	.4999	.4999	.4999	.4999	.4999	.4999	.4999
3.8	.4999	.4999	.4999	.4999	.4999	.4999	.4999	.4999	.4999	.4999
3.9	.5000	.5000	.5000	.5000	.5000	.5000	.5000	.5000	.5000	.5000

Appendix F: Important Addresses

See the introduction for the phone number of your state registration board.

American Congress on Surveying and Mapping
210 Little Falls Street
Falls Church, VA 22046

American Society of Civil Engineers
345 East 47th Street
New York, NY 10017

National Council of Engineering Examiners
P. O. Box 1686
Clemson, SC 29633

National Geodetic Survey
National Ocean Survey
Rockville, MD 20852

Training Staff, Surveying and Mapping
Bureau of Land Management
Federal Center
Denver, CO 80225

National Geodetic Information Center (N/CG17 × 2)
National Oceanic and Atmospheric Administration
6001 Executive Boulevard
Rockville, MD 20852
(301)443–8316
(for publications)

Horizontal Network Branch
National Geodetic Survey
Rockville, Maryland 20852

U. S. Department of the Interior
Western Distribution Branch
U. S. Geological Survey
Box 25286, Denver Federal Center
Denver, CO 80225
(for maps)

PROFESSIONAL PUBLICATIONS, INC. • P.O. Box 199, San Carlos, CA 94070

Appendix G: Critical Constants

0.0000001 per °F	=	coefficient of expansion invar tape
0.00000645 per °F	=	coefficient of expansion steel tape
0.6745	=	coefficient for 50% standard deviation
1.15 miles	=	1 minute of latitude
1.6449	=	coefficient for 90% standard deviation
6 miles	=	length and width of township
10 square chains	=	1 acre
15 degrees longitude	=	width of one time zone
23 degrees 26.5 minutes	=	maximum declination of the sun at solstice
24 hours	=	360 degrees of longitude
36	=	number of sections in a township
69.1 miles	=	1 degree latitude
100	=	usual stadia ratio
101 feet	=	1 second of latitude
400 grads	=	360 degrees
480 chains	=	width and length of township
640 acres	=	1 normal section
4046.9 square meters	=	1 acre
6400 mils	=	360 degrees
43,560 square feet	=	1 acre
20,906,000 feet	=	mean radius of earth

PROFESSIONAL PUBLICATIONS, INC. ● P.O. Box 199, San Carlos, CA   94070

Appendix H: Metric Conversion

1 millimeter (mm)	=	1000 micrometers
1 centimeter (cm)	=	10 mm
1 meter	=	100 cm
1 meter	=	3.2808 feet = 39.37 inches
1 kilometer (km)	=	1000 meters
1 kilometer (km)	=	0.62137 mile
1 inch	=	25.400 mm
1 foot	=	304.80 mm
1 square mm	=	0.00155 in^2
1 meter^2	=	10.76 feet^2
1 kilometer^2	=	247.1 acres
1 liter	=	0.264 U.S. gallons
1 meter/sec.	=	3.28 ft/sec
1 km/hr.	=	0.911 ft/sec

PROFESSIONAL PUBLICATIONS, INC. • P.O. Box 199, San Carlos, CA   94070

Appendix I: Surveying Conversion Factors

Multiply	By	To Obtain
acres	43,560	square feet
	10	square chains
	4046.87	square meters
acre-feet	43,560	cubic feet
	1233.49	cubic meters
chain	66	feet
	22	yards
	4	rods
day (mean solar)	86,400	seconds
day (sidereal)	86,164.09	seconds
degrees (angle)	.0174533	radians
	17.77778	mils
engineer's link	1	feet
feet (U.S. Survey)	3.048006	meters
grads	0.9	degrees (angle)
	0.01570797	radians
hectare	2.47104	acres
	10,000	square meters
inches	25.4	millimeters
labors	177.14	acres
leagues	4,428.40	acres
link—see 'engineer's link' and 'surveyor's link'		
mils	0.05625	degrees (angle)
	3,037,500	minutes
miles (statute)	5280	feet
	80	chains
	320	rods
	1.1508	miles (nautical)
square miles	640	acres
	27,878,400	square feet
minutes (angle)	0.29630	mils
	0.000290888	radians
minutes (mean solar)	60	seconds
minutes (sidereal)	59.83617	seconds
outs	330	feet
	10	33-foot chains
radians	57.2957795	degrees (angle)
	$57°17'44.806''$	degrees (angle)
rods	16.5	feet
	1	perches
	1	poles
seconds (angle)	$4.848137 \ 10^{-6}$	radians
seconds (sidereal)	.9972696	seconds (mean solar)
surveyor's link	0.66	feet
	7.92	inches
VARA (California)	33	inches
VARA (Texas)	33.333	inches
yard (U.S.)	0.914402	meters

PROFESSIONAL PUBLICATIONS, INC. • P.O. Box 199, San Carlos, CA 94070

Appendix J: Glossary of Terms

Not all words are covered in this glossary. Refer to the index for additional words that are discussed in the text.

Abstract – A summary of facts.

Abstract of title – A condensed history of the title to land.

Accessory to corner – A physical object that is adjacent to a corner. An accessory is usually considered part of the monument.

Acclivity – An upward slope of ground.

Accretion – The gradual accumulation of land by natural causes.

Acknowledgment – A declaration by a person before an official (usually a notary public) that he executed a legal document.

Acquiescense – Implied consent to a transaction, to the accrual of a right, or to any act by one's silence (or without express assent).

Adjudication – The giving or pronouncing a judgment or decree in a cause.

Adverse possession – A method of acquiring property by holding it for a period of time under certain conditions.

Affidavit – A written declaration under oath before an authorized official (usually a notary public.)

Alienation – The transfer of property and/or possessions from one person to another.

Aliquot – A portion contained in something else a whole number of times.

Alluvium – Sand or soil deposited by streams.

Appellant – The party which takes an appeal from one court or jurisdiction to another.

Appurtenance – A right, privilege, or improvement belonging to and passing with a piece of property when it is conveyed.

Assigns – Those to whom property is transferred.

Avulsion – A sudden and perceptible change of shoreline by the violent action of water.

Bayou – An outlet from a swamp or lagoon to the sea.

Bed of stream – The depression between the banks of a water course worn by the regular and usual flow of the water.

Bequest – A gift by will of personal property.

Bounty lands – Portions of the public domain given or donated as a bounty for services rendered.

Chain of title – A chronological list of documents which comprise the record history of title of real property.

Civil law – That part of the law pertaining to civil rights, as distinguished from criminal law. Civil law and Roman law have the same meaning. In contradistinction to English common law, civil law is enacted by legislative bodies.

Clear title – Good title. One free from encumbrances.

Cloud on title. – A claim or encumbrance on a title to land that may or may not be valid.

Color of title – Any written instrument which appears to convey title, even though it does not.

Common law – Principles and rules of action determined by court decisions which have been accepted by generation after generation, and which are distinguished from laws enacted by legislative bodies.

Consideration – Something of value given to make an agreement binding.

Conveyance – Any instrument in writing by which an interest in real property is transferred.

Covenant – When used in deeds, restrictions imposed on the grantee as to the use of land conveyed.

Crown – The sovereign power in a monarchy.

Cut bank – The water washed and relatively permanent elevation or acclivity which separates the bed of a river from its adjacent upland.

Decree – A judgment by the court in a legal proceeding.

Dedication – An appropriation of land to some public use made by the owner, and accepted for such use by or on behalf of the public.

Deed – Evidence in writing of the transfer of real property.

Deed of trust – An instrument taking the place of a mortgage, by which the legal title to real property is placed in one or more trustees to secure repayment of a sum of money.

Demurrer – In legal pleading, the formal mode of disputing the sufficiency of the pleading of the other side.

Devise – A gift of real property by the last will and testament of the donor.

Easement – The right which the public, an individual, or individuals have in the lands of another.

Egress – The right or permission to go out from a place; right of exit.

Eminent domain – The right or power of government or certain other agencies to take private property for public use on payment of just compensation to the owner.

Encroachment – An obstruction which intrudes upon the land of another. The gradual, stealthy, illegal acquisition of property.

Encumbrance – Any burden or claim on property, such as a mortgage or delinquent taxes.

Equity – The excess of the market value over any indebtedness.

Erosion – The process by which the surface of the earth is worn away by the action of waters, glaciers, wind, or waves.

Escheat – Reversion of property to the state where there is no competent or available person to inherit it.

Escrow – Something placed in the keeping of a third person for delivery to a given party upon fulfillment of some condition.

Estate – An interest in property, real or personal.

Estoppel – A bar or impediment which precludes allegation or denial of a certain fact or state of facts in consequence of a final adjudication.

Et al – An abbreviation for "and others".

Et Mode Ad Hune Diem – An abbreviation for "and now at this day".

Et ux – An abbreviation for "and wife".

Evidence aliunde – Evidence from outside or from another source.

Extrinsic evidence – Evidence NOT contained in the deed, but offered to clear up an ambiguity found to exist when applying the description to the ground.

Grant – A transfer of property.

Grantee – The person to whom a grant is made.

Grantor – The person by whom a grant is made.

Good faith – An honest intention to abstain from taking advantage of another.

Gradient – An inclined surface. The change in elevation per unit of horizontal distance.

Hereditament – Something capable of being inherited, be it real or personal property.

Hiatus – An area between two surveys of record described as having one or more common boundary lines with no omission.

Holograph – A will written entirely by the testator in his own handwriting.

Incumbrance – A right, interest in, or legal liability upon real property which does not prohibit passing title to the land but which diminishes its value.

Ingress – The right or permission to go upon a place; right of entrance.

Intent – The true meaning (from the written words of an instrument).

Intestate – Without making a will.

Judgment – The official and authentic decision of a court of justice.

Leasehold – An estate in realty held under a lease; an estate for a fixed term of years.

Lessee – He to whom a lease is made.

Lessor – He who grants a lease.

Lien – A claim or charge on property for payment of some debt, obligation, or duty.

Lis pendens – A pending suit. A notice of lis pendens is filed for the purpose of warning all persons that a suit is pending.

Litigation – Contest in a court of justice for the purpose of enforcing a right.

Littoral – Belonging to the shore, as of seas and lakes.

Logical relevancy – A relationship in logic between the fact for which evidence is offered and a fact in issue such that the existence of the former renders probable or improbable the existence of the latter.

Mean – Intermediate; the middle between two extremes.

Memorial – That which contains the particulars of a deed, etc. In practice, a memorial is a short note, abstract, memorandum, or rough draft of the orders of the court, from which the records thereof may at any time be fully made up.

Mortgage – A conditional conveyance of an estate as a pledge for the security of a debt.

Muniment – Documentary evidence of title.

PROFESSIONAL PUBLICATIONS, INC. • P.O. Box 199, San Carlos, CA   94070

Option – The right as granted in a contract or by an initial payment, of acquiring something in the future.

Parcel – A part or a piece of land that cannot be identified by a lot or tract number.

Parol evidence – Evidence which is given verbally.

Patent – A government grant of land. The instrument by which a government conveys title to land.

Plat – A scaled diagram showing boundaries of a tract of land or subdivisions. May constitute a legal description of the land and be used in lieu of a written description.

Power of attorney – A written document given by one person to another authorizing the latter to act for the former.

Prescription – Creation of an easement under claim of right by use of land which has been open, continuous, and exclusive for a period of time prescribed by law.

Prima facie evidence – Facts presumed to be true unless disproved by evidence to the contrary.

Privity – The relationship which exists between parties to a contract. Mutual or successive relationship to the same rights of property, such as the relationship of heir with ancestor or donee with donor.

Privy – A person who is in privity with another.

Probate – The act or process of validating a will.

Quiet title – Action of law to remove an adverse claim or cloud on title.

Quitclaim deed – A conveyance which passes any title, interest, or claim which the grantor may have.

Reliction – A gradual and imperceptible recession of water, resulting in increased shoreline, beach, or property.

Relinquishment – The forsaking, abandonment, renouncement, or gift of a right.

Remand – To send a cause back to the same court out of which it came for the purpose of having some action taken upon it there.

Riparian – Belonging or relating to the bank of a river.

Royalty – A share of the profit from sale of minerals paid to the owner of the property by the lessee.

Said – Refers to one previously mentioned.

Scrivener – A person whose occupation is to draw up contracts, write deeds and mortgages, and prepare other written instruments.

Shore – The space lying between the line of ordinary high tide and the line of lowest tide.

Sovereign – A person, body, or state in which independent and supreme authority is vested.

Squatter – One who settles on another's land without legal authority.

Statute – A particular law established by the legislative branch of government.

Statutory – Relating to a statute.

Submerged land – In tidal areas, land which extends seaward from the shore and is continuously covered during the ebb and flow of the tide.

Substantive evidence – Evidence used to prove a fact (as opposed to evidence given for the purpose of discrediting a claim).

Tenancy by entirety – Husband and wife each possesses the entire estate in order that, upon the death of either spouse, the survivor is entitled to the estate in its entirety.

Tenancy, Joint – The holding of property by two or more persons, each of whom has an undivided interest. After the death of one of the joint tenants, the surviving tenant(s) receive the descendent's share.

Tenant – One who has the temporary use and occupation of real property owned by another person (the landlord).

Tenements – Property held by tenant. Everything of permanent nature. In more restrictive sense, house or dwelling.

Testament – A will of personal property.

Testator – One who makes a testament or will. One who dies leaving a will.

Thalweg – The deepest part of channel.

Thence – From that place; the following course is continuous from the one before it.

Title policy – Insurance against loss or damage resulting from defects or failure of title to a particular parcel of land.

To wit – That is to say; namely.

Upland – Land above mean high-water and subject to private ownership (as distinguished from tidelands which are in the state.) Also used as meaning NON-riparian.

Watercourse – A running stream of water fed from permanent or natural sources running in a particular direction and having a channel formed by a well-defined bed and banks (though it need not flow continuously).

Warranty deed – A deed in which the grantor proclaims that he is the lawful owner of real property and will forever defend the grantee against any claim on the property.

Will – The legal declaration of a person's wishes as to the disposition of his property after his death.

Witness mark – A mark placed at a known location to aid in recovery and identification of a monument or corner.

Writ – A mandatory order issued from a court of justice.

Writ of coram nobis – A common law writ, the purpose of which is to correct an error in a judgment in the same court in which it was rendered.

INDEX

PROFESSIONAL PUBLICATIONS, INC. • P.O. Box 199, San Carlos, CA 94070

2   LAND SURVEYOR REFERENCE MANUAL

PROFESSIONAL PUBLICATIONS, INC. • P.O. Box 199, San Carlos, CA   94070

PROFESSIONAL PUBLICATIONS, INC. • P.O. Box 199, San Carlos, CA 94070

PROFESSIONAL PUBLICATIONS, INC. • P.O. Box 199, San Carlos, CA   94070